C. G. Warnford Lock

Mining and ore-dressing Machinery

A comprehensive Treatise dealing with the modern Practice of winning both

metalliferous and non-metalliferous Minerals

C. G. Warnford Lock

Mining and ore-dressing Machinery

A comprehensive Treatise dealing with the modern Practice of winning both metalliferous and non-metalliferous Minerals

ISBN/EAN:

Printed in Europe, USA, Canada, Australia, Japan

Cover: Foto ©ninafisch / pixelio.de

More available books at www.hansebooks.com

· MINING

AND

ORE-DRESSING MACHINERY:

A COMPREHENSIVE TREATISE DEALING WITH

THE MODERN PRACTICE OF WINNING BOTH METALLIFEROUS
AND NON-METALLIFEROUS MINERALS,

INCLUDING ALL THE OPERATIONS INCIDENTAL THERETO, AND PREPARING
THE PRODUCT FOR THE MARKET.

BY

C. G. WARNFORD LOCK
AUTHOR OF 'PRACTICAL GOLD MINING.'

E. & F. N. SPON, 125, STRAND, LONDON.
NEW YORK: 12, CORTLANDT STREET.
1890.

44850

ILLUSTRATIONS.

FIG.		PAGE	FIG.		PAGE
1	Angle of windmill sails	1	118–121	Machine-boring tools	66
2	Area of windmill sails	3	122–131	do.	67
3	Governor for windmill	3	132–138	do.	68
4	Turning contrivance for windmill	4	138–147	do.	69
5	Velocity of sails	4	148, 149	Lever-boring machine	70
6	Overshot water-wheel, without head	6	150, 151	Lever-boring machine, steam worked	71
7	Overshot water-wheel, with head	8	152, 153	Boring tackle provided with steam winch	72
8	Water-wheel for varying volume	8	154	Dolly	76
9	Slow breast wheel	9	155	Automatic sampler	76
10	Vortex turbine at bottom of fall	15	156, 157	Scoop shovels	76
11	Vortex turbine, part of fall acting by suction	16	158–172	Drilling tools	77
12	50-H.P. vortex turbine, with 276 ft. fall	17	173	Air compressor used at Rio Tinto	84
13	60-H.P. vortex turbine, with 26 ft. fall	17	174, 175	Compensating joint for air pipes	86
14	300-H.P. vortex turbine, with 300 ft. fall	18	176, 177	Water reservoirs on wheels	86
15	Girard turbine by Gilkes	18	178–187	Blasting fuses	80
16	Vertical turbine by J. & H. Gwynne	19	188, 189	Magneto battery and fuse for blasting	95
17	Horizontal turbine by J. & H. Gwynne	19	189A	Bornhardt's electric firing machine	96
18, 19	Girard turbines by Günther	20	190	Davis's magneto exploder	96
20	Jonval turbine with vertical shaft	21	191–195	Blasting machines	97
21	Jonval turbine with horizontal shaft	21	196	Ladd's frictional exploder	99
22	Priestman's oil engine	22	197–200	Tamping	99
23	Splicing wire ropes	25	201	Gelatinous cartridge	103
24	Pneumatic transmission of power	30	202, 203	Heath & Frost safety lamp	105
25, 26	Electric transmission of power	43	204	Multiple wedge	107
27–47	Stone-quarrying machinery	48, 49	205–210	Miners' shovels	108
48	Stone-polishing machinery	50	211–215	do.	109
49	Priestman's excavator	50	216–230	Miners' picks	112
50	Priestman's dredger	50	231–235	Miners' wedges	114
51	Hand-boring frame	53	236–240	Ore dressing hammers	114
52–55	Portions of hand-boring frame	55	241	Hansa shaft	117
56–59	do. do.	56	242–248	Details of walling, tubbing, and wedging cribs	118
60–67	Hand-boring tools	57	249–256	Shaft-sinking at Zollern colliery	119, 120
68–71	do.	58	257	do. Marsden colliery	123
72–77	do.	59	257a	do. do.	124
78–81	do.	60	258–263	Kind-Chaudron shaft-sinking tools	125
82–84	do.	61	264–267	do. do.	127
85	Derrick	61	268–274	do. do.	128
86–91	Machine-boring tools	62	275–277	Lowering metal tubbing for shaft	130
92–103	do.	63	278	Concreting shaft	131
104–111	do.	64	279	Lippmann's cutting tool	133
112–117	do.	65	280	Sinking through quicksand	133

b

ILLUSTRATIONS.

FIG.		PAGE
281	Winstanley & Barker's coal-cutter	135
282	do. do.	136
283	Baird's coal-cutter	138
284, 285	Gillot & Copley's coal-cutter	139
286	Hurd & Simpson's coal-cutter	140
287	do. do.	141
288	Heating and expanding air for coal-cutter	142
289	Upheaving bottom coal	142
290	Firth's coal-cutter	144
291	Garrett, Marshall, & Co.'s coal-cutter	145
292, 293	Stanley's coal-heading machine	151
294	Goolden's electric coal-cutter	152
295	Goolden's standard dynamo	152
296, 297	Water skips	155
298	Water barrel	156
299–303	Pneumatic water barrel	157
304	do.	150
305	do.	161
306	Steam capstan for raising and lowering pump rods	162
307	Cornish pump	163
308	Cornish pump body	163
309	Pump for varying level	164
310	Steam pump	164
311	Water wheel pump	164
312, 313	Hydraulic system of draining mines	165
314	do.	166
315–318	Methods of dealing with pumps during sinking	168
319–321	Davey's adjustment for pumping engines	170
322	Water supply balance valve	170
323–326	Gwynne's centrifugal pumps	171
327, 328	do. do.	172
329	Goolden electric dip pump	173
330	Ventilating box	176
331	Cornish duck engine	177
332	Water blast	177
333	Hand fan	178
334	Fabry's wheel	178
335	Cooke's ventilator	180
336	Guibal's fan	181
337	Root's blower	182
338	Hickie's air-cooling apparatus	182
339	Davis's self-acting anemometer	183
340, 340A	Shippey's electric fan	184
341	Sheet-iron lamp	186
342	Brass lamp	186
343	Spider candlestick	186
344	Davy lamp	189
345	Stephenson's lamp	190
346	Davis-Ashworth-Mueseler safety lamp	190
347	Marsaut safety lamp	190
348	Bonneted Marsaut safety lamp	190
349	Gwynne's engine and dynamo	193
350–355	Mine waggons	196
356–359	Mine waggons	199
360, 361	do.	200
362–367	do.	201
368–373	do.	202
374, 375	Self-tipping waggons	203
376	Kerr, Stuart & Co.'s waggon	204
377, 378	Dumping cradles	204
379	Frongoch skip	204
380, 381	Cornish skips	205
382, 383	Side-tipping arrangements	206
384	Improved chair and sleeper	207
385	Tramway junction	208
386–390	Sheaves and pulleys	209
391–393	do.	210
394–397	do.	211
398–402	Connections	212
403–407	do.	213
408–414	Cages	216
415–417	do.	217
418–420	do.	218
421	Cornish shackles	219
422	Safety hooks	221
423	Winks, Cowling, & Hoskin's automatic check for overwinding	222
424	Philips's safety winding appliance	223
425	Keeps	225
426–428	Pit-head framing	228
429–437	do.	229
438, 439	Horse whims	238
440	German horse whim	239
441	do.	240
442	Cornish water whim	241
443	do.	242
444	Harz water whim	243
445	Cornish system of steam winding from shallow shafts	244
446	Plan of South Duffryn Colliery	246
447, 448	Hauling in South Duffryn Colliery	248
449	Arrangement of self-acting incline	250
450	Expansion gear	271
451	Fowler's hydraulic loading and unloading	273
452, 453	Ransome's winding machinery	274, 275
454	Ransome's winding gear and frame	276
455	Hornsby's arrangement of pumping and winding engines separate	277
456	Hornsby's vertical winding and pumping engine	278
457	Hornsby's geared winding engine	279
458	Whim engine and winding cage	280
459	Wild's portable hauling and winding engine	281
460, 461	Pair of Wild's semi-portable hauling engines	282, 283
462	Wire tramway by Bullivant & Co.	285
463–465	Details of Otto's ropeway	286
466	Otto's ropeway system	287

ILLUSTRATIONS.

FIG.		PAGE	FIG.		PAGE
467	Otto's ropeway system	288	531	Rittinger's percussion table	352
468	do.	289	532	End-shake percussion table	356
469	Robey's ore breaker	290	533	Rittinger's rotating table	356
470	Calvert & Co.'s ore breaker	290	534, 535	Convex or centre-head buddle	360
471	Green's ore breaker	291	536, 537	Concave buddle	361
472	Stamp battery in section	292	538, 539	Borlase's buddle	362
473	Details of foundations and frames	293	540–542	Propeller knife buddle	363
474–476	Mortars	294	543	Munday's round buddle	364
477	Screens	296	544	Dodge's concentrator	366
478	Die	296	545	Duncan concentrator	366
479	Stamp head	298	546	Frue vanner	367
480	Shoe fastening	298	547	Halley's percussion table	370
481	Tappet or collar	301	548	Hendy's concentrator	370
482	Stamp guides	301	549	Imlay concentrator	371
483	Cam or wiper	301	550	Keeve or tossing tub	372
484	McNeill's cam	302	551	McNeill's concentrator	373
485	Stanford's ore feeder	304	552, 553	Self-acting slime frame	374
486	Tulloch's ore feeder	304	554	do.	375
487	Port Philip self-feeding hopper	304	555	Stationary slime table	276
488	Roller ore feeder	305	556	Green's crushing and dressing machinery	380
489	Hendy's ore feeder	305	557	Green's 4-compartment self-acting jigger	383
490	Cornish tin stamps	308	558	Green's trommel	383
491	Husband's pneumatic stamp	309	559	Galena and blende dressing machinery at Sentein	384
492	Pneumatic stamps	310	560, 561	Classifying apparatus at Sentein	387
493	Harvey's high-speed revolving stamps	311	562–565	do. do.	389
494	Hornsby's arrangement of gold mill	312	566–570	do. do.	390
495	Hornsby's gravitation stamp mill	313	571–574	do. do.	391
496, 497	Hornsby & Oglo's mortar	314	575–577	do. do.	394
498	Cornish crushing rolls	315	578	do. do.	396
499	do.	316	579	Dolly tub at Sentein	397
500	Rubber springs for rolls	316	580, 581	Classifying apparatus at Sentein	398
501	Green's crushing rolls	318	582, 583	Commans's dressing plant	401
502	do.	319	584	Linkenbach buddle	402
503, 504	Krom's system of crushing rolls	320	585	Tipping cradle at Dowlais	405
505	Huntington's centrifugal roller mill	325	586	do. do.	406
506, 507	Details of Huntington mill	326	587	do. do.	407
508	Arrangement of mill on Huntington system	328	588	Plated chain carrier	413
509	Edge runner mill	328	589	Marsaut washer	416
510	Globe mill	329	590	Sheppard's washer	419
511	Jordan's reducer	330	591	do.	420
512	Jordan's amalgamator	330	592	Dowlais washery	421
513	Arrangement of Jordan's reducer and amalgamator	331	593, 594	Bérard washer	422
514, 515	Hand-lever jigger	339	595–597	do.	423
516	Rittinger's jigger	340	598	Ebbw Vale washer	424
517	Huet & Geyler jig	341	599–601	Lührig's coal washer	433
518–520	Clausthal jig	342	602, 603	Lührig's fine coal jigger	434
521–524	Collom's jig	343	604, 605	Hutch and tumbler	435
525	Frongoch jig	344	606, 607	Horse-shoe washer	435
526	Settler	345	608	Creeper at Barrow	436
527	German pyramidal boxes or spitzkästen	346	609, 610	Guinotte & Briart's differential grid	437
528	Pointed box	349	611, 612	Rigg's hutches	438
529	Frongoch separator or classifier	349	613	Dry cleaning at Aldwarke Main	439
530	Triangular double troughs or spitzlutten	350	614	Roller delivery	440

ILLUSTRATIONS.

FIG.		PAGE
615	Screening and washing at Flimby	440
616	do. do.	441
617	Screen and kepper	441
618	Bell & Ramsay washer at Robin Hood	443
619–622	Bell's washer	444
623	Coppée washer at Cwm-Avon	446
624	Sheppard washer	447
625	do.	448
626	Robinson washer	449
627	Double indicator	455
628	Signalling arrangement	455

FIG.	
629	Signalling arrangement
630	Electric bell
631	Davis's clinometer
632	Louis's Davis clinometer
633	Mining dial
634	Legs with tribrach adjustment
635	Stanley's prismatic dial
636	Stanley's mine staff
637	Mining survey lamp
638	Harling's theodolite
639	Portable anemometer

MINING AND ORE-DRESSING MACHINERY.

CHAPTER I.

MOTIVE POWER.

WIND.—For giving motion to machinery, windmills have been and still are very extensively used. Engineers of the last generation devoted much attention to the construction of windmills, and brought them to great perfection. The introduction of steam-power—manageable, and always to be depended on—has, in a great measure, superseded that of wind. True, after the first cost of a windmill, the power is comparatively inexpensive; but it is so variable in intensity—sometimes, when it is not required, exerting great force, and sometimes, when it may be most wanted, totally ineffective—that it is generally preferable to apply a force, perhaps considerably more expensive in its production, but constant, steady, and completely under control. The intervention of electric storage batteries, however, is an obvious method of reducing this evil to a minimum.

Windmills are of two kinds, horizontal and vertical. The former have been very little used, being much less effective than the latter. The vertical windmill consists of an axle or shaft, nearly horizontal, mounted in bearings at the summit of a tower, with four or more blades or sails attached to it. These sails are set at an angle with the axis, so that when the wind blows directly on the face of the mill, its oblique action on the sails is resolved into two forces—one in the direction of the axis, and the other perpendicular to it, which is the direction in which the sails revolve. Numerous experiments and computations were made to determine the most advantageous angles for setting the sails, and their most effective forms and proportions. If we suppose the radius of a sail divided into six equal parts, Fig. 1, and circles traced through the points of division, the velocity of each point in revolving is proportional to the part of its circle intercepted between two radii, or proportional to its own radius. If, then, we make a series of plans of the sail at these different parts, we see that as we approach the centre we should increase the obliquity of the sail to its plane of motion, so as to allow for its slower escape sideways from the impulse of the wind. The sails accordingly are not made flat surfaces inclined equally to the plane of their revolution, but surfaces of varying inclination, somewhat like portions of screw blades, twisting as it were from a certain obliquity at their extremes

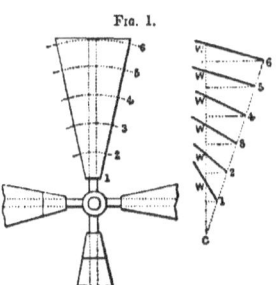

Fig. 1.

Angle of Windmill Sails.

B

in a greater obliquity at the centre. The angles found most advantageous in practice are given by Smeaton and others, as follows:—

Distance from centre					1	2	3	4	5	6
Inclination to plane of motion (Smeaton)					18°	19°	18°	16°	12½°	7°
„	„	others			24°	21°	18°	14°	9°	3°

In the angles given by Smeaton an irregularity is observed in the first, which should by theoretical reasoning be greater than the second, whereas Smeaton makes it less. The following rule may be adopted as a very near approximation. To find the angle at which the sail should be inclined to the plane of revolution at any distance from the centre:—

Rule.—Multiply 18 twice by the distance from the centre, divide the product twice by the total radius, and subtract the quotient from 23; the remainder is the inclination in degrees.

Example.—In a windmill 60 ft. in diameter, required the inclination of the sail 20 ft. from the centre.

Here 30 ft. is the total radius, and $\frac{18 \times 20 \times 20}{30 \times 30} = 8$, which, subtracted from 23, gives 15°, the angle of that point.

Were we to divide the radius 30 ft. into six equal parts, and calculate the angles at each point, we should find them correspond nearly with the means of those given by Smeaton and others.

Mills are generally made with four strong wooden arms or radii, fixed firmly in a central socket, and steadied and stiffened by tie-rods, connecting their extremities together, and with a projecting strut on the central boss. The width of each sail at the extreme should be about half of the radius, so that in a mill 60 ft. diameter, or 30 ft. radius, each sail would be 15 ft. wide at the extreme. The part of the arm next the centre for about ⅙ of the radius, that is, 5 ft. in the case supposed, is not fitted with sails because the surface there is so little effective, as well as from its short leverage as from its obstructing the wind reflected from the head of the turret behind it. The width at the inner end should be ⅓ of the radius, or 10 ft. The surface of each sail is therefore 312½ sq. ft., and the total of the four is 312½ × 4 = 1250 sq. ft.

The total area of the circle 60 ft. in diameter is somewhat above 2800 sq. ft., so that not half the surface of the circle is clothed with sails. There would be no disadvantage in extending the surface by making the sails broader or more numerous, until it became ¾ of the whole surface. Beyond this additional sail-surface is disadvantageous, for it appears to obstruct the free passage of the currents reflected from the sails, and thus clogs their motions. It is found advantageous to arrange the surface of the sail somewhat in the proportions of Fig. 2, which represents the front view of one sail.

Thus if A C is 30 ft., then A E should be 25 ft., A D or E F 10 ft., and A B 5 ft.

The covering of the surface, so as to catch the impulse of wind, formerly consisted of canvas fixed on a roller at one side of the arm, on which it could be rolled like a window-blind, or from which it could be unrolled so as to cover the whole sail, which was filled in with wooden framing to support the canvas pressed against it by the wind. Sometimes the canvas, instead of being in one sheet, was subdivided into numerous separate sheets mounted on rollers, and apparatus was provided so that the canvas might be wound on the rollers or unwound at pleasure while the mill was in motion. As the wind is exceedingly variable, and as the quantity of work required of the

MOTIVE POWER.

mill also might vary to a considerable extent, it was found necessary to provide some apparatus by which the mill might regulate itself, so that its velocity should not be excessive at one time, and too small at another. One mode of effecting this object was to apply to the machinery of a mill a governor, like that of a steam-engine. This governor consists of two heavy balls suspended from the summit of a vertical revolving spindle by jointed rods. The spindle being at rest, the balls hang close to it on each side; but on the spindle being caused to revolve rapidly, the balls, impelled by centrifugal force, fly away from the central axis. A system of levers and rods connected this apparatus with the sail-rollers, so that when the balls flew outwards from increased velocity, the sails were furled; and when they fell

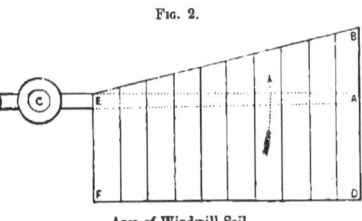

Fig. 2.

Area of Windmill Sail.

inwards from diminished speed of revolution, the sails were unfurled. The quantity of surface thus presented to the wind was adjusted to its force, and a tolerably equable velocity of the machinery was attained. In some more recent mills an ingenious contrivance for regulating the surface of sail according to the force of the wind has been successfully adopted. The sails consist of a framework filled in with louvre-boards hinged on pivot-pins near one of their edges, and all connected by levers and rods with a sliding boss on the central axis of the windmill, Fig. 3. When the wind blows strongly against the louvre-boards, it forces them out of their vertical position, and passes freely through the openings between them. The surface of the sails is thus diminished by the pressure of the wind itself. To prevent its being too much diminished, the sliding boss connected with the louvre-boards is pressed upon by a lever loaded by a certain weight sufficient to balance, as far as may be desirable, the pressure tending to force aside the louvres, and thus to keep them, to a certain extent, up to their work. When the load on the mill—that is to say, the quantity of work effected by it—is varied, the weight may be varied accordingly; and thus the effective amount of surface in the sails may be adjusted to the average force of the wind and the work to be done by it. When the wind-force exceeds or falls short of its average, the greater or less inclination of the louvres very nearly compensates for the variation.

Fig. 3.

Governor.

The sails of a windmill should directly face the wind in order to receive its most advantageous action; but, as the direction of the wind often changes, it is necessary to adopt some arrangement for varying that of the mill-shaft accordingly. The summit of the mill-tower, in which the mill-shaft is mounted, is therefore made to revolve, so that at any time the direction of the shaft may be varied and the sails presented to the wind. In many small mills this change of direction is effected by hand. A long lever is fixed to the movable cap or summit of the tower, and extends obliquely to the ground. The miller watches the direction of the wind, and by moving this lever turns the cap round to its proper position. But in large mills this would require considerable power; and, moreover, constant attention would have to be paid to the changes of the wind. Were a single change neglected the mill might be destroyed; for as the sails are made and strengthened by tie-rods to receive the wind's pressure on their face, a change of the wind to the opposite direction might throw a great strain on their back, for meeting which no provision is made. A simple mode of making the change of direction self-acting is to fit the back of the cap with a large vane, which, like that of a weathercock, would cause the sails to be presented to the wind from whatever quarter it might blow. But when mills are of considerable size the vane would require to be very large and cumbrous. The contrivance generally employed is neat and ingenious. Behind the cap, Fig. 4, on the side opposite that through which the wind-shaft passes, a framing is made to project outwards. On this framing there is mounted a small windmill on an axis transverse to that of the main arms. The cap rests on rollers fitted to the circular top of the tower so that it may move freely round; and a toothed circular rack is also fixed on the summit of the tower. A spindle, fitted with bevel-gearing so that it may be caused to revolve by the revolution of the small mill, conveys motion to a toothed pinion which gears into the circular rack. When the main mill has its face presented to the wind, the small one stands edgeways to it, and therefore remains at rest; but as soon as the wind veers it begins to act on one side or other of the small mill, and thus causes it to revolve. The pinion is thus made to travel along the fixed rack and turn the cap of the mill round until the main mill is again brought to face the wind in its new direction. This arrangement is found to be very effective, and when it is properly applied the mill requires no attention in respect of direction to the wind.

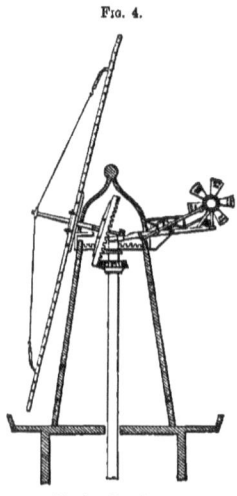

Fig. 4.

Turning Contrivance.

In estimating the velocity with which the sails of a windmill revolve, we have to consider not only the force of the wind upon them, but also the resistance to their motion occasioned by the work done by the mill. A, B, Fig. 5, may represent the edge of a surface presented obliquely to the wind, and capable of moving in the direction C, D, at right angles to that of the wind. If the surface be free and unresisted in its motion, and the wind be considered to produce its full effect upon it, the proportion of its velocity to that of the wind would be estimated by that of the line B, B', to the line B, A'; for it is clear that while the wind travels over the distance B, A', the surface moves to the position dotted, that is, over B, B'. But if the motion of the surface be resisted, its velocity in relation to that of the wind is diminished. In the case of windmill sails, we may suppose such a load of work on the mill that the velocity of the sails is not more than half what it would be were there no resistance. We may therefore assume that the velocity of the sail relatively to the wind would be expressed by the ratio of half the length of the line B, B', to the length of A, B'. Taking the wind as a gentle breeze, the velocity of which is about five miles an hour, and the inclination of the sail or angle A, B, B', half-way from the centre 18°, we should find the

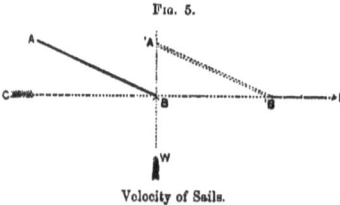

Fig. 5.

Velocity of Sails.

half of B, B,' to be about $1\frac{1}{2}$ times A, B, or the velocity of the sail, $1\frac{1}{2} \times 5 = 7\frac{1}{2}$ miles an hour—about 660 ft. a minute. If the windmill be about 60 ft. in diameter, the diameter of the middle point of the arm is 30 ft., the circumference of the circle in which that point revolves is 94 ft. and the number of revolutions made a minute is therefore $\frac{660}{94}$, about 7.

The speed of the extremities of the arms is 1320 ft. a minute, or about 15 miles an hour; three times that of the wind, which we have assumed as 5 miles an hour. Did we assume a wind of greater velocity, we should have to take into account the self-regulating arrangement, which diminishes the amount of surface exposed, and therefore prevents the mill from attaining so much increase of speed as it would without regulation. Under ordinary circumstances the speed of the outer extremities of the arms ranges from 20 to 30 miles an hour. We may assume 30 miles an hour when the wind blows at 10 miles with a pressure of about $\frac{1}{2}$ lb. on the square foot. The total surface of the sails unfurled in a mill 60 ft. diameter, is 1250 sq. ft.; we may suppose half lost by furling, leaving 625 effective. As the surface is set obliquely to the wind, the pressure in the direction of motion would be reduced from $\frac{1}{2}$ lb. to about $\frac{1}{4}$ lb. as a mean over the whole of the arms, giving a total pressure in the direction of motion of about 90 lb. The mean velocity of the arms is half that of the extreme, 15 miles an hour, or 1320 ft. a minute. We have therefore 90 lb. moving at 1320 ft. a minute, which is equivalent to a force of $90 \times 1320 = 118,800$ lb. moving at 1 ft. a minute. A horse-power is reckoned as equivalent to 33,000 lb. moved 1 ft. a minute; therefore the power of the mill we have reckoned is about $3\frac{1}{2}$ horse-power. When we double the diameter of a mill, we quadruple its power, for we quadruple its effective surface. The areas of circles are proportional to the squares of their diameters; and as the similar parts of the areas are occupied by sails, they are also as the squares of the diameters. It is not at all an easy matter to estimate the powers of windmills. The proper guide as to power, velocity, and construction is experience.

As a force applied to the movement of machinery, wind has few advantages except its little cost after the first outlay for a windmill has been made. It is chiefly available in flat countries, where there is no opportunity of obtaining the preferable power of water, and where there is little interruption to the aërial currents. In hilly countries windmills are often subject to derangement from the excessive force of the gusts of wind that often occur in such regions. In tropical countries, particularly islands and places near the sea-shore, the daily occurrence of the land and sea breezes, occasioned by the action of the solar heat on the land, provides a certain amount of wind-power, which may be almost always depended on. But in these countries, on the other hand, there often occur tornadoes or hurricanes of extreme violence, that sweep away almost everything that may oppose their progress; and thus frequently destroy windmills, and occasion renewed outlay in their reconstruction.

WATER.—Water motors render great and frequent service; for though not adequate to every emergency, they possess the advantage of requiring only the first outlay necessary to establish them, the redemption of which, with the interest accruing thereto, added to the expense of repairing, which is very small, constitute the only general costs of the motive power. Their disadvantage lies in the variations of level and volume to which a fall of water is liable; whence it follows that the power employed through its medium is not constant throughout the year; in some seasons it may be insufficient, in others greater than the requirements of the mill demand. Hence, regulating the power of water-courses becomes a matter of great importance. The causes of the variations of level and volume in a stream of water are such that, in most cases, they can be only imperfectly counteracted. The remedy consists in establishing large reservoirs in which the water may accumulate during the rainy seasons, and from which it may be drawn in nearly constant quantities.

The gross power of a water-mill is found by multiplying the weight P of the volume furnished by the stream a second, by the height H of the fall. Dividing this product by 75 kilogrammetres (the work corresponding to 1 horse-power) we get the gross power F expressed in horse-power,

$$F = \frac{PH}{75}.$$

The effective power of the mill depends solely upon the kind of motor adopted: it is the product of the gross power by the useful effect K of the motor:—

$$\text{Effective power } Fe = K\frac{PH}{75}.$$

It is therefore necessary in each particular case to choose the motor best adapted to the conditions of fall and volume in the stream to be used.

Wheels which receive the water on the top, or in a point situated between the summit and the horizontal plane passing through the axis, are called "overshot wheels." Wheels which receive the water between their centre and the bottom are called "breast wheels." Wheels which receive the water at the bottom, and upon which the water arrives with a velocity due to a height nearly equal to that of the fall are called "undershot wheels."

Overshot wheels are applicable to high falls, that is, comprised between 9 ft. and 40 ft.; above this limit their construction becomes difficult and costly.

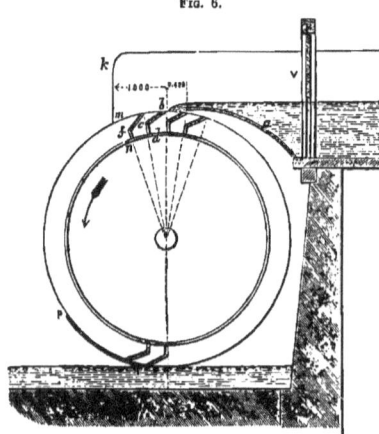

Fig. 6.

Overshot Water-wheel without Head.

When the stream has only a very small discharge, not exceeding 60 gal. a second, the canal which brings the water to the wheel is brought out to the crown of the wheel by a kind of trough, the bottom of which is cylindrical, a, nearly concentric with the wheel itself, Fig. 6. This bottom, which is usually of wood, terminates in a horizontal plank forming the overfall, which is placed at about 15½ in. short of the vertical line drawn through the axis of the wheel. The water flows over in a sheet, the thickness of which must not exceed 5¾ to 7¼ in. Therefore this system of wheel does not admit of variations in the level of the upper lade, for the smallest variations in this level would be great relatively to the thickness of the sheet of water on the overfall, and would cause considerable variation in the expenditure of water and consequently in the force of the wheel and its

MOTIVE POWER. 7

velocity; and the wheel must never dip into the tail-water, because the immersion of the buckets would prevent the efflux of the water and lessen the work of the wheel. If the level of the tail or back water varies, the bottom of the wheel must be fixed at the highest level. Overshot wheels of this kind, that is, without a head of water, are only suitable to streams that are nearly constant in their flow, and to work that offers a regular resistance. These wheels may be constructed wholly of wood, of wood and iron, or wholly of iron (cast iron, wrought iron, and plate-iron). Two cheeks k placed on each side of the trough enable several buckets to be filled every time the wheel is started. These cheeks should extend about 3 ft. beyond the vertical passing through the axis of the wheel. The sluice v fixed at the head of the trough is only for the purpose of stopping the wheel; when the wheel is going, the sluice is wholly raised, and consequently does not regulate the discharge.

When the buckets are of wood, which is usually the case, they are composed of two pieces, bc and cd, one of which is fixed in the direction of the radius, and the other in the direction of the relative velocity of the inflow of the water into the wheel. The direction of this relative velocity is found by comparing the absolute velocity with which the water arrives upon the wheel, and an equal velocity directly opposed to the linear or tangential velocity from a point in the outer circumference of the wheel. Usually the distance of two consecutive buckets apart is equal to the depth mn; this depth should not exceed 5 ft. The buckets are enclosed between rims or shroudings fixed to the arms. If the breadth of the wheel exceed $15\frac{1}{2}$ in., one or two intermediate rims are required, supported by a system of arms similar to those for the outer rims. The rotatory motion of the wheel impresses upon the surface of the water in each bucket the form of a portion of a cylindrical surface, the generatrices of which are horizontal, and the straight section of which is an arc of a circle, whose radius is expressed by $\frac{g}{w^2}$, w representing the angular velocity of the wheel. The water has a tendency to leave the wheel before the lowest point is reached; the consequence of this is a loss of work great in proportion to the height of the point p, where the anticipated discharge begins, above the level of the lower mill-race. This loss may be avoided by fixing a circular apron pq around the lower portion of the wheel from the point p. Overshot waterfalls, without a head of water, ought not to receive more than $7\frac{1}{2}$ gal. of water a second to the ft. of breadth. Their effective work varies from $0 \cdot 75$ to $0 \cdot 85$ of the gross work.

If the level of the upper mill-race and the volume of water are variable, the wheel cannot be fed by means of an overfall. Arrangements must be made by which the volume of water expended by the wheel may be varied, according to circumstances, without changing the velocity with which the water flows upon the wheel. These conditions are satisfied by constructing a vertical sluice a with a head of water h', Fig. 7, so that the distance mn from the bottom of the sluice to the floor of the pen-trough may in all cases be much less than the height h' of the head of water. A wheel-race bc, inclined to about $\frac{1}{10}$, brings the water upon the wheel; this race is provided with two side cheeks d, which extend about 3 ft. beyond the vertical line, passing through the axis of the wheel. The height h' of the head of water depends upon the total height H of the fall, and on the variations of level in the upper mill-race. It is not possible to fix absolute figures with respect to this; yet the values adopted should approximate to the following numbers:—

Values of H.	Values of h'.	Values of H.	Values of h'.
9 to 12 feet.	23 inches.	18 to 21 feet.	31 inches.
12 „ 18 „	27 „	21 „ 24 „	35 „

In this system of wheel, as in the preceding, the linear velocity measured on the outer circumference of the wheel should be about equal to that with which the water flows upon the wheel. Wheels with a head of water may receive 9 gal. and even more to the ft. of breadth a second. Their effective work is a little less than that of wheels without a head of water, and may be reckoned, as a mean, 0·75.

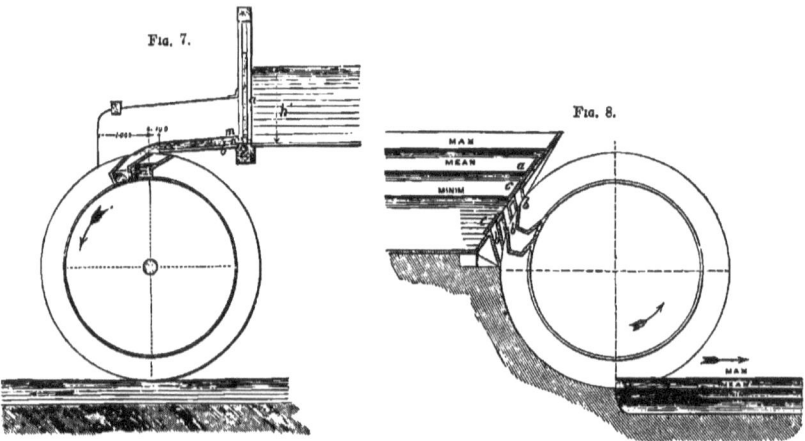

Overshot Water-wheel with Head.   Water-wheel for varying Volume.

When the level of the lower mill-race varies a little (4–6 in. at the most), and the level of the upper race and the volume of water vary greatly, the most suitable kind of wheel is that represented in a general way by Fig. 8. The upper mill-race terminates in a cast-iron pen-trough a, the inclined front of which is provided with a number of ajutages b. These may be opened or shut by two rectangular sluices c, each worked by its own mechanism. The buckets have the form shown in the figure, and the sole is provided with ventilators. One or more of the orifices is opened, according to the volume of water to be expended and the position of the level in the upper mill-race. The water is applied to this wheel at a point situate between the summit and the centre; it is known as the Wesserliug. The diameter of these wheels is usually determined by taking it equal to the height of the fall increased by 3 ft. There is nothing absolute about this rule; it is subordinate to the condition of obtaining the ready introduction of water into the wheel and a convenient form for the buckets. As this wheel moves in the direction of the water in the lower race, it may be submerged to a certain degree, 4–5 in. It may receive 18 gal. a second to the ft. of breadth, and its effective work is from 0·65 to 0·72.

The shaft of a bucket-wheel may be of wrought iron, cast iron, or wood; the arms may be of the same materials, but they are usually fixed in cast-iron sockets bolted to the shaft. When the buckets are of plate iron, they are usually curved according to a cylindrical surface.

MOTIVE POWER.

Breast-wheels include those which are enclosed in a circular breast or arc, and which receive the water at a point situate between their centre and their lowest part. Let H denote the whole fall made use of by the wheel, that is, the difference of the height of the levels in the upper and lower mill-race; h the fall utilised by the wheel, that is, the height of the point at which the water is applied to the wheel above the level of the lower race; V the velocity of the water on its arrival upon the wheel; v the velocity of a point of the periphery of the wheel; and P the weight of the volume of water expended a second. Then the useful effect or work T of the wheel is

$$T = Ph + \frac{P}{g}(V \cos. V v - v) v;$$

so that the fall utilised by the wheel is expressed by $h + \frac{v}{g}(V \cos. V v - v)$, and its duty

$$K = \frac{h + \frac{v}{g}(V \cos. V v - v)}{H},$$ of which the maximum is $v = \frac{V \cos. V v}{2}$. The duty increases as V

decreases, that is, the height of the portion of the fall taken as the generating weight of the velocity must be reduced as much as possible. Hence we have an arrangement which consists in supplying the wheel by means of a sluice that allows the water to flow upon the wheel from an overfall, called "slow" wheels. But this condition of flowing from a weir or overfall is often incompatible, either with the volume of water to be expended, or with the variations of level in the upper mill-race; hence the necessity of a sluice allowing the water to flow beneath it, that is, with a head of water. In this case the velocity V, and consequently that v of the wheel, are greater than in the preceding case. We thus obtain what are known as "mixed" or "impulse" breast-wheels.

Fig. 9 represents in elevation a "slow" with straight floats. The driving sluice is inclined so as to be placed as near as possible to the wheel. This sluice slides between two cast-iron supports fixed in the side walls, and rests against a fixed cast-iron apron called a swan-neck, to which a circular stone arc, covered with a layer of cement, forms a continuation; this arc must be constructed with care, so that the play to be left between the wheel and the arc may be reduced to within a few millimetres. The thickness of the sheet of water received by a slow wheel from an overfall should be at most 13–16 in.; with respect to the percentage of work and the ready introduction of the water into the wheel, the best thickness is about 10 in. The upper edge of the sluice should be rounded on the side of the water; often the sluice is provided on this side with a strip of sheet iron curved from left to right to guide the lower fillets before they reach the sluice, and consequently lessen the contraction. Instead of satisfying the relation $v = \frac{V \cos. V v}{2}$, most builders fulfil the condition

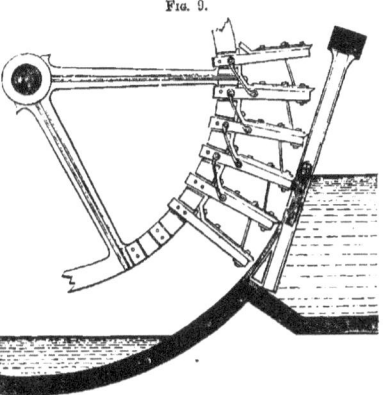

Fig. 9.

Slow Breast Wheel.

c

$v = V \cos. V v$, which is less favourable with respect to the percentage of work, but which allows of the floats being fixed in the direction of the radii of the wheel; this arrangement of straight floats simplifies the construction of wheels. To utilise, in part at least, the relative velocity of the water upon the floats, each straight float is continued by a counter-float inclined upon the float and the sole-plate. Between two consecutive floats is a ventilating aperture in the sole-plate, to enable the water to enter readily.

An absolute condition, from a theoretical point of view, which every breast-wheel must satisfy, is to be immersed in the water of the tail-race by a quantity exactly equal to the height occupied by the water in the floats that have reached the line perpendicular to the axis of the wheel. If the wheel does not dip deeply enough, there is a loss of fall equal to the half of this quantity; if the wheel dips too deeply, it meets in the water of the tail-race with a resistance which is equivalent to a loss of fall. Great care is therefore necessary in all cases to fix the position of the wheel in accordance with the variations of the volume which it is to expend, and the level of the water in the tail-race.

In undershot wheels the water arrives upon the floats with a velocity due to a head of water nearly equal to the height of the fall. When the floats are straight and radiate from the centre, the wheel is most imperfect, and its theoretical duty cannot exceed 50 per cent.; so that the practical duty does not exceed 35 or 40 per cent. of the gross work; and even to obtain this result, the wheel must be enclosed, on its lower portion, in a circular course equal in extent to the space of three consecutive floats, in order that there may not be in any case direct communication between the upper and lower races; care must be taken also to incline the sluice-gate from the wheel, and to place it as near to the latter as possible. The duty of undershot wheels with straight floats may be improved by utilising the velocity possessed by the water on leaving them. This is effected by making the floor of the course, immediately beyond the plumb-line of the wheel, a little lower than the natural level; in this way $0 \cdot 35$ or $0 \cdot 45$, or the real height of fall, may be gained. The way of doing this is to give, for a distance of 6 ft., an inclination to that portion of the race which immediately follows the circular course sufficient to enable the water to flow over it with a velocity equal to that of the wheel; from this part the race slopes about $\frac{1}{15}$ down to the natural bed of the stream. The depth of the floats should be equal to at least three times the opening of the sluice-gate; but this rule is not always sufficient; it is better to lay down the condition that the depth of the floats must be such as to keep them above the highest level of the tail-water.

Colladon's wheel with straight floats is designed to utilise the power of streams very variable in level, and offering only a very low fall. To prevent its being submerged in flood time, the axes are upon movable supports, which renders them capable of being raised or lowered at pleasure. It is a very primitive kind of wheel, having a duty inferior to that of common undershot wheels with straight floats, when well established. It is not suitable for wheels of great power, on account of the complication which is the consequence of the movability of the axis and the little rigidity which results from it.

The overshot wheels made by Green, Aberystwith, South Wales, are specially designed with the view to ensure the utmost durability, combined with moderate cost of construction, and are easily erected and taken down. The rings are of cast-iron segments, and all joints are truly planed, fitted, and well bolted together; centre pieces of cast iron, bored, slotted, and with wrought-iron bands shrunk on bosses; shaft of cast iron, turned and grooved; plummer blocks or carriages made with large base plates for resting on timber framework of wheel pit, and fitted with gun metal bottom steps; arms of best pitch pine, truly fitted to centre pieces and rings; buckets, backing, and risers of best yellow pine, planed. The whole fitted ready for erection, including all necessary bolts, arm plates, and washers (wrought-iron shaft and arms can be supplied, if desired, at a small extra cost), and one coat of paint:—

Diameter.	Breadth of Breast.	Price.	Diameter.	Breadth of Breast.	Price.
Ft.	Ft.	£	Ft.	Ft.	£
15	3	36	40	3	198
20	3	63	45	3	230
25	3	97	50	3	290
30	3	130	55	3	345
35	3	170	60	3	395

Extra charge if breadth or width is more than 3 ft. Delivered in Aberystwith Station, or on Aberystwith Quay. Packing and packing cases extra, from 3 to 5 per cent.

The only merit possessed by the preceding forms of water motor is their simplicity, making them available where a more efficient but more complex form would be undesirable, owing to inability to execute necessary repairs. Whenever possible, they are now replaced by turbines, whose great advantage is that they utilise the *vis viva* possessed by the water in virtue of the velocity with which it arrives upon the wheel, this velocity being due to a height sensibly equal to that of the fall. The water is brought upon the buckets or blades of the turning portion of the wheel, or turbine proper, by channels distributed over the whole, or sometimes over a portion only of the circumference of the turbine; these, with their various parts, constitute the fixed part of the wheel, sometimes called the distributor. Turbines may be erected upon either vertical or horizontal shafts. There are two classes of turbines with a vertical shaft. In those of the first class the water arrives horizontally upon the blades of the revolving part of the wheel through the interior of the latter, and issues horizontally, thus flowing away from the axis. The revolving blades form thus a series of vertical cylindrical channels included between two horizontal walls. In those of the second class the water enters the wheel from above and issues from below, remaining thus at a constant distance from the axis.

The practical Cullen, in his excellent treatise on the 'Construction of Horizontal and Vertical Water-wheels' (Spon, 2nd edition, 1871), remarks that if a ponderous vertical wheel be applied to a very high waterfall, its diameter will be so large, and its revolutions very few, that it must be connected with a great deal of auxiliary machinery to impart that rapid motion which is generally required. The consequence is, that through the friction occasioned by this additional machinery, considerable water-power is uselessly expended. On the contrary, the turbine being comparatively small, and its revolutions numerous in a given time, its motive power can be at once transmitted, thereby dispensing with the erection of additional shafting and wheels, and at the same time ensuring a considerable increase of power, with a machine not subject to get out of repair. Moreover, what operates as a disadvantage in the ordinary wheels, contributes to the more efficient working of the turbine; for the higher the waterfall, the smaller, and consequently the less expensive, the turbine adapted to it; also, it is applicable on falls of water so high that the ordinary wheel cannot be used. Another great property of the turbine is its constant and uniform motion, which arises from the diffusion of the impelling water over the whole of the circumference at the same instant. The turbine is capable of working under the back water as long as the surface of the fluid in the reservoir remains the highest, during which time it will produce a moving force proportional to the difference between these two levels, without a perceptible diminution of the useful effect, thereby evidencing that it is exempt from the casualties to which the vertical wheel is so often subject.

If the turbine be connected to a steam-engine during the summer months, while water is scarce, it can be made to transmit the highest obtainable power from the quantity of water by which it may be supplied, and it can be made so large as to drive all the works in winter—fuel being then dear and water abundant—saving the expense of fuel, economising the liquid that commonly runs to waste, and giving sufficient time for any repairs that might be required on the steam-engine, whereas a vertical wheel could not be made so large as to receive the extra water in winter, without lessening the effective power of the smaller quantities in summer. The turbine is capable of working under the tail water, and of discharging the largest supply of water for which it was made, and can work any less quantity without sustaining any diminution of its percentage power.

If an undershot wheel be applied to a fall of 3-4 ft., the useful effect produced will not exceed 30-34 per cent. of the expenditure. If a more favourable situation be selected, where, for instance, the waterfall would be 6-8 ft., and where the water is made to act as much as possible by its own weight, the useful effect might be 50-60 per cent.; the small percentage in the former case may be accounted for by considering the oblique direction in which the force of the stream acts on the floats, and the loss sustained by the water which escaped between the breast arc and rim of the wheel. There is another loss of power in all vertical wheels, by keeping them at a convenient height from the tail course, which is necessary in order that the water may have free room to escape from the wheel. It is, moreover, necessary that the water should descend a determinate distance to have the required velocity on entering it, and only one-half of this fall can have its full effect; whereas every inch of the entire waterfall may be made available when applied to drive a turbine, and which would yield under any fall a power of at least 75 per cent. of the water passing through it. Moreover, it can be adapted to work by the ebb and flow of the tide, though that advantage does not often affect its application in mining.

Cullen gives the following proportions for turbines:—

Q The quantity of water in cubic feet per second.
H The height of the waterfall in feet.

P The horse-power of the water at 75 per cent. $= \dfrac{QH}{700}$.

d The inner diameter of the wheel $= \sqrt{\dfrac{Q}{\sqrt[3]{H}}} + \cdot 1$.

N The number of buckets $= d \times 3 + 28$.

B The breadth of shrouding $= \dfrac{d \times 55}{N}$.

s The shortest distance between two buckets $= \dfrac{B}{4 \cdot 5}$.

D The external diameter to point of buckets $= B \times 2 + d$.

A The sectional area in inches between all the buckets $= \dfrac{Q \times 60}{\sqrt{H} \times 2 \cdot 18}$.

MOTIVE POWER. 13

h The height of buckets $\quad = \dfrac{A}{N\,8}$.
b The breadth of rim for directors $\quad = 8 \times 2\cdot 8$.
r The radius for centro of directing channels $\quad = D \times 3\cdot 6$.
v The velocity of inner circumference for low falls $\quad = \sqrt{H} \times 4\cdot 4$.
V The velocity of inner circumference for high falls $\quad = \sqrt[3]{H} \times 8\cdot 1$.
R The revolutions of wheel per minute $\quad = \dfrac{V \times 60}{d \times \frac{N}{7}}$.

U The diameter of turbine shaft in inches $\quad = \sqrt[3]{\dfrac{P \times 240}{R}}$.

Note.—$A = \dfrac{Q \times 60}{\sqrt{H} \times 2\cdot 18}$ for high falls; but $A = \dfrac{Q \times 60}{2\cdot 08}$ for falls under 38 ft. Power is gained by extending the shroud about ⅛ its breadth past the buckets when the water leaves them.

To find the Power of the Wheel at 75 per Cent.—The cubic feet of water passing through the wheel per minute, multiplied by the height of the waterfall, and divided by 700, will show by the quotient the power of the wheel. Thus, given 100 cub. ft. of water per second on a waterfall of 9 ft., required the proportions for a turbine in accordance with the foregoing rules, to be driven by 50 cub. ft., and 25 occasionally, and at the time of working with these supplies to produce at least 75 per cent. of useful effect:—

$\sqrt{\dfrac{Q}{\sqrt[3]{H}}} + \cdot 1 = 7\cdot 03$ ft., the interior diameter.

$d \times 3 + 28 = 49$, nearest number of buckets.

$\dfrac{d \times 55}{N} = 7\cdot 89$ in., breadth of shrouding to point of buckets.

$\dfrac{B}{4\cdot 5} = 1\cdot 753$ in., shortest distance between two buckets.

$B \times 2 + d = 8\cdot 345$ ft., exterior diameter.

$\dfrac{Q \times 60}{\sqrt{H} \times 2\cdot 08} = 961\cdot 53$ in., sectional area of bucket opening.

$\dfrac{A}{N\,8} = 11\cdot 175$ in., collected height of buckets.

$8 \times 2\cdot 8 = 5\cdot 806$ in., breadth of rim of directors.

$d \times 3\cdot 6 = 25\cdot 308$ in., radius for directors.

$\sqrt{H} \times 4\cdot 4 = 13\cdot 2$ ft., velocity of inner circumference.

$\dfrac{v \times 60}{d \times \frac{N}{7}} = 34\cdot 54$ revolutions per minute.

$\dfrac{77\cdot 14 \times 240}{34\cdot 54} = 8\cdot 12$ in., diameter of shaft.

$\dfrac{11\cdot 175}{2} = 5\cdot 5877$ in. high, first tier of buckets to pass 50 ft.

$\dfrac{5\cdot 5877}{2} = 2\cdot 793$ in. high for second and third tiers, each to pass 25 ft.

Or, required the number of cubic feet of water per minute, and all the other dimensions necessary to construct a turbine that will have 34 H.P. on a waterfall of 99 ft. 2 in.:—

$\dfrac{34 \times 700}{99\cdot 16} = 240$ cub. ft. of water per minute, or 4 per second.

$\sqrt{\dfrac{Q}{\sqrt[3]{H}}} + \cdot 1 = 1\cdot 029$ ft., the interior diameter.

$d \times 3 + 28 = 31$, nearest number of buckets.

$\dfrac{d \times 55}{N} = 1\cdot 826$ in., breadth of shrouding.

$\dfrac{B}{4\cdot 5} = \cdot 400$ in., shortest distance between two buckets.

$B \times 2 + d = 1\cdot 333$ ft., exterior diameter to point of buckets.

$\dfrac{Q \times 60}{\sqrt{H} \times 2\cdot 18} = 11\cdot 06$ sq. in., sectional area of openings between buckets.

$S \times 2\cdot 8 = \cdot 9288$ in. for rim of directors.

$\dfrac{A}{NS} = \cdot 888$ in. height of buckets.

$d \times 3\cdot 6 = 3\cdot 694$ in., radius of directors.

$\sqrt[4]{H} \times 8\cdot 1 = 37\cdot 478$ ft., velocity of inner circumference.

$\dfrac{v \times 60}{d \times \tfrac{22}{7}} = 715\cdot 49$ revolutions per minute.

$\sqrt[3]{\dfrac{34 \times 240}{715\cdot 49}} = 2\tfrac{1}{4}$ in., diameter of shaft.

In simpler terms, the height of the fall in feet, multiplied by the number of cubic feet of water per minute, divided by 706, will give the actual brake H.P. The H.P. required, multiplied by 706, and divided by the height of the fall in feet, will give the number of cubic feet of water required per minute. When the available quantity of water and the requisite H.P. are determined, the H.P., multiplied by 706, and divided by the quantity of water in cubic feet per minute, will give the height of fall in feet that will be required to produce the H.P. It must be remembered that these rules are based upon 75 per cent. efficiency. But when overshot water-wheels are used, the factor 706 must be altered to 815; or allowing only for 65 per cent., which is as much as can safely be relied upon, after deducting the loss of power in gaining speed by means of heavy gearing wheels. A good turbine will give 75 to 80 per cent., but in practice nothing more than 75 per cent. should be depended upon.

The old-fashioned water-wheel is, at best, clumsy and cumbrous, but in cases where the fall is less than 20–25 feet, it may be used, provided there is no scarcity of water, and that the cost of transit of so ponderous a machine is not serious; but the danger of accident to the gearing wheels, and the wear of bearings, render it out of place in most instances. It is very largely superseded by turbines, which are so much lighter, and which make so much better use of the water.

Turbines may be suitably classed under three heads:—

(1) Parallel-flow turbines, in which the water is supplied and discharged in a current parallel to the spindle.

(2) Outward-flow turbines, in which the water is supplied and discharged in currents radiating from the axis.

(3) Inward-flow turbines, in which the water is supplied from the outside of the wheel, and discharged near the centre.

In all these cases the water is caused by guide-passages, guide-blades, or jets, to enter into the vanes or partitions of the revolving wheel, in the direction of rotation desired, and if the makers know their business, good, useful turbines can be produced of either class. But there are reasons, which appear to be unanswerable, that tend to show that the best results can be obtained from the inward-flow turbine, for it does not depend on impulse only, and as it parts with the water near the centre of motion where the speed is slow, it would appear that it has a better chance of exhausting the energy of the water than those turbines in which it is thrown off the wheel at the outside, where it must be moving faster than at the inside.

The "Vortex" turbine is the invention of Professor James Thomson, of Glasgow, and as it is in a sense the father of all the others in Class 3, and as space prohibits a description of all, this form will serve better than any to illustrate the general principles and application of the inward-flow turbine.

The smaller wheels are constructed of rolled brass, and the larger ones of wrought iron, or steel plates locksmithed together. It is of great importance that the angle of the vanes where they discharge the water into the central opening should be suited to the speed, and to the quantity of water to be discharged, therefore each wheel is specially designed to suit the circumstances under which it has to work. It is in order that this may be done that wrought, and not cast, materials are used. Wheels that are cast, and not built together, must of necessity have the disadvantage of thicker, and consequently fewer vanes. When the fall is more than 10-12 ft., it is a frequent practice to place the vortex turbine at a height, not exceeding in any case 26 ft. above the tail-water, the shaft being horizontal. In such cases the water passes from the wheel into suction pipes, and the fall below the turbine is thus utilised, with precisely the same result as if it had been placed at the bottom of the fall with a vertical shaft and bevel wheel, as illustrated in Fig. 10. It is much more

FIG. 10.

Vortex Turbine at bottom of Fall.

convenient to have the turbine high and dry, and at a suitable height to take the power off by belts, as illustrated in Figs. 11 to 14. Many other turbines are placed in this position, but difficulty is sometimes experienced in the end thrust, which is developed by reaction, the water passing out at one side of the wheel. This difficulty is not met with in the vortex turbine, as half the water is discharged at each side of the wheel, and perfect equilibrium is maintained.

In all well-constructed turbines with an inward flow, the power can be adjusted by opening or closing orifices between the guide-blades, and the turbine can consequently be used when the water supply is abnormally low, or when very little power is required. Such turbines also possess, to a greater or less extent, an inherent power of self-governance, due to the centrifugal force of the water contained in the revolving wheel; but where the same speed must be maintained automatically under varying loads, a simple governor can be applied.

The following illustrations are from photographs of vortex turbines, and with the particulars given may be found useful to any readers who have not already seen such at work.

Fig. 11.

Vortex Turbine; part of Fall acting by Suction.

Fig. 11 is a vortex turbine arranged vertically, with part of the fall acting by suction. It is made by G. Gilkes & Co., Kendal.

Fig. 12 represents a 50 H.P. vortex turbine with 276 ft. fall, designed and constructed by G. Gilkes & Co., Kendal. This turbine is one of four which are in use in Wellington, New Zealand, in connection with the electrical plant for lighting that city. The wheel makes 1275 revolutions per minute. The diameter of the turbine case is $4\frac{1}{4}$ ft.

Fig. 13 illustrates a 60 H.P. vortex turbine, for 26 ft. fall, designed and constructed by G. Gilkes & Co., Kendal. In order to lessen the difficulty of carriage to the La Union Gold Mines, in Peru, this turbine was made in small pieces, none of which (the revolving wheel only excepted) weighed more than 300 lb. The speed of the revolving shaft is 175 per minute. The diameter of outer case is 9 ft.

Fig. 14 shows a 300 H.P. vortex turbine, with 300 ft. fall, designed and constructed by G. Gilkes & Co., Kendal. The normal speed of this high-fall turbine is 665. As it was designed for spinning, a simple hydraulic governor is attached. The rise and fall of the governor balls direct the flow of the water above or below the piston in the hydraulic cylinder, thereby regulating the position of the guide-blades inside the turbine case. The diameter of the outer case is $8\frac{1}{4}$ ft.

It will have been noticed in the foregoing instances that when the fall is high the speed of the turbine is great. The simple fact that the speed of the periphery of the wheel varies as the square root of the height of fall, will be found useful. An increase in the size of the wheel will of course diminish the speed, and *vice versâ*. When the fall is very high it may be found desirable to use the Girard turbine, which is one of the best outward-flow turbines. There is no limit to the size of this wheel, whereas the diameter of the vortex wheel cannot be increased beyond certain limits without materially decreasing its efficiency.

Fig. 15 illustrates the Girard turbine with a horizontal shaft, as made by G. Gilkes & Co., Kendal. In some cases the fall of water may be below the place where the power is wanted, and

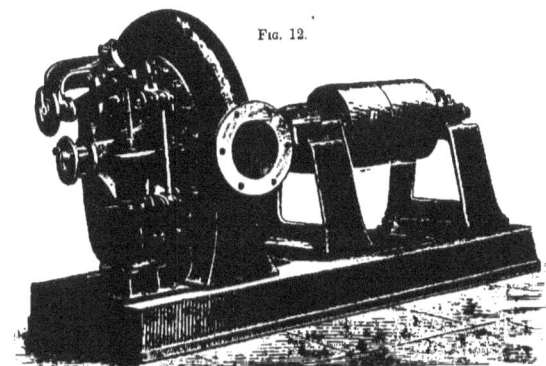

Fig. 12.

50 H.P. Vortex Turbine, with 276 feet fall, by G. Gilkes & Co., Kendal.

Fig. 13.

60 H.P. Vortex Turbine, 26 feet fall, by G. Gilkes & Co., Kendal.

the turbine consequently at some distance from its work. Transmission of the power by wire ropes or electricity is then often adopted with great success, and without any serious loss of power.

A large number of the turbines made by Messrs. John and Henry Gwynne, of Hammersmith Iron Works, and Cannon Street, London, have been used in connection with mining. Fig. 16

Fig. 14.

300 H.P. Vortex Turbine, with 300 feet fall, by G. Gilkes & Co., Kendal.

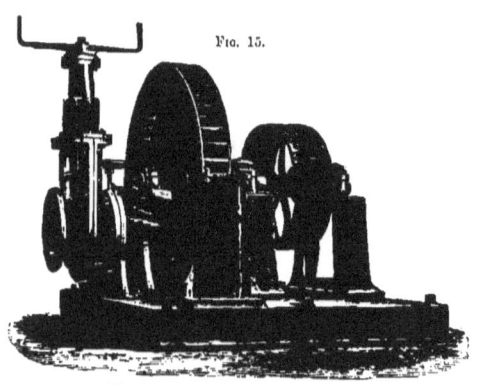

Fig. 15.

Girard Turbine, by G. Gilkes & Co., Kendal.

shows a general view of a turbine having a vertical spindle. It will be seen that the wheel is raised half out of the case to show its formation; the buckets of the wheel are formed in two halves, the lower part being a fac-simile of that shown; the lower buckets are made of rather larger area than the top, for the purpose of balancing the wheel and lessening the strain on the footstep, which object is further assisted by the wheel being made hollow, thus giving it buoyancy. The wheel

Fig. 16.

Vertical Turbine, by J. & H. Gwynne.

works in a special casing supported on a strong base plate, whereon the footstep, carrying the vertical shaft, is fixed. The lignum vitæ with which the footstep is lined can be removed and replaced in a few minutes without interfering with the shaft.

To enable the wheel to work with less than the normal quantity of water, the ports on each side are provided with slides that can be opened and closed at will. Assuming there are fourteen ports in the turbine, and seven are fitted with slides, the turbine could be worked at half power by closing all the slides, or at different powers by regulating the slides. By these means the water consumed is proportionate to the power developed, and a great saving is the result. This is of considerable advantage in dry seasons, when water is scarce and it is necessary to curtail the consumption. The arrangement ensures economy, by allowing the turbine to work at a reduced power, with corresponding reduced supply of water. As already explained, the lower end of the shaft works in the lignum vitæ footstep, and the upper end is carried by a massive thrust bearing. These turbines give an efficiency of 80 per cent.

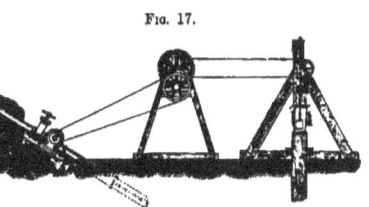

Fig. 17.

Horizontal Turbine, by J. & H. Gwynne.

Small streams of water falling from great heights may often be found in mining districts, which cannot be utilised with the common water wheels, because they would have to be of very large diameter. Messrs. John and Henry Gwynne make a small turbine with horizontal spindle arranged as in Fig. 17 to suit these falls. It is shown working stamping machinery. It requires very little foundation, and can be placed in the line of piping as shown.

Appended are short descriptions of the various types of turbines made by W. Günther, Central Works, Oldham, which are extensively adopted, both at home and abroad, and have a high reputation for durability and efficiency. As the height of fall and conditions of water supply are so varied, and different in almost each case, it will be readily understood that there can be no single type of turbine which is suitable to every want; and for this reason Günther varies the design and construction of his turbines to suit the conditions under which they have to work.

There are two classes of turbines—impulse (or action) and pressure (or reaction) wheels. To the former belong the Girard turbine, and to the latter the Jonval and numerous other types. The essential difference in principle between impulse and pressure turbines is that in the former the water leaves the guide ports with the full velocity due to the fall, and thus acts entirely by impulse; the wheel buckets are only partially filled with water, and the wheel works clear of the tail water. In pressure turbines the water leaves the guide ports at a much less velocity, and acts partly by impulse and partly by pressure; the wheel buckets are entirely filled with water under pressure, and the wheel works immersed in the tail water.

Since the water in a Girard turbine acts entirely by impulse, each jet works independently of the others, and hence the turbine will give the same efficiency, whether working with all the ports open or with a number of them closed. For this reason a turbine of this description is useful in places where the water supply is very variable, as, by closing some of the guide ports when the supply is reduced, the same efficiency is obtained and the available water power always utilised to the utmost extent. In some types of turbines the efficiency falls off when working at part gate; and in dry seasons, when only one-half or one-quarter the usual quantity is available, such turbines are often almost useless.

FIG. 18.　　　　　　　　　　FIG. 19.

Girard Turbines by W. Günther.

For low and medium falls with variable water supplies, Günther recommends Girard turbines with vertical shafts, and Fig. 18 illustrates the usual application of such turbines to falls from 12 ft. upwards. The turbine shaft is arranged with footsteps above water, so as to be readily accessible at all times. On falls above 50 ft. it often happens that a turbine as illustrated, with ports covering the entire circumference of the wheel, would be of comparatively small diameter, and run at an excessive speed. To avoid this, in such cases, the water is admitted on only a portion of the circumference, and the diameter of the wheel is increased, so as to obtain a moderate speed.

On high falls, over 60-70 ft., unless the turbine has to be placed in a deep pit, it is usually most convenient to adopt a turbine with horizontal shaft driving by ropes or belts, and Fig. 19 shows a Girard turbine so arranged. The wheel is made with partial injection and of comparatively large diameter, so as to avoid too high a speed. Experience has proved that, on account of its moderate speed, the Girard turbine can be successfully applied to very high falls.

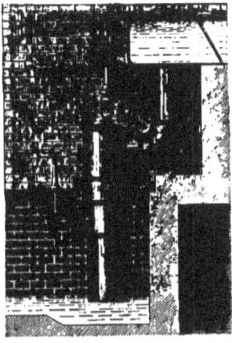

Fig. 20.

The Jonval turbine is applicable to low or medium falls where the water supply is not subject to great fluctuations, and to places where the rivers are frequently flooded, as it will work equally well even when deeply immersed in back water. Since it works drowned without loss of efficiency, it may be placed some height (not exceeding 25 ft.) above the tail water, the part of the fall below the wheel being utilised by air-tight suction pipes. Fig. 20 illustrates a Jonval turbine with vertical shaft so placed; and in Fig. 21 is shown a Jonval suction turbine with horizontal shaft driving by belting; this arrangement being of very frequent application for mining purposes. These turbines can be arranged with adjustments enabling them to give a good efficiency with a reduced water supply down to about half gate; but if the water supply is very variable, a Girard turbine is preferable.

Jonval Turbine.

OIL.—The Priestman oil engine (Fig. 22), manufactured by Priestman Brothers, of Hull, and Queen Victoria Street, London, should prove very useful to miners and others. Common

Fig. 21.

Jonval Suction Turbine.

mineral oils, flashing at 75° to 150°, are only used, and the engine is therefore perfectly safe, and may be worked by an unskilled person. No boiler, steam, coal, or gas being required, very little attention is necessary whilst the engine is working. The consumption of oil is about a pint per H.P. per hour. The engine is also very easy of transport, and is being made in fixed and portable types. They are already in use, amongst other purposes, for pumping in coal mines in restricted places where it would be impossible to use steam, and at a much less cost. We venture to think this engine will be largely used for mining work.

Fig. 22.

Priestman's Oil-engine.

CHAPTER II.

TRANSMISSION OF POWER.

THE transmission of power over short distances, as accomplished by gearing and driving belts, is not intended for discussion here, but rather that class of transmission where the distance becomes an important factor, comprising cases where (a) the power of natural sources is to be conveyed to distant points for useful application, and (b) the distribution of power from a great generating centre to a number of small independent working centres. There are four chief methods now in use—wire ropes, hydraulic pressure, compressed air, and electricity.

WIRE ROPES.—This method has been in use since 1850, and is very widely applied, though likely to be in some measure displaced by electricity as the latter becomes better known. The principle underlying this system is the conversion of force into velocity and its reconversion into force, the energy of the prime motor being transmitted in the form of velocity (say 80 ft. a second), and again resolved into force for application at the receiving station. The plant required is eminently simple, consisting of grooved wheels suitably supported and carrying an endless rope. The wheels are made of cast iron or steel, as light as possible consistent with due strength, and having a filling of leather or other soft material in the base of the V-shaped groove on which the rope runs. The speed of rotation varies from 25 to 100 ft. per second at the periphery, the limit being determined by the danger of destruction to the wheels by centrifugal force. Perfect balancing is essential. The shape of the V groove differs with circumstances. Wheels over 8 ft. in diameter are cast in sections and bolted together. Their approximate cost is as follows:—

Diameter.	Weight.	Cost.
5 feet	700 lb.	8l.
6 "	950 "	13
7 "	1100 "	16
8 "	1400 "	22
9 "	1700 "	38
10 "	2300 "	42

The figures on p. 24 concerning wire rope are given on the authority of Stahl's useful little book ('Transmission of Power by Wire Ropes,' 2nd edition, 1889, Van Nostrand) as being the American practice, the ropes being 6-strand of 7 wires each.

As it is often necessary to effect a splice in wire ropes, it will be well to reproduce the excellent directions given by Roebling, a well-known American maker. The tools needed are : pair nippers for cutting off ends of strands; pair pliers for pulling through and straightening ends of strands ; point to open strands; knife to cut core ; wooden mallet ; and 2 rope nippers, with sticks to untwist the rope. In operating : (1) heave the two ends taut, with block and fall, till they overlap each other about 20 ft.; open the strands of both ends for 10 ft.; cut both hemp cores as closely as

MINING AND ORE-DRESSING MACHINERY.

Iron Wire Rope.

Trade No.	Diameter in inches.	Price per foot.	Estimated Weight per foot in lb.	Breaking Stress, in tons of 2000 lb.	Proper Working Load in tons of 2000 lb.
		s. d.			
11	1½	2 0	3·37	36·0	9
12	1⅜	1 7½	2·77	30·0	7½
13	1¼	1 5	2·28	25·0	6¼
14	1⅛	1 1½	1·82	20·0	5
15	1	0 11½	1·50	16·0	4
16	⅞	0 9½	1·12	12·3	3
17	¾	0 7	0·88	8·8	2¼
18	11⁄16	0 6	0·70	7·6	2
19	⅝	0 5¼	0·57	5·8	1½
20	9⁄16	0 4	0·41	4·1	1
21	½	0 3½	0·31	2·83	¾
22	7⁄16	0 2¾	0·23	2·13	½
23	⅜	0 2½	0·19	1·65	..
24	5⁄16	0 2	0·16	1·38	..
25	9⁄32	0 1¾	0·125	1·03	..

Special Cast-steel Wire Rope.

11	1½	2 11	3·37	88·38	22·0
12	1⅜	2 6	2·77	67·20	16·8
13	1¼	2 1	2·28	60·67	15·2
14	1⅛	1 8	1·82	39·84	10·0
15	1	1 4	1·50	31·82	8·0
16	⅞	1 0½	1·12	24·70	6·2
17	¾	0 9½	0·88	18·48	4·6
18	11⁄16	0 8	0·70	16·32	4·0
19	⅝	0 7	0·57	12·44	3·1
20	9⁄16	0 5½	0·41	9·33	2·3
21	½	0 4	0·31	6·89	1·7
22	7⁄16	0 3¾	0·23	5·23	1·3
23	⅜	0 3½	0·19	3·93	1·0
24	5⁄16	0 2½	0·16	3·25	·81
25	9⁄32	0 2¼	0·125	2·96	·75

possible (A, Fig. 23), and bring the open bunches of strands face to face, so that the opposite strands interlock regularly. (2) Unlay any strand a, and follow up with strand 1 of the other end, laying it tightly into the open groove left upon unwinding a, and make the twist of the strand agree exactly with the lay of the open groove, until all but about 6 in. of 1 are laid in, and a has become 20 ft. long; cut off a within 6 in. of the rope (B), leaving two short ends which must be temporarily tied.

(3) Unlay a strand 4 of the opposite end, and follow up with the strand f, laying it in the open groove as before, and treating it precisely as the first (C); pursue the same course with b and 2, stopping, however, within 4 ft. of the first set; then with e and 5, c and 3, and d and 4; thus all the strands are laid in each other's places, with the respective ends passing each other at points 4 ft. apart, as in D. (4) To secure and dispose of the ends without increasing the diameter of the rope,

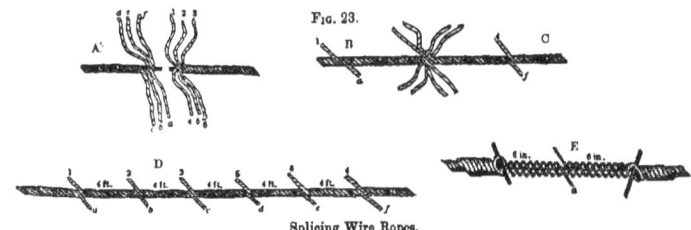

Fig. 23.

Splicing Wire Ropes.

nipper two rope slings around the wire rope, say 6 in. on each side of the crossing-point of two strands; insert a stick through the loop, and twist them in opposite directions, thus opening the lay of the rope (E); next cut the core for 6 in. on the left, and stick the end of 1 under a, into the place occupied by the core; then cut the core in the same way on the right, and stick the end of a in the place of the core, the ends of the strands being straightened before they are stuck in; loosen the rope nippers, and let the wire rope close; a wooden mallet will beat out any slight inequalities remaining. Repeat the operation in the other five places.

In the transmission of power by wire ropes, the causes which tend to thus waste a portion of the power in doing useless and even prejudicial work are—(a) rigidity of the ropes in bending to the curve of the main wheels and carrying-sheaves, which may usually be regarded as insensible, for when the wheels are made sufficiently large, the wires of the rope straighten themselves by their own elasticity after leaving the wheels; (b) friction of the journals of the wheel-shafts; (c) resistance of the air to the rotation of the wheels and to the passage of the rope through it. The friction of the journals varies directly with the pressure on the bearings, while the resistance of the air depends only on the velocity of the wheels and ropes. It must be noted that, as the pressure on the bearings depends only on the tension and deflection of the rope, which, with a given velocity of wheels, are constant, irrespective of the power transmitted, it follows that these losses are to a large extent independent of the transmitted power. When the direct tension due to the latter is small compared with the tension due to the span and deflection of the rope, these losses will become of considerable *relative* magnitude, so that it is a condition of efficiency that the system shall be worked at the highest suitable power. Under such circumstances, the efficiency of a single pair of stations has been determined to be 0·962. The efficiency of any whole system including a certain number of intermediate stations is given by Stahl as follows :—

Number of Intermediate Stations.	Efficiency of System.	Per cent. of Power Wasted.	Number of Intermediate Stations.	Efficiency of System.	Per cent. of Power Wasted.
0	0·962	3·8	3	·908	9·2
1	·944	5·6	4	·890	11·0
2	·925	7·5	5	·873	12·7

The efficiency is thus seen to be greater the fewer the number of intermediate stations.

MINING AND ORE-DRESSING MACHINERY.

A comparison of the four principal systems employed for transmitting power to distances, given in Beringer's "Kritische Vergleichung der Elektrischen Kraftübertragung" (Berlin, 1883), representing the commercial efficiency under various conditions of distance and power, all the systems being supposed to be working to the best advantage, is as follows:—

Distance of Transmission.	Electric.	Hydraulic.	Pneumatic.	Wire Rope.
300 ft.	·69	·50	·55	·96
1,500 ,,	·68	·50	·55	·93
3,000 ,,	·66	·50	·55	·90
15,000 ,,	·60	·40	·50	·60
30,000 ,,	·51	·35	·50	·36
60,000 ,,	·32	·20	·40	·13

It appears from this table that wire rope is most efficient up to about 3 miles, beyond which electric and pneumatic transmission such transmission are most efficient.

The cost of erecting and operating such transmissions varies greatly according to local circumstances. The following table gives the probable capital outlay required to establish transmission plants, not including buildings, boilers, chimneys, and cost of prime mover, as they are taken into account in the cost of producing 1 horse power per hour, but including all other expenses. From this table it will be seen that for distances less than about 1 mile the cost of the plant for wire-rope transmission is less than that for any other system, and specially so as the power to be transmitted increases in amount. For powers greater than 100 H.P., its cost is less for any distance not exceeding 3 miles.

PRIME COST PER H.P. TRANSMITTED.

Maximum HP. transmitted.	Distance of transmission.	Capital outlay per H.P.			
		Electric.	Hydraulic.	Pneumatic.	Wire Rope.
	ft.	£	£	£	£
5	300	73	40	71	6
	1500	76	64	94	30
	3000	79	94	204	50
	15000	105	348	584	296
	30000	138	594	1060	740
	60000	204	1206	2000	1188
10	300	50	29	58	5
	1500	53	44	70	22
	3000	55	63	86	46
	15000	75	214	208	225
	30000	100	406	360	448
	60000	150	784	662	910
50	300	39	16	30	2
	1500	40	20	35	7
	3000	41	30	41	14

TRANSMISSION OF POWER.   27

PRIME COST PER H.P. TRANSMITTED.—*continued*.

Maximum H.P. transmitted.	Distance of Transmission.	Capital outlay per H.P.			
		Electric.	Hydraulic.	Pneumatic.	Wire Rope.
	ft.	£	£	£	£
50	15000	54	89	86	67
	30000	67	166	143	132
	60000	97	316	258	265
100	300	31	14	25	1
	1500	32	20	29	4
	3000	34	27	33	8
	15000	44	86	65	40
	30000	57	160	106	79
	60000	85	302	187	158

COST PER H.P. RECEIVED (*Steam Power*).

Maximum H.P. transmitted.	Distance of Transmission.	Cost per H.P. received.			
		Electric.	Hydraulic.	Pneumatic.	Wire Rope.
	ft.	d.	d.	d.	d.
5	300	2·3	2·55	2·75	1·15
	1500	2·35	2·9	3·0	1·45
	3000	2·45	3·2	3·35	2·9
	15000	2·9	6·6	5·3	5·5
	30000	3·35	10·65	9·65	10·5
	60000	5·25	19·25	16·95	23·0
10	300	2·0	2·4	2·55	1·15
	1500	2·1	2·6	2·7	1·4
	3000	2·15	2·85	2·9	1·75
	15000	2·55	5·65	4·55	4·55
	30000	3·65	7·8	6·35	8·6
	60000	4·9	14·5	10·55	19·35
50	300	1·9	1·65	2·05	1·1
	1500	1·95	1·7	2·15	1·2
	3000	2·0	1·8	2·2	1·3
	15000	2·3	2·95	2·9	2·55
	30000	2·8	4·25	3·6	4·55
	60000	4·3	7·9	5·85	11·25
100	300	1·8	1·65	2·0	1·1
	1500	1·85	1·7	2·05	1·15
	3000	1·95	1·8	2·1	1·25
	15000	2·2	2·9	2·65	2·25
	30000	2·65	4·2	3·15	3·9
	60000	4·15	6·95	4·55	9·85

MINING AND ORE-DRESSING MACHINERY.

Cost per H.P. Received (*Water Power*).

Maximum H.P. Transmitted.	Distance Transmitted.	Cost per H.P. received.			
		Electric.	Hydraulic.	Pneumatic.	Wire Rope.
	ft.	d.	d.	d.	d.
	300	·35	·20	·40	·11
	1500	·36	·38	·47	·19
	3000	·37	·48	·58	·30
5	15000	·45	1·40	1·28	1·20
	30000	·52	2·53	2·43	2·53
	60000	·85	4·85	4·50	4·92
	300	·27	·25	·35	·09
	1500	·28	·30	·38	·17
	3000	·29	·37	·45	·25
10	15000	·36	·96	·89	·97
	30000	·47	1·56	1·44	1·93
	60000	·72	3·21	4·02	4·05
	300	·23	·15	·22	·09
	1500	·24	·18	·24	·11
	3000	·26	·22	·28	·13
50	15000	·29	·46	·44	·38
	30000	·31	·77	·65	·73
	60000	·55	1·44	1·09	1·63
	300	·20	·16	·22	·08
	1500	·22	·17	·23	·10
	3000	·23	·19	·24	·11
100	15000	·26	·43	·36	·28
	30000	·32	·73	·48	·48
	60000	·50	1·15	·84	1·20

These figures rather exaggerate the cost for electric transmission. The pneumatic system is well adapted for underground work, where it may assist in ventilation. It is almost always cheaper to transmit water power than to employ local steam power. Wire-rope transmission is always cheapest under ¾ mile, and electric beyond that.

HYDRAULIC.—The principal facts relating to this mode of transmitting power may be gathered from Donaldson's 'Transmission of Power by Fluid Pressure' (Spon, London, 1888.) Fluids under pressure transmit power simply as fluid pistons. When an elastic fluid is used for the transmission of power, no power can be transmitted until the minimum pressure required to do the work has been attained by actual compression. When an incompressible fluid like water is used, this piston is already formed, and the whole of the engine power expended actually transmits power. The only difference which can arise in the work due to machinery friction when expressed as a percentage of the whole work done in the pump barrel, must be caused by variation in piston friction work. If the work due to piston friction bears in both cases the same ratio to the total pressure on the piston, the whole work due to machinery friction will be the same in both cases, when the whole work done in the pump barrel is the same. In estimating, however, the value of the work due to friction in terms of the net effective result produced, measured by the product of the volume multiplied by the pressure in each case, the percentage of work due to friction in the case of air will be equal to the percentage in the case of water multiplied by the adiabatic and partial isothermal ratios respectively. Since water is incompressible, the density is constant for all pressures, and the increase of frictional resistance due to increase of pressure must be caused solely by the "skin" friction of the external film of water against the sides of the pipes, and will therefore affect the relative frictional values more in small than in large pipes. Kutter's formulæ for the flow of water are generally accepted as the most reliable, and are equivalent to the following, in which v is the velocity in inches per second and s the hydraulic inclination :—

FOR SMOOTH PIPES.

Pipes ¼ in. to 2½ in. diameter .. .. $v = 107\, d^{\cdot 9}\,\sqrt{s}$
„ 2½ „ 5 „ .. .. $v = 115\, d^{\cdot 8}\,\sqrt{s}$
„ 5 „ 10 „ .. .. $v = 134\, d^{\cdot 7}\,\sqrt{s}$
„ 10 „ 72 „ .. .. $v = 166\, d^{\cdot 6}\,\sqrt{s}$
„ 6 ft. to 400 ft. „ .. .. $v = 256\,\sqrt{d\,s}$

FOR MODERATELY SMOOTH PIPES.

Pipes ¼ in. to 2½ in. diameter .. .. .. $v = 63\, d\,\sqrt{s}$
„ 2½ „ 5 „ .. .. .. $v = 68\, d^{\cdot 9}\,\sqrt{s}$
„ 5 „ 10 „ .. .. .. $v = 78\, d^{\cdot 8}\,\sqrt{s}$
„ 10 „ 24 „ .. .. .. $v = 100\, d^{\cdot 7}\,\sqrt{s}$
„ 24 „ 96 „ .. .. .. $v = 138\, d^{\cdot 6}\,\sqrt{s}$
„ 8 ft. to 400 ft. „ .. .. .. $v = 221\,\sqrt{d\,s}$

Investigations prove that the diameters of pipes necessary to convey any assigned quantity of power in the case of water at a pressure of about 800 lb. are very much less than those required to convey the same quantity of power by means of compressed air, when the pressure to which the air is subjected does not exceed the limits usually found in practice. The extra thickness of the metal of the pipes in the case of high-pressure water is compensated by the greater size of the pipes required in the case of air and the greater cost of testing the soundness of the work. Since the engines required to utilise water of 800 lb. indicated pressure are of much less size than engines of equal I.H.P. actuated by air of 45 lb. indicated pressure, the first cost of such engines ought to be less. Engines for utilising compressed air must be capable of admitting the air during the whole stroke, and therefore ordinary stationary engines cannot be used until their valve gear has been altered. In the case of water, difficulties connected with the disposal of the exhaust water will not unfrequently arise. The first cost of the prime motors (engines and boilers) will vary with the I.H.P. required to produce the assigned effective power. The high-pressure water pumps, being of much less size than the air pumps, will cost very much less; and the accumulators, although much more costly than receivers of the same capacity, will probably cost less than the receivers, because the capacity of the latter must be about twenty times that of the accumulators. In the case of compressed air, the buildings required will be much larger also.

The application of hydraulic power to pumping deep mines is described farther on.

PNEUMATIC.—At the Newcastle (1889) meeting of the British Association, the subject of pneumatic transmission of power was ably dealt with by Prof. A. B. W. Kennedy, in a paper which is substantially reproduced below.

Compressed air has, of course, been used over and over again in rough and uneconomical fashion in connection with tunnelling and boring work, but only two practical attempts have been made to utilise it economically and on a large scale for industrial purposes. Of these two, one has been made in Birmingham and the other in Paris. The Birmingham Compressed Air Power Company has established works on a very large scale, but various causes have unfortunately combined to cause delay in the commencement of its operations, which indeed are hardly yet fairly started. In Paris, however, the transmission of power by compressed air has been in operation on a somewhat large scale and with very great mechanical success for a few years past.

In view of a recent discussion on the Hydraulic Power Company's work in London, in which some comparisons were made between power transmission by air and by water, Kennedy remarks that the two systems at present practically occupy different fields, and overlap but little. The work that each appears to do best is exactly that for which the other is least fitted. It would be a pity if there were to be any impression that two systems were antagonistic which, in point of fact, rather supplement each other. Kennedy's paper was limited to a description of the plant and methods used in Paris, and to a statement of the actual results obtained there, as determined by his own experiments on the spot. The plant and methods are by no means absolutely perfect; they are not only susceptible of, but are now receiving, considerable improvements in detail in the extensions which are being carried out.

Until about two years since, a pair of single cylinder horizontal engines by Farcot, and a beam engine by Case, sufficed for the whole work, but by that time the demand for compressed air for working motors had so increased that extension had become imperative, and the present working plant of six compound condensing engines, each working two air compressors, with the necessary complement of boilers, was put down. This plant, except the compressors, was supplied from England by Davey, Paxman & Co., of Colchester. The compressors for the English engines were made in Switzerland on the Blanchod system. The demand for power is at present so great that, at certain hours of the day, practically the whole plant, old and new, indicating considerably over 2000 H.P., is fully at work, and in consequence a duplicate main is being laid throughout, and new engines and compressors are being pushed forward as rapidly as possible.

FIG. 24.

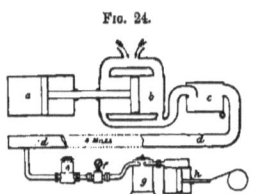

Pneumatic Transmission of Power.

The general system of working is illustrated roughly by the sketch diagram, Fig. 24, which of course is in no way drawn to scale, and it is as follows: The steam cylinders a compress the air to a pressure of 5 atmospheres (6 atmospheres absolute) or thereabout in the compressor cylinders b. The air is drawn in direct from the engine house at about 70° F., and after it has finally passed along the mains for some little distance it is again about the same temperature. It is, therefore, of the greatest importance to prevent its temperature rising during the compression, as all heat so taken up by the air represents work done in the steam cylinders of which no part whatever can be utilised. If the air were compressed adiabatically, i.e., without any cooling whatever, its temperature on leaving the compressor would be about 430° F.—a temperature higher than that of saturated steam of 300 lb. per square inch pressure. At St. Fargeau, water for cooling is allowed to run into the cylinders through the suction valve, during the suction stroke, in such quantity that the final temperature is only 150° F. So far the result is satisfactory enough, but owing, unfortunately, to the particular way in which the cooling water is utilised mechanically, the air does not get cooled until after it has been compressed, so that practically no benefit is obtained from the cooling, in spite of the extent to which it occurs. The power expended is practically equal to what would have been expended had the compression been adiabatic. The quantity of air dealt with at each revolution is 47·6 cub. feet (for the pair of double-acting compressing cylinders), which is equivalent to 3·55 lb., the quantity of water used being about 2·4 lb.

After compression, the air, now having an absolute pressure of 6 atmospheres and a temperature of 150° F., is pushed into large boiler plate receivers c, of which some are arranged to act as separators, and in these a large portion of the cooling water, which has been carried along mechanically by the air, is deposited and removed, before the air enters the mains d. The principal main is 11·8 in. in diameter and about $\frac{3}{8}$ in. thick. It is of cast iron, made in lengths perfectly plain at each end, and connected by a very simple external joint made airtight by rubber packing rings. This joint leaves the pipe quite free endwise, and also allows all necessary sideway freedom, so that accidental distortion to a quite measurable extent is entirely without effect on the tightness of the joint. The mains are partly laid under roadways and footways and partly slung from the roof of the sewer subways. They are supplied at intervals with automatic float-traps for carrying off the entrained water and the water of saturation, as they deposit.

On entering a building on its way to a motor, the air is first passed through a meter e exactly as gas would be. The quantity passing is of course too great to allow anything like an ordinary gas meter to be used; indeed, only inferential meters seem to have been at all successful. The meter actually in use in Paris is a small double cylindrical box. The air passes by a branch through to the bottom of the inner box, up through it, down outside it between the two boxes, and away through a branch at the bottom opposite the inlet branch.

The whole measuring apparatus is a little four or six-armed fan, with aluminium or nickel vanes, placed near the bottom of the inner casing, and communicating motion by a light vertical steel spindle to a clock-work register, like that of a gas meter, placed on the top. The quantity recorded is simply the number of revolutions made by the fan, or some proportional number, and this is turned into cubic meters by multiplication by an arbitrary constant, determined by direct experiment. This meter is the only type used by the Paris company, and serves in a very large number of cases as basis of payment.

After passing the meter the air is carried through a reducing valve f, by which the initial pressure in the motor is prevented from rising above a certain limit, which in practice appears to vary between $3\frac{1}{2}$ and $5\frac{1}{2}$ atmospheres absolute, according to the size of the motor in proportion to its work.

Between the reducing valve f and the motor h there is placed in all ordinary cases a small stove or heater g. This heater is simply a double cylindrical box of cast iron, having an air space between its outer and its inner walls. The air under pressure traverses this space, and is compelled, by suitably-arranged baffle plates, to circulate through it in such a fashion as to come into contact with its whole surface. A coke fire is lit in the interior of the stove, and the products of combustion are carried over the top of it, and made to pass downward over its exterior surface, inside a sheet iron casing, on their way to the chimney flue. The heater for the motor on which Kennedy experimented (which indicated 10 to 12 H.P.) was about 21 in. in diameter and 2 ft. 9 in. high over all.

The motors themselves h used in Paris are mainly of two types. Up to 1 H.P. or thereabout small rotary engines, of a form patented by Popp, are used. They start very readily, are easily governed, are provided with capital automatic lubricators worked by compressed air, are run at a very high speed, and are altogether very convenient. They use the air with little or no expansion, without previous heating, and have, of course, no pretence to economy in use of air.

The larger sized motors, up to double-cylinder engines 12 in. by 14 in., which is the largest size used, are simply ordinary Davey-Paxman steam engines, employed for air absolutely without any alteration or modification. These engines have, in most cases, automatic cut-off gear controlled by the governor, and can, therefore, easily work with the largest economical ratio of expansion for the not very high available initial pressure. In every case heaters are provided for these engines, although in some instances, where both power and refrigeration are required, they are used sparingly or not at all, in order to take advantage of the cooling due to expansion.

To come now to the experiments which Kennedy made to ascertain its efficiency. Starting from the main engines at the central station, the particular matter which he had to determine was the indicated H.P. which would be shown by a small motor three or four miles from St. Fargeau for each indicated H.P. expended by the main engines on the air which passed through that motor. The ratio thus obtained would be the total indicated efficiency of the whole system of transmission. This ratio is in reality the product of a number of separate efficiencies, and the separate determination of these formed a necessary check on the value of the total efficiency. These separate efficiencies may be summarised as follows :—

1. Mechanical efficiency of main engines, or ratio of work done in compressors to indicated work in steam cylinders.

2. Efficiency of compressors, or ratio of maximum work which could be done in a motor by each cubic foot of compressed air at 70° F. to the work actually done in compressing that air.

3. Efficiency of mains, or ratio in which the capacity of the compressed air for doing work is reduced by friction and leakage.

4. Efficiency of reducing valves, or ratio in which the capacity of the compressed air for doing work is reduced by the lowering of its initial pressure at the motor.

5. Indicated efficiency of motor, or ratio in which the actual indicated work done falls short of the maximum work which the quantity of air measured through the meter could do after passing the reducing valve.

The product of these five efficiencies is the total efficiency of transmission without the use of a heater. When a heater is used, the matter is somewhat more complicated. All the ratios given above represent what may be called mechanical efficiencies, all of them have *unity* for their maximum attainable value. It is, therefore, not possible to introduce in direct combination with them a thermodynamic efficiency (ratio of additional heat supplied to additional work done) which has for its maximum value, not unity, but 0·3 or some similar small value. This could only be done if the measurement of efficiency had started originally from the heat given to the steam instead of from the indicated H.P., and this would have given numbers having a minimum of practical value or convenience. Probably the best practical value of the efficiency of the whole transmission, when using heated air, is obtained by finding the equivalent in indicated H.P. at the central station of the coke used in the heater, and adding this to the indicated H.P. actually used. It would not be possible, by the expenditure of this or any other amount of indicated H.P. at the central station, to obtain the same results as by heating the air just before entering the motor, but that, of course, does not affect the question.

The determination of the indicated H.P. of the main engines presented no difficulty. Kennedy measured it on one pair of engines at different speeds from 21 to 44 revolutions per minute. At 31·5 revolutions per minute it amounted to 254·9, and at all speeds it was approximately 8·1 indicated H.P. per revolution per minute. The mechanical efficiency was sensibly the same at all speeds, viz., 84·5 per cent., as given in the table. There was no method available for ascertaining to what extent the real quantity of air delivered corresponded to the nominal volume swept through by the compressor pistons. The indicator diagrams showed no signs of leakage past the valve, but there are no doubt various possible leakages which would not show on the diagrams. In the absence of any direct means of determination, however, Kennedy assumed the compressor cylinders delivered their full volume, which corresponds to 348 cub. ft.* of air per indicated H.P. per hour. This air has a weight of about 25 lb. It may be pointed out that the water injected practically fills up the clearance space at the end of each stroke.

At whatever temperature the air is delivered, it must fall to about its original temperature in the long length of mains before it reaches the motors. It is therefore a simple matter to find the maximum amount of work which can be done by the air delivered per indicated H.P., for it simply amounts to the P V of the air at 6 atmospheres absolute and at 70° F., plus the work it can do in expanding adiabatically to a pressure of 1 atmosphere, and minus the work necessary to expel it from the cylinder at that pressure and at the corresponding temperature.† In the present case this work is equivalent to 0·52 indicated H.P. for an hour, so that the efficiency of the compressors is, as given in table, 61 per cent.

Kennedy determined the loss of head in the mains by a series of observations made simultaneously at known points in Paris and at St. Fargeau. The pressure gauges used having been carefully compared, and all the necessary corrections made, he found the loss of pressure to vary from 0·35 to 0·25 atmosphere, according to the distance from St. Fargeau and the amount of air passing through the pipes. The average loss may be taken as 0·3 of an atmosphere at 3 miles from St. Fargeau, when the indicated H.P. there was about 1250, and the maximum velocity of the air in the mains about 1550 ft. per minute. What proportion of this loss of head may have been due to leakage, and what the amount of leakage (if any) may have been, he had no means of determining.

In a table Kennedy gives approximate values of the loss due to fall of pressure in the mains and through the reducing valve, with various values of the total reduction of pressure. With a total reduction of half an atmosphere the combined efficiency of mains and valves is 0·96, reducing

* Here and elsewhere, unless specially mentioned, volumes are supposed to be at atmospheric pressure and at 70° F., the actual admission temperature during the experiments.

† In symbols, if suffixes 1 and 2 be used for the initial and final conditions of the air, if pressures be measured in lb. per sq. ft. and volumes in cub. ft., the maximum work possible, without addition of heat, is

$$(P_1 V_1 - P_2 V_2)\frac{\gamma}{\gamma - 1} = 3 \cdot 45 (P_1 V_1 - P_2 V_2).$$

F

the maximum possible work at the motor to 0·5 indicated H.P. per indicated H.P. at central station. Under these conditions the minimum possible consumption of air per indicated H.P. at the motor would be twice 348, or 696 cub. ft. per hour.

The motor on which Kennedy made most experiments was an ordinary horizontal Davey-Paxman engine, with a single cylinder 8¼ in. in diameter and 12-in. stroke, fitted with automatic cut-off gear. For convenience he tested it at St. Fargeau, and not in Paris, but used a pressure only of 4½ atmospheres, which pressure he found to be exceeded on branch mains 3½ miles from St. Fargeau, where he made later experiments. The position of the motor did not, therefore, put it under any conditions different from those existing in the centre of Paris. The motor, when indicating 9·9 H.P., and making about 125 revolutions per minute, used 800 cub. ft. of air per indicated H.P. per hour. The work which this quantity of air, at a given pressure and temperature, is theoretically capable of doing behind a piston, expanding down to atmosphere pressure, is equivalent to 1·27 H.P. for an hour. The indicated efficiency of the motor (the ratio expressing loss by rounding of curves, by insufficient expansion, by back pressure, &c.) is therefore 0·79. This figure gives us a check on the ratios already worked out, for if they are right, the air actually used should be $\frac{1}{0·79}$ times as great as the 696 cub. ft. already allowed for. This would be 880 cub. ft., which represents, of course, a most satisfactory check.

It will, however, be recognised that this agreement checks the figures only so far as they apply to air actually used, and would not be vitiated or in any way affected by losses by leakage. The vital measurement of all the experiments was, of course, that of the quantity of air used. The air was passed through one of the fan meters already described, readings of which were taken every quarter of an hour. After the experiments were over, air was passed through the same meter at exactly the same pressure, and in as nearly as possible the same quantity, and then passed, at atmospheric pressure, through two large standard wet gas meters. The readings of these were taken as correct, and the multiplier for the fan meter determined from them. Kennedy found, from numerous experiments on several fan meters, that this multiplier varied both with pressure and with quantity, but that the latter variation was very small within the limits of his experiments.

It will be seen that the total indicated efficiency of transmission with cold air is 0·39 (see table); in other words, that work requires to be done at the rate of 2·6 indicated H.P. at the central station per indicated H.P. at the motor. The motor was worked on a brake, and its mechanical efficiency was found to be 0·67, so that (see table) in round numbers 4 indicated H.P. were required at St. Fargeau per brake H.P. at the motor.

To examine the economy due to heating the air before using it, Kennedy used the same motor, working as nearly as possible at the same power and speed and with exactly the same pressure, but passing the air between the meter and the engine through such a heating stove as already described. He weighed all the coke used, and read the temperature every 5 minutes during a 4 hours' trial. The air was heated in passing through the stove up to 315° F., with a consumption of about 0·39 lb. coke per indicated H.P. per hour.

As the admission temperature on the cold trials was 83° F.[*] only, this corresponds to an increase of about 42 per cent. in the volume of the air, and should, therefore (had the indicated efficiency

[*] This somewhat high admission temperature was the only point in which the motor at St. Fargeau differed from those in Paris, where the admission temperature was from 69° to 71° F.

TRANSMISSION OF POWER.   35

remained the same), have been accompanied by a decrease of air consumption in the ratio $\frac{1}{1\cdot 42}$ or 0·70. The air actually used was 665 cub. ft. per indicated H.P. per hour, which is 0·75 of the 890 cub. ft. formerly required, so that the full economy is nearly realised. An air consumption of 665 cub. ft. per indicated H.P. per hour corresponds to an indicated efficiency over the whole system of 0·52; in other words, 1·92 indicated H.P. is required at St. Fargeau per indicated H.P. at the motor.

The mechanical efficiency of the motor was very much greater hot than cold, rising to 0·81. Hence about 2¼ indicated H.P. at St. Fargeau gave 1 brake H.P. at the motor.

These figures, however, take no account of the coke burnt in the heater, and are, therefore, only to be considered as *apparent* efficiencies. Allowing for the value of the coke in the manner already described, the indicated efficiency of the whole transmission is 0·47.

A shorter experiment with slightly higher temperatures and considerably larger indicated H.P. gave still more economical results, the air consumption falling to 623 cub. ft. per indicated H.P. per hour, an "apparent" indicated efficiency of 0·56. This experiment was not, however, of sufficient duration to allow of coke measurement.

As to the value of the preliminary heating, the figures given show that it caused a saving of 225 cub. ft. of air per indicated H.P. per hour, at a cost to the consumer of about 0·4 lb. of coke per indicated H.P. per hour.

Probably the stoking of the heater during the experiment was much more careful than it would be in ordinary practice, although, on the other hand, it would not be difficult to design a more economical stove. If, however, the coke consumption were even doubled, it would only amount to 72 lb. per day of 9 hours for 10 indicated H.P., the value of which might be 6d. or 7d.

The air saved under the same circumstances would be over 20,000 cub. ft., the cost of which, at the high rate charged in Paris, would be 7s. 3d.

There is no doubt, therefore, that to attain the maximum of economy, the preliminary heating of the air should be carried as far as is practicable.

Of course, heating the air serves the purpose also of preventing any chance of the exhaust pipe becoming ice-clogged. Kennedy found this to happen once or twice when working with cold air, its occurrence depending rather on the amount of moisture in the air than on the exhaust temperature, for the engine, after running freely with an exhaust of − 35° F., choked later on at + 2° F. Kennedy does not think that in any case which he met with there would have been any trouble from choking had the exhaust pipes been properly arranged. As it was, they were merely the ordinary vertical exhaust pipes of a steam engine, quite suitable for their original and for their intended purpose, but singularly unfitted for the purpose to which he was putting them.

Summarising the whole matter as regards efficiency, it may be said that the result of Kennedy's detailed investigations is to show that the compressed air transmission system in Paris is now being carried out on a large commercial scale in such a fashion that a small motor 4 miles away from the central station can indicate in round numbers 10 H.P. for 20 indicated H.P. at the station itself, allowing for the value of the coke used in heating the air, or for 25 indicated H.P. if the air be not heated at all.

Larger motors than the one tested (and there are a number of such in Paris) may work somewhat more, and smaller motors somewhat less, economically.

The small rotary motors would, of course, be much less economical. The figures given are,

F 2

however, such as can be reached by any motor of between 5 and 25 indicated H.P. if worked at a fair power for its size.

While unwilling to lay stress on possibilities which are not yet actualities, Kennedy has no doubt whatever that with mere improvement of existing methods and appliances, and without the adoption of any new or untried methods whatever, the new plant of the Paris company now being constructed can be made to have an indicated efficiency of 67 per cent. instead of 50 per cent., and to give about 0·54 effective H.P. at the motor for each indicated H.P. at the central station, in the case of such a motor as that on which he experimented.

Under these circumstances the air used per indicated H.P. at the motor would be 520 cub. ft., or 650 cub. ft. per brake H.P. He has the less hesitation in giving these hypothetical figures because the more important imperfections of Popp's transmission system arise from a very obvious cause. Nothing, indeed, can be easier than to point out various weak points in the arrangements adopted, and yet the fact remains that no one has yet carried out a compressed air transmission with anything approaching to the same success on anything like the same scale. The fact is, that the success of the system has been essentially due rather to the practical good sense with which the work has been carried through than to any special novelty in the methods employed.

The air-compressing arrangements at St. Fargeau are in no respects novel or specially perfect, they had been used over and over again before. There is no special advantage in Popp's rotary motor that may not probably be possessed by many other rotary motors; the larger motors are simply good ordinary steam engines, such as can be bought any day in open market, without the slightest alteration.

Of the fan meter it can only be said that it works well enough to allow progress to be made while it is being improved, and even of the coke stove one would not like to say very much more.

SUMMARY OF EFFICIENCIES OF COMPRESSED AIR TRANSMISSION AT PARIS, 1889, BETWEEN THE CENTRAL STATION AT ST. FARGEAU AND A 10-H.P. MOTOR WORKING WITH PRESSURE REDUCED TO 4½ ATMOSPHERES.

(The figures below correspond to mean results of two experiments cold and two heated).

1 indicated H.P. at central station gives 0·845 indicated H.P. in compressors, and corresponds to the compression of 348 cub. ft. of air per hour from atmospheric pressure to 6 atmospheres absolute. (The weight of this air is about 25 lb.)	Efficiency of main engines 0·845.
0·845 indicated H.P. in compressors delivers as much air as will do 0·52 indicated H.P. in adiabatic expansion after it has fallen in temperature to the normal temperature of the mains.	Efficiency of compressors $\dfrac{0\cdot 52}{0\cdot 845} = 0\cdot 61$.
The fall of pressure in mains between central station and Paris (say 5 kilometres) reduces the possibility of work from 0·52 to 0·51 indicated H.P.	Efficiency of transmission through mains $\dfrac{0\cdot 51}{0\cdot 52} = 0\cdot 98$.
The further fall of pressure through the reducing valve to 4½ atmospheres (5½ atmospheres absolute) reduces the possibility of work from 0·51 to 0·50.	Efficiency of reducing valve $\dfrac{0\cdot 50}{0\cdot 51} = 0\cdot 98$.

The combined efficiency of the mains and reducing valve, between 5 and 4½ atmospheres, is thus $0\cdot 98 \times 0\cdot 98 = 0\cdot 96$. If the reduction had been to 4, 3½, or 3 atmospheres, the corresponding efficiencies would have been 0·93, 0·89, and 0·85 respectively.

TRANSMISSION OF POWER.

Incomplete expansion, wire drawing, and other such causes reduce the actual indicated H.P. of the motor from 0·50 to 0·39.	Indicated efficiency of motor $\dfrac{0\cdot 39}{0\cdot 50} = 0\cdot 78$. Indicated efficiency of whole process with cold air 0·39.
By heating the air before it enters the motor to about 320° F., the actual indicated H.P. at the motor is, however, increased to 0·54. The ratio of gain by heating the air is therefore $\dfrac{0\cdot 54}{0\cdot 39} = 1\cdot 38$.	Apparent indicated efficiency of whole process with heated air, 0·54.
In this process additional heat is supplied by the combustion of about 0·39 lb. coke per indicated H.P. per hour, and if this be taken into account the real indicated efficiency of the whole process becomes 0·47 instead of 0·54.	Real indicated efficiency of whole process with heated air 0·47.
Working with cold air, the work spent in driving the motor itself reduces the available H.P. from 0·39 to 0·26.	Mechanical efficiency of motor, cold, 0·67.
Working with heated air, the work spent in driving the motor itself reduces the available H.P. from 0·54 to 0·44.	Mechanical efficiency of motor, hot, 0·81.

ELECTRIC.—The electric transmission of power was the subject of a paper by A. T. Snell, read recently at the Wigan Mining School, from which the following remarks are condensed. An electrical plant consists of four essential parts :—(a) steam or water plant used to drive the dynamo; (b) dynamo in which the steam power is converted into electrical energy; (c) conductor by which the current is carried from the dynamo to the motor; (d) motor which reconverts the electrical energy into mechanical work. The motor is simply a machine capable of giving so many H.P., and may be coupled to any required work by the ordinary methods—belting, gearing, &c. It is not necessary to understand the principles of the motor in order to successfully work an electrical plant; but a few words of explanation with reference to the conversion of energy will not be uninteresting. The motor may be considered as the converse of the dynamo. The rotation of the dynamo armature converts mechanical energy into electrical; and electrical energy supplied to a motor causes a rotation of the motor armature. In each case the phenomenon is traced to magnetism. In the dynamo we are continually doing work against the attraction produced by magnetism, and in the motor the magnetism induced by the current of electricity causes rotary motion.

The practical electrical units, the volt, ampère, and ohm are connected by the equation :—

$$C = \frac{E}{R}.$$

where C is expressed in ampères, E in volts, and R in ohms. These quantities are severally certain numbers of units of quantity, pressure, and resistance. The first two are measured directly by simply reading suitable instruments placed in the circuit; and their product, C × E, gives us the rate at which electrical work is being done. This rate is measured in units called watts. Now, 746 watts are equal to 1 H.P. The ratio between the watt, and the H.P. is thus 1 : 746. If we divide the product of ampères and volts by 746, we shall have the electrical work done in the circuit expressed in H.P.

The H.P. = 33,000 ft. lb. per min. or 550 ft. lb. per sec.

And since the ratio of the watt to the H.P. is 1 : 746,

$$\text{the watt} = \frac{33,000}{746} = 44 \text{ ft. lb. per min. or } \cdot 733 \text{ ft. lb. per sec.}$$

We can thus express electrical output in H.P. It is only necessary to read the ampères shown on the ammeter, and the volts indicated by the voltmeter, to multiply the product of these two quantities by 44, and we have the number of foot pounds per minute done by our dynamo. It is usual, however, to divide the quantity C E by 746, and thus estimate the electrical work in H.P.

The dynamo is the first point which demands especial attention. Primarily, we require a clean dry house, and a fairly steady-running engine; essentially, this is all, and the dynamo will give a maximum of work for a minimum of attention. The same remarks apply to the motor, but special stress may be laid on the compactness and large output for a given weight. One H.P. can be obtained for about 70 lb. of material in cases where weight is an object. In ordinary mining practice, however, an output of 1 H.P. for every 100–120 lb. is generally preferred. There remains, then, the conductor, or cable connecting the dynamo and motor. Essentially, this is of metal, and the circuit must have metallic continuity. Theoretically, the kind of metal does not affect the problem if sufficient cross section be allowed; indeed, this is rather a matter of cost and convenience. In practice, only copper and iron are in general use. Copper has about seven times the conductivity of iron, or is seven times as good a conductor of electricity. If we use copper for the cable, we only require one-seventh the cross section that would be necessary with iron for the same loss in transmitting a given quantity of energy. Bare copper wire does not, at the usual market prices, cost much more than iron cable of seven times the area, and is, further, less bulky and cheaper to erect; so in most cases, where bare conductors can be used, copper obtains the preference. With covered or insulated wires copper is universally employed, since the smaller bulk requires less insulation. The choice of conductor depends principally on the question of prime cost. If, however, a high electrical pressure is employed, we are compelled to use a high-class insulated cable, with a copper conductor, not so much owing to the cost as to the necessity of presenting a high resistance to leakage of the current. We may note here that electricity is always trying to shorten the path between the positive and negative brushes of the dynamo. There is no such thing as absolute insulation in practice, and hence a certain quantity of the current is always wasted. The amount is negligible in a good installation, being a fraction of $\frac{1}{10}$ per cent. at the most. It varies directly with the pressure, and inversely with the insulation resistance, hence the necessity for good insulation in large power plants.

The efficiency of the conversion of the dynamo and motor for medium size machines may be taken as averaging 90 per cent. Hence, if there were no loss in the conductor, the ratio between the work done by the motor and that on the belt of the dynamo would be about 80 per cent. But energy cannot be transmitted by the cable without some loss. The law which regulates this loss is very simple. The H.P. lost in a conductor carrying an electric current is equal to the square of the current in ampères, multiplied by the resistance in ohms, and divided by 746. The resistance can be taken approximately from any of the wire manufacturers' tables, if we know the gauge. The resistance of a given wire, however, is directly proportional to its length. Also, the resistance is inversely proportional to the area of the conductor, or a wire of 1 sq. in. cross section has one-half the resistance of one only $\frac{1}{2}$ sq. in. In practice the problem usually presents itself this way. Given a certain distance between dynamo and motor, determine the best size of cable to transmit a definite quantity of energy. This presents several practical points which somewhat complicate the simple theory.

To give some idea of the magnitude of the loss to be expected in the cable, we will refer to the Normanton plant. The conductor is about 1000 yd. in length, and is composed of 19 strands of

No. 16 B.W.G. copper wire. The energy transmitted is roughly 50 H.P., and the loss is approximately 2¼ H.P., or 5 per cent. If the cable were 1 mile in length, the loss would be about 8·75 per cent. It will be interesting to give particulars of a few installations, commencing with a small pumping plant erected at St. John's Colliery, Normanton, in August 1887. The pump delivered about 39 gal. per minute through 530 ft. head. The results obtained were so satisfactory that the owners decided to put down a larger plant, to deliver 120 gal. per minute through 900 ft. of vertical head. This larger plant was started in February 1888, and has continued running about 22 hours a day since then. The run now exceeds 18 months, and during that period there have been no breakdowns traceable to the electrical details, all stoppages being due to mechanical defects, incidental generally to the engine or pumps. The engine is semi-fixed, compound, and is rated by the makers at 30 nominal H.P. It has indicated during the past 18 months on an average about 80 H.P. The dynamo is driven off the fly-wheel by a link belt 14 in. wide, curved to fit the pulley. The belt speed is about 2750 ft. per minute. The dynamo is designed to give 600 volts and 70 ampères. The motor is of similar design to the dynamo, but the field magnets are of lighter construction. The pumps are differential with two 6-in. and two 4½-in. rams. The suction is made by the two large rams only. On the in-stroke the 6-in. rams deliver water partly into the rising main and partly into the small rams. On the out- or suction-stroke the 4½-in. rams deliver into the column, and so the discharge is fairly continuous. When doing full duty, the pumps make 25 revs. per minute. The rising main is about 450 yd. long, and is composed of 4-in. cast-iron pipes. An air vessel about 5 ft. high is fitted at the lower end near the delivery clacks. The piping is too small for 120 gal. per minute. It was designed for a feeder of about 50 gal. The water travels in the column at about 250 ft. per minute, and there is nearly 10 H.P. lost in friction. This heavy loss is against the total efficiency of the plant. The cable is built up of 19 strands of No. 17 B.W.G. copper wire, insulated and lead covered. It is about 1000 yd. long, and has a resistance of about 0·5 ohm. About one week after the erection, the electrical quantities for the full load averaged about 65 ampères and 603 volts, with 450 revs. of the dynamo and 134 revs. of the engine per minute. The motor run at about 450 revs. per minute, and the pumps at 25 revs. Under the above conditions the water delivered was measured, and found to be 118-120 gal. per minute. The suction has a rise of about 14 ft., and the column is 860 ft. in vertical height. Thus the theoretical H.P. in the water is about 32. The engine was indicated at the same time, and found to be 80 H.P. The efficiency of the system, that is the ratio between theoretical work in the water and I.H.P. of engine, was therefore equal to $\frac{32}{80}$ = 40 per cent. These tests have been repeated from time to time, and the readings show a gradual decrease of the losses due to friction, particularly in the engine and pumps. At the last test the current had fallen to about 62 ampères, and the engine only indicated about 73 H.P. The efficiency had thus increased to about 43 per cent.

The percentage of losses, as calculated from the indicator cards, are:—

	H.P.	Per Cent.
Engine friction	= 6·9 =	9·4
Belt and dynamo friction	= 4·8 =	6·5
Leads and motor	= 6·7 =	9·4
Motor, belt, gearing, and pumps empty	= .10·2 =	14·0
Load of 117 gal. through 890 feet	= 31·5 =	43·1
Water friction in pumps and rising main	= 12·9 =	17·6
	73·0	100·0

The engine doing above load indicated 73·0 H.P. The friction of the water in the pumps and rising main is arrived at by subtracting the sum of the theoretical H.P. in the water delivered, and the total friction, from the total I.H.P. of the engine; or $73 - (31·5 + 28·6) = 12·9$. This is not quite correct. The friction diagrams are necessarily taken with no load on the pumps, and hence are all slightly lower than is actually the case when the load is on. From the known efficiency of the motor, the loss in the pumps and rising main has been found to be not less than 10 H.P.

As an example of electricity applied to drive a single-rope hauling engine, Snell refers to a plant at Llanerch Colliery, near Pontypool, Monmouthshire. This plant has now been in use for about 5 months. The dynamo is driven by a horizontal engine with an 18-in. cylinder and 3 ft. 6 in. stroke, making at full speed about 50 rev. per minute. The steam pressure varies between 60 lb. and 50 lb. at the main boilers, and the speed is controlled by a Pickering governor. The engine drives a countershaft by a 9-in. × ⅜-in. link belt, and the dynamo is run off this shaft by a similar but lighter belt. Provision is made for running a second dynamo for lighting and other purposes. The engine-house is built with sandstone quarried on the spot; the pit shaft is 250 yd. deep, and the motor-house is about 750 yd. in-bye on the main intake. The copper cable is composed of 19 strands of No. 18 B.W.G., insulated and covered with lead. The shaft is rather wet, and the road is also damp, but no special trouble has been caused by this; the roofs and sides are not particularly good. The "falls" have caused some trouble with the cable, but it is fully expected that the arrangements now made will meet the case. The motor is coupled to a countershaft on the hauling engine bed by a link belt. A pinion on this shaft engages a wheel mounted on the drum axis. The drum is thrown out of gear by a lever to let down the empties, as is usual with single rope haulage. In fact the haulage engine is an old type machine, previously driven by a single engine, and converted for its present use by the removal of the crank shaft and cylinder. The rope is steel, and is ¾ in. in diameter. Two drifts are worked; one with a grade of 1 : 8, the other about 1 : 12. They are both at present some 300 yd. long, but will increase in length as the "face" recedes. The trams used are built entirely of iron; they average about 7 cwt. empty, and carry about 22 cwt. of coal, so that each loaded tram represents a rolling load of about 29 cwt. The wheels are 12 in. in diameter, and each pair is keyed to the axle. The rails are of the girder type, and weigh about 28 lb. per yd. The general condition of the road is well up to the average of colliery practice. The rolling friction, as measured by a dynamometer on a level part of the road, averages nearly 70 lb. per ton. The drift with the 1 : 8 grade has two "partings" in continual use at present, and hence there is a great deal of shunting, stopping, and starting. The signals are given by a "rapper," and the motor is controlled by a lever similar to the regulator in ordinary use on steam engines. The notches correspond to different rates of motor power and speed. The trams can thus be moved a few inches at a time, if necessary, and the speed can be varied from a foot per second to the desired maximum. An ammeter is placed in the circuit near the lever, so that the brakesman can tell the number of trams on the rope, and can also judge whether it is going right with the "journey." If, for instance, a tram leaves the metals, the ammeter needle registers more than the maximum current for the full load; the driver at once stops the motor and brakes the drum; the tram is then set right, probably before any damage is done. The power absorbed by the drum and rope averages about 5 H.P. The ohmic resistance of the cable is 1·25 ohms. The electrical losses are small. The heaviest loss is that incurred by the friction of the drum and rope; this could probably be reduced by designing special gearing. Accepting things as they are, the efficiency between the maximum work on the rope and the probable I.H.P. of engine is about 50 per cent. The strain on the rope on the 1 : 8 grade was measured by a dynamometer, and the result averaged for one and two tram readings. It is fairly represented by a pull of 4·5 cwt. per tram. Taking the grade as 1 : 8, the resistance due to gravity is 280 lb. per ton. The average resistance measured by dynamometer on the level was 70 lb. per ton. These figures show the high efficiency that a properly-designed electrical plant is capable of. If we take the dynamo conversion at 85 per cent., and the engine efficiency at the same figure, the total efficiency for six trams (i. e. the ratio between work in the rope and the I.H.P. of the engine) will be equal to 67 × ·85 × ·85 = 48·5 per cent. The motor is also arranged to run a set of three-throw pumps. When these are running the hauling drum is thrown out of gear.

Experiments made in the laboratory of the Compagnie Electrique, Paris, on the transmission of motive power to great distances by means of electricity, initiated by Hippolyte Fontaine, comprised the electrical transport of an initial power of about 100 H.P., against a resistance of 100 ohms, with an efficiency of 52 per cent. Transmission is easily effected where the power does not exceed 30 H.P., nor the distance 1¼ mile. On the contrary, there is considerable difficulty for higher powers or longer distances, especially longer distances; and it is necessary to reduce the intensity of the current, and augment its electromotive force, in order to obviate the loss of the greater part of the disposable energy in the line. A 4-valve steam engine of Farcot's, nominally of 60 H.P., working with steam of 5 atmospheres, develops 95 H.P. at the fly-wheel, which is 16¼ ft. in diameter, and making 55 turns per minute. By a band from the fly-wheel, an intermediate shaft, making 180 rev. per minute, is driven, from which 4 dynamos are worked, by means of two 6½-ft. pulleys, and 4 friction wheels, constructed of compressed paper, 15¼ in. in diameter, 10 in. wide, mounted on Gramme dynamos, and driven by the pulleys by contact. Each machine oscillates on a pivot placed below it, so that the weight of the machine itself determines the pressure of the friction-wheels on the pulleys. By means of a fast-and-loose coupling the friction-wheels can be promptly placed in or out of contact with the pulleys. The whole system, pulleys comprised, is contained in a space 11½ ft. by 12 ft. The intermediate shaft is about 20 ft. from the fly-wheel shaft. The receiving apparatus is simple. The dynamos are mounted end to end, on a stone foundation, and are connected together with rubber couplings; they make 1200 rev. per minute. They occupy a space of about 23 ft. square, which includes the connection for the brake. Following are the results of experiments made in July 1887 :—

Mechanical power disposable on the periphery of the fly-wheel of the steam-engine .. 95 H.P.
Mechanical power delivered to the brake from the receiving apparatus .. .. .. 50 H.P.
Resistance of the intermediate conductors (this resistance is that of a copper wire about
  ¼ in. in diameter, 77½ miles long) .. .. 100 ohms.
Number of volts at the origin of the conducting line .. .. 6700 volts.
Intensity of the current .. .. 8 ampères.
Ultimate efficiency .. .. 52·52 per cent.

At St. John's Colliery, Normanton, 39 gal. per minute are raised 530 ft., equal to 6·3 H.P. work done. This is with an old girder engine, which indicates 14·2 H.P.; the efficiency, therefore, is 44·4 per cent. This plant commenced working in the latter part of 1887. In the early part of 1888, so satisfied were the owners of the colliery with its working, that they laid down what is at present the largest electrical pumping plant in England. 120 gal. per minute are raised 900 ft., equal to nearly 33 H.P. The engine has not been indicated, but there is an output of 53 H.P. from the generator, so that the efficiency, comparing the actual work done with the output of the dynamo, is about 62 per cent. The current averages 66 ampères with an electromotive force of 600 volts.

At Allerton Main Colliery very small quantities of water are being dealt with at inaccessible points, the power being taken to the motor in secondary batteries, charged at the surface and conveyed in the colliery tubs. The manager of these collieries has also in work a coal-cutting machine. The motor is carried upon the bedplate of the machine, a rotary motion being transmitted to the shaft carrying the cutter bar through gearing. The current is conveyed to the coal face from the dynamo machine, placed on surface, by flexible cable.

Underground haulage at Zankerode Colliery, in Saxony, has been successfully and economically working since 1882. The distance from the dynamo to the working level is 300 yd., and the total length of the line is 700 yd. The current is carried by well-insulated conductors to ang irons which run along the roof of the roadway, and from which the current is taken to the motor carried upon the locomotive. A full train is 15 tubs, each containing 10 cwt., and the locomotive weighs 30 cwt. The speed varies from 5 to 7 miles per hour. The total cost of plant was about 800l., and the cost of working about $\frac{3}{4}d$. per ton. At Paulus and Hohenzollern Colliery, in Upper Silesia, a similar plant has been working since 1884. The distance from the dynamo to the working level is 250 yd., and the length of the line 820 yd. The current is conveyed as at Zankerode. The full train is 16 tubs, each containing 10 cwt., and the locomotive weighs 42 cwt. The cost is about $\frac{1}{2}d$. per ton, or one-half cheaper than horse transport. A similar arrangement is also working in the salt mines of New Stassfurt, and at the Salzberg works. At New Stassfurt tail-rope haulage has been applied.

At the Phœnix Gold Mines, Skipper's Creek, New Zealand, dynamos driven by 2 water-wheels generate about 52 H.P. A No. 8 B.W.G. copper wire, on telegraph poles, conveys the current over a mountain 800 ft. high a distance of some 3 miles, to a motor which drives 20 stamp heads (each of 8 cwt.) at the rate of 70 blows per minute, and is said to be powerful enough to drive 30 heads. The loss in transmission is stated to be about 3 H.P. only.

The most extensive set of electric mining plant in existence is now working at Big Bend Tunnel Camp, California. A tunnel 16 ft. by 12 ft., and $2\frac{1}{4}$ miles long, is cut from the Feather River through the mountain. A permanent dam is built across the river just below the head of the tunnel, by which the river is diverted from its channel and made to flow through the tunnel, thus drying up the bed of the river. A canal 2 miles long is constructed from the tail end of the tunnel, by which a fall of 300 ft. is obtained. Here powerful water-wheels are fixed, driving the dynamos. The working electromotive force is 1000 volts. The conductors are double, metallic, and extend a distance of 18 miles, delivering electricity at 14 different points in the circuit where power is required for winding, pumping, &c. Some 10 to 20 motors, varying from 5 to 50 H.P., are worked by branch conductors from these various stations. The potential at the motors varies from 500 to 700 volts. It is apparent, where water power can be thus applied, the saving, when compared with the use of fuel, is very considerable, especially so in metal or diamond mining districts, where coal is expensive. The ease with which any power can be conveyed over hill and dale, and into the intricacies of the mine, are factors of no mean importance.

The subjoined table shows the approximate cost of various sized mining motors :—

TRANSMISSION OF POWER.

Horse Power of Motor.	Speed.	Volts.	Ampères.	Approximate Weight.	Price of Motor.	Size of Lead-covered and Double Insulation Cable recommended.	Approximate Cost of Cable per Mile.	Price of Dynamo.
					£	B.W.G.	£	£
2	1200	200	9·5	3 cwts.	30	$\frac{7}{18}$	60	34
4	1200	200	18·5	4½ „	40	$\frac{7}{17}$	80	45
6	1200	200	27·5	6 „	50	$\frac{7}{16}$	100	57
9	1000	250	31·5	10 „	75	$\frac{19}{15}$	120	84
12	850	300	35·	15 „	100	$\frac{19}{14}$	120	112
16	800	350	39·	1 ton.	125	$\frac{7}{16}$	150	142
20	750	400	41·5	1½ „	150	$\frac{7}{16}$	150	170
25	700	450	46·5	2 tons.	175	$\frac{19}{14}$	175	200
30	650	500	49·5	2½ „	200	$\frac{19}{13}$	175	225

At the Trafalgar Colliery a very small pumping plant, started in December 1882, developed, in May 1887, into three sets of plant, doing the greater part of the underground pumping of the

Fig. 25.

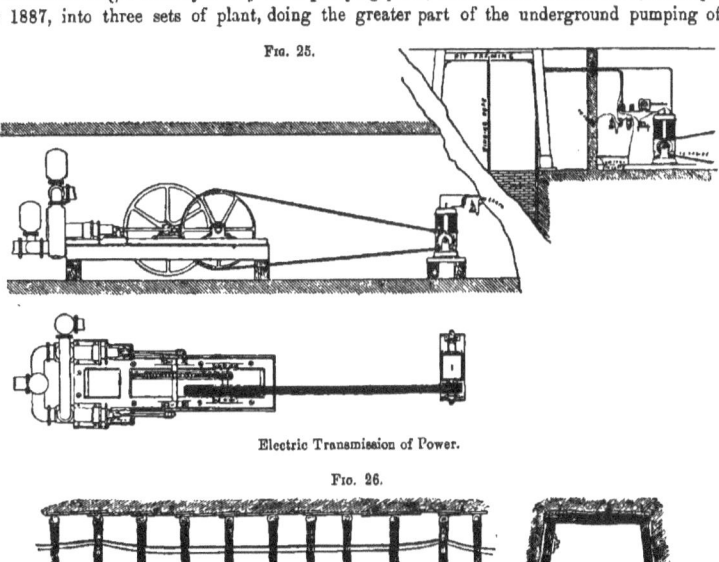

Electric Transmission of Power.

Fig. 26.

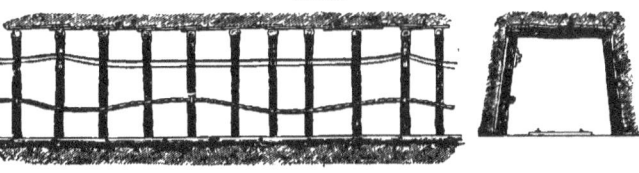

Electric Transmission of Power.

colliery. Following is a brief description of the last installation, by Frank Brain. The pump, a double-throw 9 in. plunger with 10 in. stroke, is fixed 2200 yd. from the generator and 1650 yd. from the bottom of the pit shaft. The pipe main is 7 in., and at a maximum speed of 25 strokes the pump lifts 120 gal. per minute 300 ft. high. It is geared 6 to 1, the small pinion being driven by a belt from the motor. Current is conveyed to the motor by a conductor of nineteen No. 16 wires, insulated and carried on earthenware insulators. An old 4 in. iron wire rope serves for the return current. The current is 43 ampères, and the E.M.F. 320. The cost of engine and electric plant was 644l. The weekly cost of maintenance, allowing 15 per cent. for interest and depreciation on plant, 7l. 17s., or ·002 of a penny per horse-power per hour, and the economy effected about 500l. per annum. The power lost and the useful effect in the different stages is as follows :—

Received by		Loss in		Useful Effects.
Generator	23·00	Steam-engine	22 per cent.	78 per cent.
Cables	18·44	Generator	20 „	80 „
Motor	14·99	Cables	20 „	80 „
Pump	11·99	Motor	20 „	80 „
Water	10·36	Pump	14 „	86 „

The efficiency obtained throughout is only 35 per cent., but it will be noticed that the engine, which is an old one, loses 6·49 horse-power, or 22 per cent. alone. This plant, worked daily since May 1887, without interruption, is giving every satisfaction. Figs. 25 and 26 are views of the pump, showing its connection with the motor, and of the leads along the underground roads.

CHAPTER III.

QUARRYING.

THE mode of working underground stone quarries near Bath is somewhat peculiar, and is thus briefly described by Dr. C. Le Neve Foster, H.M. Inspector of Mines in Wales.

The beds of freestone which are worked occur in the Great Oolite, and vary from 8 or 9 to 18 or 24 ft. in thickness; the dip is slight, being only 1 in 33. The bed of stone which it is proposed to work, is reached by an inclined plane, and then a main heading is driven out, 15 or 16 ft. wide, with "side holes" at right angles as wide as the roof or ceiling will admit with safety, say 20–24 ft., leaving pillars 10 ft. square and upwards. If rock is unsound, it is left as a pillar, and this may cause some irregularity in the plan of the mine.

The first process in removing the stone consists in excavating the "jad," a horizontal groove at the top of the bed, which is cut in for a depth of 5 ft. and width of 20–25 ft. The jad is cut out with a pick, which is not set quite at right angles to the hilt. This form enables the workman to cut right into the corners. The first pick weighs 7 lb., the second 6 lb., and the third 5 lb. This last has a hilt 5 ft. long, so that the man may cut the jad to a full depth. Projecting pieces of roof are broken down by the "jadding iron," a long bar. After the jad has been excavated with the pick, two vertical cuts are made with a saw, and a piece, called the "wrist," is wedged up from the bottom or off from the side. When the "wrist" has been removed, the blocks are simply cut out with saws. These saws are 6–8 ft. long, by 10–12 in. wide. The first saw used in the jad has to be narrower, and is called the "razor saw." The heaviest saw weighs 56 lb., and the handle can be used entirely below the eye when working near the roof.

When set free by sawing on all four sides, the block can easily be detached by wedges driven in along a plane of bedding. The blocks are lifted off by cranes, and either loaded at once on to trucks, or stacked inside the quarry, after having been roughly dressed with an axe or with a saw. A workman can saw 15 sq. ft. of the softest beds in an hour. The men work in gangs, and the ganger is paid at a certain rate per cub. ft. of stone delivered on the trolleys at his crane. The payments are fortnightly; stock is taken every alternate Wednesday, and the men paid on the following Friday. The men make 20s. to 28s. a week; the ordinary hours are from 6 a.m. to 5 p.m., with two hours for meals. Good pickers cutting out the jad can earn as much as 1s. per hour while at work, but at this rate they will not work more than 5 or 6 hours a day. The trade price of the stone is 11d. per cub. ft. at the railway station. The royalties are calculated in various ways:—
(a) At per superficial yard of land, irrespective of the depth of the stone from the surface or its quality; (b) at per superficial yard of land, but varying according to the aggregate thickness of the beds; (c) at per ton of 16 cub. ft. of stone for sale; (d) at a proportion of the selling price of the stone delivered at the G.W. Railway. Owing to false bedding and other irregularities, a bed of

stone which is 20 ft. thick will only yield on an average one-half of blocks fit for the market; the other half is left in the quarry.

About ten years ago a Belgian company was formed to work the old Roman marble quarries of Schemton in Tunis. Though the marble, of various colours and structure, was estimated at more than 253,165,800 cub. ft., working was discontinued on account of the expense. Lately the company has been organised to work the quarry by the "Helicoidal-wire" system, by which not only can the blocks be subdivided, but also the marble extracted from the mountain side.

Power from a 60-H.P. engine is transmitted by teledynamic cable to the highest point of the quarry, whence it is distributed to the several working places by three helicoidal cords, each composed of three steel wires twisted spirally, and running at the rate of $14\frac{3}{4}$ ft. per minute. The cord cuts the marble into slabs by penetrating into the rock at the rate of $5-5\frac{1}{2}$ in. per hour for hard marble, sand and water being allowed to flow constantly into the groove. By changing the direction of the cord, by means of pulleys with adjustable axes, their bearings being fed down as the stone is penetrated, the same cord can be made to serve several working places. The marble, cut to the required dimensions without being touched by the chisel, is brought down in tramways to a workshop, where the blocks may be still further subdivided by the helicoidal wire so as to be reduced to the required dimensions. The workshops are connected with the Bona-Guelma Railway by a tramway, $2\frac{1}{4}$ miles long, made by the company.

An installation of the Société Anonyme Internationale du Fil Hélicoïdal in the grounds of the Brussels Exhibition of 1888, exemplified the principal applications of this new method of working quarries. The endless wire cord is sent by the driving pulley to a tension truck at the end of the yard, and, guided by pulleys with universal joints, is diverted at given points for sawing a mass of concrete and a block of marble, while there are also the following appliances:—A frame, in which the usual blades are replaced by cords for sawing slabs; a finishing apparatus; and a drill, driven by teledynamic rope, for sinking the shafts by which the cord carriers are introduced, the whole being driven by a 14-H.P. engine.

In most quarries, especially those of marble, it is less important to extract the greatest quantity of stone, than to obtain blocks of the form and size desired with as little waste as possible, and this is accomplished in a high degree by the helicoidal cord; while, manual labour being superseded by a regular mechanical operation, there is no need for skilled workmen, but only a few boys to tend the apparatus. A still further saving of labour is effected by the mass being subdivided into blocks of the desired size on the spot where it is quarried.

The rapidity of the operation naturally depends on the hardness of the stone, but it may be put roughly at ten times as great as that by old methods; while concrete, and such rocks as cannot otherwise be worked, yield to the helicoidal cords. At the Exhibition, the same cord which sawed a block of marble also cut simultaneously a mass of concrete composed of quartz and flint pebbles.

Quarries in France, Algeria, Tunis, Italy, Spain, Germany, Russia, and Finland, have been provided with the new apparatus, while it is exclusively used in the marble quarry of Traigneaux, near Philippeville, Belgium. Here the trench, nearly 2 ft. wide, which was formerly, as it is still generally in other quarries, made by hand, is superseded by vertical cuts with the helicoidal cord on all faces not detached, and a horizontal cut underneath the mass to be extracted. If the mass be not detached on any side, it is necessary to run two cuts, 2 ft. apart, along one of the faces.

In order to permit the cord to descend, it is also necessary to sink shafts at all the angles of the mass where not detached, in order to receive the pulley carriers; and this work is now performed by the drill invented by Thonar, at the same time preserving the cores for use as columns. It is usual to make three contiguous shafts, and break down the intervening angles; but the number and size of the shafts may be made subservient to the diameter of columns most in demand. The drill, driven by teledynamic cable, requires 3-3½ H.P., and descends at the rate of about 4 in. per hour in Belgian marble.

The endless helicoidal cord, composed of three steel wires, varies from 100 to 300 yd. in length, and receives its longitudinal motion from a fixed engine, the requisite tension being preserved by a weighted truck on an incline. The downward feed is given by screws in the pulley carriers, turned either automatically or by hand; and the helical twist of the cord causes the rotary motion, which is demonstrated by the even wear of the wires. The cord serves as a vehicle for conveying the sand and water, the former of which is the real agent in cutting the stone.

The diameter of cord found most suitable for quarrying is less than ¼ in., running at a speed of 4 yd. a second, while smaller diameters and quicker speeds are adopted for subdividing the masses. A cut of more than 4 in. per hour is obtained for lengths of 3-4 yd. in Belgian marble. in Quenast porphyry, which it had not before been found possible to saw, a cut of 1-1½ in. per hour is obtained.

For quarrying 2 H.P. is found sufficient. If the cord should break, it is readily spliced; and a cord of average (150 yd.) length will produce 40-50 sq. yd. of sawn surface before wearing out, when it may be used for fencing. The sawn surface, plane if not smooth, is readily finished by the application of an amalgam of emery with lead, tin, and antimony, used in a machine like that for polishing glass.

Fig. 27 shows a plan of part of the Traigneaux Quarry, where the process is exclusively employed for extracting the marble from the rock, as well as for reducing it to blocks and slabs; and Fig. 28 is a vertical section of the same. The size of the mass being sawn out at A is exaggerated for the sake of clearness. At B a perforator—which will be described further on—driven by a teledynamic cord, is sinking a shaft for letting down the pulley carriers C. D are posts carrying grooved pulleys for distributing the cords, a double universal joint permitting the pulleys to assume of themselves the direction of cord necessitated by each case. E are weighted trucks on inclined planes, for keeping the cords at the proper tension. The 30-H.P. engine, which is seen to the left of the illustration, not only serves to give motion to the cords for sawing and the teledynamic cable for working the perforator, but also to drive a three-speeded winch, capable of hauling a block of 25 tons, as shown at F, up the incline of one in seven. Fig. 39 is a side elevation, showing at G such blocks being further subdivided on the surface, also by wire cord.

Enlarged details of the above appliances are given. Fig. 30 shows an elevation, Fig. 31 a plan, Fig. 32 a horizontal section, and Fig. 33 a vertical section of the perforator. This consists of a hollow cylinder of boiler plate, shod at the bottom with a serrated steel cutter, slightly thicker than the plate, so as to clear an annular space in the rock. Sand and water are allowed to run into the hole to assist the action of the cutter, and this agent will in all probability be relied on exclusively in future, in connection with a collar of soft iron superseding the steel cutter. In the event of clogging, the cylinder is raised by the winch which forms part of the apparatus, and the cores are extracted in the same manner. The latter serve as columns, and the perforator may be of any diameter, so as to produce the size of columns most in demand; but as their diameter is

48 MINING AND ORE-DRESSING MACHINERY.

decreased their number must be increased, so as to produce a shaft of about 2 ft. 6 in. diameter, to receive 10 pulley carriers. With a speed of 140 rev. per minute the advance is about 1 in. per hour. Fig. 34 shows the pulley carriers inserted in the shafts, with the tension truck and distributing post, while Figs. 35 and 36 give the arrangement of guide pulleys, with universal joint

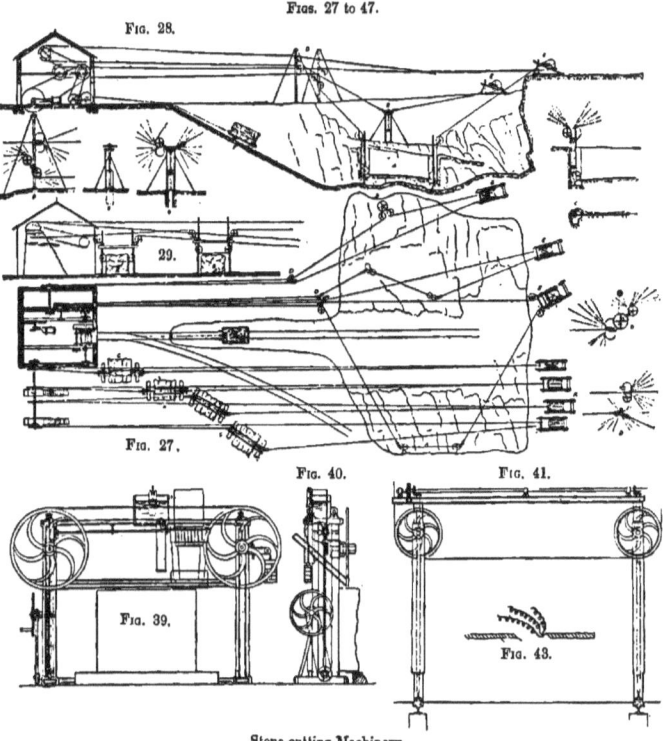

Figs. 27 to 47.

Stone-cutting Machinery.

bearings. Figs. 37 and 38 illustrate the general arrangement of a frame for sawing a block into slabs; and Figs. 39, 40, 41, and 42 give the details. The pulleys are loose on the shaft, so that each cord shall be independent of the others, and the distance between them is adjusted by collars. Fig. 43 shows the method of splicing the wire cord; the splice is a very long one, and the ends of the wires all break joints. Fig. 44 gives plan and section of the footstep, on which runs the shaft A; Fig. 45, one of the four uprights of a frame for sawing slabs, with screw for feed. Figs. 46 and 47 show the method of mounting groove pulleys, f, on the plain drums K, by means of screws v.

QUARRYING.

The surface produced by sawing with the cord is truer than that by blades, but not so smooth. However, the highest degree of polish is given by clamping down the slab to the reciprocating machine, shown by Fig. 48, and subjecting it to the action of the rapidly-revolving plate set with amalgams of emery and various metals, iron being used for marble.

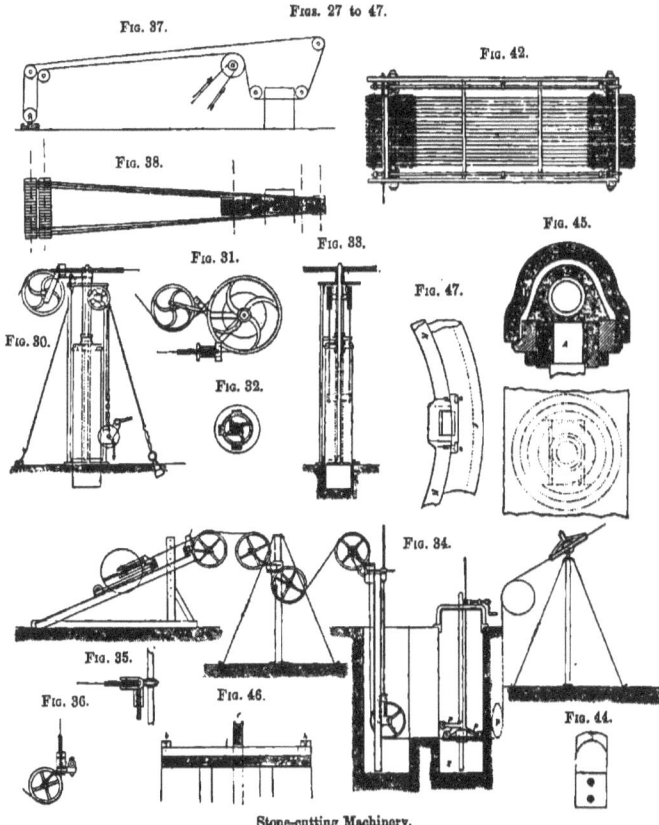

Figs. 27 to 47.

Stone-cutting Machinery.

For an annual production of 14,000 cub. ft. of marble sawn to size, the Traigneaux Quarry only employs 30 hands all told, in addition to the 5 boys who tend the apparatus and give the feed.

For excavating and top-stripping in mines, Priestman's patent excavators (Fig. 49) are very suitable and useful machines.

They are made by Priestman Bros., of Hull, and Queen Victoria Street, London, in various sizes, and supplied with buckets or grabs to lift different kinds of material. They may be fixed upon wheels and axles for running on rails, or fixed upon a barge and used as a dredger for lifting alluvial

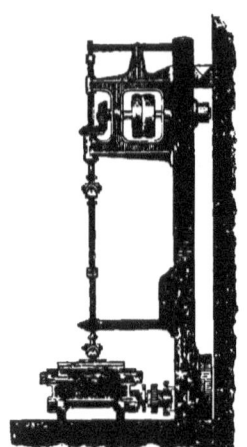

Fig. 48.

Stone-polishing Machinery.

deposits from the beds of rivers, &c. For top-stripping purposes, these machines have long been used at the Frodingham Iron and Stone Mines with excellent results, the cost of working being about 1d. per yard. A large number of machines have also been supplied for various public works,

Fig. 49.
Priestman's Excavator.

Fig. 50.
Priestman's Dredger.

for lifting blasted rock, &c.; and the fact that the crane may be used for ordinary lifting purposes, by detaching the bucket or grab, is a great point in favour of these machines.

For dredging the beds of alluvial rivers, Priestman Bros. have recently sold several machines.

These (Fig. 50) deposit the material lifted into the washers, &c., which are usually placed upon the barge with the dredger. Mr. White, engineer to the French Nechi Gold Mining Company, who use a dredger, speaks very highly of it, and says, in a letter written to Messrs. Priestman, "I know of no other system of dredge so adapted to the general requirements of such work as your machine."

The Siam Gold Fields Company have lately purchased two large machines, each capable of lifting 2 tons at a time. A French company have also purchased a machine capable of lifting 1 ton at a time, for dredging a river in Italy, and we think there is a likelihood of these dredgers and excavators being very largely used.

CHAPTER IV.

PROSPECTING MACHINERY.

THE work of examining the rocks of a locality for the purpose of discovering the mineral deposits of commercial value contained in them, is known to miners as "prospecting." The machines used in these operations consist mainly of boring tools and the apparatus needed to work them. A detailed description of the tools and the methods of boring generally adopted in Europe will be found in André's 'Mining Engineering,' and an exposition of the conditions under which earth boring may be successfully prosecuted is also given in that work. The most important of these tools and apparatus are, however, illustrated below.

HAND-BORING.—A set of hand-boring apparatus consists of the "head-gear," by means of which the tools are worked from surface; the "rods" which are used to connect the tools with the head-gear; and the tools by which the perforation is made. Of the last, there are two kinds: "cutting" tools, which are used to penetrate the rock, and "clearing" tools, which are used to remove the débris that collects in the bore-hole. Besides these, a set of "extracting" tools is required to extract broken tools from the bore-hole, in case of a fracture occurring.

Head-Gear.—The head-gear consists of a boring-frame, or shear legs, with the accessory parts and appliances for raising, lowering, and turning the tools. The use of the boring frame is to furnish an elevated point of support from which the rods attached to the tools may be conveniently suspended. The rods have to be very frequently raised for the purpose of changing the cutting tool and clearing the bore-hole; and it is obvious that the time required for the performance of this operation will, in a great measure, depend upon the height of the frame. If this were equal to the depth of the bore-hole, the rods might be withdrawn at one lift. If it were equal to half the depth, one half the length of the rods might be withdrawn at one lift; but this half would then have to be disconnected from that remaining in the bore-hole by unscrewing the joints, and the latter half subsequently raised by a second lift. A high frame saves much labour and time, by increasing the length of the "offtake," as it is called; for deep borings, a high frame is indispensable. The height found most convenient in practice is 45–60 ft. Whatever the height adopted, it must be a multiple of the lengths of which the rods are made up, so as to bring the joint to be unscrewed a convenient distance above the top of the bore-hole.

The support furnished by the top of the boring frame is provided with a pulley, usually of cast iron and grooved to receive a rope. This rope is attached at one end to the rods, and at the other to a windlass, by means of which the rope is drawn in and the rods raised. The windlass is in most cases fixed upon the frame at a convenient height above the ground, and is worked either with an intermittent motion by levers and ratchet-wheel and pawl, or with a continuous motion by winch handle, as in the case of a common drawing well. The former method is unsuitable for any but small depths, and even in such cases is inferior to the latter. Sometimes the windlass is arranged with a vertical axis, and worked with horizontal bars, like a capstan. When the depth (and, consequently, the weight of the rods) becomes great, the windlass is worked by intermediate gearing, which may be so contrived as to increase the speed when a large portion of the weight has been taken off. In very deep borings, a steam engine may be used to work the windlass. In all cases, a break is attached, to regulate the descent of the rods.

The "sludger" is the clearing tool generally employed, and being used independently of the rods, it is usually provided with a special pulley and windlass of smaller dimensions. This windlass is, like the large one, fixed to the boring frame; but for convenience, upon the opposite side, and furnished with a sufficient quantity of rope. The pulley is so contrived that it may run out exactly over the bore-hole when about to be used, and back again out of the way of the rods when done with. The sludger, being swung over the hole, is lowered rapidly by its own weight, its descent being checked by a brake upon the windlass. To raise it, the windlass is turned by winch handles, and these are also made use of to produce the reciprocating or "pumping" motion required to fill the sludger. But sometimes this motion is derived from the oscillating lever by winding the rope two or three times round the head.

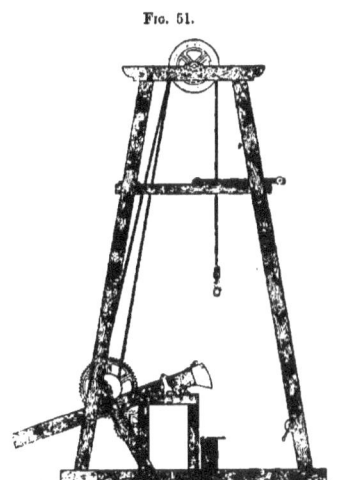

Fig. 51.

Hand-boring Frame.

For deep borings by hand power, the frame shown in Fig. 51 is very suitable. It consists of two pairs of shear-legs, of 12 in. by 9 in. scautling, set into the projecting ends of the side pieces of a strong rectangular wooden framing, constructed of balks, 12 in. by 9 in. for the side pieces, and 9 in. by 9 in. for the end pieces. The timbers of each pair of legs have a slight inclination towards each other, being 3 ft. 8 in. apart at the bottom, and 14 in. at the top, in a vertical height of 30 ft. These timbers are stayed at intervals of 3 ft. by horizontal wooden ties, each 9 in. by 6 in., mortised into them, and keyed on the outside with wooden keys. The two pairs of legs are connected at the top by two cross pieces, into which they are mortised: these pieces carry the pulleys. One pair of the shear-legs is provided with stout diagonal timbers fixed between them and the bottom framing, for the purpose of carrying the windlass, which is moved by spur gearing, and furnished with a ratchet-stop. The barrel of this windlass is 18 in. diameter, and the proportion of the driving to the driven gear is 1 to 3. The two top cross-bars carry a horizontal wrought-iron axle, upon which two independent cast-iron guide-pulleys run loosely. The use of the two pulleys is to save time in raising and lowering the rods. To effect this object, the ends of two ropes are led over the pulleys and coiled in contrary directions upon the barrel of the windlass. By this arrangement, one of the ropes is always down, in readiness to be attached to the rods the moment the offtake has been removed, without the labour of uncoiling it from the windlass.

For the purpose of raising and lowering the sludger, a pair of traverses, 9 in. by 6 in., is fixed across from one pair of shear-legs to the other, at a distance of about 8 ft. below the top traverses supporting the pulleys. These pieces, which are mortised and keyed into the shear-legs, are intended to carry another and smaller pulley, mounted on a cast-iron frame capable of motion between horizontal wooden slides provided for the purpose, and fixed upon the traverses. The slides are made to project beyond the shear-legs, and are furnished with a roller, as shown in the figure, for the purpose of carrying the rope out clear of the frame. The end of the rope, after being led over the pulley and the roller, is brought down and wound upon a smaller windlass fixed upon the shear-legs opposite those carrying the larger windlass. The sludging windlass is provided with a brake, to regulate the descent of the tool; such brakes should be self-acting, the power being obtained preferably by means of a weight. The rope used for the sludger will be $\frac{3}{4}$-1 in. diameter, according to the dimensions of the tool. Hempen rope is usually employed, but aloe fibre, allowing of smaller dimensions, has often been used with advantage.

Next to the boring-frame, the most important part of the head-gear is the oscillating or "rocking" lever. It is by means of this lever that the requisite motion is communicated to the rods when working the cutting tools. It consists of a piece of straight-grained ash, provided with an iron axle, upon which it turns as a fulcrum. This axle is supported upon a wooden framing, composed of four upright pieces, fixed at the bottom in two cross timbers, inserted for that purpose into the framing of the shear-legs, and connected in pairs at the top by two cross pieces, into which they are mortised. The two inner upright pieces are connected in the same way, to afford a support for the lever axle. The height of the support thus obtained is about 5 ft. 6 in. The dimensions of the lever will be determined by the weight of the rods, and will therefore vary with the depth of the bore-hole. The same conditions will determine the proportions of the iron axle and its attachments. This axle is fixed upon the lower side of the lever by means of straps and bolts, in the manner shown in Figs. 52 to 55, an iron carriage being bolted down to the framing to carry the axle. As these parts will be subjected to severe strains, the materials should be of good quality, and the dimensions ample. The proportion of the shorter to the longer arm of the lever will be determined by the weight of the rods and the length of the stroke: 1 to 4 and 1 to 5 are the usual proportions; but in some instances as much as 1 to 9 has been adopted. The total length of the lever will depend somewhat upon the proportion of the arms, but in most cases 10–12 ft. will be found to be a convenient depth. And with a proportion of 1 to 4, or 1 to 5, and a bore-hole of considerable depth, say 500–700 ft., a scantling of 9 in. by 7 in. will be sufficient. The diameter of the axle in such a case should be $2\frac{1}{4}$ in. The length of the stroke should, as far as is practicable, be proportioned to the hardness of the rock which is being bored through. For moderately soft clay, 6 in. may be sufficient; but compact limestone may require 24 in., or even more.

To allow several men to work at the end of the longer arm of the lever, a cross-bar is affixed to it. This cross-bar should be of tough ash, of circular section, and of such a diameter as to be conveniently grasped by the hand; it should be fixed, by means of iron straps, upon the upper or upon the lower side of the lever, and never passed through it or notched into it. Instead of the cross-bar, straps of iron provided with a hook at each end may be fixed across the upper side of the lever, leaving the hook projecting over the edge. Short pieces of rope, with a ring on one end, and a piece of wood, of circular section and about 8 in. in length, on the other end, may be used instead of the cross-bar, by placing the ring over the hook and grasping the piece of wood to pull by.

In this way, four men can work with two hooks on each side, or six men with three hooks. One advantage gained by this method of working the lever is the directness of the strain. With the cross-bar, a preponderance of force on one side—and such a preponderance must always exist, since the men will never be exactly equal in strength—produces a torsional strain great in proportion to the amount of the preponderating force and the leverage of the cross-bar. To steady the lever, the longer arm is sometimes made to move between guides.

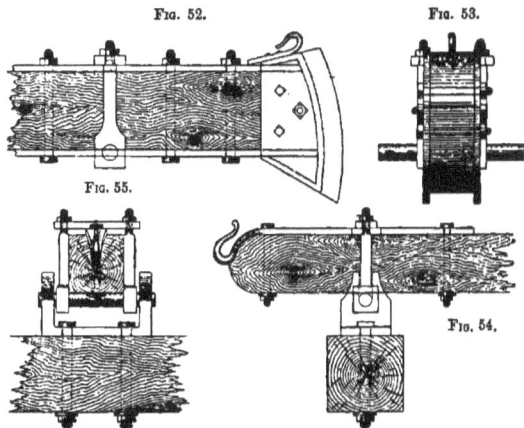

Portions of Hand-boring Frame.

The head of the lever should be formed of a sector of a circle, the centre of which is the point of support. This is needed to raise and lower the rods in a straight line. Usually the sector-head is of cast iron, as in Fig. 52. Above the head a stout hook is firmly fixed by means of bolts. In this hook the rods are hung, a short piece of chain, or preferably of flat hempen rope, furnished with a ring at one end, and a swivel-head and hook at the other, being required when the lengthening stirrup is used. As the head of the lever partly overhangs the bore-hole, the axle must be so set in its bearings that the lever may be withdrawn when it becomes necessary to use the sludger. The usual manner of providing for this requirement is shown in Fig. 54, where the construction of the carriage allows the lever to be readily lifted off its bearings.

For the purpose of suspending the rods from the oscillating lever or the pulley, the top length of the rods terminates above the surface of the ground in a stirrup, the construction of which allows the rods to be turned round during the operation of boring, and to be lowered as the boring progresses. Several forms of stirrup are in use, but the most convenient is that represented in Fig. 56. This stirrup keeps the upper end of the rod always at the same height above the ground, a necessary condition for the perfect working of the lever, and it enables the borer to see exactly the progress that is being made at the bottom of the hole.

An essential part of the surface apparatus is the bore-hole guide-tube, shown in elevation and

MINING AND ORE-DRESSING MACHINERY.

in plan in Figs. 57, 58, and 59. This tube is of wood, and for bore-holes of ordinary dimensions is about 12 in. diameter and 6 ft. in length. The diameter of the bore of this tube is the same as that of the bore-hole, so that the thickness of the wood is about 4¼ in. for a 3-in. hole. The guide-tube should be inserted into the bore-hole to a depth that will leave about 10 in. of its length above the

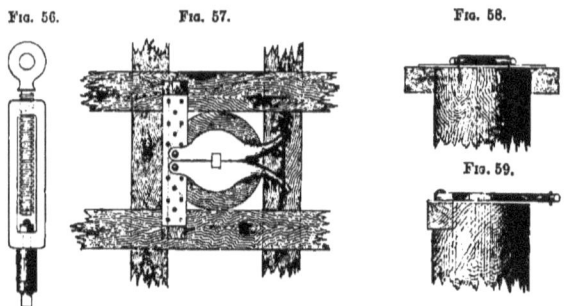

Fig. 56.   Fig. 57.   Fig. 58.

Fig. 59.

Portions of Hand-boring Frame.

surface of the ground, and firmly held in its position by four pieces of timber, 9 in. by 6 in. in section. These pieces are laid upon the ground in pairs, one piece on each side of the tube, the pairs being at right angles to each other. The ends of the pieces forming each pair are then pressed partially together, to make them tightly clasp the tube, and are held in a state of tension by iron straps across the ends, and these ends are firmly fixed to the ground. Sometimes they are fixed to the framing in which the shear-legs are set; but this practice is not to be recommended, as the vibration of the framing tends to produce injurious effects. When a staple is sunk, they are, of course, set at the bottom of the staple. Various means of fixing these timbers may be employed, the only necessary condition being that there shall be no liability of their becoming loose during the progress of the work. The upper surfaces of these timbers are 1 in. below the top of the tube. The aperture of the tube is provided with a pair of iron shutters, opening and closing horizontally, as shown in Fig. 57. Each shutter is notched to form a square aperture 1¼ in. wide, through which the rods may freely move from joint to joint when the shutters are closed. The use of these shutters is to prevent anything from falling down the bore-hole. Instead of this kind, flap shutters may be used. They consist of two semi-circular iron discs hinged upon the tube, and opening vertically like a clack-valve, a notch in each forming the aperture for the rods, as in the preceding kind.

The top-length of the rods terminates in a swivel-head, by which it is suspended from the rocking lever. For the purpose of adjusting the height of the head to the requirements of the lever, several short lengths are needed, varying from 1 ft. to 3 ft., which are screwed on as the boring progresses, the shorter lengths being removed and a larger one substituted at each change.

These lengthening pieces, one of which is represented in Fig. 60, are all provided with a screwed socket. Sometimes they are furnished with an eye through the shank just below the head, through which a piece of wood is passed to form a lever, by means of which the rods are turned round during the operation of boring. This arrangement is common in Belgium; but in England it is more usual to employ the brace-head or tiller for this purpose.

The tiller may be of wood or of iron. When of wood, it consists of a piece of ash, 4 in. diameter, square in section in the middle, and rounded off and reduced in size towards the end, as shown in Fig. 61. The middle portion is provided with a notch 1 in. square to receive the rod, one side of the notch being formed by an iron plate turning on a bolt at one end and fixed at the other by a screw. On withdrawing this screw, the plate drops and leaves the notch open. Another screw through the centre of the plate is provided for the purpose of fixing the tiller upon the rods. When the tiller is of iron, it is constructed as in Figs. 62 and 63. It consists of two portions, each

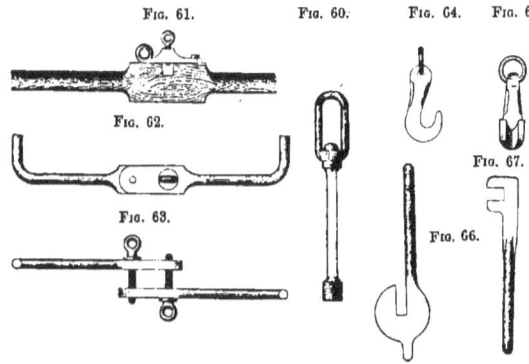

Hand-boring Tools.

18-24 in. in length, joined by two screws. The ends of the tiller are turned up to afford a convenient hold for the workmen.

For the purpose of raising and lowering the rods, a "lifting dog" is required. This consists of a claw-hook, through the shank of which a ring is passed, by means of which it is attached to the rope. When in use, the claw is placed under the head on the shoulder of the top length of rod, and the latter is hauled up or lowered by means of the windlass. The lifting dog is represented in Figs. 64 and 65.

Another instrument required for raising or lowering is the "nipping fork" or "tiger." When the rods have been hauled up as far as the height of the shear-legs will allow, they must be supported in that position while being unscrewed. For this purpose, the nipping fork (Fig. 66) is placed upon the top of the guide-tube beneath the joint in the rod, and the latter is lowered till the joint rests upon the fork. In like manner, in lowering, the rods are let down till the lifting dog rests upon the fork; the next offtake is then screwed on, and the lifting dog hanging from the other pulley is placed under the shoulder of the top length, and the rods are slightly lifted thereby to allow the lower dog to be removed. When only one lifting dog is used, after the first offtake has been removed, a short swivel-head lengthening piece must be screwed on to each subsequent offtake, to afford a hold for the dog. Provided the shutter of the guide-tube is sufficiently strong,

I

it may be made to fulfil the purpose of the nipping fork. For screwing up and disconnecting the rods, a kind of wrench, called a "hand dog" (Fig. 67), is required.

Rods.—The rods by which the excavating tools are worked from the surface consist usually of bars of iron 1 in. square, for boring of ordinary dimensions. Other sections have been employed, notably the circular and the octagonal; but the greater simplicity of the square section, and the advantage which it possesses of allowing the application of keys and spanners to any portion of its length, have caused it to be preferred to the more complex forms. Ordinary forms are shown in Figs. 68 and 69.

FIG. 68.   FIG. 69.   FIG. 70.   FIG. 71.

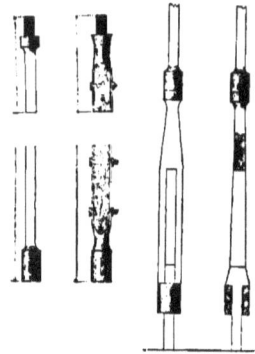

Hand-boring Tools.

The rods are generally made up of 10- or 15-ft. lengths, and the several lengths of rod are connected by a screw-joint. Other modes of connection have failed in practice, leaving the screw-joint in universal use. All the joints should be identical in every respect, so that any two lengths may be connected together; and in making up the rod, care should be taken always to have the socket on the lower end of each length, to prevent rubbish from being jammed into it. The enlarged portion at the joint serves as a point of support to suspend the rods from during the operations of raising and lowering. A grave objection to the employment of iron rods for deep borings is their great weight. Two methods have been proposed of overcoming this difficulty: the first is the substitution of wooden rods for iron ones when the depth is great. Such rods, with their iron connections, lose the greater part of their weight in water, and thus are not exposed to the danger mentioned.

They have been successfully employed in many instances. These rods are made of sound, straight-grained pine, in lengths of 25 and 35 ft., have a square section of not less than 2½ in. side, with the angles slightly planed off. They are connected by iron screw-joints in the same manner as the iron rods, each end being provided with an iron joint-piece forming a socket into which the rod is bolted, as shown in Fig. 69. A fatal objection to wooden rods for small borings is the necessity for a large section. Less than 2 in. side could not be used, and for great depths 3–4 in. would be required. But when the bore-hole is of sufficient diameter, they may be employed with advantage in some cases, and are frequently adopted for deep borings on the continent of Europe.

The second method proposed possesses greater advantages, inasmuch as it removes the difficulty without abolishing the iron rods. It consists in forming the rods of two distinct portions: a short and massive part at the bottom, to which the cutting tool is attached, and on which alone the force of the shock is expended; and the rod proper, which is used solely to raise the former part, and which, not being attached to it in an invariable manner, is not exposed to the shocks occasioned by the percussive action of the former. As this method allows all the advantages attending the use of iron rods to be retained, it is by much the more important of the two, and has, consequently, been very generally adopted. The forms of construction by which the latter method has been carried out have undergone numerous modifications since its first introduction. But, as in the case of the tools, the lessons of experience have led to the abandonment of all or most of the recent devices in favour of the extremely simple "sliding joint," which is the only one that can be relied upon for borings of an ordinary character. Its construction is shown in Figs. 70 and 71. The lower part, to which the cutting tool is affixed, terminates upwards in a head, moving in a slot in the lower extremity of the upper portion constituting the rods proper. When the rods are raised, the tool is lifted by this head, which then rests upon the bottom of the slot. On dropping the rods to produce the percussive action of the tool, the latter falls with the former till it comes in contact with the rock at the bottom of the hole, when it is abruptly arrested and thereby subjected to a violent shock. But the rods continue the descent by allowing the head of the arrested portion to slide up in the slot, and by that means the shock is confined to the part carrying the tool. As soon as the head of this part begins to move up the slot, the end of the lever at surface to which the rods are attached comes in contact with an elastic stop, which is capable of bringing the rods to rest within the space allowed by the play of the slot. In this way the descent of the rods is gradually arrested, and injurious shocks are avoided, without diminishing in any degree the action of the cutting tool. Another advantage is the possibility of considerably reducing the section of the rods, which are required only to raise the part to which the tool is affixed.

Tools.—The simplest and commonest form of cutting tool is the "flat chisel" or "straight bit" (Figs. 72, 73). It is made of the toughest iron, and steeled at its cutting edge with the best material. The length is usually 18 in.; at the upper end it is provided with a thread by means of which it is screwed to the rods. This form of tool is applicable to all but the softest and the hardest strata. The ease with which it may be re-sharpened is a quality that commends it to the choice of the practical man. For penetrating very hard rock, the form shown in Figs. 74, 75, and variously known as the "diamond-point," "drill," or "V chisel," is used. This differs from the straight bit only in the form of its cutting edge. For boring through gravel, Fig. 76 is used: it consists of two cutting edges at right angles to each other, one of them being curved towards its extremities; it is known as the T chisel. The "auger" borer (Fig. 77) is employed in boring through plastic clay and loose sand; it is similar to that used for boring in wood. The bottom is partially closed by the lips, which are turned down to a greater angle than in the case of wood augers. The clay auger is made of greater lengths than the chisel bits, but it terminates upward in the same form. Usually it is half cylindrical, as in the figure; but sometimes it is made wholly cylindrical, with the exception of a length of about 6 in. at the bottom, where it is left open to allow of the admission of the material which is being bored through. When used in clay, this tool, on being raised to surface, carries the "core" with it.

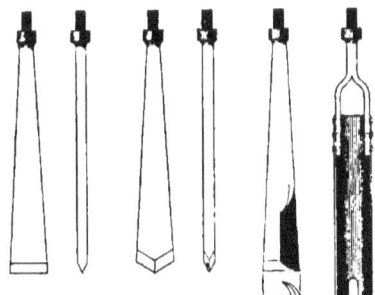

FIG. 72.   FIG. 73.   FIG. 74.   FIG. 75.   FIG. 76.   FIG. 77.

Hand-boring Tools.

For clearing the bore-hole of the débris of the rock chipped off by the cutting tool, a "sand-pump," or "sludger" (Fig. 80) is used, so called, because it removes the débris in the form of sludge or mud. It consists of a wrought-iron cylinder, a little less in diameter than the cutting tools, the lower extremity of which is furnished internally with a ball-valve, of metal, its weight being proportioned to the degree of fluidity of the matters to be extracted. It is made to rest upon a conical seating formed by an annular piece riveted to the cylinder. The sludger is worked by jerking it up and down in the bore-hole on the end of a rope. During the descent of the tool, the valve is raised by the water in the hole, and as it sinks by its own weight into the débris, the latter passes above the valve. During the ascent of the sludger, the material which has entered acts, with the water, to close the valve. By this means, the escape of the sludge is prevented, though a large portion of the water passes out through the accidental interstices caused by small pieces of stone upon the valve seating. The action of the sludger is very effective, as much as a cubic yard of sludge being sometimes removed at one time by a large tool. When the operation of "pumping" the sludger has been continued sufficiently long to clear the hole, it is raised and its contents removed by turning it upside down.

The materials brought up by the sludger show the nature of the stratum that is being passed through. But as these materials are in a divided state, being reduced to small fragments by the action of the chisel, they indicate but little of the physical condition of the rock-bed, and nothing whatever of its dip. Moreover, their indications concerning the nature of the bed are hardly trustworthy, inasmuch as particles of the higher beds are continually falling from the sides of the bore-hole. As it is highly desirable that full information on all points should be obtained when boring in search of minerals, and especially on the dip of the beds, their physical character, and their geological age as evinced by contained fossils, it becomes necessary to have recourse to special tools for that purpose. The use of such tools is to bring up a solid core of the rock, and to bring it up in such a condition that the lines of stratification will show the dip of the bed. To obtain this result, the core must be marked relatively to the north point before it is broken from the rock. This is effected by means of a chisel with an eccentric cutting edge. Having previously cleared out the hole, this chisel is lowered, care being taken, by suitable marks on the rods, and a fixed plumb-line, that it be not turned in the least degree during the operation. When it has reached the bottom of the hole, two or three light blows are struck without turning the rods, and it is again raised. A special tool (Fig. 78) composed of a number of chisels set in a ring, is then lowered, and worked with light blows in the same manner as the common chisel. By this means an annular space is cut round the marked core. When this space has been cut nearly to the depth of the chisel, the tool is raised, and another special extracting instrument (Fig. 79) is let down. This instrument drops over the core, and by means of a wedge thrust in by the weight of the rods, exerts a sufficient lateral pressure to break it off. The core is held between the wedge and a spring fixed on the inside of the instrument for that purpose, and in this way it is raised to the surface. The inclination of the lines of stratification may then be observed relatively to the mark upon the upper end, and the direction and amount of the dip determined. By repeating this series of operations, a complete section of the strata may be obtained.

FIG. 78. FIG. 79. FIG. 80. FIG. 81.

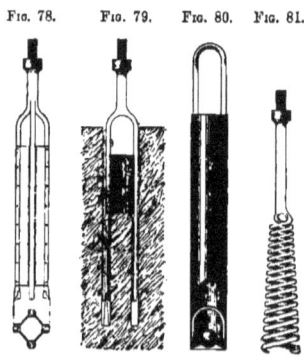

Hand-boring Tools.

It not unfrequently happens that the rods become fractured, and the tools are consequently left at the bottom of the bore-hole. In such a case, "extracting" tools are made use of to remove the fragments from the hole. When the rods have parted either in, or immediately above a joint, the portion in the bore-hole may be seized by the "crow's-foot" (Fig. 82). This is lowered on the end of the rods into the bore-hole, and turned round till it has grasped the broken rod beneath the joint, when the whole may be raised without difficulty. When the fracture has occurred immediately below a joint, or near the middle of a length, the crow's-foot cannot be used, because the long portion above the joint would catch in the side of the bore-hole on being lifted by the tool. In such a case, the "bell" or "horn-socket" is used to recover the last portion. Another extracting tool is the "wad-hook" (Fig. 81); this is attached to the end of the rods, and lowered as far as the broken portion, when it is turned round till it has taken a firm hold of the rod. It may be used to extract a broken bit, pebble, or any substance that may have accidentally fallen into the bore-hole.

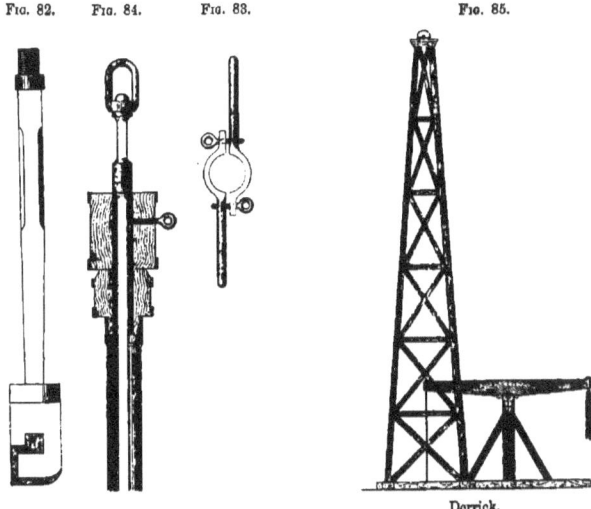

Fig. 82. Fig. 84. Fig. 83. Fig. 85.

Derrick.

Tubes.—In earth boring it is very frequently necessary to line a portion, or the whole, of the bore-hole with iron-tubes. These tubes are provided with screwed plug and socket for joining in the same way as the rods. Other forms of joint are used, but the screw joint, though somewhat more expensive, is to be preferred; the tubes are generally in 10-ft. lengths. To line the bore-hole, a length of pipe, the outside diameter of which is equal to that of the hole, is inserted, and driven down till its socket is nearly level with the top of the guide-tube; another length is then screwed on by means of the pipe-clamp (Fig. 83), and driven down in like manner. The lower end of the first tube is steeled, and the edge sharpened, to enable it to penetrate readily. The operation of forcing the tubes down the bore-hole is one demanding great care. Proper provision must be made for keeping the tubes perfectly vertical during the driving, to avoid the danger of fracture arising from transverse strains and indirect shocks. The driving is effected either by blows or by pressure. When the former method is adopted, a block of wood, bound with an iron hoop to prevent crushing, and having a hole through the centre sufficiently large to allow the free passage of the rods, is placed upon the socket of the upper length of tubing to receive the blow, the object being to prevent the fracture of the tube by interposing an elastic medium between it and the instrument with which the blow is given. The latter consists of another block of wood bored and bound in the same manner as the first, and constituting a kind of "monkey," to be used as in pile-driving. This monkey is fixed by pressure-screws upon an upper length of rod, as shown in Fig. 84. Several lengths of rod are then screwed on to give weight, and passed through the hole in the lower block, and allowed to hang down the tube and the bore-hole ; a rope is then attached to the head of the rod, carried over the pulley at the top of the shear-legs,—the ordinary rope having been lifted off for the occasion, —and wound with one turn upon the windlass. Frequently it will be found convenient to use the sludger pulleys and windlass for this purpose. To work the monkey, two men turn the windlass, a third man holding the end sufficiently taut to enable the former, by means of the friction, to raise the rods with the monkey attached. The drop is occasioned by releasing the end of the rope. The descent of the tubes may be assisted by giving them a partial turn after each blow by means of the clamp-lever represented in Fig. 83.

MACHINE BORING.—The advantages consist mainly in the substitution of steam power for manual labour, and the use of a rope instead of rods in the bore-hole ; by the adoption of this flexible suspending medium, which has been used for centuries by the Chinese, much of the time expended in raising and lowering the tools is saved.

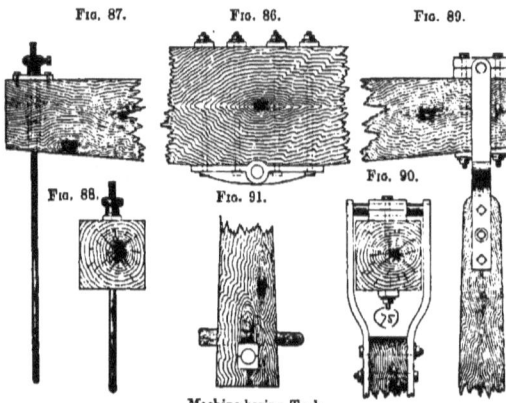

Machine-boring Tools.

The head-gear consists of a boring frame or "derrick," Fig. 85, usually 60–70 ft. high; the "working beam," or rocking lever, Figs. 86 to 88, by means of which the jigging motion is given

PROSPECTING MACHINERY.

to the tools; the "pitman bar," or connecting rod, Figs. 89 to 91, by which the outer end of the working beam is connected to the crank of the band-wheel; the "sampson post," Figs. 92 to 95, upon which the working beam is hung; the "band-wheel," Figs. 96, 97, upon the shaft of which is the crank to which the pitman bar is attached; the "jack-frame," upon which the band-wheel is supported; the "sand-pump reel," Figs. 98, 99, fixed upon the jack-frame and worked by friction from the band-wheel; and the "bull-wheel," or windlass, Figs. 100, 101, upon which is wound the rope from which the tools are suspended. The connections consist of : "a temper-screw," or stirrup by means of which the rope is suspended from the inner end of the working beam; a rope, to which the tools are attached; a "rope-socket," Fig. 102, by means of which the attachment is made; a "substitute," Fig. 103, a short bar of round iron having a box at one end and a pin at the other, used in the place of, or with, the sinker bar; the

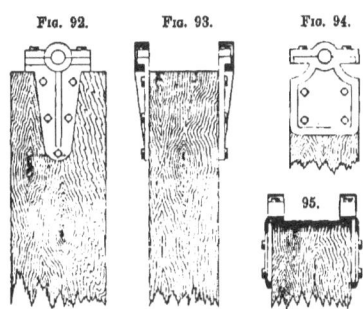

Machine-boring Tools.

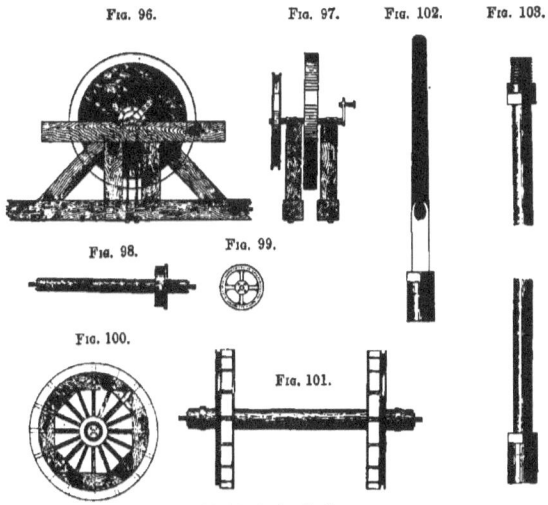

Machine-boring Tools.

"sinker bar," a bar of round iron 10-12 ft. long, terminating, like the substitute, in a box at one end and in a pin at the other, the use of which, when screwed on to the jars, is to give additional weight to the tools; the "jars," or sliding joint, which are screwed on to the sinker bar, and used to lift the cutting tools and to let them drop; and the "auger-stem," by means of which the tools are attached to the jars, is in form like the sinker bar, but longer, 20–24 ft. The cutting tools consist of: two straight "bits," for cutting away the rock; and a flat and a round, or a half-round reamer, to follow the bit in order to enlarge the hole, and to keep it true and round. The clearing tool used is the "sand-pump," or sludger. Besides these, two wrenches are required for screwing up and unscrewing the connections. The engine used is generally of the portable class, and of about 25 H.P.

The derrick is a tall framework, in the shape of a pyramid. It was formerly built of rough poles, or hewn timber, the bottom being 10–12 ft. square, the poles, four in number, being erected one at each corner, 30 ft. in height, converging towards each other, forming a square at the top of $2\frac{1}{2}$ ft., with girths and braces at suitable distances to make the structure sufficiently strong for the work required of it. Derricks are now built of sawn lumber or planks, 2 in. thick, and 6–8 ft. wide, the two edges being spiked together, forming a half square on each corner of the foundation, which is 14–16 ft. square, and in some localities more. The derrick is put up in sections, being braced transversely as it goes up, in order to secure the strength necessary, until it reaches the proper height, which for deep holes is about 56 ft.; for shallow ones, less height and a lighter derrick is required; at the top it forms a square of 2–3 ft.

On the top of the derrick is put a strong framework for the reception of a pulley (Figs. 104, 105), over which the drill-rope passes. The floor of the derrick is made strong by cross sleepers, covered with planks or boards. A roof for the protection of the workmen is laid with boards across the girths, 10–12 ft. above the floor. In cold weather the sides are boarded up. The bull-wheel (Figs. 100, 101) is a shaft of timber, 6–8 ft. long, fastened like the shaft of a common windlass, and 6–8 in. diameter, the ends of the shaft being banded with iron, and a journal of inch-iron driven into each end for it to revolve upon. Mortices are made through this shaft 8–10 in. from each end, for the arms of the wheel. The wheels are usually made 6–8 in. thick on the face, with strips of plank sunk into and spiked on to the outer surface, for the double purpose of receiving the rope-belt and connecting it with the band-wheel for drawing up tools, tubing, &c., out of the well, and for the workmen to take hold of with their hands when working it without the help of the engine. The bull-wheel is placed on the side of the derrick next to or opposite the band-wheel and engine, as the workmen may desire. The drill-rope is coiled on this shaft between the wheels, one end being passed from it over the pulley on the top of the derrick, and attached to the tools.

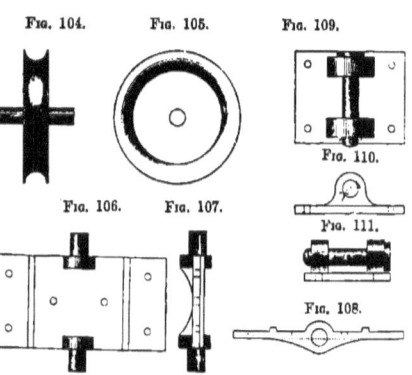

Fig. 104.   Fig. 105.   Fig. 109.

Fig. 106.   Fig. 107.

Fig. 110.

Fig. 111.

Fig. 108.

Machine-boring Tools.

The sampson post is of hewn timber, 12–15 in. square, and usually 12 ft. in height, erected on heavy framed timbers which cross each other, and are bedded firmly in the ground, and having a mortice to receive the tenon on bottom of post; there is also a brace on each side, reaching nearly to the top of the post. On the top of this are irons (Figs. 106 to 108) fitted to receive the working beam, which is balanced on the top of the sampson post, admitting of the rocking motion required in drilling and pumping. The working beam is a piece of timber, 20–26 ft. long, 8–10 in. square, at each end, 8 by 14–16 in. in the middle, with iron attachment in the centre, fitting to a similar one on the sampson post. To the end over the bore-hole is an iron joint, for attaching the temper-screw when drilling, and sucker-rods when pumping. On the other end of the working beam is attached an iron joint (Figs. 109 to 111), for attaching the pitman-bar, which connects the same with the crank, or band-wheel shaft. The band-wheel, shaft, crank, and spider are shown in Figs. 112 to 117.

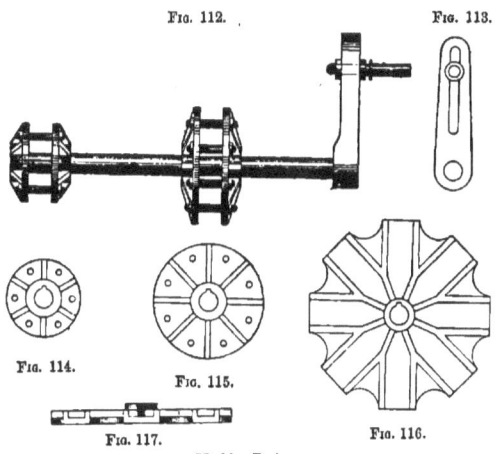

FIG. 112.   FIG. 113.
FIG. 114.
FIG. 115.
FIG. 117.   FIG. 116.
Machine Boring.

The band-wheel is usually about 6 ft. in diameter, with a 6-in. face, and is placed upon a strong frame, called the jack-frame. This is secured in position by two heavy timbers, bedded into the ground, with jams sunk into them to receive the sills of the jack-frame, to which they are keyed fast. The engine is usually placed 8–12 ft. distant from the band-wheel, and connected by rubber or other belting. The belting in general use is 6 in. in width.

When all the parts of the apparatus have been placed in position, the iron guide-pipe or driving pipe (Fig. 118) is first driven down to the solid rock. This acts as a conductor, and prevents earth or stones from falling into the pit or hole while the drilling is going on. It is generally of cast iron, 6–8 in. diameter, having walls of about 1 in. in thickness, and is in lengths of 9–10 ft. The driving of this pipe requires the utmost skill, since the pipe must be forced down through all obstructions, often to a great depth, while it must be kept perfectly vertical; the slightest deflection from the vertical ruins the well, as the pipe acts as the conductor for the drilling tools. The process of driving is simple, but effective: Two slide-ways (Figs. 119, 120), made of plank, are erected in the centre of the derrick to the height of 20 ft. or more, 12–14 in. apart, with edges in toward each other, and the whole made secure and plumb. Two wooden clamps or followers are made to fit round the pipe, and slide up and down on the edges of the ways. The pipe is erected on end between the ways, and held vertical by these clamps, and a driving cap of iron is fitted to the top. A battering ram is then suspended between the ways and arranged to drop perpendicularly upon the end of the pipe. The ram is of timber, 6–8 ft. long, and 12–14 in. square, banded with iron at the lower or battering end, with a hook in the upper end to receive a rope. When the whole is in position, a rope is attached to the hook, passed over the pulley of the derrick, and led down to and passed round the shaft of the bull-wheel. When everything is in readiness to drive the pipe the machinery is put in motion: a man standing behind the bull-wheel shaft grasps the rope attached to the ram and coiled round the bull-wheel shaft, holds it fast, and takes it up in his hands, thus

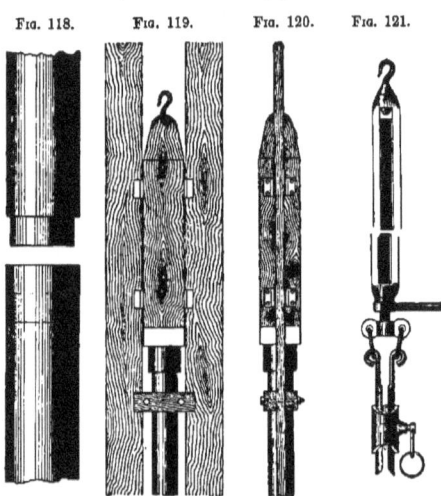

Fig. 118.   Fig. 119.   Fig. 120.   Fig. 121.

raising the ram to its required elevation; he then lets it fall upon the pipe. Thus by repeated blows the pipe is driven to the requisite depth. When one joint of pipe is driven, another is placed upon it, the two ends are secured by a strong iron band, and the process is continued as before. The pipe has to be cleaned out frequently, both by drilling and sand-pumping. Where obstacles, such as boulders, are met with, the centre-bit is put in requisition, and a hole two-thirds the diameter of the pipe is drilled through the same. The pipe is then driven down, the edges of the obstacle being broken by the force applied, the fragments falling into the space cleared by the bit. When this cannot be done, the machinery and derrick are moved sufficiently to admit of driving a new set of pipes. It sometimes happens that the pipe is broken, or diverted from its vertical course by some obstacle. The whole string of pipes driven has, in such a case, to be drawn up again, and the work commenced anew; if this is not possible, a new location is sought.

After the pipe is driven, drilling is commenced. The drilling rope, which is generally $1\frac{1}{4}$-in. hawser-laid cable, of the required length (500–1000 ft.), is coiled round the shaft of the bull-wheel, the outer end passing over the pulley on the top of the derrick down to the tools, and attached to them by a rope-socket. When connected, these are 30–40 ft. in length, and sometimes more, weighing 800–1600 lb., according to the depth required to be reached. The process of drilling, until the whole length of the tools is on, and suspended by the cable, is slow. When the depth required to suspend the tools is reached, the attachment between the working beam and the drilling

PROSPECTING MACHINERY.

cable is made by means of a temper-screw (Fig. 121) suspended from the end of the working beam, and attached to the rope by a clamp. The temper-screw, which is provided with a coarse thread, is 2–3 ft. in length; it works in a thin iron frame, and is furnished with a wheel at the lower end of the screw for the driller to let out as required. As the drill sinks down into the rock, the screw is let down by a slight turn of the wheel by the driller, some allowing a full revolution every few blows of the bit, others once only in a few minutes, according to the hardness of the rock which is being drilled through.

The "jars" (Fig. 122) play a highly important part in the work of drilling. They are two long links or loops of iron or steel, sliding in each other. Drillers always allow about 4–6 in. play to the jars, which they call the "jar," and by this they can tell when to let down the temper screw. With the downward motion, the upper jar slides several inches into the lower one; by the upward motion, it is brought up, bringing the end of the jars together with a blow like that of a heavy hammer on an anvil, making a perceptible jar. Experienced drillers can, as soon as they take hold of the rope, tell how much "jar" they have on.

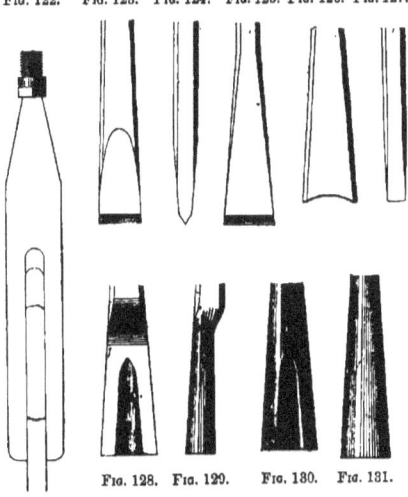

FIG. 122.　FIG. 123.　FIG. 124.　FIG. 125.　FIG. 126.　FIG. 127.

FIG. 128.　FIG. 129.　FIG. 130.　FIG. 131.

In drilling, the tools are alternately lifted and dropped by the action of the working beam in its rocking motion. One man is constantly required in the derrick to turn the tools as they rise and fall, to prevent them from becoming wedged fast, and to let out the temper-screw as required. This is one of the most important duties of the work, requiring constant attention to keep the hole round and smooth. The centre-bit is run down the full length of the temper-screw. The centre-bit (Figs. 123 to 125) is about 3¼ ft. long, with a shaft 2⅓ in. diameter, and a cutting edge of steel 3¼–4 in. wide, with a thread on the upper end by which it is screwed on to the end of the auger-stem. The reamer (Figs. 126–131) is about 2½ ft. long, having a blunt instead of a cutting edge, with a shank 2½ in. diameter, terminating in a blunt extremity 3½–4½ in. wide by 2 in. thick, faced with steel. The weight of heavy centre-bits and reamers averages from 50 to 75 lb. each. The centre-bit is followed by the reamer, to enlarge the hole and make it smooth and round.

The sediment, or battered rock, is taken out after each centre-bit, and again after every reamer, by means of a sand-pump. That now in use is a cylinder of wrought iron, 6–8 ft. in depth, with a valve at the bottom and a bail at the top, to which a ½-in. rope is attached, passing over a pulley suspended in the derrick some 20 ft. above the floor, and back to the sand-pump reel, attached to the jack-frame, and coiled upon the reel-shaft. This shaft is propelled by means of a friction pulley, controlled by the driller in the derrick by the rope attached. The sand-pump is usually about 3 in. diameter. Some drillers use two, one after the centre-bit and a larger one after the reamer, the two being preferable. When the sand-pump is lowered to a requisite depth, it is filled by a churning process of the rope in the hands of the driller, and is then drawn up and emptied. This operation is repeated each time the tools are drawn up out of the well, the pump being let down and drawn up a sufficient number of times to remove all the drillings. The fall of the tools is 2-3 ft. This labour goes on, first tools and then sand-pump, until the well is drilled to the required depth.

Several kinds of regulating tools are occasionally used. For the purpose of keeping the hole straight, the "winged" substitute (Fig. 132) is often used. To straighten a crooked hole, recourse is sometimes had to the "hollow reamer" (Fig. 133); sometimes the "star reamer" (Fig. 134) is employed for the same purpose. If skill and care, however, are exercised in the execution of the boring from the beginning, these tools are seldom needed.

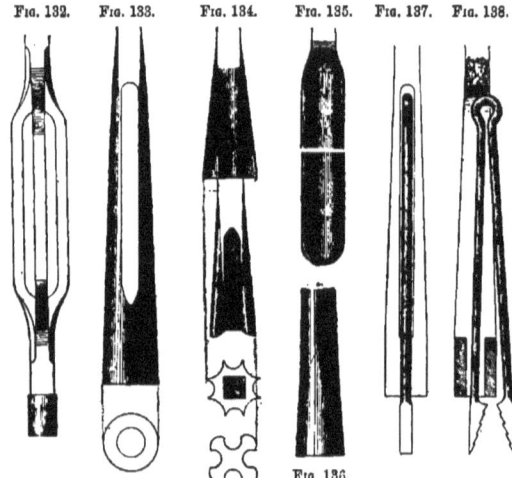

FIG. 132.　　FIG. 133.　　FIG. 134.　　FIG. 135.　　FIG. 137.　　FIG. 138.

FIG. 136.

The rate of boring with the machinery described varies from 2 in. an hour in very hard rock to 10-12 in. in shale. This rate is greatly in excess of that attainable by the system of rods.

In its passage, the drill not unfrequently dislodges gravel or fragments of hard rock, that have a tendency to wedge it fast in the hole. When a bit or reamer becomes so firmly imbedded as to render its removal impossible by jarring or breaking it in pieces, the well is abandoned. Sometimes a bit or reamer breaks, leaving a piece of hard steel fastened securely in the rock several hundred feet below the surface. Where the fragment is small, it is pounded into the sides of the well, and causes no further annoyance. When it is larger, the difficulty is not unfrequently insurmountable. The bit or reamer sometimes becomes detached from the auger-stem by the loosening of the screw from its socket; this is often heightened by the workman not being aware of its displacement, and for an hour or two pounding on the top of it with the heavy auger-stem. Various plans are resorted to in order to extract the fastened tool, and a large number of implements have been devised for "fishing up" the same. Many persons have become so expert and successful as to adopt this as a regular calling. The first instrument used is an iron with a thin cutting edge, straight, circular, or semicircular, acting as a spear, or to cut loose the accumulation around the top, and along the sides of the refractory bit or reamer, so as to admit a spring socket, that is lowered by means of the auger-stem over the top of it, and lays hold of the protuberance just below the thread. If the socket can be made fast, the power of the bull-wheel and engine is brought into requisition, and in a great number of cases it is brought to the surface.

In the jarring and other operations rendered necessary in cases of this kind, the entire set of tools, 40–60 ft. in length, may become fastened; and cases are of frequent occurrence where two and even three sets of tools have become fastened in a bore-hole as they were successively let down to extricate the first ones. A most effective instrument now commonly used for the extraction of broken tools consists of a number of heavy iron rods, similar to an auger-stem, weighing about 10 tons; to the end of the rods is attached a socket, or bell, which is lowered over the head of the tools and secured fast to them, the joints of the rods being provided with left-handed screws. When a set of tools have become fast each separate piece may by this means be unscrewed and raised to surface. The rods are lowered and raised from the top by jack-screws.

A running stratum often occasions much difficulty. Sometimes in passing through a bed of soft clay the material will flow into the hole suddenly, and bury 10 or even 20 ft. of the tools. In

FIG. 139. FIG. 140. FIG. 142. FIG. 143. FIG. 144. FIG. 145. FIG. 146. FIG. 147.

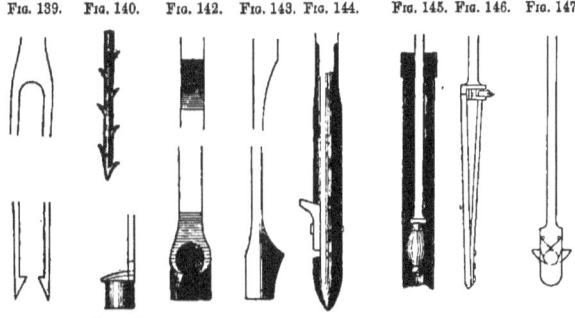

FIG. 141.

such a case a cutting instrument is attached to rods, and the rope is severed by it above the sinker bar. The cutting tool is then replaced by a spear-pointed instrument, with which, by means of a light set of tools, the substance imbedded round the tools is forced out. When they are sufficiently loosened, efforts are made to jar them out, an extra pair of jars being used for this purpose. Instead of the spear, the "spud" or spoon" (Fig. 135) is frequently used, being simply half a hollow reamer. The "horn-socket" (Fig. 136) is a tapering iron tube, designed to be dropped over and wedged upon the head of a lost tool. The "slip-socket" (Figs. 137, 138) is intended for the same purpose, but is provided with dogs or teeth to fall out and catch the tool under the collar. A similar kind of tool is the "grabs" (Fig. 139). The "rope-grabs" (Fig. 140) are for grappling the rope or cable; the "rope-knife" (Fig. 141), for severing the rope in the bore-hole. The "hook" (Figs. 142, 143) is used for grappling lost tools that are leaning against the side of the hole. The "slip-spear" (Fig. 144), to extract tubing.

Tools for Extracting Tubes.—When the bore-hole has been completed, and the end for which it was undertaken attained, it becomes desirable to recover the tubes used to line the hole. Also when more sets of tubes are required than anticipated, and the diameter of the bore-hole has consequently been so reduced that farther progress is impracticable, it becomes necessary to withdraw the lining and to enlarge the hole from surface. Withdrawing the tubes is always difficult, and when the hole is deep is seldom altogether successful. But in most cases a large proportion of the tubing may be recovered if suitable means are employed. These means consist of tools for disconnecting and lifting the several lengths of tubing, or for lifting them altogether.

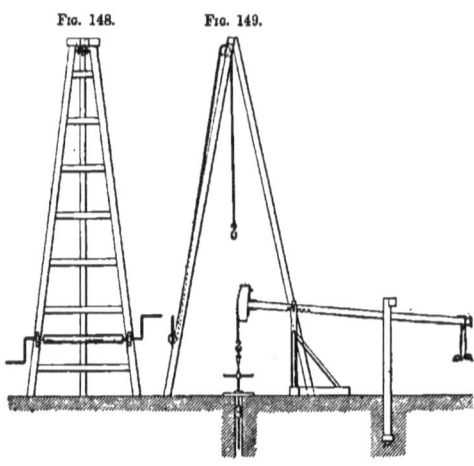

Fig. 148. Fig. 149.

Lever-boring Machine.

Of the former kind, the simplest and most effective is the screw-plug. This instrument consists of a conical plug having its lower end slightly less in diameter than the bore of the tube, and its upper end slightly greater, and provided with a left-handed steeled screw-thread. This plug terminates upwards in a shank and screw-socket for the purpose of fixing it to the rods. The latter, which are constructed specially for this purpose, are of large section, and are connected by left-handed screw-joints. The screw-plug is lowered at the end of these rods into the end of the tube, and turned slowly round till the thread has bitten. When the plug has obtained a firm hold of the tube, the latter will be unscrewed by the continued left-handed motion of the former, and may be lifted by it. The same operation is repeated for each length of tubing.

Of tools designed to lift the whole length of tubing, the best is that known as "Kind's plug" (Fig. 145). It consists of a block of oak of an ovoid form fixed upon the end of an iron rod, which passes through the centre of the plug, holding it by means of a nut, and terminating upwards in a screw-plug for the purpose of attaching it to the ordinary boring rods. The diameter of this wooden plug at its largest part is slightly less than that of the tube, so that a little amount of play is allowed between it and the sides of the tube. When it is required to raise the tubes, the plug is lowered to the desired depth, and one or two shovelsful of coarse, gravelly sand, washed and sifted, are thrown down upon it. This sand fills the space between the sides of the tubing and the plug, and the latter is thereby firmly wedged in. The rods being then hauled up, the tubing is raised with them. If it be desired to make the plug leave go its hold on the tube, it is only necessary to lower it below the lining, when the sand will run out.

When the tubing is too firmly held by the friction against the sides of the bore-hole to allow of it being raised altogether, and it is deemed undesirable to have recourse to the spare rods required for the screw-plug, Kind's plug may be used in conjunction with another kind of tool to raise the tubing in portions. The use of the latter tool is to cut through the lining so as to divide it into portions capable of being raised at once. Numerous tools have been invented for this purpose. One of the simplest is represented in Figs. 146, 147. By suspending this tool at the requisite and fixed height in the bore-hole, on the end of the boring rods, and turning it round, the cutting edge, which is pressed by a spring against the sides of the hole, cuts through the lining, the severed portions of which may then be raised by Kind's plug in the manner described above. The cutter is so constructed that it may be readily withdrawn from the cut and raised to surface.

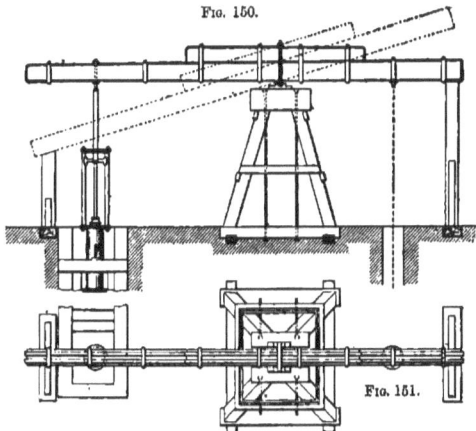

Lever-boring Machine, Steam Worked.

Figs. 148, 149 show, in side and end elevation, a useful set of boring tackle, consisting of a simple lever worked by hand. Triangular shear-legs are erected over the bore-hole, and provided with a windlass for the purpose of drawing the rods.

Figs. 150, 151 show the same kind of machine, worked by means of a direct-acting steam cylinder instead of by hand.

Figs. 152, 153 show, in side and end elevation, a boring tackle provided with a steam-winch for raising and lowering the rods and tools. The shear-legs are 50 ft. high, and are provided with

72 MINING AND ORE-DRESSING MACHINERY.

two stagings $a\ b$ for facility in handling the rods. The steam-winch is provided with a strong, flat hemp rope c, going from the main drum d and over the top sheave, to lift or lower the boring rods; and a light round rope e going from a larger drum f over a lower sheave, to be used for the

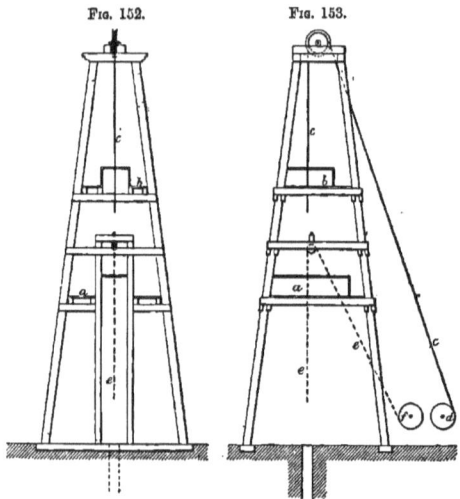

Fig. 152.   Fig. 153.

Boring Tackle, provided with Steam Winch.

purpose of raising and lowering the clearing scoop. The actual jumping of the tool is done in the same way as in Fig. 150. With a speed for the driving pinion of 120 revs. per minute, the velocities when running at high speed are 136 ft. per minute for the scoop rope, and 132 for the rod rope; and when working at low speed, 20 ft. per minute for the rod rope.

In the annexed table is shown the cost of some trial borings for ironstone in the Barrow district, made with the aid of a steam winch and free-falling tool, similar to Figs. 152 and 153.

Depth of Hole.	Diameter of Hole.	Cost per Yard.	Cost of Labour Alone.	Time Occupied.
yards	inches	s. d.	£ s. d.	weeks
126	6 to 2	7 10¼	49 10 0	15
124*	,,	9 7	59 8 0	18
50	,,	9 10¾	24 15 0	7½
63	,,	9 5	29 14 0	9
76½	,,	6 0½	23 2 0	7
88	,,	8 3	36 6 0	11
48	,,	11 0	26 8 0	8

* The strata passed through in this hole were as follows, proceeding from the surface downwards:—45 ft. pinder, 75 ft. red sand, 3 ft. white sand, 30 ft. red sand mixed with clay, 150 ft. red sand, 30 ft. red and white sand, 6 ft. white sand, 6 ft. shale, 4 ft. ore, 6 ft. clay, 1 ft. ore, 2 ft. stone, 9 ft. ore, 2 ft. black shale, 3 ft. stone—total, 372 ft.

PROSPECTING MACHINERY.

The approximate cost per set of boring tools, including rigger, rope, and ordinary shear legs, and windlass for depths of 300 ft. and upwards is:—

	£		£
To bore 30 ft.	20	To bore 250 ft.	75
,, 50 ,,	36	,, 300 ,,	120
,, 100 ,,	45	,, 500 ,,	155
,, 150 ,,	58	,, 800 to 1000 ft.	195
,, 200 ,,	70		

The old-fashioned systems of boring by pulverising the rock have almost universally given place to the diamond drill, which cuts its way, and permits a solid core to be drawn out of the hole, representing exactly the strata bored through.

The general principle of boring with the diamond drill is the same, the different machines, by comparatively slight changes, being applicable to any kind of rock-drilling. For deep boring, or prospecting mineral lands, a machine is used with a double oscillating cylinder engine, mounted on an upright, or horizontal, tubular boiler. The capacity of the engine varies according to the depth and size of hole requiring to be bored. These machines have a screw shaft made of heavy hydraulic tubing 5-7 ft. in length, with a deep screw cut on the outside. The shaft also carries a spline, by which it is feathered to the lower sleeve gear. This gear is double, and connects by its upper teeth with a bevelled driving gear, and by its lower teeth with a release gear, which is a frictional gear, and is fitted to the lower end of the feed shaft, to the top of which a gear is feathered, fitting to the upper gear on the screw shaft, which has one or more teeth less than the upper gear on the feed shaft, whereby a differential feed is produced. This frictional gear is attached to bottom of feed shaft by a friction nut, thus producing a combined differential and frictional feed, which renders the drill perfectly sensitive to the character of the rock through which it is passing, and maintaining a uniform pressure upon the same. The severe and sudden strain upon the cutting points incidental to drilling through soft into hard rock with a positive feed is thus avoided. The drill-rod, made of heavy lap-weld tubing, passes through the screw shaft, and is held firm by a chuck at the bottom of the screw shaft. To the lower end of this tubular boring-rod the bit is screwed, and to the upper end a water swivel, to which connection is made with the steam pump. By means of this pump a constant stream of water is forced down through the hollow drill-rod, thereby keeping the bit cool and the hole bored clear of sediment, which is forced by the water-pressure up the outside of the rods to the surface. The hollow bit is a steel thimble, having three rows of diamonds (bort or carbon) imbedded therein, so that the edges of those in one row project from its face, while the edges of those in the other two rows project from the outer and inner periphery respectively. The diamonds set in the first-mentioned row cut the path of the drill in its forward progress, while those embedded upon the outer and inner periphery of the tool enlarge the cavity around the same, and permit the free ingress and egress of the water as above described.

The screw shaft, being rotated and fed forward, rotates the drill-rod and bit; and as the bit passes into the rock, cutting an annular channel, that portion of the stone encircled by this channel is of course undisturbed; the core barrel passing down over it preserves it intact until the rods are withdrawn, when the solid cylinder thus formed is brought up with them, the core-lifter breaking in at the bottom of the hole, and securely wedging and holding it in the core-barrel. Where a core is not required, the perforated boring-head can be used, the detritus being washed out by the water introduced through the drill-rod, the same as when boring with the hollow bit. In order to run

L

the screw shaft back after it has been fed forward its full length, it is only necessary to release the chuck and to loosen the nut on the frictional gear, which allows the gear to run loose; then the screw shaft will run up with the same motion which carried it down, but with a velocity sixty times greater; that is, the speed with which the screw shaft feeds up is to the speed with which it fed the drill down as 60 to 1, the revolving velocity in both cases being the same. By tightening up the chuck and nut on the frictional gear, the drill is ready for another run. The drill-rods may be extended to any desired length by simply adding fresh pieces of tubing, the successive lengths being quickly coupled together by an inside shoulder nipple coupling, made of the best forged iron, and having a hole bored through the centre to permit the passage of the water. In order to withdraw the drill-rods, they are uncoupled below the chuck; the swivel head, which is hinged, is unbolted and swung back, thereby moving the screw shaft to one side, and affording a clearance for the rods to be raised by the hoisting gear on the machine without moving the drill. By the erection of a derrick of sufficient height, it will be necessary to break joints only once in every 40–50 ft.

The advantages claimed for diamond drills over the steel or percussive system of drilling are:—

1. They drill rock faster than is possible in any other way; not only boring more feet in any given time—as a day or an hour—but accomplishing far greater results in the aggregate of a month's or a year's practical use.

2. They perform a given amount of work more cheaply than it can otherwise be done, saving at least one-third the entire cost of heavy excavations (including the blasting and removal of material) as compared with hand labour, besides economising time in a much larger ratio.

3. They are extremely simple, both in construction and operation, and seldom need repairs, the very best material and workmanship being used in their construction; and workmen of ordinary intelligence are perfectly competent to operate with them successfully. These machines, it will be remembered, are not subject to the constant and destructive shocks of concussion against the rock, which disable the best percussive machines so often, and render them so expensive to keep in repair, nor have they any delicate parts or nice adjustment to be carefully watched. Every part is equally simple and durable. The diamond teeth are the only part of the tool which comes in contact with the rock, and their hardness is such that more than 2000 ft. have been drilled by the same points with but little apparent wear. The cost of resetting the diamonds, so as to present new points, is very slight, and no special skill is required for the operation. Other repairs are seldom needed.

4. They produce holes uniform in diameter—not three-cornered or funnel-shaped, as must necessarily result from percussion drilling, but perfectly cylindrical from top to bottom—a feature of great importance in blasting, as the force of the explosive material is thereby fully utilised, and its practical effects are greatly increased. They are not deflected from a right line by seams and crevices, nor impeded in their progress by the hardest rock.

5. The great advantage of being able to bore blast holes to the "bottom of grade" is well understood by rock contractors. For grades of over 20 ft. in depth, this is the only machine with which holes can be readily bored. They bore as rapidly and cheaply at a depth of 50 to 100 ft. as at the surface, while at a depth of 500 to 600 ft. there is but little appreciable difference. With the prospecting machines, thousands of holes, from 300 to 1500 ft. in depth (many of them horizontal), have been bored through solid rock.

6. All the drills are adapted to bore holes at every angle of either a vertical or horizontal circle, and in a shaft or tunnel they bore as close to either side-wall as a hand-drill can be turned.

The peculiar shape of the boring-bit, holding a cylinder of solid rock inside, prevents the drill from running out of line; hence the hole bored, however deep it may be, is perfectly straight.

7. They are not only adapted to shafting, tunnelling, well-boring, submarine blasting, and all kinds of rock-excavating in mines, quarries, railroads, &c., but in their application to "prospecting" they accomplish most important results, otherwise wholly unattainable. By their use only can mines be penetrated to a depth of 1000 ft. and upwards through solid rock, vertically or horizontally, and perfect samples of mineral taken out the entire distance, disclosing the character and value of the mine by means of a single drill-hole. It should also be observed that the samples so obtained are not disintegrated fragments of rock, but continuous solid cylinders, showing clearly the stratification and character of the ground so prospected.

Wherever these drills are used, it is necessary to arrange for a regular supply of diamonds. The two kinds of diamond used in drills are known as "bort" and "carbons." The bort is a real diamond, which, owing to imperfections, is useless as a gem; being nearly globular in shape, it is generally set in the outer edge of the drill, as being less likely to catch in irregularities of surface in the rock. The carbon is a black stone of very varying shape and usually sharp-edged. The bort is much the harder, and resists greater pressure. It is considerably dearer, costing about 42s. a carat, while carbons may be had for 26s. a carat. About 6-8 weeks' constant work suffices to wear out the setting of a drill. The working capacity is about 8 ft. a day in quartz and granite, to 10 ft. in sandstone and slate.

STAMPING.—When the vein of mineral matter has been reached by the borings, it is necessary to reduce some of it to a fine state of subdivision for testing. A primitive yet efficient apparatus for this purpose, and such as may be erected by the prospector himself without the aid of much engineering skill, is shown in Fig. 154, and goes by the name of a "dolly." On the end of a solid log a, fixed in the ground and standing about 4 ft. high, is cut a square hole b, about 6 in. across, in which are firmly fitted wrought-iron bars about $\frac{1}{2}$ in. apart, $\frac{1}{4}$ in. thick, and 3 in. deep, made thinner beneath, so that whatever enters above will fall through. A wooden box c is placed round this to keep the ore from jumping away. A square block of wood d, about 3 or 4 ft. long, shod with wrought iron, and small enough at the lower end to work in the box, forms the stamper. It is hung on the end of a long pole e, the spring of which keeps it on the swing without too much labour. It is worked by laying hold with the hands of a wooden pin f on each side of the stamper, and pulling it down, its own rebound and the spring of the pole taking it up again. The iron bars might be replaced by an old stamp die. The gold is caught on the table g.

SAMPLING.—The fine stuff must next be properly sampled, which is well performed by the automatic sampler, shown in Fig. 155: a is a sheet-iron funnel by which the falling ore is thrown together into a small stream, whence it falls on b, and is immediately scattered over the entire width of the box and falls into funnel c, which, from the shape of its mouth, discharges a flat stream 12 in. by 1 in., and thoroughly mixed. The trough d, 1 in. wide, is placed directly under c at a sharp inclination, and carries off constantly $\frac{1}{16}$ of the stream, the remainder going directly into the sack. The trough d leads by a tube into a box, and the portion conveyed there constitutes the sample. Thus when a 10-ton lot is run through the mill, a 1-ton sample is found in the box. This is then dumped on a clean floor and mixed by shovelling over a conical pile, every shovelful going to the apex of the pile. This is then thrown by shovelfuls over a scoop-shovel, constructed as in Fig. 156, or sometimes over one like Fig. 157. What is caught in the scoop, viz., about $\frac{1}{16}$ of the main sample, is dumped alternately in two heaps, making a double sample, called original and duplicate. The two samples are then each cut down separately by throwing again over the scoop-shovel, sweeping up carefully each time, until they weigh about 10 lb. each, when they are taken to a small pulverisor and ground to about the size of coarse sand. Each sample is then thoroughly mixed by rolling on

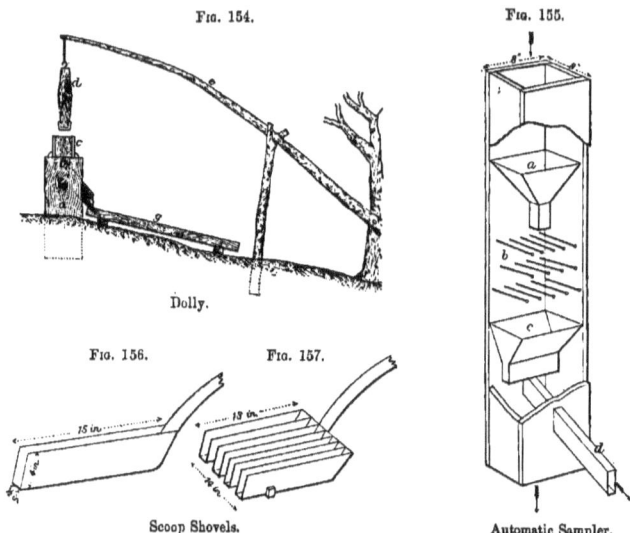

Fig. 154.

Fig. 155.

Dolly.

Fig. 156.   Fig. 157.

Scoop Shovels.

Automatic Sampler.

an oil-cloth and then cut down on an ordinary tin splitter, like Fig. 157,—of rather deep and wide troughs, however—to about 2 lb., when it is ground on a rubbing-plate till all passes a 20-mesh sieve. It is then cut down on a finer splitter with $\frac{1}{2}$-in. troughs to about $\frac{3}{4}$ lb., and this is finally ground on the plate till all passes an 80-mesh sieve.

CHAPTER V.

EXCAVATING MACHINERY.

DRILLING TOOLS.—Hand-drilling tools consist essentially of the drill and the hammer. The drill is an iron or a steel rod terminating at one end in a cutting edge, and at the other end in a flat face to receive the blow from the hammer. Thus the parts of a rock-drill are the bit or chisel edge, the stock, and the striking face. Formerly the stock was always of iron, but now it is usual to make the whole drill of cast steel, the superior solidity of texture in steel rendering it capable of trans-

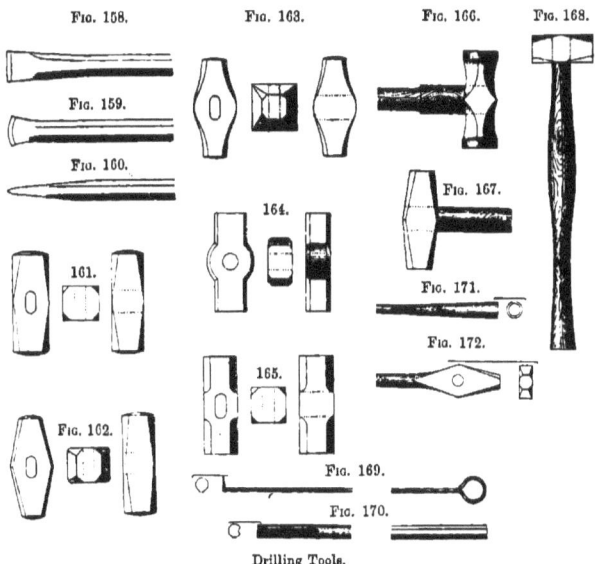

Drilling Tools.

mitting the force of a blow more effectively than iron. The cutting edge of a drill demands careful consideration. To enable the tool to free itself well in the hole, and also to avoid introducing unnecessary weight into the stocks, the bit is made wider than the latter, sometimes as much as 1 in. In hard rock, the liability of the edge to fracture increases as the difference of width. The edge of the drill may be straight, as in the flat chisel for deep drilling, or slightly curved. The straight edge cuts its way somewhat more freely than the curved, but it is weaker at the corners than the curved, a circumstance which renders it less suitable for very hard rock. It is also slightly more difficult to forge. Figs. 158 to 160 show the straight and curved bits, and the angles of the cutting edges for use in rock. The width of the bit varies, according to the size of the hole required, from 1 to $2\frac{1}{2}$ in.

The stock is octagonal in section, and is made in lengths varying from 20 to 42 in. The shorter the stock, the more effectively does it transmit the blow, and therefore it is made as short as possible; for this reason several lengths are employed in drilling a blast-hole, the shortest being used at the commencement of the hole, a longer one to continue the depth, and a still longer one, sometimes, to complete it. To ensure the longer drills working freely in the hole, the width of the bit should be very slightly reduced in each length. It has already been remarked that the diameter of the stock is less than the width of the bit; this difference may be greater in coal drills than in rock or "stone" drills; a common difference in the latter is $\frac{3}{8}$ in. for the smaller sizes, and $\frac{1}{2}-\frac{3}{4}$ in. for the longer. The following proportions may be taken as the average adopted:—

Width of the Bit.	Diameter of the Stock.	Width of the Bit.	Diameter of the Stock.
1 in.	$\frac{5}{8}$ in.	$1\frac{3}{4}$ in.	$1\frac{1}{4}$ in.
$1\frac{1}{4}$,,	$\frac{3}{4}$,,	2 ,,	$1\frac{3}{8}$,,
$1\frac{1}{4}$,,	$\frac{7}{8}$,,	$2\frac{1}{4}$,,	$1\frac{1}{2}$,,
$1\frac{1}{2}$,,	1 ,,	$2\frac{1}{2}$,,	$1\frac{5}{8}$,,

The striking face of the drill should be flat. The diameter of the face is less than that of the stock in all but the smallest sizes, the difference being made by drawing in the striking end. The amount of reduction is greater for the larger diameters, that of the striking face being rarely more than $\frac{7}{8}$ in.

Hammers are important tools in the hands of the miner. The distinction between a sledge and a hammer is founded on dimensions only; the hammer, being intended for use in *one* hand, is made comparatively light, and is furnished with a short handle; while the sledge, being intended for use in *both* hands, is furnished with a much longer handle, and is made heavier. Sledges are used for striking the drill in making blast holes, for driving wedges in rock and in coal, and for breaking up large masses of the latter. The blasting sledge and the wedge-driving sledge, being employed under different conditions, require different forms and dimensions. The striking face of the blasting sledge should be flat, to enable the striker to deliver a direct blow with certainty upon the head of the drill; and to facilitate the directing of the blow, as well as to increase its effect, the mass of metal composing the head should be concentrated within a short length. To cause the sledge to fly off from the head of the drill in the case of a false blow being struck, and thereby to prevent it from striking the hand of the man who holds the drill, the edges of the striking face should be chamfered or bevelled down till the diameter is reduced by nearly one-half. This requirement is, however, but seldom provided for. When used for wedge driving, the head of the sledge is very frequently required to follow the wedge into the cleft, and to enable it to do this, the head must be made long and of small diameter, that is, the mass of metal composing the head must be distributed throughout a greater length. The striking face should be rather convex than flat to avoid a sharp edge, which would soon be battered off by coming into contact with the edges of the rocks in the cleft. A longer handle or helve is also needed for the wedge-driving than for the blasting sledge.

The head of a sledge is of iron; it consists of a pierced central portion called the eye, and two shanks or "stumps," the steel ends of which form the striking faces or "panes." The form of the head varies in different localities, but whatever the variation may be, the form may be classed under one of four types or "patterns." A very common form is that shown in Fig. 161, and known as the "Bully" pattern. By varying the width, as shown in Fig. 162, we obtain the "broad bully," the former being called for the sake of distinction the "narrow" bully. Another common form is the "Pointing" pattern, represented in Fig. 163. The form shown in Fig. 164 is designated as the "Bloat" pattern; and that given in Fig. 165 the "Plug" pattern. Each of these forms possesses peculiar merits which render it more suitable for certain uses than the others. The same forms are used for hammers. The eye is generally made oval in shape, but sometimes, especially with the bloat pattern, it is made circular, as shown in Fig. 164. The weight of a sledge head may vary from 5 to 10 lb., but a common and convenient weight is 7 lb. The length of the helve varies from 20 to 30 in.; a common length is 24 in. for blasting, and 28 in. for wedge-driving sledges. The average weight of hammer heads is about 3 lb., and the average length of the helve 10 in.

All the forms of sledge heads may be used for wedge-driving purposes, but that which is generally employed, especially for coal wedging, is the pointing pattern. The modification made in the form illustrated is merely in the length of the head. A common length of a coal-wedging sledge is 12 in., with a diameter of about $2\frac{1}{4}$ in. in the thickest part. The stumps are tapered down to about $1\frac{1}{2}$ in. at the panes, and the angles of the stumps are taken off by a chamfer, beginning near the eye and gradually increasing to form an octagonal section at the panes.

Fig. 166 represents a blasting sledge used in South Wales. The stumps are octagonal in section, and spring from a square block in the centre. The panes or striking faces, however, are circular and flat. The length of the head is $8\frac{3}{4}$ in., that of the helve 27 in., and the weight of the tool complete 7 lb.

Fig. 167 represents a blasting sledge used in North Wales. The central block is an irregular octagon in section, formed by slightly chamfering the angles of a square section, and the stumps are chamfered down to form a regular octagon at the panes, which are flat. The length of the head is $7\frac{3}{4}$ in., that of the helve 22 in., and the weight of the tool complete 6 lb. 7 oz.

The sledges used in the north of England have shorter heads, and are lighter than the foregoing. Fig. 168 represents one of these blasting sledges. The head is nearly square in section at the centre, and the panes are flat. The length of the head is 5 in., that of the helve $24\frac{1}{2}$ in., and the weight of the sledge complete 4 lb. 14 oz.

Drills, as before remarked, are used in sets of different lengths. The sets may be intended for use by one man or by two. In the former case, they are described as single-hand sets, and contain a hammer for striking the drills; in the latter case, as double-handed, and contain a sledge instead of a hammer for striking.

Besides the drill and the hammer, other auxiliary tools are used in preparing the hole for the blasting charge. If the hole is inclined downwards, the débris or "meal" made by the drill remains at the bottom of the hole, where it is converted into mud or "sludge" by the water there present. This sludge, as in the case of deep boring, has to be removed as the work progresses, to keep the rock exposed to the action of the drill. The removal of the sludge is effected by a simple tool called a "scraper" (Fig. 169). It consists of an iron rod ¼–½ in. diameter, and of sufficient length to reach the bottom of the hole. One end of the rod is flattened out on the anvil and made circular in form, and then turned up at right angles to the stem. The disc thus formed must be less in diameter than the hole, to allow it to pass readily down. When inserted in the hole, the scraper is turned round while it is being pressed to the bottom; on withdrawing the instrument the sludge is brought up upon the disc. This operation, two or three times repeated, is sufficient to clear the hole. The other end of the scraper is sometimes made to terminate in a ring for convenience in handling. Instead of the ring, however, at one end, a disc may be made at each end, the discs in this case being of different diameter, to render the scraper suitable for different size holes. Sometimes the scraper is made to terminate in a spiral hook or "drag twist." The use of the drag is to thoroughly cleanse the hole before inserting the charge. A wisp of hay is pushed down the hole, and the drag end of the scraper is introduced after it, and turned round till it has become firmly entangled. The withdrawal of the hay by the drag wipes the bore-hole clean. Instead of the twist drag, the "loop" drag is frequently employed. This consists of a loop or eye, through which a piece of rag or tow is passed. The rag or tow is used for the same purpose as the hay, namely, to thoroughly cleanse and dry the hole previous to the introduction of the charge.

When the charge has been placed in the hole, and the fuse laid to it, the hole needs to be tamped, that is, the portion above the charge has to be filled up with some suitable substance. For this purpose, a rammer, stemmer, or tamping iron (Fig. 170) is required. It consists of a metal bar, the tamping end of which is grooved to receive the fuse lying against the side of the hole. The other end is flat, to afford a pressing surface for the hand, or a striking face for the hammer when the latter is needed. To prevent the danger of accidental ignition from sparks caused by the friction of the metal against silicious substances, the employment of iron stemmers has been prohibited by law. They are usually made of copper, or of phosphor-bronze, the latter substance being more resisting than the former.

Sometimes in wet ground it becomes necessary to shut back the water from the hole before introducing the charge of gunpowder. This happens very frequently in shaft sinking. The method employed in such cases is to force clay into the interstices through which the water enters. The instrument used for this purpose is the "claying-iron" or "bull" (Fig. 171). It consists of a round bar of iron, called the stock or shaft, a little smaller in diameter than the bore-hole, and a thicker portion, called the head or pole, terminating in a striking face. The lower end of the shaft is pointed to enable it to penetrate the clay, and the head is pierced by a hole about 1 in. diameter to receive a lever.

Clay in a plastic state having been put into the hole, the bull is inserted and driven down by blows with the sledge. As the shaft forces its way down, the clay is driven into the joints and crevices of the rock on all sides. To withdraw the bull, a bar of iron is placed in the eye, and used as a lever to turn it round to loosen it; the rod is then taken by both hands, and the bull lifted out. To allow the bull to be withdrawn more readily, the shaft should be made with a slight taper, and kept perfectly smooth. As the bull is subjected to a good deal of heavy hammering on the head, the latter part should be made stout. This tool is very serviceable, and should always be at hand in wet ground when gunpowder is employed.

Another instrument of an auxiliary character is the beche (Fig. 172), used for extracting a

EXCAVATING MACHINERY.

broken drill. It consists of an iron rod of nearly the diameter of the hole, and hollow at the lower end. The form of the aperture is slightly conical, so that the lower end may easily pass over the broken stock of the drill, and being pressed down with some force, may grasp the stock in the higher portion of the aperture with sufficient firmness to allow of the two being raised together. When only a portion of the bit remains in the hole, it may often be extracted by means of the drag-twist end of the scraper.

A set of coal-blasting gear will include a drill, 22 in. long, with cutting edge straight and $1\frac{1}{2}$ in. wide, and weight $2\frac{1}{2}$ lb.; another drill, 42 in. long, with straight cutting edge $1\frac{7}{16}$ in. wide, weight 4 lb. 10 oz.; the hammer weighs 2 lb. 14 oz., length of head $4\frac{1}{2}$ in., and that of handle $7\frac{3}{4}$ in. A single-hand stone set includes shorter drill, 22 in. long, cutting edge strongly curved, and $1\frac{1}{2}$ in. wide, and weight 3 lb. 10 oz.; longer drill, 36 in. long, cutting edge $1\frac{7}{16}$ in. wide, and curved as in the shorter drill, and weight 6 lb. 5 oz.; hammer weighs 3 lb. 6 oz., length of head 5 in., and that of handle 10 in. A double-hand stone set comprises first or shortest drill, 18 in. long, $1\frac{3}{4}$ in. wide on the cutting edge, and weighs $4\frac{1}{4}$ lb.; second drill, 27 in. long, $1\frac{1}{4}$ in. wide on the cutting edge, and weighs 6 lb.; third, or longest drill, 40 in. long, $1\frac{3}{8}$ in. wide on the cutting edge, and weighs $9\frac{1}{4}$ lb. The cutting edges of all these drills are strongly curved; sledge weighs about 5 lb.

Machine drills penetrate rock in the same way as the hand drills already described, namely, by means of a percussive action. The cutting tool is, in most cases, attached directly to the piston rod, with which it consequently reciprocates. Thus the piston with its rod is made to constitute a portion of the cutting tool, and the blow is then given by the direct action of the steam or the compressed air upon the tool. As no work is done upon the rock by the back stroke of the piston, the area of the forward side is reduced to the dimensions necessary only to lift the piston, and to overcome the resistance due to the friction of the tool in the bore-hole. The piston is made to admit steam or air into the cylinder, and to cut off the supply and to open the exhaust as required, by means of tappet valves, or other suitable devices; and provision is made to allow, within certain limits, a variation in the length of the stroke. During a portion of the stroke, means are brought into action to cause the piston to rotate to some extent, for the purposes that have been already explained. To keep the cutting edge of the tool up to its work, the whole machine is moved forward as the rock is cut away. This forward or "feed" motion is usually given by hand; but in some cases it is communicated automatically. The machine is supported upon a stand or framing which varies in form according to the situation in which it is to be used. This support is in all cases constructed to allow of the feed motion taking place, and also of the cutting tool being directed at any angle. The support for a rock drill constitutes an indispensable and a very important adjunct to the machine; for upon the suitability of its form, material, and construction, the efficiency of the machine will largely depend. The foregoing is a general description of the construction and mode of action of percussive rock-drills. The numerous varieties now in use differ from each other, as already remarked, rather in the details of their construction than in the principles of their action, and the importance of the difference is, of course, dependent upon that of the details. Assuming the necessity for a high degree of strength and rigidity in the support, a primary condition is that it shall allow the machine to be readily adjusted to any angle, so that the holes may be drilled in the direction and with the inclination required. When this requirement is not fulfilled, hand-labour will have to be employed in conjunction with it, and such incompleteness in the work of a machine constitutes a serious objection to its adoption.

Besides allowing of the desired adjustment of the machine, the support must be itself adjustable to uneven ground. The bottom of a shaft which is being sunk, or the sides, roof, and floor of a heading which is being driven, present great irregularities of surface, and as the support must of necessity, in most cases, be fixed to these, it is obvious that its design and construction must be such as will allow of its ready adjustment to these irregularities. The means by which the adjustment is effected should be few and simple, for simplicity of parts is important in the support as well as in the machine, and for the same reasons. A large proportion of the time during which a machine drill is in use is occupied in shifting it from one position or one situation to another; this time reduces in a proportionate degree the superiority of machine over hand labour in respect of rapidity of execution, and it is therefore evidently desirable that it should be shortened as far as possible. Hence the necessity for the employment of means of adjustment which shall be few in number, rapid in action, and of easy management.

For reasons similar to the foregoing, the drill support must be of small dimensions, and sufficiently light to allow of its being easily portable. The limited space in which rock drills are used renders this condition, as in the case of the machine itself, very important. It must be borne in mind that after every blast the dislodged rock has to be removed, and rapidity of execution requires that the operations of removal should be carried on without hindrance. A drill support that occupies a large proportion of the free space in a shaft or a heading is thus a cause of inconvenience and a source of serious delay. Moreover, as it has to be continually removed from one situation to another, it should be of sufficiently light weight to allow of its being lifted and carried without difficulty. In underground workings, manual power is generally the only power available, and therefore it is desirable that both the machine and its support should be of such weight that each may be lifted by one man. Of course, when any endeavour is made to reduce the weight of the support, the necessity for great strength and rigidity must be kept in view.

In spacious headings, such as are driven in railway tunnel work, supports of a special kind may be used. In these situations, the conditions of work are different from those which exist in mines. The space is less limited, the heading is commenced at surface, and the floor laid with a tramway and sidings. In such a case, the support may consist of a massive structure mounted upon wheels to run upon the rails. This support will carry several machines, and to remove it out of the way when occasion requires, it will be run back on to a siding. But for ordinary mining purposes, such a support is unsuitable.

Air-compressing Machines.—Its easy conveyance to any point of the underground workings, and its ready application at any point; the improvement which it produces in the ventilation; the complete absence of heat in the reservoir and conducting pipes, a condition which tends greatly to their preservation; these and numerous other advantages, when contrasted with the defects of steam under like conditions, give to compressed air a special value as a means of transmitting force in underground operations. In applying a motor fluid underground, it is rather a question of distributing small forces over a large number of points, than of concentrating a large force at one or two points. This is particularly the case when it becomes necessary to employ hauling engines, coal-cutting machines, and portable rock-drills, the positions of which are daily changing. Great, however, as the merits of compressed air as a medium for transmitting force are, it possesses defects of an important character. These defects lie chiefly in its inherent and essential qualities as a gas, whereby a loss of the force to be transmitted is occasioned. The amount of the loss due to this source is necessarily considerable; but when due precautions are not taken to keep it near its minimum limits, it may assume very grave proportions.

The most important source of loss of work in compressing air is the accumulation of heat. The means of remedying the evil lies in the application of a suitable medium of abstraction. Hitherto water has shown itself to be the most effective and convenient body for such a purpose, and numerous modes of applying it have been devised in order to obtain the best possible results. To favour the action of the water, the velocity of the piston should be kept low. This would appear to be a necessary condition of efficiency.

Another source of loss of work, the consequences of which increase in importance with the degree of compression, is the clearance space at the end of the cylinder. Those consequences will, on reflection, become sufficiently apparent. Suppose a cylinder in which the compression is carried to six atmospheres. When the piston arrives at the end of its stroke the clearance space contains air compressed into one-sixth of its volume at atmospheric pressure; and it is obvious that when the piston recedes, this air must expand into six times that volume, that is, into its volume at atmospheric pressure, before any air from the surrounding atmosphere can enter the cylinder; or, in other words, before the suction valve can open. Thus a volume of air is lost at each stroke, equal, at atmospheric pressure, to that assumed by the compressed air in the clearance space when, by expansion, it has dropped to the same pressure. To remove altogether the necessity for a clearance space, columns or cushions of water have been employed. These fulfil the purpose required very satisfactorily; but it must be borne in mind that they are themselves a source of loss of work, by the inertia which they oppose to the motive force. It should be remarked that the contents of the clearance space includes the air in the receiver behind the valve, which air returns into the cylinder as the valve closes. This is called the slip of the valve, that is, the quantity of air which the valve as it returns to its seat allows to slip back into the cylinder. When the lift of the valve is great, this quantity may be considerable; and when the lift is slight, the resistance from friction, due to the contracted passages, may also be considerable.

Leakage of the valves and pistons, and the friction of the moving parts, constitute sources of loss of greater or less importance according to the degree of perfection attained in the construction of the machine and the condition in which it is maintained. As these sources of loss are greatly dependent for their existence upon design, workmanship, and supervision, they are capable of being reduced to narrow limits. It is, however, needful to remark here that the loss of work due to the friction of the air in the valve-ways, and to the influence of the contracted vein, is by no means insignificant.

There is yet another source of loss of motive force, the influence of which is very great, and which increases with the degree of compression. This source of loss exercises an important bearing upon the question of economy relatively to this mode of transmitting force, and is, therefore, deserving of careful consideration. As the air has to be compressed by the application of force, it is clear that the fraction of that force which remains after the important deductions have been made for the losses already described, cannot be fully recovered without working the air expansively down to the pressure of the atmosphere. As this is in all cases impracticable, there must be always a loss of work, the value of which may be determined from the degree of expansion adopted. In the case of machine rock-drills, which work without expansion altogether, the loss is necessarily very great, and, when high pressures are used, may become enormous.

Compressed air has to be conveyed in pipes and tubes from the reservoir into which it has been forced, to the machines in position at the various points where operations are being carried on, through distances in many cases considerable. During this transmission, a loss of work is occasioned by the friction of the air in the pipes. Numerous and exhaustive experiments have been made to determine accurately the value of the loss thus occasioned. From the results of these experiments the following three conclusions have been deduced, namely: 1, that the resistance is directly as the length of the pipe; 2, that it is directly as the square of the velocity of flow; and 3, that it is inversely as the diameter of the pipe. Upon these results and conclusions formulæ have been established whereby the value of the loss of force may be ascertained with ease and accuracy. These formulæ show that for pipes of the diameters usually employed for this purpose, and for distances not exceeding one mile, the loss of motive force due to the friction of the air in the pipe is insignificant when the velocity does not exceed 4 ft. a second. And it can be shown that even this loss is notably diminished, and in some cases entirely annulled, by the increased head due to the depth of the shaft, when the compressed air is employed in mines. The influence of this head may often be taken advantage of to diminish slightly the diameter of the pipes, and thereby to effect a considerable economy of cost. This source of loss of motive force is of small moment when compressed air is applied to ordinary mining operations, so long as the velocity is kept below the limit already mentioned.

Fig. 173.

Air Compressor, built by Harvey and Co., for Rio Tinto Mines.

It becomes apparent, from a consideration of the foregoing facts, that compressed air as a medium of transmission is, under the most favourable conditions, exceedingly wasteful of the motive force, and that the waste may become enormous if means are not employed to keep it near its minimum limits. The application of such means involves conditions which can be satisfied only when the machinery employed for the compression is designed to form a portion of the permanent plant of a mine. Even with machinery of this character as at present constructed, not more than

30 per cent. of the motive force remains to be utilised when the necessary deductions for loss have been made; and calculations for practical purposes ought, therefore, to be based upon this, or a smaller proportion.

Fig. 173 represents a portable air-compressor constructed by Harvey and Co., Limited, of Hayle, Cornwall, and 186, Gresham House, London, for service in the Rio Tinto Mines.

Air Receivers.—As machines driven by compressed air in underground workings do not run continuously, the consumption of air is irregular; consequently it becomes necessary to store the air as it is received from the compression cylinders, in a reservoir of sufficient capacity to annul the effects of the irregularity existing between the production and the consumption. The minimum capacity of this reservoir should be twenty times the average consumption a minute, when only one compressor is employed; ten times when the compressors are two in number; and five times when there are three compressors. The form of the receiver or reservoir is a matter of small importance. Frequently an old boiler is used for this purpose. Every receiver should be provided with a pressure-gauge, in order that the pressure of the contained air may be readily ascertained at any moment. The outlet from the receiver should be provided with a cock, for the purpose of cutting off communication between it and the conduit pipes. There should also be a discharge cock at the bottom of the receiver, for the purpose of removing the water which is carried in by the air, and which accumulates at the bottom. It is hardly necessary to remark that the receiver should occupy a situation in which it will not be liable to be exposed to heat, the action of which upon the contained air would occasion a loss of work.

Air Conduits.—For the conveyance of compressed air from the reservoir to the points at which the machines are required both cast-iron and wrought-iron pipes are used. When cast iron is the material employed, the pipes are cast with a flange, and the joint is made by means of a rubber washer inserted between the flanges. In some cases, one flange is provided with a groove, and the other with a corresponding annular prominence, the edges both of this projection and of the groove being bevelled. To form the projection, the flange is turned down, and the groove is also cut in the lathe. The ring of rubber is placed in this groove, which is $\frac{1}{2}$ in. broad and $\frac{1}{4}$ in. deep.

Wrought-iron tubing is in many cases preferable, especially on account of its lighter weight. This tubing is usually manufactured in 14-ft. lengths, and with an internal diameter of $3\frac{1}{4}$ in. Such a length will weigh about 80 lb. The joints in this case are made by means of flanges $7\frac{1}{2}$ in. diameter, strongly brazed upon the ends of the tube, and pierced with four bolt-holes $\frac{3}{4}$ in. diameter. In order to obtain the requisite degree of rigidity, great care is needed in putting on these flanges. Before applying them, a groove is cut in the lathe in the thickness of the flange, and a copper ring of $\frac{1}{32}$ in. on the side is inserted. This ring is, by the process of soldering, rendered solid with the flange and tube, and by that means a perfectly air-tight joint is made. In joining the tubes, a rubber washer is inserted between the flanges.

In certain situations, where rapid oxidation of the metal is likely to occur, an air-conduit may be composed of both cast- and wrought-iron pipes, each material being confined to the locality for which its qualities render it the more suitable. All air-pipes should, previously to their being put into use, be tested by hydraulic pressure up to ten atmospheres, and also by air pressure under water up to seven or eight atmospheres. By such means any latent defects in them are rendered apparent, and confidence in their resisting and retaining powers is created.

When a considerable length of air-piping is laid down, means must be provided to allow of expansion and contraction under the influence of changes of temperature. In underground workings, the temperature is sufficiently constant to justify a neglect of such precautions; though even there it is well to be prepared against a sudden rise of the temperature. But in the shaft and at surface, the conditions are different, and rollers and compensating joints become a necessity. A form of compensating joint that has been proved to act satisfactorily in practice is shown in Figs. 174, 175. It consists of a tube of copper, of the same dimensions as the iron tubes, and bent into the form shown. The flanges of this tube are applied in the same manner as those of the iron tubes. In practice, it has been found necessary and sufficient to insert one of these compensating joints at intervals of about 100 yd.

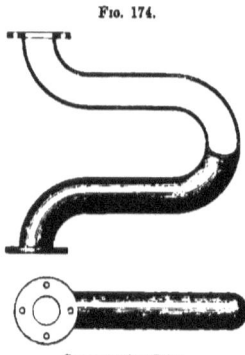

Fig. 174.

Compensating Joint.

The tubing used to connect the machine drills with the fixed iron tubing is always flexible. The material used is rubber, and great thickness is given to the tubing to ensure strength. As a protection from injury caused by friction against the rough surfaces of the rock, such tubing is covered with coarse canvas. The internal diameter of the largest size used is 2 in., and that of the smallest about ¾ in.

Water Reservoirs.—In rock drilling by machinery, the application of water greatly facilitates the operation, by keeping the bottom of the hole free from débris, and hence it is highly conducive to rapidity of execution. It has been proved by experience that the rate of drilling in a dry and in a wet hole varies as $1 : 1\cdot 5$; that is, it takes $1\frac{1}{2}$ times as long to drill a hole dry, as to drill a hole with the assistance of water. Thus it is possible to reduce the time of drilling by one-third. Moreover, the great heating produced in the bit by the blows upon hard rock causes a rapid deterioration of the tool, and hence it becomes necessary to change and to repair it more frequently. Another great objection to dry drilling is the production of much fine dust, which annoys the workmen and destroys the packing and the rubbing surfaces of the machinery. Hence a supply of water is desirable, and where the holes are inclined upward it becomes necessary to inject the water into them with considerable force. For this purpose water reservoirs are required. These are made of galvanised iron, and filled through a funnel cock, which renders them air-tight when closed. A piece of tubing communicates with the air pipe, from which the requisite pressure is obtained. The water is directed into the drill-hole as a strong jet by means of another piece of flexible tubing provided with a nozzle. When the drill is carried on a support running on rails, the reservoirs may be conveniently fixed to the support, as in Figs. 176, 177.

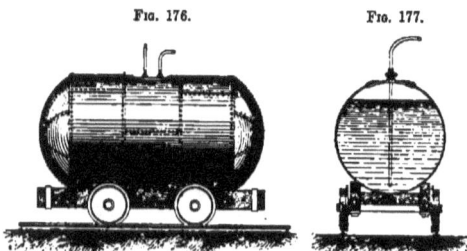

Fig. 176. Fig. 177.

Water Reservoirs on Wheels.

EXCAVATING MACHINERY.

Annexed is a statement of the work done each week in a 3 months' trial at North Skelton mines of percussive and rotary drills, to compare the capabilities of the two systems:—

Week ending	Number of Places Drilled.	Percentage of Wide Places.	Percentage of Narrow Places.	Number of Holes Drilled.	Number of Feet and Inches Drilled.	Stone Got.		Shifts Worked this Week.	Average Number of Holes Drilled per Shift.	Average Length of Holes.		Average Stone Got per Hole.		Average Feet and Inches Drilled per Shift.		Average Time Occupied in Moving Machine from Place to Place, in Minutes.	Excess in Quantity Obtained over Percussive Drill.	Excess in Number of Holes Drilled over Percussive Drill.	Excess in Feet and Inches Drilled over Percussive Drill.	
						tons	cwts.	8 hrs. 6 hrs.		feet	ins.	tons	cwts.	feet	ins.		tons	cwts.	feet	ins.
1882.					feet ins.															
Nov. 11	18	67·23	27 77	263	1032 8	417	17	5 1	43 8	..	..	..	..	..	..	19·00	..	..	..	..
"	23	39·13	60·87	321	1245 8	409	13	5 1	53·5	3	10	1	5	207	7	15·88	..	..	58	213 0
Nov. 18	..	..	..	245	.. ..	..	..	5 1	40·6	..	..	..	..	..	..	..	..	..	..	..
"	27	40·73	59·27	344	1385 1	642	0	5 1	57·3	4	0½	..	..	230	10	12·33	..	..	..	..
Nov. 25	16	56·25	43·75	239	958 9	535	12	4 1	47·8	4	0	2	4½	191	9	19·21	..	..	..	..
"	25	40·00	60·00	307	1239 8	624	12	4 1	61·40	4	0½	2	0	247	11	12·00	89	0	68	280 7
Dec. 2	17	75·00	25·00	246	968 7	497	5	4 1	49·00	3	11	2	0	193	8	25·33	..	..	..	..
"	21	50·00	50·00	342	1333 7	624	18	4 1	68·40	3	11	1	14	266	8	12·76	127	13	96	365 0
Dec. 9	16	61·00	39·00	251	983 9	580	14	4 1	50·20	3	10	2	6	196	9	25 87	..	..	..	..
"	24	52·00	48·00	369	1488 9	729	6	4 1	73·80	4	0	1	19	297	9	10·09	148	12	118	505 0
Dec. 16	17	49·00	51·00	264	1014 3	566	15	4 1	52·80	3	10	2	3	202	10	24·25	..	..	..	..
"	25	53·00	47·00	372	1498 2	808	6	4 1	74·40	4	0	2	3	299	7	11·66	241	11	108	483 0
Dec. 23	21	66·00	34·00	303	1174 6	797	9	5 1	50·50	3	10½	2	12	195	9	25·83	..	..	..	..
"	29	39·60	60·40	467	1887 6	1066	4	5 1	77·80	4	0½	2	5	314	7	10·26	268	15	164	713 0
Dec. 30	17	75 00	25·00	255	1003 5	583	6	4 1	51·00	3	7	2	5	200	8	25·56	..	..	..	..
"	23	59·00	41·00	353	1450 1	883	1	4 1	70·60	4	1½	2	10	290	0	15·90	299	15	98	446 8
1883.																				
Jan. 6	17	77 50	22 50	280	1089 4	678	7	5 1	46·06	3	10¼	2	8	181	6	..	..	..	..	..
"	23	47·00	53·00	390	1576 3	835	11	4 1	78·00	4	0	2	2¾	315	3	13·40	157	4	110	486 9
Jan. 13	20	86·00	14·00	332	1311 4	772	1	5 1	56 33	3	11	2	8	218	6	18·83	..	..	..	..
"	29	51·00	49·00	522	2110 6	1116	11	5 1	87·00	4	0½	2	2¾	351	9	13·04	344	10	190	799 2
Jan. 20	20	70·00	30·00	290	1175 5	763	18	5 1	50·00	3	11	2	7	195	11	21·36	..	..	..	..
"	30	52·00	48·00	501	2051 7	1049	13	5 1	83·50	4	1	2	1¾	341	11	14·03	345	15	202	876 2
Jan. 27	20	60·00	40·00	313	1220 10	802	10	5 1	52·18	3	11	2	11	203	5	22·22	..	..	..	..
"	30	53·00	47·00	518	2135 1	1186	0	5 1	86·30	4	1	2	5¼	355	10	12·80	383	10	205	914 3

The work done by the percussive drill is represented by the *smaller* figures, and the work done by the rotary drill by the *larger* ones. When the trial commenced, the men working the rotary drill were strange to the mode of working the North Skelton stone, and did not therefore get on at all well for the first few weeks.

The yield for a set of 15 or 16 holes, put in, in a wide place, varies from 22 to 34 waggons of stone. The yield for the same number of holes in a narrow place, varies from 13 to 26 waggons of stone. Each waggon carries 30 cwt.; and if the figures showing the percentage of wide and narrow places drilled by each machine are noted, it will be seen that a greater percentage of narrow places have been drilled by the rotary drill than by the percussive, which (as compared with working the larger quantity from wide places) reduces the yield of stone per hole, and increases the cost of powder per ton.

If the results obtained during the first four weeks of the trial are carefully compared with those obtained during the last four weeks, the figures will, in these respects, speak for themselves. Speaking of the cost of repairs during this trial, the Percussive Drill cost over £6, whilst the Rotary was under 5s.; in fact, the difference between the value of the wiping hay used for cleaning the holes out in the case of the percussive and that needed by the rotary, would more than cover all the repairs required by the latter. There is a wide difference to the men between the working of these two classes of drills. In the percussive class, there is twice as much hard work required from the men as in the rotary, and in point of comfort there is no comparison, the machinery in the rotary case doing all the work, and the men are kept quite dry; but in attending to the percussive, they are working very wet during the whole shift, and have to do much more laborious work in the fixing of the drill in each place, as well as during its general working. The water needed for drilling with the percussive drills has to be led out of the places going to the dip, and in all the places the state of the roads is made very disagreeable and objectionable for the men and horses travelling upon them. In the use of the percussive drills, a second costly main of pipes, with all necessary connections and cocks for conveying the water, under a heavy pressure, into every place, has to be provided and kept in order, involving double cost in both pipes and skilled labour.

Since the date of this competitive trial, further improvements have been added to the rotary drill, by which a 4-feet hole can be put in with 1000 revolutions less of the engines, and the holes cleaned of the small drillings, thus saving a considerable amount of skilled labour, reducing the consumption of the compressed air and the wear and tear on the drilling machine by 71 per cent.

Both the machines here referred to have been continued at work at North Skelton every day, and (working two shifts) the rotary is now daily averaging 324 tons of stone, whilst the quantity from the percussive is not much more than one-third of this.

BLASTING.—The means by which the charge of explosive matter placed in the drill-hole is fired constitutes a very important part of the set of appliances used in blasting. The conditions which any such means must fulfil are: (1) that it shall fire the charge with certainty; (2) that it shall allow the person whose duty it is to explode the charge to be at a safe distance away when the explosion takes place; (3) that it shall be practically suitable and applicable to all situations; and (4) that it shall be obtainable at a low cost. To fulfil the second and most essential of these conditions, the means must be either slow in operation, or capable of being acted upon at a distance. The only known means possessing the latter quality is electricity. The other means in common use are those which are slow in operation, and which allow thereby sufficient time between their ignition and the explosion of the charge for a person to retire to a safe distance. These means consist generally of a train of gunpowder so placed that the ignition of the particles must necessarily be gradual and slow. The old, and still commonly employed, mode of constructing this train was as follows: An iron rod of small diameter, and terminating in a point, called a "pricker," was inserted into the charge and left in the bore-hole while the tamping was being rammed down. When this operation was completed, the pricker was withdrawn, leaving a hole through the tamping down to the charge.

EXCAVATING MACHINERY.

Into this hole a straw, rush, quill, or some other like hollow substance filled with gunpowder was inserted. A piece of touch-paper was then attached to the upper end of this train, and lighted. When the train became ignited, the powder being confined in the straw, except at the upper end, burned slowly down and fired the charge, the time allowed by the touch-paper and the train together being sufficient to enable the man who applied the match to retire to a place of safety. This method of forming the train does not, however, satisfy all the conditions mentioned above. It is not readily applicable to all situations. Moreover, the use of the iron pricker may be a source of danger; the friction of this instrument against silicious substances in the sides of the borehole or in the tamping has in some instances occasioned accidental explosions. The danger, is, however, very greatly lessened by the employment of copper or phosphor-bronze instead of iron for the prickers. But the method is defective in some other respects. With many kinds of tamping, there is a difficulty in keeping the hole open after the pricker is withdrawn till the straw, can be inserted. When the holes are inclined upwards, besides this difficulty, another is occasioned by the liability of the powder constituting the train to run out on being ignited. And in wet situations, special provision has to be made to protect the trains. Moreover, the manufacture of these trains by the workmen is always a source of danger. Most of these defects in the system may, however, be removed by the employment of properly constructed trains. One of these trains or "fuses" is shown in Fig. 178.

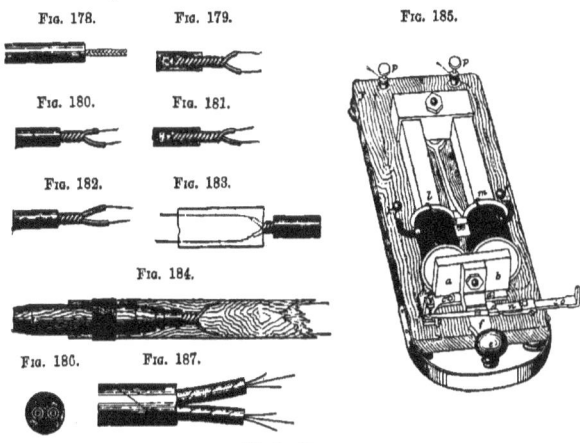

Blasting Fuses.

Safety Fuse.—Many of the defects pertaining to the system were removed by the introduction of the "safety fuse." The train of gunpowder is retained in this fuse, but the details of its arrangement are changed so as to fairly satisfy the conditions previously laid down as necessary. It consists of a flexible cord composed of a central core of fine gunpowder, surrounded by hempen yarns twisted up into a tube, and called the countering. An outer casing is made of different materials according to the circumstances under which it is intended to be used. A central touch thread, or in some cases two threads, passes through the core of gunpowder. This fuse, which in external appearance resembles a piece of plain cord, is tolerably certain in its action : it may be used with equal facility in holes bored in any direction; it is capable of resisting considerable pressure without injury; it may be used without special means of protection in wet ground; and it may be transported from place to place without risk of damage.

In the safety fuse, the conditions of slow burning are fully satisfied, and certainty is in some measure provided for by the touch thread through the centre of the core. As the combustion of the core leaves in the small space occupied by it a carbonaceous residue, there is little or no passage whatever left through the tamping by which the gases of the exploding charge may escape, as in the case of the straw trains. Hence results an economy of force. Another advantage offered by the safety fuse is, that it may be made to carry the fire into the centre of the bursting charge if it be desired to produce rapid combustion. This fuse can also be very conveniently used for firing charges of compounds other than gunpowder, by fixing a detonating charge at the end of it, and dropping the latter into the charge of the compound. This means is usually adopted in firing the nitro-glycerine compound, the detonating charge in such cases being generally contained within a metallic cap. In using this fuse, a sufficient length is cut off to reach from the charge to a distance of about an inch, or farther if necessary, beyond the mouth of the hole. One end is then untwisted to a height of about $\tfrac{1}{4}$-in., and placed to that depth in the charge. The fuse being placed against the side of the bore-hole, with the other end projecting beyond it, the tamping is put in, and the projecting end of the fuse is slightly untwisted. The match may then be applied directly to this part. The rate of burning is about 2 ft. a minute. Safety fuse is sold in coils of 24 ft. length. The price varies according to the quality and the degree of protection afforded to the train.

Electric Fuses.—The employment of electricity to fire the charge in blasting rock offers numerous and great advantages. Those who have most carefully studied the matter are most earnest in praise of the method and its economical value. A very little thought will make apparent the greater effect which can be produced by firing simultaneously a number of blasts instead of firing them singly, while a little experience will teach that even in firing single blasts by this apparatus, much can be gained. One advantage gained in firing single holes is that in case of missfire, which can rarely happen by this method, no time is lost in waiting, as in the case of firing by safety fuse there would be, before approaching the work. There is no "hanging" fire. Another advantage is that the explosion of the electrical fuse in the centre of the charge throws the fire through the whole body of the powder, igniting it all at once and by detonation giving the same charge by far greater explosive effect, as has been fully demonstrated by experiment.

An electric fuse consists of a charge of an explosive compound suitably placed in the circuit of an electric current, which compound is of a character to be acted upon by the current in a manner and in a degree sufficient to produce explosion. The mode in which the current is made to act depends upon the nature of the source of the electricity. That which is generated by a *machine* is of high tension, but small in quantity, whereas that which is generated by a *battery* is, on the contrary, of low tension, but is large in quantity. Electricity of high tension is capable of leaping across a narrow break in the circuit, and advantage is taken of this property to place in the break an explosive compound sufficiently sensitive to be decomposed by the passage of the current. The electricity generated in a battery, though incapable of leaping across a break in the circuit, is in sufficient quantity to develop a high degree of heat. Advantage is taken of this property to fire an explosive compound by reducing the sectional area of the wire composing a portion of the circuit at a certain point, and surrounding this wire with the compound. It is obvious that any explosive compound may be fired in this way; but for the purpose of increasing the efficiency of the battery, preference is given to those compounds which ignite at a low temperature. Hence it will be observed that there are two kinds of electric fuses, namely, those which may be fired by means of a machine, and which are called "tension" fuses, and those which require a battery, and which are known as "quantity" fuses.

In the tension, or machine fuses, the circuit is interrupted within the fuse case, and the priming, as before remarked, is interposed in the break; the current, in leaping across the interval, passes through the priming. In the quantity or battery fuses, the reduction of the sectional area is effected by severing the conducting wire within the fuse case, and again joining the severed ends of the wire by soldering to them a short piece of very fine wire. Platinum wire, on account of its high resistance and low specific heat, is usually employed for this purpose, but iron is sometimes used. The priming composition is placed around this fine wire, which is heated to redness by the current as soon as the circuit is closed.

The advantages of high tension lie chiefly in the convenient form and ready action of the machines employed to excite the electricity. Being of small dimensions and weight, simple in construction, and not liable to get quickly out of order, these sources of electricity are particularly suitable for use in mining operations, especially when these operations are entrusted, as they usually are, to men of no scientific knowledge. Moreover, as the means of discharging the machine may be removed until the moment when it is required, this mode of firing offers greater security than the battery. Also by employing a current of high tension, a large number of shots may be fired simultaneously in single circuit with greater certainty than is obtained with a battery, unless the power of the latter be accurately calculated for the number of fuses in circuit and the thickness of the platinum wire used. Another advantage of high tension is the small effect of line resistance upon the current, a consequence of which is that mines may be fired at any distance from the machine, and through iron wire of very small section. A disadvantage of high tension is the necessity for a perfect insulation of the wires.

When electricity of low tension is employed, the insulation of the wires needs not to be perfect, so that leakages arising from the injury to the coating of the wire are not of great importance. In many cases bare wires may be used. Other advantages of low tension are, the ability to test the fuse at any moment by means of a weak current, and an almost absolute certainty of action. On the other hand, the copper wires used must be of comparatively large section, and the influence of line resistance is so considerable that only a small number of shots can be fired simultaneously when the distance is great. Moreover, as the number of fuses is increased, the power of the battery must be augmented by adding to the number of its cells, so that for ordinary mining operations the battery becomes large and unportable. But the chief disadvantage of the battery lies in the fact of its requiring a liquid to excite the current, and the consequent careful attention and delicate handling which the elements require. This defect may, however, be removed to some extent by a suitable form of the battery.

The details of the construction of an electric fuse may be varied greatly. The central part consists, in one class of tension fuses (Figs. 179, 180), of a piece of guttapercha covered wire doubled, and the two portions twisted tightly together. The loop thus formed is stripped of the insulating material and cut through to make the interruption. Around these wires is a casing of a water-resisting compound, and over this is another casing of paraffin paper. Above the terminals, that is, the severed ends of the wire, is a chamber of peculiar form to receive the priming compound. This chamber, after the priming has been put in, is closed with the compound. The whole fuse thus constructed is sufficiently small to pass easily into the tube of a guncotton or a dynamite detonator. The electric current, in leaping across the interruption at a, passes through that portion of the sensitive compound which occupies the interval, and fires it. In a quantity fuse (Figs. 181, 182), a similar mode of construction is adopted; but in this fuse, the interval between the ends of the wire is made wider, and is bridged over by a piece of fine platinum wire. The current in passing heats this wire to redness, and thereby fires the priming compound which is placed round it. This fuse is also capable of being inserted into the tube of a guncotton or dynamite detonator.

Tension fuses have hitherto been, except when newly made, somewhat uncertain in their action, especially when used in warm climates. The defect is due mainly to the influence of moisture, particularly when combined with heat, and to the shocks and vibrations to which the fuses are necessarily subjected during transport. Heat and moisture cause decomposition to take place in the priming, especially when the latter consists of unstable elements; and shocks and vibrations loosen the mass of priming between the terminals, an effect that is very destructive to its sensitiveness.

The quantity fuses, illustrated, possess the following important advantages :—(1) The diameter of the platinum wire is such as will give a high degree of sensitiveness, a condition which renders small battery power sufficient. (2) The length of the platinum wire is kept absolutely uniform, a condition of very great importance when several fuses are to be fired simultaneously. (3) The priming composition is perfectly protected from atmospheric influences and from moisture, so that these fuses are peculiarly suitable for blasting in very wet ground, and for submarine work. And (4) The size and form of the fuse are such as to admit of its easy introduction into the charge.

The wires which lead from the fuse up through the tamping above the charge, and called for that reason the shot-hole wires, must be "insulated," that is, covered with some material capable of preventing the escape of the electricity. Various materials are used for this purpose; the best is rubber, but its expensive character is a serious obstacle to its common use for shot-hole wires. Guttapercha is the next most suitable material, and, as it is comparatively cheap, it is largely employed in blasting under water, and in very wet ground. When guttapercha-covered wires are used, they are simply a continuation of those within the fuse, as shown in the drawings. But even guttapercha is too expensive for ordinary use. A much cheaper substitute is found in paper. When this material is used as an insulator, the wires are cemented between two strips about ½ in. broad. These "ribbons," as they are called, are then dipped into some resinous substance to protect them from water. They are attached to the bared ends of the guttapercha-covered wires projecting from the fuses. Fig. 183 shows a ribbon in which the positions of the wires are indicated by dotted lines. For ordinary blasting operations, these ribbons are very suitable. Another insulating material employed is wood. A lath of this material, ¼ in. thick and ¾ in. broad, receives a narrow groove along its edges. Into these grooves the wires are placed, and the fuse with its detonator is fixed to the end of the stick. The stick with the wires and fuse attached is called the "blasting stick" (Fig. 184). The rigidity of the stick is found by miners to afford great facility for placing the fuse in the charge, an advantage that leads them to prefer this means of insulation.

Electrical Machines.—For the excitation of electricity in a state of high tension, two kinds of machines are used. In one kind, the current is excited by the motion of an armature before the poles of a magnet; in the other kind, the electricity is excited by friction and stored in a condenser, to be discharged by special means. The magnetic machines are the more simple in construction and the more constant in their action. But they generate only a small quantity of electricity, so that only a small number of fuses can be fired simultaneously in single circuit. By using the divided circuit, however, a very large number may be fired, if the machine be of the rotative class. The friction machines generate a larger quantity of electricity, and therefore are capable of firing a larger number of fuses in single circuit than the magnetic machines. But they are far less constant in their action than the latter, a defect of grave practical importance. For the excitation of electricity in a state of low tension, batteries are employed in which chemical means are made use of to generate a current. These apparatus require a liquid to promote their action.

Breguet's Magnetic Firing Machine.—The simplest form of magneto-electric machine in common use is "Breguet's Exploder" (Fig. 185). It is not of the rotative class, and is, therefore, unsuitable for firing a large number of fuses in divided circuit. But if not more than five or six be required to be fired simultaneously, it will be found to be most convenient. The induction coils are placed upon bars of iron constituting a continuation of the arms of the magnet. The latter is fixed upon a base of wood, $x\,y$. Against the bars of soft iron upon which the coils are placed presses the armature $a\,b$ fixed upon the lever $c\,d\,e$, which turns about a horizontal axis $c\,d$. When this lever is pressed down by a blow upon the knob e, the armature is withdrawn from the coils, but remains parallel with them. The lever is, moreover, provided with a spring $f\,g$, which descends with the point f on the lever. While the armature $a\,b$ is in contact with the coils, the free end g of the spring presses against the lower end of the screw h, the support of which is in communication with the wire $i\,k$. But when the lever is depressed, the spring descends also, until, at a certain point, it is separated from the screw. When this separation takes place, the short circuit is interrupted, and the current is then forced to pass into the long circuit in which the fuse is placed. The point at which interruption shall take place is regulated by raising or lowering the screw h. This point should be a little above that at which the spring stands when at the end of its stroke, because then the intensity of the current is at its maximum. When the hand is removed from the knob e, a spring beneath the lever, aided by the attraction of the magnet, forces the armature back into contact with the poles $l\,m$. A safety bolt $n\,o$ is pushed under the lever to prevent accidental discharges. The whole of the apparatus, with the exception of the knob l and the terminals p, is enclosed with a wooden case. The leading and the return wires being fixed to the terminals, and the fuses included in the circuit, the latter are fired by striking a sharp blow upon the knob e.

For firing a large number of fuses, Siemens' dynamo-electric machine is the most suitable. This is a rotative machine, and is, therefore, adapted to the divided circuit.

Friction machines possess the advantage of generating a larger quantity of electricity than the magnetic machines, with a tension sufficient to carry the current over wide interruptions in the circuit. By reason of this property, they are capable of firing a large number of fuses in single circuit, an advantage that renders them very suitable for use in industrial operations. Unfortunately, however, they are inconstant in action, in consequence of the wear of the rubbing surfaces, and, chiefly, the influence of moisture in the atmosphere. Magneto-electric machines are also affected by this influence, but in a lower degree. The difficulty lies in effectually excluding the air so as to isolate the atmosphere inside the machine from that on the outside. This difficulty is created by the necessity which exists for communicating with the inside. One of the rubbing surfaces must be set in motion from the outside, and through the aperture traversed by the winch handle usually adopted for this purpose the air enters. And in addition to this, there is generally some contrivance for effecting the discharge, which contrivance requires a second opening into the machine. In some of the American frictional machines, this second opening is avoided by means of an arrangement whereby the discharge of the condenser is effected by simply reversing the motion of the winch handle. Some machines are constructed to discharge themselves as soon as a sufficient quantity of electricity has been generated. This, besides the advantage of rendering a communication with the inside unnecessary, constitutes an important quality when the machine is used by an inexperienced or a careless operator, since it cannot be made to deliver an insufficient charge, nor can it be injured by overcharging.

Cables.—Cables are used to connect the fuses with the machine or the battery. They generally consist of several strands of copper wire well insulated with guttapercha or rubber, and protected from injury by a coating of hemp or tape. Two cables are needed to complete the circuit; the one which is attached to the positive pole of the machine or the battery (through which the electricity passes out) is distinguished as the "leading wire," and the other, which is attached to the negative pole (through which the electricity returns), is described as the "return wire." The most convenient form of cable is that which contains both the leading and return wires under one hempen covering, as in Figs. 186, 187.

Cable Box.—It is convenient to have the cable wound upon a reel, contained in a wooden box; the reel may be turned by a winch handle from the outside. The cable upon the reel is in metallic connection with two brass eyes on the upper edges of the box. The machine may thus be connected, without removing the cable from the reel, by simply attaching a connection to those eyes. The lower portion of the box has a drawer divided into several compartments for fuses, connecting wire, small tools, and necessaries.

Fig. 188 shows a magneto machine, used in America. To blast 12 holes at a time, the machine costs about 5l.; the platinum fuses range from about 16s. a hundred with 4-ft. wires to 2l. with 10-ft. wires; leading wire, $\frac{1}{4}d$. a foot; connecting wire, 2s. a lb. Following are directions for operating. Use a fuse with the wires attached, of such length that the ends may protrude from the surface after the hole is charged, the fuse head being in the centre of the charge. Tamp with dry sand, or in such a manner that the wires may not be cut or the insulating covering upon them be injured. When all the holes to be fired at one time are tamped, separate the ends of the two wires in each hole, joining one wire of the first hole with one of the second, the other, or free wire, of the second with one of the third, so proceeding to the end or last hole. If the wires attached to the fuses should not be long enough, use connecting wire for joining. All connections of wire should be by hooking and twisting together the bare and clean ends, and it will be best if the parts joining be bright. The charges having all been connected as directed, the free wire of the first hole should be joined to one of the "leading wires," and the free wire of the last hole with the other of the two leading wires. The leading wires should be long enough to reach a point at a safe distance from the blast, say 250 ft. at least. All being ready, and not until the men are at a safe distance, connect the leading wires, one to each of the projecting screws on the front side of the machine, through each of which a hole is bored for the purpose, and bring the nuts down firmly upon the wires. Now, to fire, taking hold of the handle of the magneto-electric battery, shown in Fig. 188, lift the rack (or square rod toothed upon one side) to its full length, and press it down, for the first inch of it stroke with moderate speed, but finishing the stroke with all force, bringing the rack to the bottom of the box with a solid thud, and the blast will be made. Fig. 188a shows the fuses which are to be used with the magneto machines. The cut shows in section a platinum fuse nearly of actual size. These fuses are kept in stock with wires of 4, 6, and 8 ft. in length. Fuses with longer wires are made to order. The length of wire should be equal to depth of holes drilled. Enough leading wire is needed with each machine to make two leaders of sufficient length to reach from the blast to a safe distance for the person to stand who shall operate the machine : 500 ft. is the quantity usually sold, but in some cases 1000 ft. are used. Connecting wire is sold in coils of 1 or 2 lb. weight (about 225 ft. in each coil), and is used in connecting the fuses to each other where several charges are to be fired simultaneously. A small quantity only of this will be needed if the fuse wires are picked up after each blast, as they can be twisted together and used in place of connecting wire.

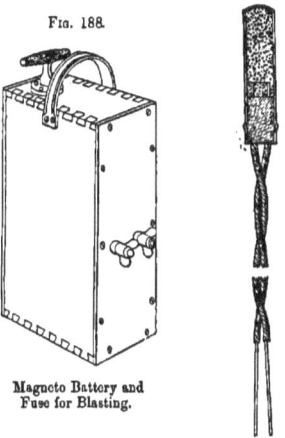

Fig. 188.

Fig. 188a.

Magneto Battery and Fuse for Blasting.

Bornhardt's electrical firing machine is designed to excite electricity by means of friction, the electricity so excited being stored in a receiver known as a Leyden jar, whence it can be discharged suddenly by pressing a knob. The electricity thus discharged is of very great power, sufficient to leap over, in the form of a vivid spark, any small breaks which may occur in the circuit or path taken by the fluid. It is this property which is taken advantage of to fire the detonator. Small wires are led into a specially prepared detonating powder contained in the fuses. On connection being made with the firing machine, the current passes along the wires, leaps across the break, which occurs in the circuit in the form of a spark; this spark explodes the fulminate, which in its turn explodes the charge of dynamite or other explosive near it. The passage of the electric spark is so rapid that it may be carried through any number of charges up to 20, all of which it will explode simultaneously. The machine (Fig. 189) is composed of a thin plate wheel a of ebonite, which can be rotated rapidly by turning the handle. The top part of the wheel a is in contact with a rubber b, formed of cat's skin, carried by the arms c. It is the friction of this skin on the surface of the wheel which excites the electricity. d are two ebonite rings furnished on the inside with metal points and connected with the Leyden jar e, in which the electricity is stored after having been collected from the surface of the generating wheel a by the points on the rings d. Metal rings fg project through the front edge of the machine; it is to these rings that the ends of the wires forming the circuit are to be attached when everything is prepared for firing. h is the firing knob by which the discharge is made. The knob being quickly pressed causes the arm i to be thrown into the position shown in dotted lines, making contact with the top of the Leyden jar and completing the circuit, allowing the charge stored in the jar to pass along the wires, and in its passage firing the prepared shot holes simultaneously.

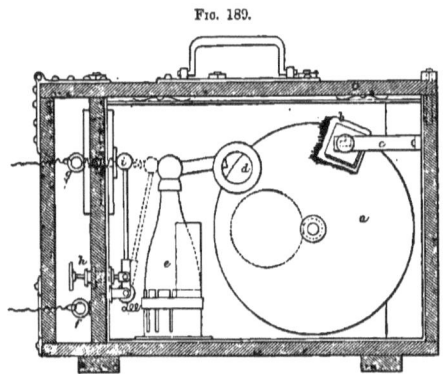

Fig. 189.

Bornhardt's Electrical Firing Machine.

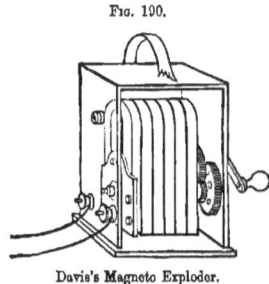

Fig. 190.

Davis's Magneto Exploder.

Davis's magneto exploder is shown in Fig. 190. The complete apparatus includes a wood cover, handle and terminals for leading wires: dimensions, 9 in. by 6¼ in. by 5¼ in. This small magneto exploder is a practical and useful instrument for ordinary purposes when not more than 6 or 12 fuses are required to be exploded simultaneously. Its weight is only about 12 lb., and its efficiency is not impaired either by damp or dust. Price 13*l*. 13*s*.

Davis's dynamo tension exploder is sufficiently portable for all practical purposes, weighing about 55 lb., while at the same time its power to generate a charge of electricity renders it available under nearly all conditions for the ignition of a mine with certainty. Perhaps the most important advantage this apparatus has over all others is the large number of shots that can be fired simultaneously. Price 32*l*.

C. W. Kinder, who for some years conducted important blasting works in China, at first used Siemens' dynamo exploders, on account of their great durability and freedom from being affected by moderate dampness. They are undoubtedly best, provided suitable fuses can readily be purchased in an absolutely fresh condition. Under the description of fuses, it will be noticed why Kinder was forced to discontinue the employment of this machine, and to fall back upon another type.

The dynamos are of two kinds, commonly known as "high tension" and "quantity," the first giving a short and intensely hot spark which ignites a chemical priming; the latter a continuous current which heats a platinum wire to incandescence. The first cost of these machines is greater than others, but their durability, and excellent construction, make them the cheapest in the end. The 'high-tension" exploder (Figs. 191, 192) consists of a small Siemens dynamo enclosed in a suitable wooden box. The peculiar cam-motion k below the commutator c is to enable the electro-magnets m to become excited by a current generated from their own residual magnetism; the lever l

is thus held down during two revolutions of the handle; on its being allowed to rise by the notch cut in the cam, the strong current accumulated in the magnets is permitted to flow to the line by means of the terminals t; or should these not be in communication, by a suitable conducting medium, the current is absorbed in a mica protector p, placed under the bed of the machine. In case of the

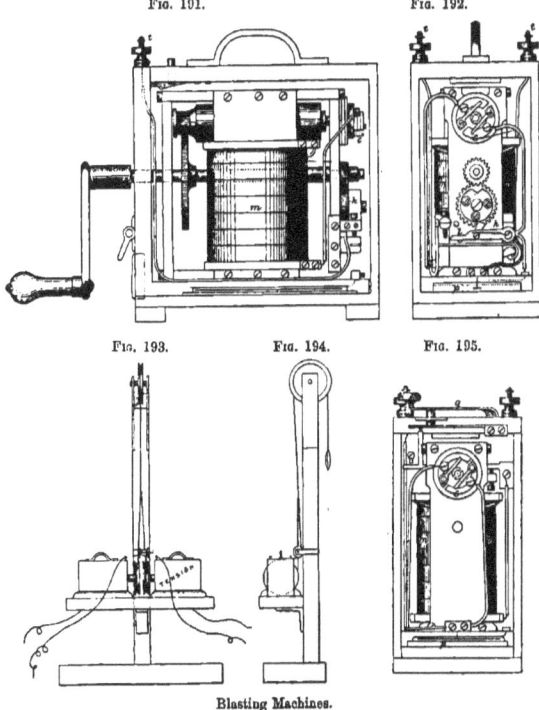

Fig. 191.   Fig. 192.

Fig. 193.   Fig. 194.   Fig. 195.

Blasting Machines.

handle being incautiously turned too often, this protector will be pierced by the spark generated, at once giving a vent to the current and saving the machine coils from total destruction. With ordinary care, the machine requires little attention beyond slight cleaning and lubrication of the working parts. In case of the spark becoming feeble during damp weather, the cloth pads in the protector are taken out and dried; this operation needs sufficient care to prevent injury to the mica plate, but is easily done by means of a small screw-driver. The box should be constantly locked while in service, as the curiosity of the men to find out how the "fire" is made is very great, and has led to trouble. Kinder several times experienced a severe shock at the moment of firing, and at one time the sinkers became so afraid of the "box" that they refused to use it; the difficulty was easily got over by wrapping a piece of rubber sheeting around the handle.

When many fuses have to be fired in circuit, the handle is rotated slowly half a dozen times before attaching the leading wires to the terminals (this is to additionally excite the magnets). When all is ready, the handle is turned very gently till a click is heard; two quick revolutions will then fire the charge. In some cases four turns are necessary, but more than this should never be given.

A simple device for firing two series of holes simultaneously by two machines is shown in Figs. 193, 194. It consists of an upright post and a shelf, on which the exploders are placed facing each other; disks are then attached in lieu of handles, with cords wound upon them. These cords unite and go over a top pulley, the end hanging free. On this being pulled, the exploders are charged and discharged with great rapidity; and if the disks are properly marked and fitted, the explosions are simultaneous. This system acted perfectly for upwards of 50 holes, resulting in a great saving of time and certainty of ignition, due to the moderate number of fuses attached to each machine.

The "quantity" exploder (Fig. 195) strongly resembles the one above described; but electrically there is a considerable difference, as the machine has to produce a continuous flow of electricity through the fuses until they become sufficiently heated to cause ignition. To procure this the magnets m are wound with coarse wire, and the cam arrangement is dispensed with. The key q, at the top of the machine close to the terminals, enables the magnets to be considerably excited by the passage of the current through the resistance coil of fine wire placed at p. On the depression of the key, the coil p is cut out of circuit, and the current proceeds to the line; if there is no communication between the terminals, all generation of electricity will cease; while if the key be left up, the coil will be fused by careless turning, instead of more important parts of the machine being destroyed.

The great cost of fuses for this type of exploder renders its adoption very limited; but as it can be used with very defective insulation, and the platinum fuses keep well in any climate, it is a valuable appliance, especially in very wet shaft-sinking. As many as 22 holes have been fired simultaneously; but the machine needs skilful handling, and the circuit should invariably be tested with a galvanometer.

The machine is used thus:—the left hand is placed on the top of the box, with the thumb over the firing key k. The handle is turned a few times moderately quickly, and the key is then pressed; great resistance is felt, which must be overcome by applying more power, when the explosion takes place.

The ingenuity shown in the construction of both the above machines cannot be too highly commended; and it is to be regretted that rival machines, as hereafter described, are not as well constructed.

The Silverton battery consists of a box containing 6 cells, each 9 in. by 16 in. by 3 in., constructed on the Leclanché principle; the porous jars being replaced by hard felt. It is used in place of the last-named exploder, but cannot fire more than 10 fuses under favourable conditions. The pressure of the firing key is all that is required to cause ignition, so that its actual manipulation renders it the simplest means of blasting. The high cost of the fuses, great weight, bulk, and moderate powers, stand in the way of its frequent employment.

Of Bornhardt's frictional exploder, Kinder says that the number of turns required to cause a spark to leap from one brass stud to the other, is a sure test of the condition of the machine. At Kaiping, China, it averaged five to six in winter, and nine to twelve in summer, owing to the excessive dryness of the former, and dampness of the latter season.

After testing the machine, which must invariably be done before firing, in order to judge how many turns are necessary, the leading wires are attached to the terminals, and the handles rotated the required number of times; a sharp push of the key will discharge the jars and fire the fuses. The usual number of holes varies from 12 to 27; the turns in winter being 15-27, and in summer 25-35. Excellent results are now daily got by two of these exploders. In winter they want cleaning about once a month; and the fur pads are taken out, well brushed and dried in the sun. In summer this must be done somewhat oftener, care being taken to choose a bright sunny day. This exposure to a drying atmosphere will in a few minutes cause it to give a good spark with 5-7 turns, when it should be quickly screwed up and kept in a dry room until wanted. In each instrument, two paper packets of charcoal are placed to absorb moisture.

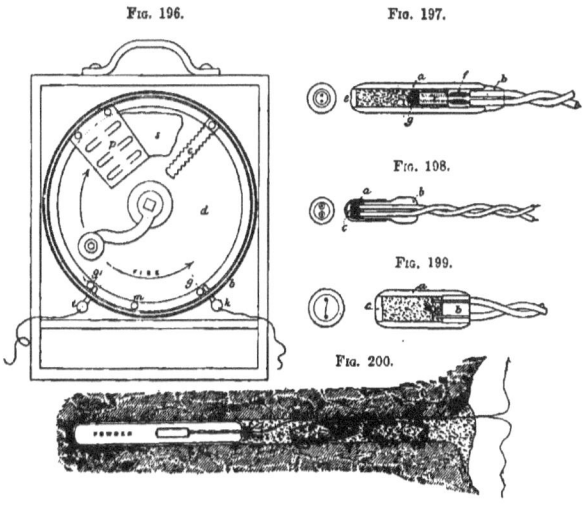

FIG. 196. FIG. 197.
FIG. 198.
FIG. 199.
FIG. 200.

Ladd's exploder; Siemens's fuse; Austrian fuse; low-tension platinum fuse; and tamping.

Ladd's frictional exploder (Fig. 196), considering its light weight and bulk, is the most powerful exploder in existence. It consists of a teak box containing an ebonite drum b, in which the whole of the essential parts are enclosed. A single ebonite disk d, 12 in. diameter, is employed in conjunction with a leather pad p, coated with amalgam. The collector c is of sheet brass almost touching both sides of the plate, s being a piece of oil silk attached to the pad. The condenser, which acts the part of the Leyden jars of the Bornhardt machine, is placed below, and consists of disks of tin-foil, separated by a sheet of ebonite. This part of the apparatus is allowed to rotate about one-quarter revolution, governed by the terminal screws $t\,k$, and the two stops g connected respectively with the tin-foil sheets of the condenser. In charging, the stop m comes against the terminal t, thus cutting out all connection with line; to discharge, the handle is brought sharply back, causing the stops g to assume the position shown. This machine, although very effective, is somewhat troublesome, owing to the use of amalgam, and the rapid deterioration of the oil silk; its high price is also an obstacle, and one which the simple construction does not account for. The firing movement, which is copied from the American machines, requires only one hole in the case to be protected from the influence of the atmosphere, and permits of great simplicity of construction. It is unfortunate that makers do not take more care in the manufacture of frictional exploders, and thus prevent many of the difficulties which arise solely from bad workmanship. In the Bornhardt type, the revolving gear is too weak, and the wooden collectors are so badly put together, that they come to pieces when moderately shaken, while the wire points injure the disks from want of steadiness and suitable attachment to the sides of the casing. The disks should permit of easier removal for polishing, which would conduce to keeping the machine in better order. No kind of exploder can be kept long below ground without its electrical efficiency being impaired; this especially applies to all frictional exploders.

The leading-wires used by Kinder have been of three kinds. The three-strand type is undoubtedly the best, and in the long run is cheapest, owing to its excellent soft rubber insulation and great flexibility which prevents it being easily damaged by the explosion. A return wire is invariably employed when many shots have to be ignited; but if proper care be taken to keep the wires close to the roof of the drift very little injury will result. The machines are generally supplied with two leading wires of 70 yd. each. These, with care, will last several months; but they often need slight repairs, which are easily effected by means of gutta-percha and tape soaked in Stockholm tar. The reels are made of wood, the inside ends of the wires projecting through the hollow axle; and, if the leading wires are tied together at intervals of 16 ft., no difficulty will be experienced from kinking or entangling on a single reel. Naked wires hung on insulators have been employed; but although well enough for low tension, they cannot give good results in the damp atmosphere of a mine, when any form of high tension fuse is used.

The fuses sent out with the Siemens tension-machine were unable to withstand the effects of the voyage to China, and were therefore useless. In appearance they closely resembled the Austrian pattern, but were primed with a mixture of nitro-glycerine, chlorate of potash, and charcoal. After this failure, which almost put a stop to electric blasting, it was decided to procure Abel fuses made by Ladd; these arrived after considerable delay, and proved to be excellent when unaffected by long storage or an unusually damp season. This fuse is remarkable for its good insulation and sensitiveness; being thoroughly well made it is not easily injured by rough handling in the mine. The priming compound can only be made by experienced chemists, as the chemicals have to be specially manufactured by a very delicate process.

The construction of the fuse is shown in Fig. 197. The wires being cut to a suitable length according to the depth of the hole, the ends are bared, and a small piece of double copper wire, made expressly for the purpose, is attached, thus forming the tip e. The joints at f are carefully insulated with gutta-percha, and the end is inserted in the paper cap g, containing the priming mixture. A small wooden tube a is now placed over the end, cemented at b, and after being charged with powder, the whole is sealed with mastic at e; or, if for use with dynamite, a detonator acts as a plug.

The Austrian fuse, Fig. 198, is primed with a composition allied to that of Abel; but the construction is totally different. No special wire tip is used, but instead, the long wires are brought together at the end by means of a sulphur cylinder b cast around it. A metal cap a containing a small piece of gun-cotton at c, together with the priming, is placed over the tip, thus completing the fuse. Although the Austrian fuses appear incapable of standing a sea voyage to the East, and do not permit of very rough handling, it is certain that on the Continent, when fresh from the maker, they give excellent results if care be taken in the tamping.

Owing to a very larger number of Abel fuses having been damaged, it became necessary to re-prime them. After numerous experiments it was found that several easily-made compositions acted well when used with a frictional exploder; but none could be found to give good results with the dynamo. This was very disappointing, as the Siemens machine had to be laid aside, and the more troublesome Bornhardt exploders used instead. They have been in daily use for over two years, and have given great satisfaction.

The fuses now employed are manufactured at the works, the Abel type being adopted, as it is superior, although slightly dearer in first cost than those made on the Austrian system. The priming consists of a mixture of equal parts by weight of chlorate of potash and black sulphide * of antimony. These are carefully ground together in a small mortar until no white streaks are visible. The hands must be protected, so that in case of explosion due to the presence of grit, they may not be injured. The addition of charcoal makes the composition far less liable to ignition from friction, but at the same time more sensitive to moisture. A quantity of 2·5 grams is sufficient for over 30 fuses. The priming described is one highly recommended by French chemists for electrical purposes, and with slight modification is largely used by artillerists in the British service for friction-tubes. After the wires have been arranged, and just previous to capping, they are tested by means of a current from a dynamo exploder, and if a fair spark results, the fuse is finished by putting on the paper cap about half filled with priming. A coating of shellac dissolved in spirits of wine greatly assists in the preservation of the cap. At one time these fuses were issued for service without any wooden shell; but it has been found by experience that the preservation of the priming is favourably influenced by keeping it surrounded with fine gunpowder, no doubt due to its absorbing any moisture which would attack the otherwise unprotected priming.

A Chinese boy, receiving about 20s. a month, keeps all the electric gear in order, besides making over sixty fuses a day. The wire tipping takes up most of the time, and requires considerable dexterity, which is easily acquired after a few days' practice. In the Abel fuses, it was found not only that the priming became caked from moisture, but that the joints were short-circuited by rust, destroying the insulation. This latter circumstance is certainly an argument in favour of the use of copper wires exclusively.

The low-tension platinum fuse, Fig. 199, is for use with quantity-dynamos and the galvanic battery. The long iron wires are similar to those used for the Abel fuse, the bare ends being thrust through two holes bored in a little cylindrical plug of wood b; the bridge of fine platinum-wire is placed between the points around which a small piece of gun-cotton is tied. A wooden shell a is

* The red sulphide is quite unsuitable.

fastened over the end, and filled with gunpowder, which is sealed with cement at *e*. In spite of apparent extreme simplicity, their manufacture is very costly and difficult; for unless the resistance of each fuse is identical, simultaneous firing is impossible. As the bridge wires weigh only 0·21 gr per yard, it will be readily understood what great delicacy is required at the hands of the fuse-maker. The wires should be invariably twisted together, and the fuse kept straight, as coiling for packing injures the gutta-percha insulation, especially in very dry situations. Iron wires are preferable to copper, as they remain stiffer in the hole, permitting of easy tamping, and are besides far less liable to kink. Copper wire is exclusively used on the Continent. It is less trying to the fingers of the blaster, and the connections are more neatly made. Unless fuses can be readily obtained shortly after manufacture from trustworthy makers, it is wisest to adopt the method above described, and only make use of the freshly primed article, thereby becoming independent of delay in transport, long storage, and risk of receiving old stock.

It is much to be regretted that the Abel priming is so exceedingly difficult to make; and it is to be hoped that before long some compound will be found, which, while giving equally good results, will possess more stability, or permit of easier manufacture by less skilled hands. The requirements of a good fuse are as follows:—The priming should be unaffected by ordinary atmospheric moisture; it should be sufficiently sensitive to permit of not less than 30 holes being fired simultaneously; incapable of being exploded by induction currents; incapable of detonation from blows or pressure; and so made as to be easily tested before use. True the low-tension platinum fuse fulfils most of these conditions, but the great cost is prohibitive of its use except for very important and special purposes.

The gunpowder used by Kinder, which was invariably enclosed in paper cartridge-cases, was manufactured at the Imperial Arsenal, Tientsin. The cases are made by boys at the works, and consist of three thicknesses of Japanese paper dressed with a composition consisting of 12 lb. of tallow, mixed with 10 lb. of rosin and 6 lb. of beeswax. When the electric fuse is used, it is placed about one-third from the top of the cartridge, filled round with powder and the mouth drawn up tight by a piece of scrap fuse-wire. These cartridges have been 10 hours under deep water and have remained perfectly good; the tough Japanese paper being a cheap and excellent substitute for the rubber skins often used in wet shafts. The tamping consisted of pellets of a mixture of soft and hard clay (Fig. 200), a large stock of which was constantly kept in readiness. When the holes are all charged, two of the best hands remain to connect the fuse-wires, while only one person is supposed to be present at the last moment when the leading wires are attached. This is not only for safety, but because a clear view is absolutely necessary. The fuses are invariably coupled up in direct circuit, which practically is quickest and most reliable, and easily understood by an ordinary native miner. Kinder personally tried the various group systems, but could get no better results underground, although an improvement was noticeable in office experiments. It is useless to introduce any extra complications so long as 30 holes can be fired by the direct method. Missfires occur at intervals, but they rarely cause serious damage to a blast, as they almost always happen in those fuses situated near the leading-wire ends, and consequently in the outside holes. Immediately after a blast, an examination should be carefully made for unexploded cartridges, which are sometimes blown out.

Although electric blasting is much safer than any system of time fuse, yet after a few successful rounds have been fired, the men who before feared electricity more than dynamite, become too careless in its use, and thus accidents have happened which would not occur with reasonable care. Mowbray, in his account of the Hoosac tunnel, refers to two fatal accidents which were not due to this cause; for at that time the danger incurred by handling the wires under certain conditions was not understood. Premature explosion was caused in each case by the victims, who wore rubber boots, becoming charged with electricity due to the use of highly compressed air employed for the ventilation. These accidents led to the following orders being issued :—" Let a blaster, before he handles these wires, invariably grasp some metal in moistened contact with the earth, or place both hands against the moist walls of the tunnel. Before taking the leading wires to the electric-fuse wires, let the bare ends of the leading and return wires be brought first in contact with themselves, and then in contact with the moist surface of the tunnel, and before inserting the armed cartridge, let him unite both of the uncovered naked wires and touch with them a metal surface having good ground connection. Above all, do not ventilate, by allowing a free blast of air through a rubber connecting-pipe, until after the electric connections have been made and the blast fired." Jutier gives an account of a severe accident, owing to the blaster having laid a packet of dynamite on the top of a frictional exploder; on firing the blast, this dynamite exploded, resulting in the death of three men standing near. Although several attempts have been made to explode dynamite by means of an electric spark, none have succeeded, as the dynamite always burned away. Nevertheless, it is wisest to place no explosives near any electrical machine capable of generating strong currents. The same writer also mentions a premature explosion due to the careless tamping of a charge, whereby the electric-fuse priming was detonated by pressure of a hard block of compressed powder.

Gelatinous cartridge.—A clause in the Coal Mines Regulation Act forbids blasting in fiery or dusty mines, unless (amongst other restrictions) the explosive is so surrounded by water or other substance, that it cannot inflame gas or dust. Though the water cartridge, if properly handled, may comply with this rule, there is always a danger of the case which contains the water receiving some damage on being pushed into the shot-hole, and the water consequently running out, and the steeper the mine, the greater the chance of this occurring. To do away with this objection, Heath and Frost, of the Sneyd Colliery, Burslem, have patented a method whereby the water is gelatinised, and being thus rendered less mobile than ordinary water, does not so readily tend to flow out at any hole there may be in the case containing it. A section of the charge immersed in the gelatinous surrounding is given in Fig. 201. Gelatine dynamite and tonite (cotton powder) may both be used in this method, but for coal getting tonite is preferred. The tonite is made up in charges of various strengths as required, a detonator and fuse are attached, and the charge is inserted in the case or bag containing the gelatinous surrounding, the neck of the bag being tied tightly round the fuse to prevent leakage. (The bags are of thin mackintosh, and can thus be filled at bank and sent down ready for the fireman). The whole is then ready to be placed in the shot-hole, stemmed up, and fired in the ordinary way. The effect of the explosion on the gelatinous substance seems to be to liquefy it, as seen from examination of the sides of the hole after the shot has been fired.

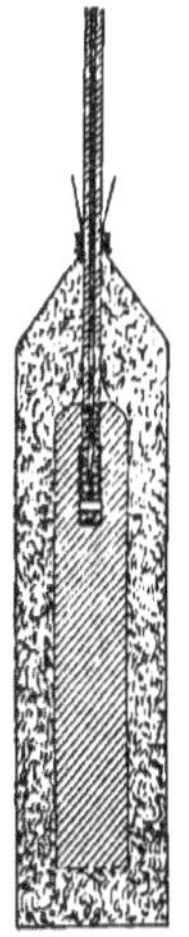

FIG. 201.

Gelatinous Cartridge.

Below are results of some experiments conducted by Mr. A. R. Sawyer, Inspector of Mines, with a view to test the safety of the cartridge. The experiments were made in a heading specially driven for the purpose, at the Sneyd Colliery. "The heading was 7 yd. long, several holes, ranging from 3 ft. 1 in. to 3 ft. 10 in. in length, and 2½ in. diameter, were bored at the back in different directions. Pipes conveying fire-damp from the mine reached to within 7 in. of the back, the supply of gas was regulated by a well fitting stop-cock. A brattice was placed at the mouth of the heading to enable a known percentage of firedamp to be retained. An agitator, which could be moved from a distance, was fixed to the roof of the heading, coal dust was placed on it and on the floor, and was agitated previous to the discharge. The shots were ignited by means of a fuse, which had necessarily to be over 7 yd. long, and took eight minutes to burn. The percentage of firedamp present was ascertained by means of a Mueseler lamp. An entrance was effected into the "crut" as soon as the smoke allowed, and the presence of firedamp was conclusively demonstrated in each case with a Mueseler lamp, showing that it had not been ignited. To show what would have occurred had the firedamp been ignited, two experiments were made in the same way with powder. Better colliery explosions on a small scale could not have been witnessed."

No. of Experiment.	Nature of Explosive.	Quantity in Ounces.	Amount and Nature of Stemming.	Per Cent. of Fire-damp previous to Lighting Fuse.	Result.	Per Cent. of Fire-damp after Explosion of Shot.
1	Cotton powder.	5	10 in. of coal dust, not solid.	Explosive.	Shot blew out well, fire-damp not exploded.	Explosive at mouth of heading.
2	Do.	5	Do.	Do.	Shot blew out well into apparently a mixture of 8 to 10 per cent. of gas well mixed with dust.	5 per cent. at mouth and 7 per cent. at back of heading.
3	Gelatine dynamite.	4½	Do.	Do.	Shot blew out.	6 per cent. at back.
4	Do.	4½	None, shot on floor, and covered with coal dust.	Do.	The concussion blew the brattice 12 yards away.	Explosive in heading.
5	Powder.		On floor.	Do.	Large flame, which filled heading and reached 7 yards outside, accompanied by clouds of smoke. Timber at mouth of crut set smouldering.	None.
6	Do.		10 in. of coal dust.	Do.	Same as above, but louder report; flame filled crut completely. Much smoke and flame far beyond heading. Brattice blown away.	None.

Experiments were also made with the cartridge in the same crut, the charge (5 oz. tonite) being stemmed with gunpowder, and a mixture of coal dust and gunpowder. The stemming was simply blown out, there was total absence of flame or sparks, and the gas was not ignited.

EXCAVATING MACHINERY.

The ordinary method of lighting fuses, where lamps are used, is by means of a thin wire pushed through a hole in the glass, or the gauze (if a Davy) on to the flame, and

Fig. 202.

Fig. 203.

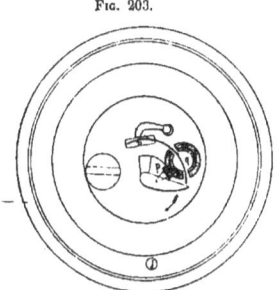

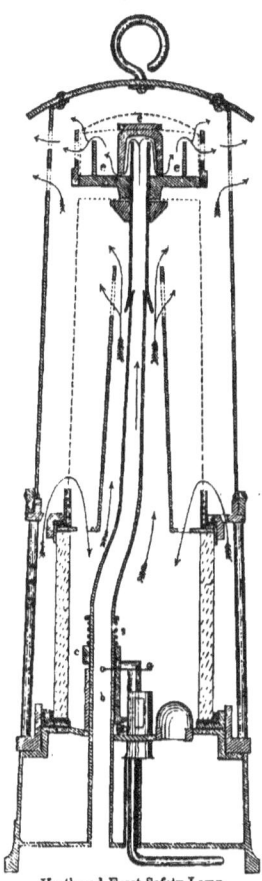

Heath and Frost Safety Lamp.

then drawn quickly on to a piece of touch-paper which is applied to the fuse. When a fuse is first ignited, it throws out a volume of sparks, and in order that no flame or spark may appear during the whole process of shot-firing, Heath and Frost have patented, for use in conjunction with the cartridge, the lamp shown in Fig. 202, Fig. 203 being a plan showing the arrangements on the oil can. (For the sake of clearness, the wick pricker is omitted in Fig. 202). The action of the lamp is simple. The fireman having, stemmed the shot-hole, turns the firing wire by means of the trigger, through the holes in the fuse tube and into the flame, and having cut the fuse to a good, clean end, pushes it up the tube till it touches the wire, and holding it tightly there, draws the wire back again over the fuse. If the wire is hot enough, it ignites the fuse, the smoke and sparks from which go up the tube and condense in the top arrangement, the gauze at the top of which is double, the smoke following the course shown by the unwinged arrows in Fig. 202. He holds the fuse in its place until smoke appears at the bottom of the tube (some 6-7 seconds). By this time the sparks have ceased, and the fuse can be withdrawn and left to do its work. Should the fuse not become ignited, it must be drawn down the tube a little, the wire again turned into the flame, and the process repeated. A shutter (annular shaped) kept down by a spring, covers the holes in the fuse

P

tube through which the wire passes, and thus prevents the fuse from extinguishing the lamp. The shutter is pushed up automatically by a cam on the trigger, to allow the wire to pass through the holes. In Figs. 202, 203, the wire is shown in position ready to be drawn over the fuse. The lamp weighs about 3½ lb. It is made either with or without a shield, and costs about 8s. 6d. The top of the oil can, inside the glass, slopes towards the centre, where is a small hole, so as to let no oil collect there. Should the lamp be objected to by firemen on account of its weight, which after all is only about 1 lb. heavier than the Belgian Mueseler (one of the lightest lamps), another, lighter one, might be carried by them to examine with, the firing lamp being kept in readiness at some convenient place. References—a, firing wire; b, fuse tube; c, shutter for air hole; d, cam working shutter; e, condenser; f, spring for shutter; winged arrows show air current; unwinged arrows show course of fuse smoke.

The Tonite or cotton powder manufactured by the Cotton Powder Company, Limited, 116, Queen Victoria Street, London, is strongly recommended for its safety in transit, storage, and manipulation, and the form of waterproof cartridges makes it very handy, saves times, and avoids spilling. It has all the advantages of the other high explosives known, and is used in the same manner; but has none of their disadvantages. It is also ready and available at any climatic temperature, and gives off very little smoke when fired. One of the greatest advantages of this powder is that the holes need not be so large nor so deep as those required by ordinary gunpowder, thereby saving a great deal of labour and hastening the work. It can be used in places where gunpowder would utterly fail, such as in soft beds, between two layers of rocks, or inserted in fissures, without any boring whatever. It will work well in damp holes.

The charges may be taken to have a density of about 1·50, and are particularly suited for any work where the maximum power is required; and in very hard rock offering difficulties to the drill, it is important that the charge should occupy the smallest possible space. The cartridges should invariably be stored in a dry place until they are required to be used. A sound ordinary fuse, to fit the detonators, and tolerably damp proof, is all that is usually needed for ordinary blasting. Cut it clean and cap it with a tonite detonator free from any of the sawdust in which the detonators are packed. Nip the open end of the detonator so as to make it fast to the fuse. Ordinary detonators will not explode tonite. The detonators should be kept dry. The Company sells special knives, by means of which the detonators can be properly fixed on the fuse. To use them, put the detonators on the fuse first, then nip the open end once between the handle and spike, but not too close to the fulminate.

Take a cartridge of about ¼ in. less diameter than the bore-hole; open the neck so as to admit of the detonator which is attached to the end of the fuse being freely introduced down the tube, and being pushed down as far as possible. The neck of each cartridge is furnished with a piece of wire, which must be twisted firmly round the fuse, so as to make both fast together. The cartridge is then ready for use. Make sure that the bore-hole be large enough to let the charge to the very bottom, but it must not be too large, or else power will be wasted. When used in wet holes, the neck of the cartridge should be protected by tar or grease, to prevent water getting to the detonator or interior of the cartridge.

Where more than one cartridge is necessary to charge a mine, put in the hole as many cartridges as necessary (without detonators), and press them gently one after the other, so as to leave no space between; then introduce the cartridge containing the detonator, press it down carefully, on account of the detonator inside, and tamp with clay or sand in the ordinary way. The rammers used for loading the holes should be scooped out somewhat in the shape of an auger, so as not to interfere with the fuse while tamping.

Should a miner find that the cartridges he has in stock do not fit the bore-hole, he can cut or break them in pieces, and press them down any dry hole. In this case, a priming charge must be made of one of the top parts of such cartridges, and put in the hole last, but this must be pressed in carefully, on account of the detonator. As the action of the powder takes place from cartridge to cartridge through the paper casing, this latter need not be removed.

Tonite is sent out in cases of 50 lb. weight only. The following sizes are always in stock :—

	Diameter.					
	1 In.	1¼ In.	1½ In.	1¾ In.	1⅞ In.	2 In.
Weights	oz. 2	oz. 2½	oz. 3	oz. 4	oz. 6	oz. 6
	2½	3	4	6	8	8
	4	5	6	8	12	12
	5	6	8	12	16	16

Wedges.—Fig. 204 illustrates a patent multiple wedge, for bringing down rock and coal without the use of explosives. The cost is No. 1, 1¾ in. diameter, 2 ft. 6 in. long, 2l. 10s.; No. 2, 2 in. diameter, 3 ft. long, 3l. 10s. But there is scarcely any wedge able to hold its own as a means of breaking down coal. The cause is much the same as that which has been the means of limiting the use of the lime-cartridge. The want of success is due almost entirely to the fact that it is difficult to get combined, a face of coal which will break down easily, a roof which will separate freely, and a coal which will break off well, conditions which are generally required, whether the wedge or the lime-cartridge is used, both being slow means of applying force to break down coal. There is

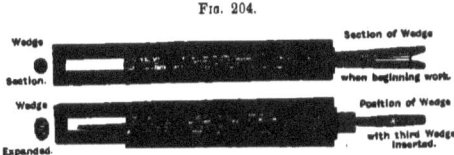

Fig. 204.

Multiple Wedge.

one wedge now in use with great success in Belgium and the North of France. It consists of two long steel wedge-pieces, which are placed in the shot-hole, the thick end inwards, and a third long wedge is driven between the two. The wedge is not employed in England on a large scale, but in France and Belgium it has been largely adopted. The objection to it is that whilst with the lime-cartridge or any other means of breaking down coal simple ordinary explosive force is applied, with the wedge a considerable quantity of "elbow-grease" is required, and a man has to take 5–10 minutes in striking the centre wedge in order to get the coal broken down.

Tools.—The hand tools used by miners comprise chiefly shovels, picks, wedges and hammers.

Shovels.—These consist of a metal plate for lifting and carrying loose material, and a wooden handle for manipulating it. The plate is always of iron with steeled front edge; the handle or helve is of ash, circular in section, and terminating in a crutch handle or hilt in preference to a D or eye. To reduce stooping in use, the handle is set at an angle with the plate. The sizes of shovels are distinguished by the width of the plate, measured in its widest part; they vary from 10 in. for heavy work, to 16 in. for light work, such as shovelling coal or loose earth. The strain upon the shovel when in use is mainly thrown upon the crease and the top strap, and it is at this part that they yield by the parting of the strap. Strength in the strap and the crease is, therefore, a requirement in a shovel.

The form of shovel used for gravel is that shown in Fig. 205. The plate is 10 in. wide, and the "mouth," or entering part, is pointed so as to form two edges. This form renders it very suit able for entering closely compressed or heavy ground. The handle is 30 in. long, and is set at an angle of about 150° with the surface of the plate.

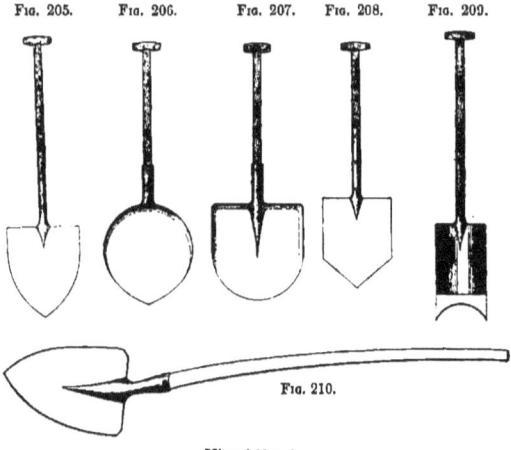

Fig. 205. Fig. 206. Fig. 207. Fig. 208. Fig. 209.

Fig. 210.

Miners' Shovels.

Fig. 206 represents a "frying-pan" filling shovel, as used in the north of England and in some other districts. The plate is nearly circular, with a short point, and the edges are turned up to give it concavity. The breadth is 14 in. and the length 16 in.; the handle is 24 in. long, and is set at an angle of 142° with the plate. The weight of this tool is 7 lb. 14 oz.; it is well adapted for loading coal into tubs, and it is very extensively employed at collieries.

Fig. 207 represents a "round-mouthed" filling shovel, which is very generally employed for shovelling loose stuff not too heavy. The plate is 16 in. wide and 15 in. long; the handle is 23 in. long, and is set at an angle of 147° with the plate. The weight of this tool is also 7 lb. 14 oz. Fig. 208 represents a "sinking shovel," $11\frac{1}{2}$ in. by 14 in. the handle of which is 23 in. long.

For use in clay ground, the "clay spade" (Fig. 209) is used. The plate of the spade is long and narrow, and has a square mouth. Sometimes the plate is curved so as to form a portion of a cylinder, as shown in the figure. When of this form it is often called a "grafting spade." The clay spade is used by forcing it into the ground with the foot placed upon the shoulder, and to form a convenient tread a piece of iron is riveted upon the shoulder. This tool is much used in soft or clay ground.

The long-handled shovel used in Cornwall is shown in Fig. 210. Fig. 211 is a bulling shovel, and Fig. 212 a hoe for tin dressing, also used in Cornwall. Fig. 213 is a cast-steel square coal shovel. Fig. 214 is a square pronged coke fork, and Fig. 215 a coal screen.

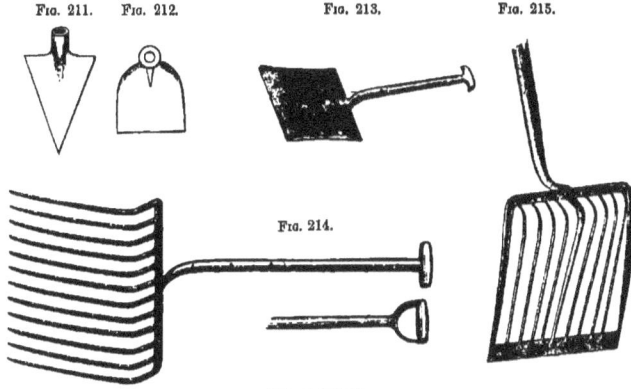

Fig. 211.   Fig. 212.   Fig. 213.   Fig. 215.

Fig. 214.

Miners' Shovels.

The shovel is the only tool which is never made at the mine; it is always purchased ready made, and when broken it is seldom capable of being repaired. The cost of gravel shovels, 10 in., 11 in., and 12 in. wide, is 25s.–35s. a dozen; all steel, from 60s. to 70s. Frying-pan, and round-mouth filling shovels cost, according to size, 35s.–48s. a dozen; all steel, round mouths, 45s.–60s.; and clay spades, 12 in. × 6½ in., about 35s. a dozen.

Picks.—The pick, mandril, or hack, as it is variously named, is the most important tool of the miner. Its use is to loosen masses of rock, or to chip away small fragments. It consists of an iron head formed of two arms, and a wooden handle or helve fitted into an eye in the middle of the head or stem. The arms are steeled at the tips, and are either pointed or chisel-edged, according to the work required of the tool. When pointed, the point is formed by a square taper. Such wedge-shaped extremities enable the arm to penetrate the joints of fissured rocks, or between the laminæ of shaly rocks. When the tip of one arm of the pick has been forced into the rock, it is used as a lever to fracture the mass by pressing or prizing upon the helve. Thus the action of the pick combines that of the hammer, the wedge, and the crowbar or lever. It acts as a hammer, in delivering a blow; as a wedge, in penetrating and disrupting the rock; and as a lever, in forcing out large masses. These several actions must be borne in mind when considering the form and construction of a pick. With the chisel edge it is very frequently used to chip off fragments of rock, as in dressing the sides of an excavation. In this case it combines the action of the chisel and the hammer.

In using the pick as a lever, the strain is thrown on the helve in the eye, and the helve yields in that part by "wincing," that is, by a crushing of the fibres. To provide against this wincing, the bearing surface at each end of the eye should be made as long and as wide as possible. It is obvious that the sharper the edges of the feather, that is, the widened portion of the helve that fits into the eye, the greater will be the tendency to wince. Wedging the helve very tightly into the eye, so as to make it press against the cheeks, also lessens the liability of the fibres to yield. Many devices against wincing have been adopted, the most effective of which, however, consist in lengthening the eye in the direction of the helve, in flattening the edges of the feather, and in providing the helve at that part with an iron strap or ring.

The pick-head is usually made of wrought iron. It consists of a central part called the "eye," made to receive the helve, and two shanks or stems. The sides of the eye are spread out to form cheeks, against which the sides of the helve may be firmly wedged. Generally the shanks are square in section, and their size varies in dimension from $\frac{3}{4}$ in. in light picks, to $1\frac{1}{4}$ in. in heavy picks near the eye, diminishing gradually towards the point. Sometimes the section of the shank is $1\frac{1}{4}$ in. × 1 in., or $1\frac{1}{2}$ in. × $\frac{7}{8}$ in., the longer side being in the direction of the helve, to give greater strength for prizing. Frequently, when the section is square, the edges are chamfered down, and in some cases the chamfering is carried so far that the section approaches the octagonal form. The ends of the shanks are steeled, and brought, as before remarked, either to a point or a cutting edge. The weight of the head varies from 2 lb. to 7 lb., according to the nature of the work to which the tool is to be applied, the difference of weight being caused by the larger section and the greater length of the shanks required for certain purposes. The helve is of ash, and consists of two portions, the haft and the feather; the latter portion is inserted into the eye, and fixed by wedging. The length of the helve also varies, according to the nature of the work to be performed, from 24 to 34 in.

Pickheads are made straight, curved, or anchored. Straight-headed picks assist the reach, and are more suitable for getting into corner work than the curved or the anchored forms. They are always preferred for long-reaching or over-hand work. When curved, the head is said to "sweep," and such a form is preferred for under-hand work, the sweep causing the tool to fall into its work better than it would do if the head were straight. The degree of curvature is always slight. Sometimes, instead of curving the shanks, they are made straight and converging to the eye. This form is described as the anchored, and is very common in the north of England.

The tips of the shanks are sharpened on an anvil, and tempered to the requisite degree of hardness. The form of the cutting edge will be determined by the nature of the work to be performed. For hard ground the four-sided pyramid point is generally the most suitable. The rate of taper in such a case will also be determined by the character of the work. A quick taper or "bluff" point is stronger than a slow taper or "slim" point; but if the point is very bluff it will not penetrate the rock readily. When the tool is required to work in a narrow slit, it is obvious that the point must be slim, even if the nature of the rock is such as to require a bluff point, since the pickhead cannot be turned sufficiently to enable the bluff point to catch the side of the cut; and such a circumstance would soon cause the sides to come together, or "cut out," as it is termed. As the bluffness of the point under such conditions is mainly dependent upon the length of the head, the latter is usually shortened to increase the bluffness. This relation between the rate of

EXCAVATING MACHINERY. 111

taper of the point and the length of the head is evident, for the shorter the head the more obliquely it may be turned in a narrow cut. The pyramid point is very generally used for holing coal; that is, for cutting a narrow slit in the seam; but the conditions existing in this case seem rather to require a chisel edge. The operation of holing consists in *chipping*, and for such a use the point is not suitable. It is somewhat remarkable that this form of cutting edge should still be used by hewers. With the exception of this case, whenever the pick is to be used for chipping the rock, the chisel edge is adopted. The chisel edge is also suitable for penetrating the joints of rocks, or between their laminæ, as before remarked, so as to disrupt them by acting as a wedge, or to dislodge them by acting as a lever.

Picks may be divided, according to the nature of the work to which they are applied, into three classes, and described as "stone picks," "holing picks," and "cutting picks." The first of these are used in rock only, and to render them suitable for such heavy work they are made very strong and heavy. Holing picks are used for undercutting coal, and are used either in the coal or in the underclay. In using them, they are swung horizontally. Cutting picks are swung vertically for downward cutting, and are used for cutting or shearing off the coal at the side of the stall or face, so as to divide the seam on each side after it has been "holed," for the purpose of causing it to fall. To avoid wasting the coal, these side cuts are made as narrow as possible. Cutting picks have a slim point, and are sometimes made slightly heavier than the holing picks. Various forms are given to picks by Continental nations, but the following are almost exclusively employed in Great Britain and America.

Fig. 216 represents a holing pick in common use in South Wales. The head is straight, and 18 in. long from tip to tip. The helve is 33¼ in. long, and the weight of the whole tool, fitted as shown, is 3 lb. 8 oz. The points of this pick are somewhat bluff.

Fig. 217 is a cutting pick used with the former. The head is straight as in the holing pick. The length, however, is somewhat less, being 17 in., and the helve is only 20½ in. long. The weight of this tool complete is 2 lb. 14 oz. The shanks of the pick in this case taper directly from the centre to the points, which it will be observed are slim.

The stone picks used in the same districts have curved heads, and are of considerably larger dimensions. Fig. 218 represents a "bottom pick," that is, a pick used for cutting the floor or thill of the coal seam. The head is 21¼ in. long, and the helve 30½ in. The weight of this tool is 3 lb. 3 oz. The shanks in this case are provided with a chisel head 1 in. wide, one edge being horizontal and the other vertical.

Fig. 219 represents a stone pick, the head of which is 24 in. long, and the helve 30½ in. The shanks are octagonal in section, and terminate, one in a wedge point and the other in a chisel edge. The weight of this tool is 9 lb. 5 oz.

Fig. 220 represents a holing pick as used in North Wales. The head is 18 in. long, and the upper has a strong curvature or sweep. The cheeks are V-shaped, and the shanks terminate in chisel edges. The length of the helve is 28 in., and the weight of the whole tool 2 lb. 10 oz. The cutting pick used with this holing tool is of a similar form, but has less sweep. It is also slightly heavier and has slim points.

Fig. 221 is a heading pick used in the same locality. The head is 16¼ in. long, and has a top sweep only. The cheeks are V-shaped, and the shanks taper regularly from the eye. The helve is 27½ in. long, and the weight of the tool is 3 lb. A somewhat heavier form of this pick is used for dead work, and is called a "driving," or "metal driving" pick. It has a head 17¼ in. long, a helve 27½ in. long, and weighs 3 lb. 10 oz.

Fig. 222 represents the form of coal picks common in the north of England. The head is 17½ in. long; the lower side is straight, but the upper side forms two inclined planes. In plan, the head is a regular lozenge-shaped figure, diminishing gradually from the eye to the points.

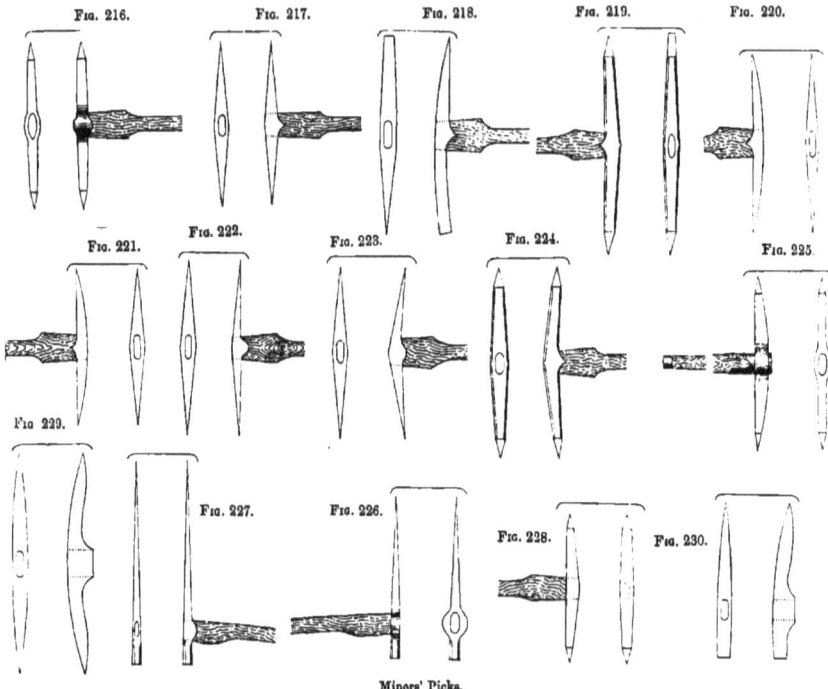

Miners' Picks.

The cheeks are semicircular and very small. The length of the helve is 32 in., and the weight of the tool is 4 lb. 5 oz.

Fig. 223 is similar to the preceding, except that the head is anchored, a form much in favour in the northern coal fields. The shanks in this case meet at an angle of 155°. The length of the head is 18 in., that of the helve 32 in., and the weight of the whole is 4 lb. 5 oz.

Fig. 224 represents a stone pick of the same district. The head is slightly anchored, and

EXCAVATING MACHINERY. 113

is provided with taper-shaped cheek-pieces. The angles are deeply chamfered or bevelled, so as to give an octagonal section. Sometimes, however, the section is square. The shanks terminate in four-sided pyramidal points. The length of the head is 19¼ in., that of the helve 30 in., and the total weight of the tool 7 lb. Frequently these stone picks are made stronger, the length of the head being increased to 23 in., and the weight to 8 lb.

An improved form of pick is shown in Fig. 225. This is made of cast steel throughout, and is known as the "interchangeable" pick. The merits claimed for this pick are, that being of solid cast steel, it will never require to be re-steeled, and will last longer than the ordinary pick; that as the helve is very endurable, and capable of being readily affixed to and removed from the head, one helve is sufficient for a number of tools; and that being thus interchangeable, when the pick requires to be resharpened, the helve need not be sent with the head to the fire, where it is liable to become shrunken from exposure to the heat. By sending only the head, not only does the helve escape damage, but the labour of carrying it is saved, and as one helve is sufficient for several picks, the labour of carrying helves is, under all circumstances, greatly lessened. To strengthen the helve, as well as to facilitate its easy application to the eye of the head, the feather is reduced, and a ferrule or hoop is affixed, as shown in the figure. By this means, the liability to wince is removed, or at least very materially diminished. The cost of these helves is 1s. 6d. each; that of the picks about 9d. per lb. for the lighter, and 8d. per lb. for the heavier kinds.

The pick commonly used in Cornwall, and in some other metal mining districts (Figs. 226, 227), is known as the "poll pick." It has one stem and one stump called the "poll." The face of the latter is steeled to form a pane, like a sledge, to render it suitable for striking blows. The pick is generally forged out of 1¼-in. iron, and weighs, without the helve, about 4 lb. Sometimes the head is made quite straight. This tool is a favourite one with metal miners. Possessing the features of both the pick and the sledge, it may be used for the purposes for which those tools are intended. It is commonly used for driving in wedges, and not unfrequently it is employed as a wedge by striking it on the poll end. The pick shown in Fig. 224 is for use in hard ground; it has the following dimensions: Length of pick end, 12½ in.; length of poll end, 3 in.; length of eye, 2·2 in.; width over eye, 3·1 in.; width of poll end, 1·2 in.; width of pick end, 1·1 in.; thickness, or depth, of poll end, 1·2 in.; thickness of pick end, 1·1 in. The length of the helve is 26 in.; the point is set at an angle of 85° to the helve. Total weight, 8½ lb. The pick shown in Fig. 227 is for use in soft ground. It dimensions are: Length of pick end, 16·5 in.; length of poll end, 3 in.; length of eye, 2·1 in.; width over eye, 1 in.; width of poll end, 0·8 in.; width of pick end, 0·8 in.; thickness of poll end, 0·8 in.; thickness of pick end, 0·9 in. The length of the helve, which is set at an angle of 83°, is 26¼ in. The total weight is about 2 lb. 10 oz.

Fig. 228 represents a "slitter" pick, used for slitting out mineral veins. It is double armed, one end being worked up to a point and the other to a horizontal cutting edge 0·4 in. wide. The head is 15·7 in. long, and the handle 29 in. The weight of this tool is about 3 lb. 10 oz.

Fig. 229 shows a Californian "drifting" or quartz pick. It is used chiefly in narrow drifts where there is not much room to swing the tool; also in working out the "gauge" or "salvage" from quartz veins. A common size used weighs 3½–4 lb., exclusive of the helve. A notable improvement of construction will be observed in the eye, which is "raised" or lengthened to give a large bearing surface to the helve, an important condition in picks that are used much for prizing.

Q

Fig. 230 shows a "poll" pick from the same locality. This pick has the same form of eye as the preceding. A size most commonly used is about $16\frac{1}{2}$ in. long, and weighs about 5 lb. The poll pick is a favourite tool among the Californian miners.

Wedges.—The wedge constitutes an important instrument in the hands of the miner. Large numbers of them are employed in every mine, as many as a dozen being sometimes required by one miner. They are used to break down large masses of hard coal, to force out blocks of rock by driving them into the joints, and to dislodge masses of rock that have been loosened by blasting. In jointed or vughy rock, they often do great service. Wedges are made of iron, and are steeled at the edge. In length, they vary from 6 to 18 in.; but a common size is 12 in. Their thickness is generally about 1 in., and their breadth $1\frac{3}{4}$ in. These dimensions, however, are frequently varied slightly.

Fig. 231 represents a coal wedge used in South Wales. The penetrating side forms a slender rectangular pyramid; the striking side is of an irregular eight-sided section, tapered from the base of the wedge. In side elevation, the breadth diminishes uniformly from the striking face to the

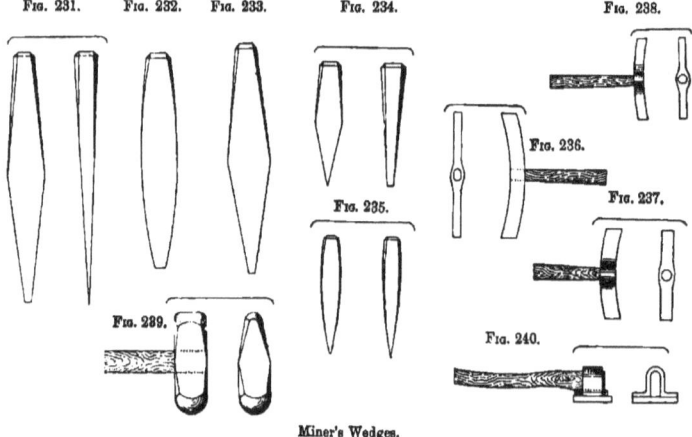

Miner's Wedges.

point. The length is $13\frac{1}{4}$ in.; in central section, the breadth is $1\frac{7}{8}$ in., and the thickness $\frac{7}{8}$ in. On the striking face, the breadth is $1\frac{1}{4}$ in., and the thickness 1 in. The weight of the wedge is 3 lb. 14 oz.

Fig. 232 is a coal wedge used in North Wales. The tapering sides of this wedge are bounded by curved lines, instead of straight ones, as in the preceding example. The length is $11\frac{1}{4}$ in., and in the greatest section, the breadth is $1\frac{3}{4}$ in., and the thickness $\frac{7}{8}$ in. The weight of the wedge is 3 lb. 9 oz.

Fig. 233 represents a wedge used in the north of England. The sides are straight, like those of South Wales. The length is 12 in., and the greatest section, or base of the wedge, 6 in. distant

EXCAVATING MACHINERY. 115

from the point, is a rectangle, $2\frac{1}{4}$ in. broad by $\frac{7}{8}$ in. thick. The striking face is an irregular octagon, 1 in. broad by $\frac{3}{4}$ in. thick. The point is cut off to a rectangle $\frac{1}{4}$ in. in the side. The weight of the wedge is 4 lb.

Fig. 234 is a stone wedge, from the same locality. The length is $6\frac{1}{4}$ in.; the wedge end is $3\frac{1}{2}$ in. long, and is drawn in from a rectangular section $1\frac{1}{2}$ in. wide and $1\frac{1}{8}$ in. thick. The opposite end is drawn in by a tapering eight-sided section to a striking face $\frac{7}{8}$ in. diameter. The weight of this wedge is 2 lb. 1 oz.

A wedge terminating in a point instead of a chisel edge is called a "gad." Gads are much used in metal mining for working jointy or vughy ground, or rock which has been fissured by a blast. They are of various sizes; the common lengths are from 6 to 12 in. in length. Fig. 235 shows a Cornish gad. It is 6 in. long, 0·9 in. tapered to 0·8 in. broad, and 0·6 in. thick: it has a central swell in breadth, but tapers uniformerly in thickness from poll to point. The weight is about 10 oz.

Ore-dressing Hammers.—Besides the sledge, which has been already described, other hammers are used for breaking up ore. The "cobbing" hammer, used for dressing ores by hand, is shown in Figs. 236 to 238; in this the arms curve upwards from the centre. In Fig. 236 the head is 13·1 in. long, and has an elliptical eye or socket 1·1 in. long; breadth across the eye, 1·6 in. The striking faces are rectangular, being 1·7 in. deep by 0·6 in. broad; the depth at the centre is 1·3 in. The arms taper in breadth from 0·8 in. at the centre to 0·6 in. at the faces. The helve is 9 in. long; the total weight is about $4\frac{3}{4}$ lb. Fig. 237 is a similar tool, of somewhat smaller dimensions. The arms are more strongly curved than those of the preceding hammer, the depth of the curve at the centre being 0·7 in.; they are of the same breadth throughout. The total weight is about $3\frac{1}{4}$ lb. Fig. 238 is a still smaller tool, in which the arms are less curved than in the preceding ones. The length of the head is 8·1 in.; that of the eye is 0·9 in.; the breadth across the eye is 1·5 in. The striking faces are 1·1 in. deep and 0·6 in. broad. The depth of the curve of the top surface is 0·3 in.; the total weight is about $2\frac{1}{4}$ lb.

The "spalling" hammer, Fig. 239, is used for breaking up pieces of ore for sorting previous to stamping or crushing. The head is of the pointing pattern, but has hemispherical ends; it is almost identical in form with the common road-metalling hammer. The weight of the head varies from 2 to 3 lb.; the length of the handle from 26 to 30 in.

The "bucking iron," Fig. 240, is a tool that is also used for dressing ores by hand. It consists of a rectangular iron striking plate, having an eye or stirrup welded on to its upper surface to receive the helve. In the tool illustrated, the striking plate is 5 in. long by 4 in. broad, by $\frac{3}{4}$ in. thick. The eye, or stirrup, is $3\frac{1}{4}$ in. high, and 1 in. broad. The helve, which is wedged into the stirrup, is 16 in. long; and the weight of the tool complete is about 6 lb.

CHAPTER VI.

SHAFT-SINKING MACHINERY.

WITH increasing depth of modern mines the difficulties of shaft-sinking have multiplied, and the rude system so long in vogue is no longer admissible. The methods employed are governed to some extent by the character of the strata and the amount of water encountered.

The Westphalian coal beds are overlaid by the chalk formation, in which the upper chalk and its flints is wanting; and whilst the upper and lower greensand are invariably present, the stratum between these, called "Gault" in England, is here represented by a very white argillaceous chalk or marl. Underneath the lower greensand there is almost invariably a thin bed of small gravel, indurated into a concrete, and containing 12–18 per cent. of iron ore. This is called in German "Bohn-erz" or bean-ore, and serves admirably, by reason of its density and structure, to shut off from the coal measures below the water contained in the marl formation above. The chalk formation is traversed, in various directions and at various levels, by fissures and clefts, some horizontal, some oblique, and some vertical, or nearly so; others siphon-like, and mostly connected with each other. These are without any large cavernous openings, such as are common in mountain limestone; but through their free communication with each other and their ample sources of supply, they are capable of delivering very large quantities of water per minute.

From the account of the methods pursued in this formation by W. T. and T. R. Mulvany, the following particulars are derived :—The system followed, up to that time, by the Germans was to sink through the quicksands and the "Thon-mergel," or soft marl lying over the solid marl, by means of a sink wall built on a wooden or iron crib-shoe; and then to sink the shaft down to the stone head, where they laid a foundation, and then walled the shaft up to bank with bricks, hydraulic lime, and "trass" mortar. In cases where feeders were large, they sought to make a foundation in the marl itself for the walling, which, in order to shut off the water, was built in that case of great thickness. This very expensive, and at the same time very slow system, was only partially successful, even in cases of moderate depth, and with comparatively small feeders of water. At all the shafts sunk by the Mulvanys, they adopted the English system of sinking and pumping; hanging and guiding the pumps with ground-spears and crabs, and using cast-iron wedging-cribs and tubbing to make the shafts water-tight.

The Shamrock Colliery gave some difficulty with the quicksand, on attempting to drain it by an open drain into an adjacent valley, in order to procure solid foundations for engines, chimney, &c. Finally they adopted for the shaft the German system of a sink wall 20 in. thick, with iron shoe attached to wooden cribs; by which means the shaft was sunk into the "Thon-mergel," or upper marl, at a depth of 26-28 ft. The foundations for the high chimney, engine, boiler, &c., were simply built on broad solid masonry platforms, with the excellent German hydraulic mortar, upon the quicksand. The sinking was proceeded with through the water-bearing marl to a depth of about 126 ft., the quantity of water not being excessive. About 60 cub. ft. per minute was pumped with 18-in. sinking sets and ground spears, by a twin horizontal winding engine and spur gear.

SHAFT-SINKING MACHINERY. 117

The water was completely tubbed off with English cast-iron tubbing. From this point to the coal measures, and down to the first working level or gallery, at 876 ft., the shaft was sunk very rapidly without any pumps whatever. Below the tubbing the shaft was walled in, at convenient lifts of 10 to 12 fathoms, with 10-15-in. brick walling set in hydraulic lime.

Fig. 241 shows a working section of the Hansa No. II. shaft from surface to bottom; and Figs. 242 to 248 show details of the walling, tubbing, and foundation or wedging cribs. Attention is called to the difficulties encountered in getting down the sink-wall to the depth of 50 ft.; to the improved system of laying the wedging cribs with pass-pipes and self-acting valves, Figs. 246 to 248, for escape of the compressed air and gases; to the facilities afforded by the system in dealing with feeders, when met with in horizontal fissures; and to the greater difficulties encountered when, as shown in the 7th and 8th lifts of tubbing, and in the lowest part of this shaft, a nearly vertical fissure happens to come within the area of the sinking. The great feeder of water was, as expected, met in a horizontal fissure at the bottom of the 5th lift of tubbing, and a similar feeder at the lower part of the 6th lift. These were effectually tubbed off at the depth of 295 ft. But upon continuing the sinking of the 7th lift, the area of the shaft encountered a vertical fissure, which let in the whole of the marl waters shut out in the upper lifts, both in larger quantity, and with the increased pressure due to the greater depth. This vertical fissure continued within the periphery of the shaft throughout the 7th lift; and, though showing a tendency to lead out of the shaft, it still continues to the present bottom. The upper feeders had been easily dealt with; but the accumulated supplies brought together by this vertical fissure gave, even after long pumping, and when the men were working at the full depth of 351 ft., a supply of water to be pumped of over 470 cubic ft. per minute; and this notwithstanding the wedging off of a portion of the total feeder (exceeding 600 cub. ft. per minute), by means of pinewood wedges driven into the fissure itself.

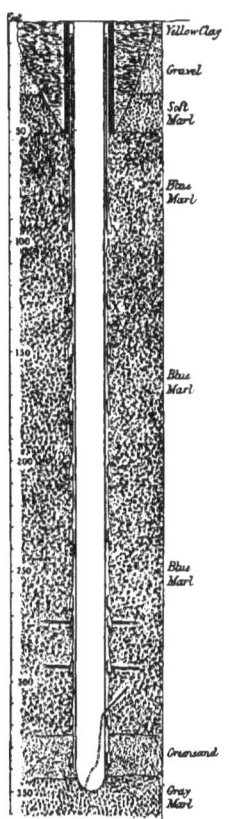

Fig. 241.

Hansa Shaft.

To deal with these feeders, Mulvany had at first a horizontal single-cylinder winding engine, 40 in. diam., and about 6 ft. stroke. This was at the south side of the shaft, and worked direct off the main crank by spears of considerable length, to which were attached two wrought-iron quadrants. To these were hung an 18-in. and 19-in. set of pumps, with a stroke of about 5 ft. At a later period another horizontal winding engine, with a single cylinder 28 in. diam., was erected to the north of the shaft, and worked an 18-in. set with about 4 ft. stroke off the back end of the piston rod. Subsequently a direct-acting vertical engine, with 36-in. cylinder, and about 6-ft. stroke, working a 21-in. set of pumps, was erected over the shaft. Upon meeting with the vertical fissure, however, this engine proved insufficient, and was at a later period removed; and in its stead was erected a direct-acting engine with 72-in. cylinder and 11 ft. stroke, which worked two 21-in. sets of pumps. All these pumps were hung in with ground spears, but with wind-bores resting on the bottom; and delivered direct to the surface. After meeting with the vertical fissure, the buckets and clacks were all obliged to be changed, and that frequently, at bank; the buckets having each time to be drawn, and the clacks fished—a tedious and laborious operation from such a depth. Upon getting into the more sandy part of the marl, near the present bottom, these changes became more frequent, the leathers wearing much faster, owing to the high speed, increased height of column, and action of the sand. This is an evil for which it is most desirable to find a remedy.

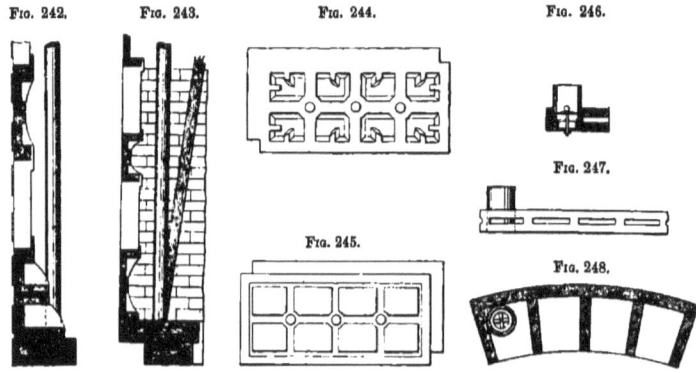

FIG. 242. FIG. 243. FIG. 244. FIG. 246.

FIG. 247.

FIG. 245.

FIG. 248.

Details of Walling, Tubbing, and Wedging Cribs.

The Zollern Colliery had been commenced by sinking two large round shafts, 24-25 ft. diam., intended for brick walls of great thickness, as at that time applied by German mining engineers for damming back the water. These shafts were sunk to the level of the first water feeder, which was met at 182 ft. from the surface, or 139 ft. 6 in. below an adit which had been constructed for carrying off the water from the pumps. The general section Fig. 249, and the enlargement of the bottom, Figs. 250, 251, show clearly the condition in which Mulvanys found both shafts as sunk down to feeder No. 1; and Figs. 252, 253, show the manner in which they finished them, down to the feeder No. 2 in shaft No. I., and to the feeder No. 1 in shaft No. II. This latter shaft they subsequently completed down to 943 ft. for coalwork, pumping, and ventilation. It will be seen from Fig. 254, that in shaft No. I. the German engineers had 10 sets of pumps firmly built into the shaft, with an enormous mass of timber framing; according to the system of that time the wind-bores were movable, or telescopic, so that they could be removed on firing shots or changing; and the pumps were lengthened by common pump pipes, each of 1 lachter or (6 ft. 10 in.) in length, added on below in the shaft. Thus the space, even in shafts of such great dimensions, was so encumbered with timber as to render sinking, even with moderate quantities of water, a very slow, expensive, and difficult operation. When the feeder No. 1 was first met with, and even before it was widened out

SHAFT-SINKING MACHINERY.

by the constant flow of water, it must have yielded 600 cub. ft. per minute. Under such circumstances, and with the inability in some of the pumps to change either buckets or clacks, for packing,

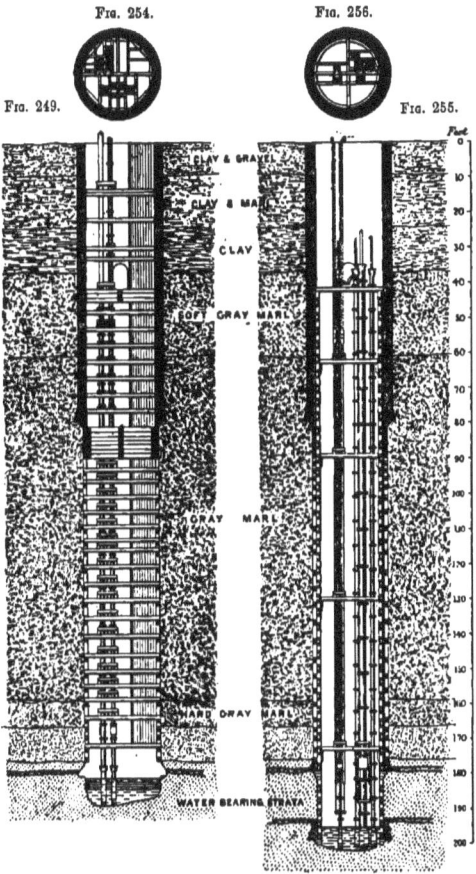

Fig. 254. Fig. 256.

Fig. 249. Fig. 255.

Zollern Shaft.

at the surface, it is only wonderful that the engineers succeeded, even in course of time, by continuous pumping and partially exhausting the feeder, in sinking the sump, and in preparing, as shown at bottom in Fig. 249, the foundation for the great walling below the first feeder.

120   MINING AND ORE-DRESSING MACHINERY.

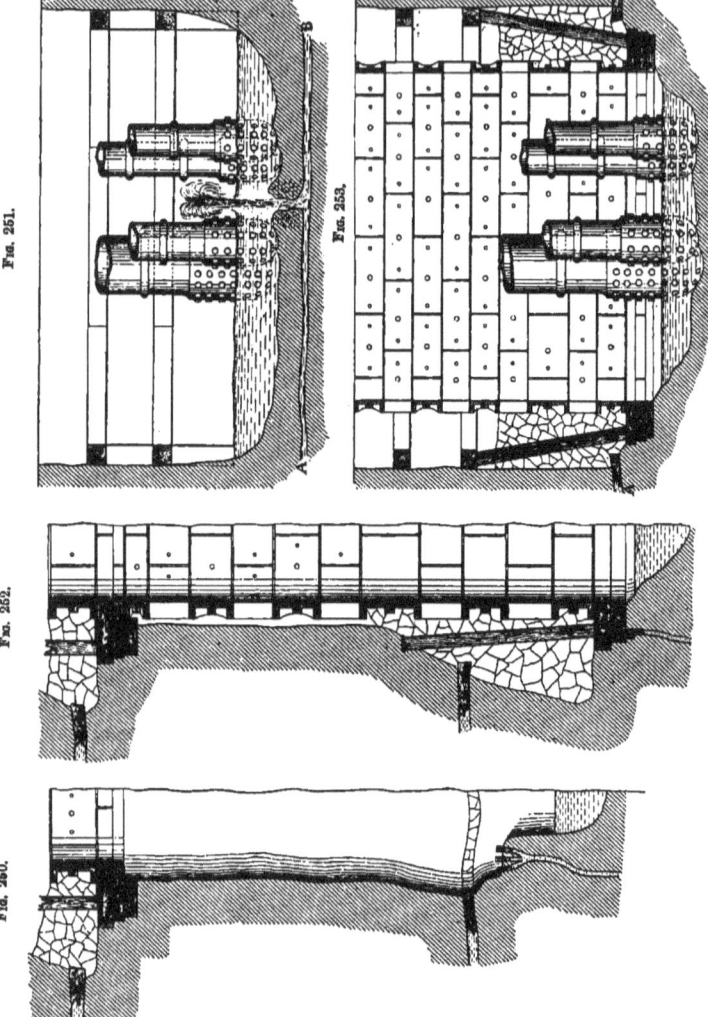

Fig. 251.
Fig. 253.
Fig. 252.
Fig. 250.

Shaft-sinking at the Zollern Colliery.

The Mulvanys, having acquired the colliery, commenced preparations for recovering shaft No. 1. They encountered great difficulties in the beginning; but by hanging in one large set of pumps, 32 in. diameter, they so far lowered the water as to enable them to take out the German pumps and timber, and then to hang in other large sets of 18 in., 19 in., and 20 in. diameter, as shown in Figs. 255-256; and after wedging off part of the supply of water coming from the horizontal cleft or fissure, they were enabled to commence cutting out the foundation for the wedging cribs, designed for the tubbing of a shaft 17 ft. 6 in. diameter. They adopted this dimension as that most suitable, according to the extensive experience they had obtained in the opening out of such large coal-fields, where the coal formation with its numerous beds is likely to reach 2500-3000 ft. depth below the surface.

In sinking shafts through heavy feeders of water, the sinkers must invariably work in water, both while drilling the holes for blasting in the sump, under the thick cast-iron wind-bores of the pumps, and subsequently while breaking up and removing the materials blasted. Now, whilst it is necessary to maintain the sump at such a depth that the wind-bores can get their full supply of water, without drawing air, yet on the other hand men cannot work efficiently if the water be more than knee-deep, or say 2 ft. Consequently the pumps must be worked by a short quick stroke; and the strokes of the several pumps must, as far as possible, be so timed as to keep the water regularly and steadily down to the proper level in the shaft. Yet this, with large feeders, is often most difficult to do, and the missing of a stroke or two often causes the men to be up to their waists in water: especially when only one shaft is being sunk, or when the two shafts are small. Again, in case of buckets or clacks suddenly failing, the men have at times to scramble for their lives up the pumps, or up wire-rope ladders provided to afford a means of escape. Such work requires the utmost perseverance and courage on the part of all concerned; the men must be relieved every 4 or 6 hours, and, in addition to receiving adequate wages, should be encouraged during each shift by a premium on every inch they lower the sets of pumps; the work must of course be continued day and night, and in extreme cases on Sundays and holidays also, without intermission. The natural feeders, when met with, must be wedged off, both to assist in reducing the quantity of water, and to prevent it from dashing out, with all the force due to the pressure of its source, over the bodies and heads of the sinkers. Under all these difficulties, whenever a solid homogeneous layer of marl, free from fissures, is found below the feeder, a perfectly smooth, level, and carefully made bed must be cut out and chiselled off, as a foundation for the wedging cribs, upon which a length of tubbing is subsequently built.

Notwithstanding these difficulties, and many other sources of care and anxiety in all the details of the work, Mulvanys maintain that, whenever the water can be pumped during the sinking of the shafts, the system of shutting off the water from the shafts by tubbing, both for the present and for all future time, is the best that can be adopted; and this for the following reasons:—

(a) Every portion of the work is seen and inspected, and can be properly treated and proved as it progresses from the surface downwards.

(b) The nature of the strata, and the separate quantities of the feeders, &c., are not only seen, but, upon closing each lift of tubbing, the feeders, so far as regards the space occupied by the shaft, are restored to their natural channels; while by the wedging cribs they are at the same time shut off from communicating with each other. This is an important matter, because, in case of accident, the repairs of any one lift of tubbing, or the removal of a broken segment, can be effected without interfering with the water in other lifts of tubbing.

(c) The cast-iron tubbing constructed in segments, as shown in Figs. 242 to 245, allows of sinking large shafts as easily as small ones.

(d) Such segments can be constructed either with brackets, or with large openings, or with taps of suitable strength and dimensions; thus giving the means of attaching pumps, or building in buntons for standing sets, or pipes for the supply of water, either to bank for surface purposes, or down the pit for hydraulic power, &c.

In short, by this system, one is master of the work as it proceeds, and it can therefore be carried out more quickly than by any other system, and in the great majority of cases at less expense on the whole.

The old German plan of attempting to shut out the water with brick walls is of course exploded; and the only system with which to compare the tubbing system in its present improved state is that known as the "Kind-Chaudron." This system is very ingenious, and has many merits, especially for boring out small trial shafts, where water is known to exist in the overlying measures; or in unexplored countries for viewing or examining the underlying minerals. It might be used for sinking auxiliary upcast shafts, for the constant discharge of gases from the goaf or broken, lying to the rise of the colliery; instead of allowing these gases to accumulate below, and upon a fall of roof, depression of barometer, or other accidental cause, to flow back into the working parts of the mine and produce explosions. The Kind-Chaudron system is also unquestionably to be recommended, when the supply of water in the upper measures, above the coal or other minerals, is practically speaking unlimited; as, for instance, in open strata, communicating directly with the sea, inland lakes, or large rivers—in other words, where the water cannot be pumped. This, of course, virtually includes all cases where, even under the best system, and with suitable means, it would not pay to pump the water and exclude it with tubbing.

Mulvanys' experience of the Kind-Chaudron system as applied at Dahlbusch in Westphalia was not favourable, because the quantity of water was not great, whilst the time occupied in sinking was much greater than with the tubbing system. Again, looking to collieries where the water-bearing strata are all near the surface, it will be seen that in carrying out the Kind-Chaudron system in such cases, and so keeping the water in the shaft the whole time that the boring is proceeding, one would be liable to be deceived into carrying the boring and the cast-iron "cuvelage" or casing down the whole depth to the coal-measures; though in fact the marl may be completely free from water below a certain level, so that one might sink and wall the shaft without pumps. Speaking generally, the objection which Mulvanys have to the Kind-Chaudron system, at least as applied to Westphalia, are as follows:—(a) The work is, so to say, carried out in the dark, with the shaft full of water; (b) all the feeders of water are brought into connection with each other, both surface feeders and under feeders; (c) the shafts are by the very nature of the machinery limited to a very small diameter, too small for the great depth of the coal-measures; and this limitation of diameter restricts the engineer to the use of a very thin bed of beton or cement, at the back of the cast-iron "cuvelage."
(d) The whole "cuvelage" being necessarily joined into one length, from the surface to the foundation (which itself is formed by a boring machine in the dark, i. e. under water), the pressure of the accumulated feeders is brought to bear over the whole height of the tube; and in case of any accident to any part of this cast-iron envelope, it would be liable to vent into the shaft the whole of the accumulated feeders. (e) Another objection, in the Westphalian district, is the risk to which the coal-measures, otherwise dry, are exposed of having the feeders in the marl let down to much

lower levels, by boring through the Bohn-erz, or layer of iron-ore, which shuts them off. Indeed, this has been done in many cases with the ordinary trial bore-holes, where they have not afterwards been efficiently stopped. There are perhaps cases where, for special purposes or for works of temporary duration in shallow depths, this system may with advantage be applied even in parts of the Westphalian district; but the number of such cases is probably small.

Fig. 257.

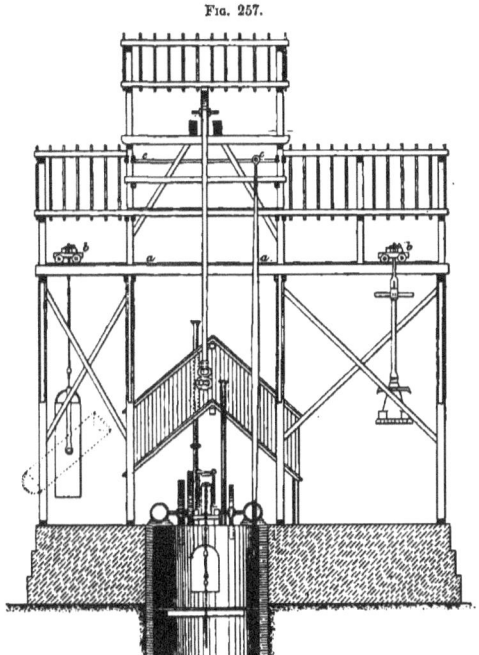

Shaft-sinking at Marsden.

The Kind-Chaudron method was employed by J. Daglish in sinking two shafts at Marsden, and the following remarks are condensed from his description of the operations, published in the Minutes of Proceedings of the Institution of Civil Engineers (James Forrest, Esq., Secretary):—A substantial headgear was erected, strongly framed together with timbers (Figs. 257, 258). The whole of this is covered in with wood cleading, so that the workmen are always protected from the weather. At 37 ft. from the ground, two rails are laid on stout balks a of timber, which carry travelling carriages b, on which the heavy tools are run backwards and forwards. At 52 ft. from the ground, similar rails on longitudinal balks of timber c, support small carriages (Fig. 259) for carrying the boring-rods, this great height being necessary in order to obtain sufficient length of rods. This system of carrying and moving the tools on traversing carriages enables the operation to be conducted with a very small amount of manual labour.

Fig. 258.

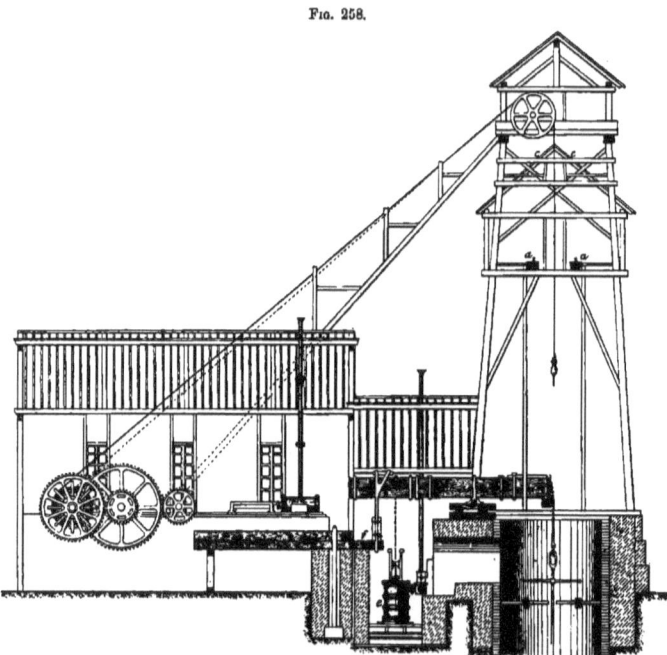

Shaft-sinking at Marsden.

The Kind-Chaudron process consists of two distinct series of operations :—(a) Those connected with the boring out of the shaft, on a system closely resembling that first adopted by Kind many years ago for boring deep holes for artesian wells. (b) That of lowering down the shaft a watertight lining or tubbing.

The first process, therefore, at Marsden was the boring of a centre hole in No. 1 pit, 4 ft. 11 in. diameter, by a small trepan or chisel (Fig. 260). This trepan, 7 tons in weight, is attached to the massive wooden lever d (Fig. 258) by rods of the best pitch pine, 5 in. square (Fig. 261) and 58 ft. long, with iron terminations, having tapered screws. One end of each rod is fitted with a male screw (Fig. 262), and the other with a female screw. The screws have coarse threads carefully cut, so that, after having entered, a few turns are sufficient to screw the joint quickly home.

The lever d is attached on the opposite end to a steam cylinder (Fig. 258), 39 in. diameter, actuated by a single valve only on the top side. The valve is worked by hand; the rods are lifted by the pressure of the steam on the top side of the piston, and they fall by their own weight when

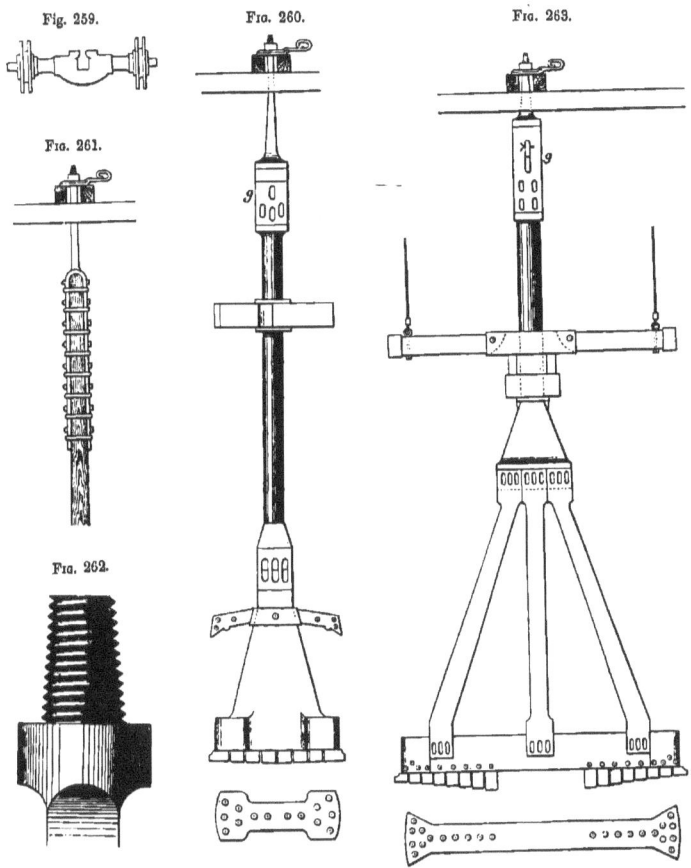

Kind-Chaudron Shaft-sinking Tools.

the valve is opened to the atmosphere. The length of stroke is regulated by the machinist, and varies from 6 to 18 in., according to the hardness of the rock. An important adjunct to the lever is the spring-beam f, against which the lever strikes at the termination of each stroke. The number of strokes per minute varies from 9 to 18. In very hard rock comparatively few and light blows only can be given. When the rods are suspended at the end of each stroke, they are turned through an angle of 2° to 4° by four workmen holding a crosshead lever, walking round the top of the pit, similarly to an ordinary boring.

An essential part of the boring tools is the sliding piece g (Fig. 260, 263), by which the trepan is connected to the rods through the medium of a slot 12 in. long. This permits the trepan to strike the bottom without communicating a severe shock to the rods, which continue their ascent until arrested by their buoyancy in the water, aided by the spring beam striking against the inner end of the lever. Except for the play thus allowed, it would be impossible to strike even a light blow without fracturing the rods.

An apparatus called the freefall (Fig. 264) is sometimes also attached. On the descent of the rods, the trepan is caught up by a pair of jaws h which are locked by a wedge. The wedge being withdrawn by means of a large disk of wood i at the commencement of the return stroke, permits the trepan to fall nearly 2 ft. without being detached from the rods. This apparatus was attached to the small trepan in boring the No. 2 small pit between the depths of 284 and 334 ft. A disk, 5 ft. 2¼ in. diameter, gave most satisfaction, the diameter of the small pit being 6 ft. 6¾ in.

After the boring has been continued about 3 hours, in moderately hard rock, the trepan is withdrawn, and the sludger (Figs. 265, 266), with a capacity of 4 cub. yd., or 10 tons, is lowered. The sludger is sometimes attached to the lever, and worked up and down by the rods, and at other times by the rope only. The débris rises into it through the valves in the bottom, it is then withdrawn and emptied. The emptying of the sludger, and the unshipping of the lever, to allow of the rods being removed, are effected by ingenious and time-saving arrangements. After the centre-boring is advanced 30-40 ft., the large trepan (Fig. 263), 16 tons in weight, is put in, and the large pit is similarly bored, the débris falling into the small pit, which requires to be frequently cleared out. This was the process in the first instance adopted at Marsden, but it was afterwards modified. In every new sinking by this system slight variations are found in the character of the rock, which entail modifications in its application. At Marsden the rock proved to be harder than in any locality where the system had been previously in operation.

During the boring out of No. 1 small pit no difficulty was found in raising the débris with the ordinary sludger; but in boring the large pit it would not rise into the sludger, but became solidified at the bottom of the small pit. This was probably due to the particles being larger than those produced in the boring of the small pit. To remedy this, at first clay was thrown down the pit, and the small trepan was again introduced to loosen the débris, and mix it with the clay, which could then be withdrawn by the ordinary sludger. But the process was a long one, the re-boring taking quite as much time as the original boring. It was therefore determined to lower the sludger into the small pit, release the rods, and leave it there to catch the débris as it fell. Accordingly the sludger was lowered to the bottom of the small pit, and left there, as shown in Fig. 267. On attempting to withdraw it, however, it was found that the mud which had settled in the water at the bottom of the pit, or which had passed the sides of the sludger, embedded it so far that great violence had to be used to extract it, which would have certainly, sooner or later, resulted in serious accidents. Arrangements were then made to suspend the sludger on the edge of the small pit at the top by claws (Fig. 268), and the two inner of the interior teeth of the large trepan were removed to avoid striking these claws. This plan succeeded imperfectly, and on several occasions when the claws were struck, the sludger fell down the small shaft, and was only extracted with difficulty, and with a liability to accidents.

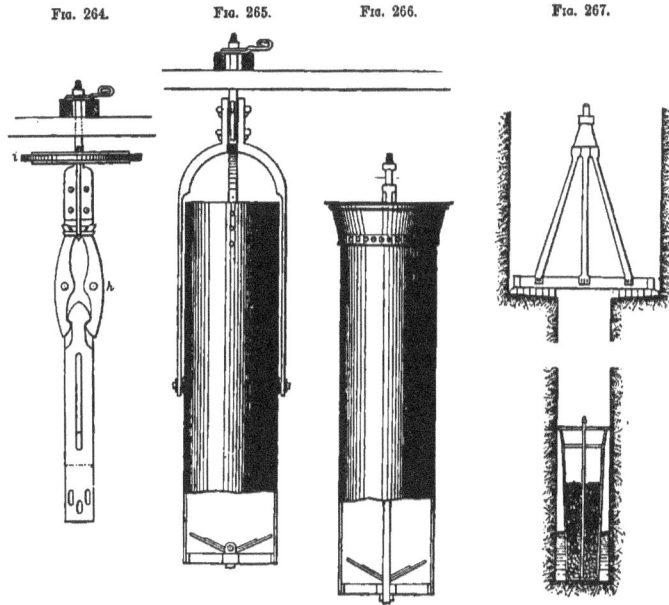

Fig. 264.   Fig. 265.   Fig. 266.   Fig. 267.

Kind-Chaudron Shaft-sinking Tools.

A successful attempt was then made to form a ledge within the smaller pit, by taking out all the teeth but the two outer, and the sludger was thus suspended about 1 ft. from the top of the small pit. This operation, however, entailed so many changes of the teeth, &c., that it was attended with great loss of time. But having found the correct principles on which to proceed, it was not difficult to devise a plan for leaving a suitable ledge within the smaller pit. To effect this, the outside tooth of the small trepan on each side was enlarged 3 in.; the tool was again introduced, and the small pit bored to a diameter 6 in. wider than previously, leaving a ledge of 3 in. all round (Fig. 269), on which the sludger was suspended by an angle-iron ring. In No. 2 pit a third trepan was used, having a diameter of 6 ft. 6 in. By this trepan the small shaft was bored to a depth of 383 ft.; not only through the limestone, but 50 ft. into the coal-measures, and 6 ft. 6 in. below where it was intended to place the moss-box of the tubbing, and therefore below, and entirely clear of, all future operations with the large trepan. The smallest trepan was then introduced, and the boring continued 32 ft. 9 in. farther, leaving a ledge of stone $9\frac{1}{4}$ in. in width all round; on this ledge a cast-iron ring was deposited, to form a permanent bed for the hanging sludger to rest on. This arrangement acted perfectly, never having been the cause of the slightest accident throughout the sinking of the second shaft. The cast-iron ring was adopted in the second pit, because the weight of the sludger soon wore away the ledge of stone by being suspended from it. At first the hanging sludger was lowered into its seat by the regular screw, which was left

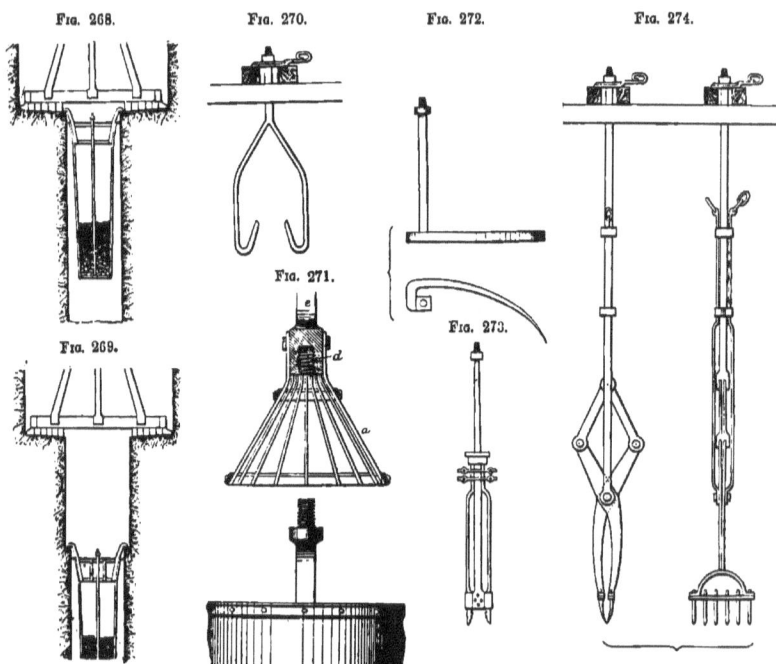

Kind-Chaudron Shaft-sinking Tools.

slightly slack, all the other screws of the rods, as they were lowered in, being tightly screwed home. When the sludger was deposited on its bed, by turning the rods backwards, the slack joint yielded, and the rods were unscrewed at this point and drawn away. It did, however, happen occasionally that some of the other screws became detached, and then the remaining rods and sludger had to be fished up. A double hook (Fig. 270) was next adopted for lowering the hanging sludger into place; it was simply fastened on to the bow of the sludger, and when the latter was lowered and rested on its bed, the rods were let down a few inches farther, and turned half round, so as to free the hook entirely from the bow; they were then drawn away, leaving the sludger in place. The bottom of the rods, where they are attached to the sludger by a female screw, is fitted with a small inverted funnel (*a*, Fig. 271), to guide the male screw *b*, which is attached to the sludger *c*, into the female screw *d* at the end of the rods *e*, as they are lowered; an arrangement successfully carried out through the whole of the boring of both pits, without failure or difficulty, even at a depth of nearly 400 ft.

No small part of the success of this process arises from the ingenious arrangements, and forms of tools, for picking up material at the bottom of the shafts, and for taking hold of broken spears, &c., which, from the character of the operations, must be of frequent occurrence. These are termed "safety tools," and consist of the following apparatus :—

(1) The catching hook (Fig. 272), which, on being swept round the shaft below the top of the broken spear, guides the spear into the angle made by the hook and its rod, where a properly-shaped recess is formed, into which the ironwork of the spear falls, and can by this means be retained and withdrawn.

(2) The spear catcher (Fig. 273) is a fish-head, with a pair of serrated jaws, which on touching the top of the broken rod, and the wooden chock keeping the jaws open being forced out, the teeth press firmly against the ironwork of the spears, enabling them to be withdrawn.

(3) The grappling tongs (Fig. 274) being a pair of large rakes, which can be opened and shut by levers worked by ropes. By moving and working this across the bottom of the shaft, any pieces of material larger than 2 in. square can be extracted with ease.

The most important part of the process, and that attended with the greatest risk, is that of lowering into the shaft the metal tubbing. At the Marsden sinking, the dimensions of each ring or cylinder were as follows (Fig. 275 shows position of moss-box before compression, and Fig. 276 shows position of moss-box after compression, with false bottom removed, and foundation tubbing and wedging cribs in place) :—

	No. 1 Pit.			No. 2 Pit.		
	ft.	in.		ft.	in.	
Internal diameter	12	$7\frac{1}{2}$		13	$8\frac{1}{2}$	
External "	12	$9\frac{1}{8}$		13	$11\frac{1}{4}$	
Thickness of top cylinder ..	0	1		0	$1\frac{3}{8}$	
" bottom " ..	0	$1\frac{5}{8}$		0	$1\frac{7}{8}$	
Height of each cylinder	5	0		5	0	
Total height of tubbing	280	0		285	0	
	tons	cwt.	qr.	tons	cwt.	qr.
Weight of top cylinder	5	4	0	6	10	1
" bottom "	7	0	0	8	19	2
Total weight, including bolts and lead joints ..	400	0	0	450	0	0

The flanges of each top cylinder are $3\frac{1}{2}$ in. wide by 2 in. thick; and between every two rings is placed a plain leaden wedge $4\frac{7}{8}$ in. wide, by $\frac{1}{8}$ in. thick, covered on each side with red lead. The cylinders are attached to each other by sixty $1\frac{3}{16}$ in. bolts of best iron. The whole of these cylinders are alike, save in varying thicknesses, excepting the bottom three pieces. The bottom pieces *ab* are telescopic, with outside flanges cd, 6 in. and $7\frac{3}{8}$ in. respectively; the bottom piece b was suspended from the upper piece by rods in No. 1 pit, and in No. 2 pit, by an internal flange, which permits of the second piece a sliding downwards on the outside of the first piece. Whilst being lowered, the outside flanges of the bottom pieces, which are called the moss-box (and which are the only two cylinders with outside flanges), are 5 ft. apart, and the interval is filled with tightly compressed

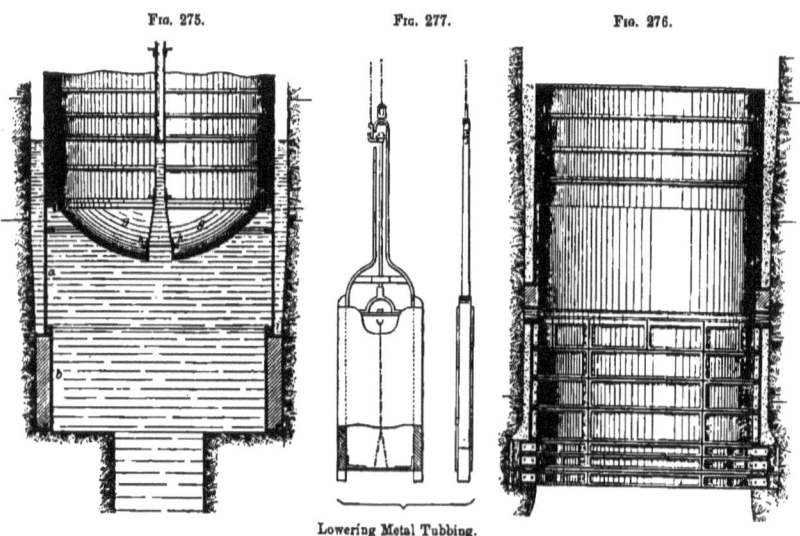

FIG. 275.   FIG. 277.   FIG. 276.

Lowering Metal Tubbing.

moss. When the lowest piece rests on its bed, at the bottom of the pit, the remainder of the cylinders continue to descend, compressing the moss with the whole weight of the tubbing, namely, over 400 tons. In the middle over the third cylinder from the bottom, there is an extra internal flange e, $3\frac{1}{2}$ in. wide; on which is screwed, by 64 bolts, a flat ring or circle of cast iron f, $5\frac{3}{4}$ in. broad. This ring admits of the false bottom g being withdrawn up the interior of the tubbing to the surface, when the operation of lowering the tubbing has been completed. A massive dish-plate g, of cast metal $1\frac{1}{4}$ in. thick, is bolted to the bottom, having a flange h on the upper side, for attaching the column of pipes. The object of the false bottom is to float the tubbing whilst it is being lowered. After carefully securing together by their respective flanges and attachments three pieces of tubbing intended for the bottom, they are lowered to the level of the water by an arrangement of screw-rods worked by six powerful winches, with two men to each; additional cylinders and central pipes are then added one by one, causing the whole of the tubbing to sink until it floats by the displacement of the water. In the Marsden No. 2 pit the tubbing floated when cylinder No. 9 was attached. The rods are thereupon removed, and as each additional cylinder is added, a certain quantity of water is run inside to cause the tubbing to sink. In the Marsden No. 2 pit the addition of cylinder No. 10 caused the

SHAFT-SINKING MACHINERY.

tubbing to sink 1 ft. 9 in., and of cylinder No. 56 at the top 1 ft. 1 in. In both pits this operation was completed without leakage, either at the joints of the cylinders, or of the central column of pipes. The work, however, requires great care and watchfulness, being attended with risk, as any leakage would cause the tubbing to sink to the bottom. In the deep sinking at Ghlin near Mons, the depth bored is 1026 ft., with an internal diameter of $14\frac{1}{2}$ ft. The thickness of the tubbing at the top being 1 in., and at the bottom $3\frac{5}{8}$ in., the total weight being taken at 1772 tons, at a cost of 12l. per ton, brings the cost of the tubbing alone for the two pits to more than 40,000l. The bottom of the hard rock was bored through at a depth of 931 ft., and below this, before reaching the impervious coal-measures (in which the moss-box will be laid at a depth of 1030 ft.), 80 ft. of running-sand, gravel, and clay were bored through, and a wrought-iron tube was inserted to protect the sides until the main tubbing is lowered down.

FIG. 278.

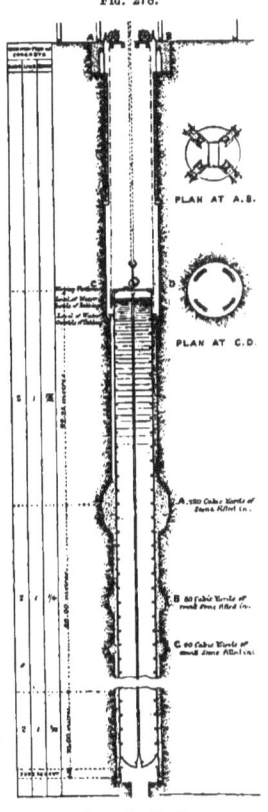

Concreting consists in filling with concrete the annular space between the exterior surface of the tubbing and the sides of the shaft, from the moss-box upwards to the top of the tubbing i (Figs. 275, 276). The concrete is lowered simultaneously all round the pit by four rectangular boxes 3 ft. long, 18 in. broad, and $4\frac{1}{2}$ in. wide, shaped to the radius of the pit (Fig. 277). A large gullet was passed through in No. 2 pit at a depth of 56 yd. from the surface, the width of which was nearly the whole diameter of the shaft. When concreting at this point, 120 cub. yd. of small stones and concrete were filled in, and 80 and 40 cub. yd. at smaller gullets lower down (Fig. 278), without sensibly raising the level of the concrete.

The absolute time taken from commencing to finishing the boring was 17 months in No. 1 pit, and 19 months in No. 2 pit. There was, however, a delay of several months in No. 2 pit on account of the tubbing not being ready; the depth of boring was also 40 ft. greater. The time occupied in lowering the tubbing and concreting, &c., was $3\frac{1}{2}$ months in No. 1 pit, and 4 months in No. 2 pit. The total time taken to complete each pit was $20\frac{1}{4}$ months in No. 1 pit, and 23 months in No. 2 pit. The average distance bored in No. 1 small pit in the limestone was 1 ft. $3\frac{1}{4}$ in. per shift of 12 hours, and in the coal-measures 1 ft. $8\frac{3}{4}$ in. In No. 1 large pit in the limestone it was $7\frac{3}{4}$ in., and $8\frac{1}{4}$ in. in the coal-measures. In the small No. 2 pit the average distance bored in the limestone was $10\frac{1}{2}$ in. per shift of 12 hours, and in the coal-measures 1 ft. 4 in. In No. 2 large pit in the limestone it was $8\frac{1}{2}$ in., and $9\frac{3}{4}$ in. in the coal measures. The terms of the contract were that no payment had to be made to the Kind-Chaudron Company for the patent right and superintendence unless the following conditions were fulfilled—that the tubbing when completed

Concreting Shaft.

s 2

should not be more than 6 in. out of the perpendicular, and not let pass more than 40 gal. of water per minute. On the formal examination by the engineers of the Whitburn and Kind-Chaudron Companies, it was found that in No. 1 pit the tubbing was only 1 in. out of the perpendicular, and let pass about 1 gal. of water per minute, and this only at the wedging joint below the moss-box. In No. 2 pit the tubbing was only 2 in. out of the perpendicular, and no water passed. In both cases the tubbing itself from top to bottom was absolutely dry.

COST OF BORING AND TUBBING NOS. 1 AND 2 PITS, MARSDEN COLLIERY, BY THE KIND-CHAUDRON PROCESS.

	Construction											Working									Summary		
	Preparing and Erecting Baraque*		Tools†	Lining Tube		Tubbing		Patent‡		Total		Boring		Repairing Tools		Tubbing		Sundries, Salaries, &c.		Total		Labour	Materials	Grand Total
	Labour	Materials	Materials	Labour	Materials	Labour	Materials	Labour	Materials	Labour	Materials	Labour	Materials	Labour	Materials/Stores	Labour	Materials	Labour	Materials	Labour	Materials			
	£	£	£	£	£	£	£	£	£	£	£	£	£	£	£	£	£	£	£	£	£	£	£	£
No. 1 Pit (12 ft. diam.)	694	654	588	175	644	4,491	1,062	869	7,439	2,033	577	825	282	457	3,708	722	7,023	1,581	7,892			9,020		16,912
No. 2 Pit (13 ft. diam.)	694	654	588	Nil	Nil	5,916	1,037	694	8,195	1,765	278	660	261	125	1,723	216	4,273		755		5,967	8,950		14,917

* The baraque was originally erected at No. 1 pit; it was taken down and rebuilt at No. 2 pit. The total cost is divided over both pits.
† The original cost of the tools was 2060l.; after the completion of the pits they were sold for 984l. The difference is divided over both pits.
‡ Patent right (including plans and specifications).

In order to overcome the difficulty of the straight cutter making the blows more closely together near the centre of the shaft, and not sufficiently closely together near the periphery, Lippmanns used a different description of cutting tool (Fig. 279). The drill is in the shape of a double V fastened together. There are two blades at that part of the tool which cuts round the circumference of the shaft, and only one blade cutting in the part of the shaft which is at the centre, so that there are more blows in cutting the stone near the periphery than at the centre. Owing to the angle at which the blades are placed, the stone is cut into checks or squares, so that it is broken up more certainly into small pieces. Not only is there that advantage, but, owing to the breadth of the tool, when it strikes the ground there is less liability for it to be deflected sideways if it happens to strike upon hard stone. In some cases a shaft sunk with simply a straight cutter is sunk not quite perpendicularly, owing to that tendency to deflection; and Lippmanns claim that their shape of cutter overcomes this. Of course it is a great advantage if only one hole has to be bored, because it appears that the shaft can be bored all at once as quickly as the enlargement can be done after boring the centre, as in the ordinary Kind-Chaudron method. The rate at which the work was done by Lippmanns in sinking a pit in Westphalia, was about 10 in. every 24 hours, compared with 7 in.

Lupton has described another way of overcoming a great body of water without pumping, where it was necessary to sink through a bed of very fine running quicksand on the seashore. The compressed air process was adopted, as commonly used in sinking the cylinders of bridge founda-

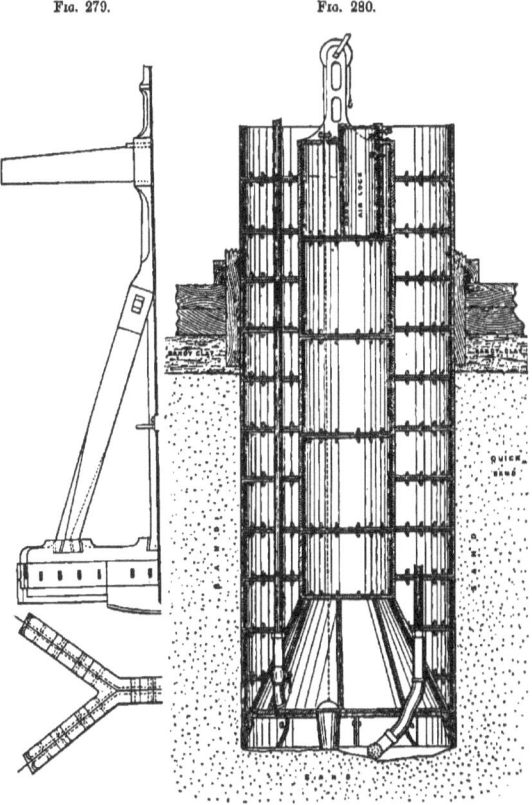

Fig. 279.   Fig. 280.

Lippmann's Cutting Tool.   Sinking through Quicksand.

tions. The shaft was sunk to a depth of 100 ft. below the water level by compressed air. There was a cast-iron cylinder 13 ft. diameter, in which air was pumped at sufficient pressure to force out the water, the pressure of the air being exactly measured by the depth of the water. The tide coming in round the pit, the level of the water was above the level of the land. By means of that process the sinking was easily performed, the workmen standing upon the sand and excavating it just as if they were on the surface and the tide was out. No doubt it is an operation that demands some care in managing the men, who are subject to pains in their shoulders, but if treated with care they do not suffer any great injury. Fig. 280 shows the cylinder as it descended through the sand. Inside the outer cylinder was a smaller cylinder, 6 ft. diameter, connected with a larger one at the bottom by a conical piece or bed. The pressure was inside the smaller cylinder, and in the conical piece at the bottom. At the top of the smaller cylinder was an air lock, through which the men and materials could enter and leave. The pit was sunk through sand, boulder clay, and gravel, into the red marl, where a firm joint was made with a solid stratum. The cost was about the same as with the Kind-Chaudron method.

Chavatte made a sinking through 107 ft. of wet running ground, and 334 ft. of chalk down to the top of the coal-measures under water, without any assistance from Chaudron, to avoid the payment of a premium of 75,000 or 80,000 francs. In sinking through the running ground he commenced with a ring of masonry about 6 ft. in depth and 15 ft. interior diameter. Inside of that, he placed a sheet-iron cylinder and then commenced with what the Germans called a "sackbohrer," working the shaft down in that way, and pressing the cylinder down by means of screws. He had lowered four cylinders of that kind, one inside the other, telescope fashion, before he reached the bottom of the running measures, 107 ft. in depth. At last, having reached that depth, he had not to line the shaft in any way with a temporary lining, the ground being good, but he proceeded to bore it out. He then bored out the remaining 334 ft. with ordinary tools, such as those of Kind and Chaudron, with a diameter of 14 ft, and he introduced the same kind of cast-iron tubbing. Instead of using the moss-box, he placed a ring in the form of a truncated cone at the bottom of the cast-iron cylinder, and floated it down in the same way as was done by Chaudron. He had contracted the boring somewhat at the bottom, so that when the truncated cone came to rest on the bottom, it crushed away a little ledge that had been left, and made what was thought to be a watertight joint. No dependence, however, was put upon that joint, but concrete was inserted in the most careful manner. Chavatte thought that he had introduced a better method of concreting than that used by Chaudron. He did not use four windlasses, but only two, and on each windlass there were two ropes, one of which let down a box of concrete while the other drew up the empty box, so that instead of men being employed alternately pulling up an empty box and letting down a full one, the full box helped to draw up the empty one. Chavatte thought that in the Chaudron process sufficient care was not used in making the concrete tight enough, and that if that were done there would be fewer failures.

In a shaft where it is necessary to place an "up-over" crib so as to tub off water in a certain stratum, theoretically the pressure behind the tubbing is simply due to the head. It has been found in many cases that the tubbing has cracked and blown, and it is difficult to account for it. Made no matter how thick, the inevitable result is that it becomes cracked, or some catastrophe happens. William Coulson, of Durham, suggested the idea of putting a small safety-pipe up the shaft to allow any gas or air that might accumulate in the stratum behind the tubbing to escape. It has never been explained, how it is that the air or gas behind the tubbing can possibly have a greater pressure than is due to the hydrostatic head; yet it is so, and ever since the insertion of that small pipe, about 2 in. diameter, and carrying it through the "up-over" crib, bringing it sufficiently high in the shaft above the water-level, no failure has occurred.

CHAPTER VII.

COAL-CUTTING MACHINERY.

The labour of hewing coal by hand is very severe. The necessity for under-cutting to a great depth in a narrow groove, and the constrained attitude of the hewer, especially in thin seams, combine to render his occupation the most laborious of any connected with coal getting. It is also evident that the force of the hewer, exerted under such unfavourable conditions, must

Winstanley and Barker's Coal-cutter.

be very wastefully applied, and, therefore, is not employed according to the requirements of economical production. Besides this, even the proportion of the force which is made effective is improperly utilised, since it is made productive of a large quantity of small coal. When holing to the usual depth of 3 ft., the average height of the cut, even with skilful hewing, is not less than 9 in.; and when it is necessary to hole in the seams such an excavation destroys a large proportion of the coal. Another important circumstance is the relation of capital to labour. To lessen the dependence of production upon hand labour, it is highly desirable that machinery should be applied to the undercutting of coal seams. Moreover, the same change is called for by the constantly and rapidly increasing demand.

It would seem to be a comparatively easy matter to design and construct machinery capable of performing the work of undercutting the seams effectively. Experience has, however, shown that the difficulties are greater than they appear. Numerous attempts have been made to overcome them. Following is a description of those machines which have shown good results in continued practice.

Winstanley and Barker's machine (Figs. 281 and 282), like most other coal cutters, is driven by compressed air, conveyed down the pit shaft and along the main roads and drawing roads in iron

FIG. 282.

Winstanley and Barker's Coal Cutter.

pipes, and from the end of the drawing road to the machine in a rubber hose-pipe 2 in. diameter. The frame is about 6 ft. long, and is supported on flanged wheels, which run on the ordinary tramway of the mine, the gauge being varied as required. On the front part of the frame are two oscillating cylinders, 9 in. diameter and 6 in. stroke, provided with ordinary slide valves. The piston rods are connected to an upright crank-shaft, on the bottom end of which is a driving pinion, shrouded at the top, and having only 5 teeth, which gear into the teeth of a spur-wheel, which is also the cutting wheel, and is 3 ft. 6 in. diameter; the driving power is thus applied with the greatest mechanical advantage, that is, directly on the circumference of the cutting wheel. The cutters are fixed in the circumference of the wheel, one in every cog or tooth, their points projecting 1 in. beyond the teeth.

The cutting wheel revolves at the end of an arm consisting of a broad flat plate, at the opposite extremity of which is a toothed segment or quadrant, actuated by a worm and hand-wheel, whereby the arm carrying the cutting wheel can be turned partly round in its bearing in the frame of the machine. Before the machine commences to hole in the coal, the cutting wheel is under the back part of the frame, as shown dotted in the plan, almost touching the straight face of the coal; and on starting the engines, the attendant, by turning the hand-wheel and worm, causes the cutting wheel gradually to hole its way into the coal, until the arm is at right angles with the frame of the machine. In this position, the cutter is holing about 3 ft. in depth from the face of the coal; and it can be placed in any position to hole less than this depth if required. As soon as the cutter has worked into the coal to the full depth, the machine is drawn along the face of the coal as it holes or cuts its way, throwing out the small coal or slack between the tram rails upon which the machine runs. The thickness of the holing or groove cut out is 3 in., but this can be reduced by using a thinner cutting wheel. There is no traverse motion on the machine, as it is considered simpler to draw it along the face by means of a small crab, turned by a lad at the end of the working face. When the holing of the entire length of the face is completed, the cutting wheel is brought back to its original position underneath the frame of the machine, by means of the worm and hand-wheel, and is ready for beginning to hole at the commencement of the new face as soon as the coal already holed has been removed.

The chief advantages in this machine are, that the swivelling movement of the arm carrying the cutter enables it to cut or hole its own way into the coal, the depth of cut increasing from nothing to about 3 ft.; and by the same movement the cutter is brought back underneath the frame of the machine when not at work. When the cutter is in this position, it can be taken through narrow parts of the mine, without the necessity of removing the cutter, the space required for the machine to pass being only the width of the cutting wheel, which, with the cutters, is 3 ft. 8 in. The two cylinders driving one crank in a horizontal plane, and the star-wheel on the lower end of the crank-shaft gearing directly into the teeth of the cutter, constitute the simplest form of coal cutter which can be imagined. Moreover, the rotation of the centre of the cutter round the axis of the crank-shaft, so as to enable it to start its own cut anywhere, is a feature of the highest importance.

Baird's machine is driven by air compressed by the winding engine of the upcast pit. The air-cylinder is 24 in. diameter and 24 in. stroke. It compresses to 45 lb. per sq. in., and works 24 strokes per minute. The air is compressed into a boiler, with a safety valve loaded to 45 lb., and is taken down the shaft in 6-in. cast-iron pipes. The same pipes are continued, mostly along the floor of the main waggon way, where they would appear to be somewhat liable to accidents, for about 1000 yd. underground, when they are reduced to 3 in., and finally to 2 in. of flexible tubing, which supplies one single coal-cutting machine. There is, of course, a certain amount of leakage in the joints, and the difference between pressure in the reservoir at bank, and that at the machine when working, is about 2 lb. per sq. in. The machine (Fig. 283) has one air-cylinder a, $8\frac{1}{4}$ in. diameter by 12 in. stroke, which propels the machine and also drives the cutters. The engine runs at about 240 revolutions per minute; but this speed is considerably reduced by the gearing of the machine, and the cutters themselves move very quietly. The machine is exceedingly compact, and the parts are fairly easy of access; it is covered by a sheet-iron case when at work, to shield the gear from injury, and also to facilitate the moving forward of the rails and sleepers on which it runs. It is necessary to provide a special road, formed of short pit-rails r, in 4-ft. lengths, fitting into cast-iron sleepers s. In consequence of the shortness of the lengths of rails, all the sleepers are joint-sleepers, and the rails and sleepers are regularly taken up behind the machine and passed forward along the sheet-iron cover before mentioned, to the man in front, who lays them down in readiness.

The pressure upon the rails, due to the direction of the cut in the chisel faces, tends to draw them *towards* the coal face. This is clearly the right direction for the pressure, as it is very easy to set wooden chocks and wedges against the sleeper ends, when necessary, to prevent them from being drawn too near the face itself, while it would not be so easy to wedge the road up against the packs and props in the goaf. The coal cutter is very heavy, weighing in all 25 cwt. The shape of the cutting edges of the teeth is, no doubt, the result of experiments, but the angle of the

Fig. 283.

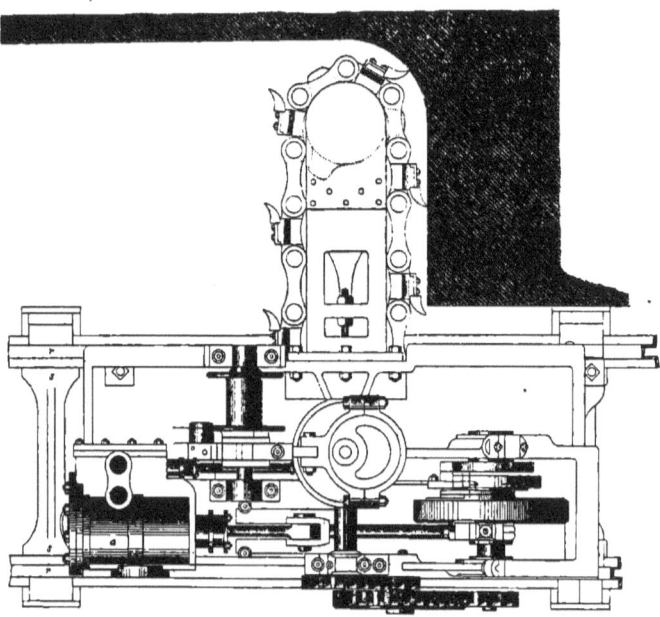

Baird's Coal-cutter.

cutting edge would appear to be too great for obtaining the best results. The average speed of progress of the machine is 1 ft. per minute, 3 ft. under. It does not, however, often happen that the distance cut exceeds 40 ft. in the hour, including stoppages; the whole distance of 120 yd. being, nevertheless, traversed with ease in the night's work. Three hands are employed at the machine, two men and a boy. The coal cutter works in the seam itself, and about 4 in. from the bottom, so that "round coal" is made from the small piece of the seam left below. The general design is strong and solid, and it is in many ways the right sort of tool for the rough usage of the pit. The working parts are also easy of access. On the other hand, its weight is very great. The single cylinder of considerable size, is obviously not so advantageous as the double cylinders of the other cutters, although the machine would appear to give no trouble from this cause. The gauge of the wheels of the machine does not fit the gauge of the pit, and thus a special road has to be employed. The adoption of a self-acting "feed" is a great advantage. It is a great drawback that it is obliged

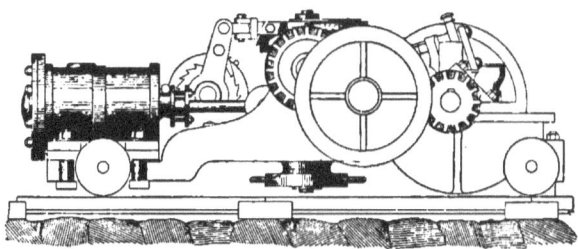

Gillot and Copley's Coal-cutting Machine.

to be started from a "loose-end," and cannot be entered anywhere, like the Winstanley cutter. The endless chain form of cutter offers great advantages in working over a very uneven floor; and by keeping a spare chain, with the cutters ready fixed, some time may be saved when it is necessary to shift them.

Gillot and Copley's machine (Figs. 284, 285), consists of two horizontal steeple-engines, with cylinders 7 in. diameter and 12 in. stroke, driving a horizontal crank-shaft, carrying on either end a

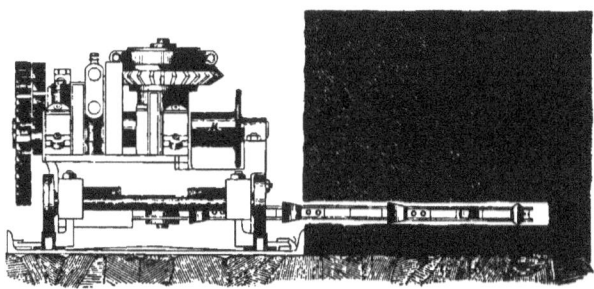

Gillot and Copley's Coal-cutting Machine.

bevelled pinion; this gears into an inclined bevelled wheel, and another bevelled pinion, on the end of this intermediate shaft, drives the circular cutter, which is *fixed* in position, not capable of rotation as in Winstanley's machine. It is made of cast steel, and very thin. The cogs, by which it is driven, are formed by slotting out holes right through the substance of the cutter, and thus making

T 2

what may be termed a lantern "crown-wheel." In the periphery are fixed 20 steel teeth by means of set-bolts, the teeth being alternately single and double. The flexible pipe, for the supply of compressed air, can be fixed on to either end of the machine, as there is a sort of "stand-pipe" fixed on to the top for this purpose. The cutter itself will fix on to either side of the frame, the gear being made reversible for this, so that the machine can be made to cut either right- or left-handed. It has no propelling gear, but is made to traverse by a chain and a winch at the far end of the face, worked by a boy. The machine requires two men to tend it, and a boy to haul at the winch. The air-compressing engine has one cylinder 18 in. diameter by 4 ft. stroke, and works at 35 lb. pressure 50 strokes per minute. The air-cylinder is 16 in. diameter, 4 ft. stroke, and pumps into an old boiler as receiver. The pipes are 4 in. diameter for the first 400 yd., after that 2-in. gas-pipe. The coal-cutter makes 90 strokes per minute. With fair working, 30 yd. per hour may be reckoned on. The wages come to $1\tfrac{3}{4}d.$ per yd. cut, including the time spent in preparing, &c., and also engineman's wages at the air-compressor, but no time in laying the road for the machine. The machine will turn out 10 tons of coal to 8 tons by hand labour, or a saving of 25 per cent. on the coal. There is also the saving of $1\tfrac{3}{4}d.$ against $7d.$, or $5\tfrac{1}{4}d.$ per yd. in the cost of holing. The machine can be reversed with great ease, so as to cut either right- or left-handed. In this respect it has a great advantage over both the Winstanley and the Baird.

FIG. 286.

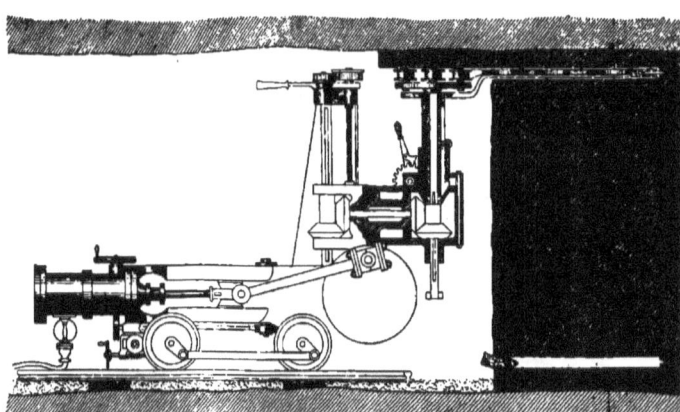

Hurd and Simpson's Coal Cutter.

Hurd and Simpson's machine is adapted for undercutting the coal by means of an eccentric wheel with cutters at its edge, driven by two specially constructed air-engines, of 6 in. cylinder and 12 in. stroke, the whole being carried upon a suitable bogie on wheels made to run upon a tramway line along the face of the coal. Figs. 286, 287, show a modification of the machine for cutting in any direction. The arm carrying the cutter is made to turn upon an axle attached to the framework of the machine; this centre can be raised up or down, and the power communicated to the cutters by means of the bevel wheels. This machine, which can cut in both the roof and thill, and can also nick at the ends, is peculiarly adapted for narrow work. The cutters can start from the face and work themselves into the nick, and are so arranged that each group of three cuts the top, centre, and bottom of the groove or nick. The cutter wheel is driven by a bevel wheel, the teeth of which, in fact, are cast with it, but are placed underneath to protect them from dirt. The cutter wheel is carried by a thin, but strong, steel arm, whose hidden end is provided with a wheel through which, by means of gear, motion is communicated to it by hand, and it is made to enter the coal or to withdraw from the groove, or take a direction in front of, or at either side of the machine. The cutters being eccentric to the wheel, they act with greater effect during one half of the revolution than at the other; and while the smaller radius of the eccentric is towards the coal, the machine is drawn forward by a self-acting hauling rope, or chain, which is wound round a drum and actuated by the machine. The leading end of the machine is kept in position on the rails when at work, by a roller fixed to a differential lever, with self-acting adjustment to adapt itself to the inequalities on the face of the coal, and this arrangement prevents the machine from getting off the rails when undercutting. These machines are sufficiently portable and compact to run on the ordinary rails into any part of the pit, and can be taken up or down the shaft in the ordinary cages used for winding the coal.

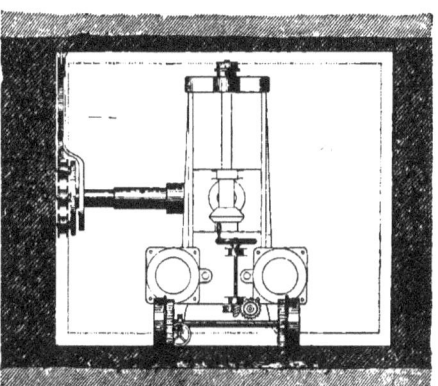

Fig. 287.

Hurd and Simpson's Coal Cutter.

Fig. 288 shows the method adopted for heating and expanding the air supplied to the machine. The air coming from the compressor is made to pass through a retort a containing a perforated crucible b, made of saponite or other suitable material, charged with ignited fuel (charcoal and scrap iron); a check-valve c is provided to prevent the return of the heated air which passes through d to the machine.

Fig. 289 shows an apparatus for upheaving the bottom coal after the top portion has been undercut and removed, and consists of a cast-steel wedge-shovel a, which is forced forward by the screw b, and the screw is worked round by the lever c and catch d acting in the toothed wheel e; the catch d can be reversed, so that the wedge-shovel can be withdrawn as well as pushed forward. The end of the screw b works in a socket which abuts against one of the props f, and the adjustable stay g serves to increase the resistance to the pushing of the screw.

With these machines above 150 yd. can be undercut in 10 hours, an amount of work which is at least 30 times as great as an experienced workman can do in the same time; but this is a minor advantage, compared with the great safety it affords the miners. By these machines the coal is undercut at night, and in most cases falls without a shot or a wedge being required, so that the miner begins to fill and send out his waggons at once without the necessity of holing the coal himself, which, when the weight is on the face, is a dangerous operation, and necessitates great vigilance on the part of the workmen. As a rule, the miners know to a few seconds, by the sound,

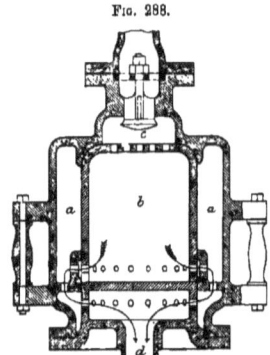

Fig. 288.

Heating and Expanding Air.

Fig. 289.

Upheaving Bottom Coal.

when the coal will part, but yet they are sometimes caught. Again, in nearly all cases the speed of the machine is so great that the coal can be undercut and got before the weight of the roof gets on to it. Thus, seams formerly known for their bad roofs are comparatively safe. Taking the average of coal seams, the usual mode of working involves a reduction of one-third of the quantity gained, into slack, which, of course, is a very serious loss; whereas by the machine the average loss in a 3 ft. seam does not amount to more than the $\frac{1}{18}$ part of the whole. In seams where there is a thin band of stone or dirt, the machine can be made to hole in such band, the débris of which can be cleared away before the main coal is brought down, which would enable the coal to be brought to bank cleaner and with much less trouble. In perfected machines, the speed of work with a pressure of air of 20 lb. to the sq. in. may be reckoned at 30 yd. per hour in medium hard coal, the groove made being 3 in. wide and a yard deep. Taking stoppages into account for removing and adjusting the machine, the average may be taken at one-third less. If it is convenient to have a working pressure of 50 lb. to the sq. in., the cutting rate can be increased to 1 yd. per minute. This extra rate, however, is not economical, on account of the additional wear upon the cutters, although the machines are of sufficient strength to resist the increased rate without breakage or heating. With respect to the replacement of the cutters, they will run 6-8 shifts of 9 hours each, without sharpening, in very hard stone coal, with an air pressure of 20 lb. to the sq. in., and the average rate of progress 7 yd. an hour. The usual price charged for undercutting a medium hard coal by contract is 1s. 6d. per yard, the contractor finding the necessary machines, and one workman to each, and the miners laying the roads and preparing the faces, which must not be less than 30 yd. in length.

If worked in pillar and stall, the rates are increased in proportion to those paid to the miner; this system is not only expensive but very dangerous, and involves costly ventilating arrangements; and it may be predicted that it will certainly go out of use by the adoption of machinery; first, because it is much easier to ventilate a straight face; and secondly, because the coal when undercut comes down before the roof has had time to settle, and this to such an extent that at many places where pillar-and-stall work is carried on, under the impression that the roof is so bad that faces of 30 yd. could not be maintained with safety, it has been found that when the seams had been struck to boundaries, and worked up half board and half endways on, and undercut by machinery, they have been worked with an open face of 1000 yd. in length, with more safety and better ventilation than before. In reference to the work done by the heading, tunnelling, or straight work machine, already described, the contracts are based on different terms, and with reference to the forward yardage only. For example, the contract price for making 3 cuts 1 yd. deep in medium hard coal, that is to say, two side cuts and one bottom cut (or, if preferable, a top cut) in a heading 5 ft. 6 in. high, by 9 ft. wide, is 10–15s. a yd. forward. When these cuts are made, if necessary, a 1½-in. drill is adjusted to the machine, and in two minutes a shot-hole 3 ft. deep is made. The machine is then run back a short distance, when a shot is placed and fired. During the whole time ventilation is kept up by a slight outlet of compressed air, which arrangement saves the expense of bratticing. The average time occupied in making the cuts, with 20 lb. pressure of air, in such a heading, is 63 minutes, which is about 5 times the speed of driving it by manual labour, which only gives at best about 3–4 yd. in 24 hours. The air-compressing machinery has been designed to obviate the very great expense of laying down the pipes to transmit the compressed air from bank to the machines in the face. The compressing machinery here shown can be placed conveniently near to the face of the coal, and can supply air to machines to undercut a large area without being moved.

Firth's machine (Fig. 290) is constructed for working a pick by means of a bell-crank lever, so as to give an action similar to that of the ordinary pick employed in hand work. The pick a is fixed in a socket in one of the arms of the bell-crank lever b, the other arm of which is worked direct by the piston-rod of the horizontal cylinder c. The slide-valve d, for the admission and discharge of the compressed air by which the machine is driven, is an ordinary slide, worked by a tappet roller e upon the piston rod; the machine is thus self-acting as regards the strokes of the pick, which is started to work as soon as the compressed air is turned on by the stop-cock f in the supply pipe g. The machine is mounted upon four wheels running upon the ordinary rails of the colliery, and is advanced the requisite distance after each blow of the pick by a hand-wheel h, connected by gearing with the two hind pair of carrying wheels. The two pairs of wheels are coupled together, in order to render the full adhesion available for the forward motion of the machine; and by this means it is found that sufficient adhesion is obtained without the necessity of laying down a special rack-rail for the feed motion. As the return of the pick after each blow is made by means of the self-acting tappet motion working the slide-valve, it is necessary that the tool should go to the full extent of its stroke at each blow before it can be withdrawn again. The amount of feed between the blows has therefore to be regulated by the attendant, according to the hardness of the seam of coal in which the machine is cutting, so that the pick shall complete an entire cut at each blow. In the event, however, of the pick being advanced too far at any blow, so as to put too much work upon it, and stop it before the stroke is completed, it is only necessary to draw the machine back again by means of the hand-wheel h, until the pick is released from the cut; the unfinished stroke is then completed, and the pick goes on working again the same as before the stoppage. In order to allow of altering the height at which the pick performs the holing in the coal, the socket i carrying the pick is made to slide vertically upon the shaft of the bell-crank lever b, the height of the socket being adjusted by the forked arm k, controlled by a screwed rod and handle.

FIG. 290.

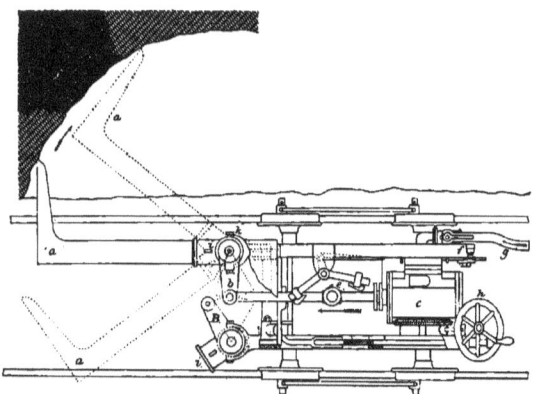

Firth's Coal-cutting Machine.

One of these pick machines worked the whole of the undercutting in the West Yorkshire Coal and Iron Company's colliery at Tingley, near Leeds, holing a seam of coal 3 ft. 8 in. thick; and the compressed air for driving it was supplied by an air-compressing engine at the surface, with steam cylinder of 20 in. diameter and 3 ft. stroke, working an air-cylinder of 18 in. diameter and the same stroke, and compressing the air to about 50 lb. per square in. pressure. The depth of the pit is 170 yd., and the air is conveyed down the shaft and along the mine in 2½ in. cast-iron pipes, with a 1¼ in. wrought-iron pipe laid up the bords to the working faces, and then a 1¼ in. flexible tube to the coal-cutting machine. Small air-vessels are placed at intervals of 500 yd. along the air main, for the purpose of maintaining the pressure of the air at the machine when working at a considerable distance in the mine; the machine is worked at a distance of as much as a mile from the shaft.

In a trial of this machine, it was found that a pick of 75 lb. weight cutting a groove to a depth of 24 in. in from the face, gave about 74 blows a minute. At the colliery, the coal was got by the long wall system of working, the machine working along a straight face of 50 yd. at one of the banks. The time occupied by the machine in undercutting a length of 56 ft. was 25 minutes, including all stoppages for clearing rubbish out of the hole, and for backing the machine when the pick occasionally made an incomplete stroke. The machine was then run back to the starting point, and set to work again with a longer pick of 90 lb. weight, completing the previous cut to the final depth of 3ft. 9 in. from the face. With this pick the

COAL-CUTTING MACHINERY. 145

blows were about 60 a minute, and the half-length of 28 ft. was undercut in 17 minutes, including all stoppages. The time occupied in running the machine back and changing the pick was 16 minutes. The machine in this case was working at a distance of about a mile from the bottom of the shaft. From this trial it appears, that in undercutting to a depth of 24 in. in a single course, the work done by the machine was at the rate of 30 sq. yd. an hour; and in undercutting in two courses, to the total depth of 3 ft. 9 in., the work was done at the mean rate of about 15 sq. yd. an hour, including the time required for running the machine back and changing the pick. The width or height of the groove cut out by the pick is 2 in. at the inner extremity, widening out slightly towards the face of the coal. It is necessary to stop the machine at intervals, in order to clear out the rubbish left in the hole; and the rails in front of the machine have also to be cleared of the material thrown out by each return stroke of the pick. Two men are required to attend to the machine, one working the hand-wheel for the advance of the machine, and the other clearing away the stuff. The picks are fixed with movable cutting points, which are detached for sharpening.

Fig. 291.

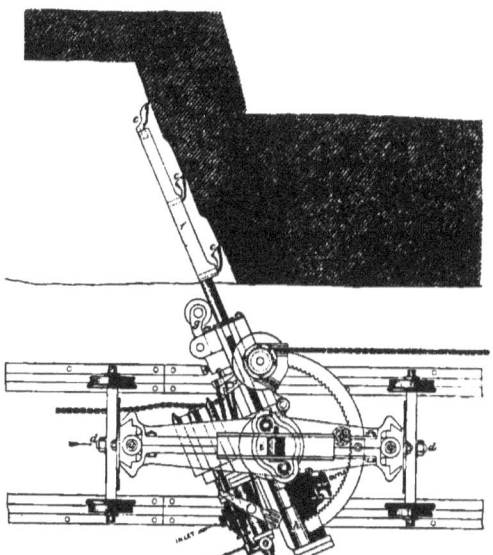

Carrett, Marshall, and Co.'s Coal Cutter.

Carrett, Marshall, & Company's machine (Fig. 291) has a different principle of action. It works after the manner of a hand-plane, cutting into the coal as a scoop cuts into cheese; it is said to be capable of working effectually in a space only 2 ft. high, and of accomplishing more in one minute

U

than 700 blows from a pick can in the same time. It is about 2 ft. high, weighs 1 ton, has 4 legs of adjustable length, and is provided with a holding piece adjusted so as to touch the roof of the drift and hold the machine firmly to its work. The motor is water, under a pressure of about 20 atmospheres (300 lb.), and supplied through a 2-inch pipe at the rate of 30 gal. per minute. This water pressure acts vertically on a 5-in. piston pressing against the roof, and horizontally on one about the same size, reciprocating 18 in. and 15-20 times in a minute. There is a pressure of 5000 lb. against the roof, and the same pressure acting horizontally, forcing three steel cutters, shaped like cheese scoops, into the coal. These cutting tools are 3 in. wide, and penetrate 4 ft., with a power equal to 3 horses, or 18 men; and this is effected by a consumption of 50 lb. of coal per hour to feed the boiler of the engine, which makes the water pressure, and pumps the same over and over again.

The machine in operation fixes itself dead fast upon the rails during the cutting stroke, and releases itself at the back or return stroke, and traverses forward the requisite amount for the next cut without any manual labour. Should the tools be prevented from making the full stroke at one cut, they will continue to make more strokes at the same place, until the maximum depth is attained, when the machine will move itself forward the required amount for the next cut. Thus, at one operation, a uniform straight depth is attained, parallel with the rails, inducing an even fracture when the coals are brought down, and thereby a straight line for the new coal face. There is no percussive action, either against the roof or into the coal, but simply a concentrated pressure, producing a steady reciprocating motion at 15 strokes per minute. There is, consequently, no dust or noise, and little wear and tear. For the same reason, when cutting pyrites, the tools throw out no sparks, and the workman can hear any movement in the coal or roof.

The required height from the line of rails in the "boling," "kirving," or "baring," varies in different mines; it follows that the hydraulic cutting cylinder, and its direct-action cutting tools, have sometimes to be arranged above the carriage, and sometimes beneath the main carriage, or close down upon the rails. The main carriage has 4 wheels far enough apart to allow the machine to be placed longitudinally when being transported from place to place. The screws a are for raising and lowering the carriage and its cylinder and cutting tools. The pinion b and the segmental rack c regulate the desired angle of the tools cutting into the coal face, and the two nuts d at each end of carriage regulate the angle required, when necessary that it shall not be in the same plane as the rails. e are the cutting tools, f the cutter bar, g a guide roller for the same; h is the main cylinder, with a self-acting hydraulic valve motion, which passes a portion of its water alternately above and below the piston of the holder-on, which thus rises and falls without percussion, and follows the uneven line of the roof of the mine, so that the required stability is given to the machine for the time being, an instant before the cutters enter the coal. The "holder-on piece" can be any length necessary to bridge over gaps in the roof; it is loose on the pin i and droops at its leading end to enable it to ride over the varying projections in roof. The traverse motion is actuated by a pin j, which connects the cutting bar with the piston rod, and at the termination of each end of its stroke actuates the lever k in both directions, which operates on the pawl l, which causes the chain pulley to revolve on the chain, made fast ahead by an anchor-prop between floor and roof. Although the length of stroke of each cutting tool is 18 in., the practical cutting length is 16 in., and, consequently, the three cutters jointly give a total effective depth of 4 ft. at each stroke of the machine, finishing the work as it goes along. The mechanism employed consists of a hydraulic reciprocating engine, adjustable to any height or angle, having a self-acting valve-motion. The cylinder is 4½ in. diameter, and lined with brass, and the piston is made tight with ordinary hydraulic leathers, which can easily be removed. Within the piston rod is attached the cutter bar of steel, carrying the tools or cutters. These can be varied in number to suit the depth to be holed at one operation. The cutting tools are of double shear steel, can be easily made, and are very strong, and can be removed and replaced in a few moments; they can be readily sharpened on an ordinary grindstone. The cutter bar is also movable, when transporting the machine from place to place, for which purpose the main cylinder is, for the time being, placed longitudinally with the rails.

The machine is about 3 H.P., and weighs 1 ton, and will work either right or left. It is self-acting in all movements, and will ascend steep gradients; being simple in all its parts, it is not liable to get out of order, is easily managed by an ordinary miner, and can be transported from place to place, on the ordinary rails, about the mine. It undercuts "holes," or "kirves," with a man and boy as attendants, and completes the work with once going over, at the rate of 15 yd. per hour, and at any angle and height from floor rails, being suitable for either "dip" or "rise" workings, and is capable of cutting the thinnest seams. The pressure of water which actuates this apparatus can be obtained either from the stand-pipes in the pits, or from pumps attached to any existing engine, or from an engine or pumps specially made for the purpose. The quantity necessary is only what is sufficient to fill the circuit of the pipes, using it over again when desirable. Each machine uses 30 gal. per minute, or about 300 lb. pressure, according to the hardness of the coal or mineral to be operated upon. In cutting the shale of the Cleveland ironstone band a somewhat greater pressure is found to be necessary. There is no limit to the pressure of water that may be used, nor the distance it may be forced without loss of power, beyond that due to its friction along the pipes. The same water pressure is also applicable to work pumps and rotary engines for hauling, &c., and other requirements in the mine, at a distance from the engine power. In cases where there is a fall of water, say of 100 lb. pressure, it can be "intensified" by a self-acting machine to 400 lb. pressure, to work the coal cutter, but sacrificing ¾ of its bulk, which is set free. In arranging the engine and pumps required to make a "continuous stream" of water pressure for working these machines, it is preferable to have two steam cylinders, so that there be no dead water. They are constructed to work one, two, or four machines. Pipes, if for one machine, are of 2 in. bore, wrought iron, a superior quality of gas-pipes strong enough to stand 500 lb. pressure. These pipes are screwed together in the ordinary manner, and adapt themselves readily to the irregularities of the floor of the mine. A flexible pipe 1½-in. bore, suitable for the same pressure, allows the machine to traverse.

The danger of firing charges of gunpowder in an atmosphere laden with explosive gas has led to several attempts to produce a machine which should be capable of breaking down or "falling" the coal after it has been undercut, more easily and quickly than the ordinary wedge driven in by hand. A hole is first bored in the upper part of the seam which has been "holed" or undercut, as for a shot; an expanding bar from the machine is then inserted into the hole, and the bar expanded vertically, so as to bring upon the coal a force tending to break it down. One of the most successful of these machines, which are usually actuated by hydraulic power, is that known as "Bidder's."

The principal machine consists of a small hydraulic press, weighing about 60 lb., and of 15 tons power. To this press is attached a pair of steel tension straps, bent in form of a tuning-fork, and which are connected with the press by a collar. At the end of these straps is first placed a clearance box, about 4 in. long, and upon each side of the straps expanding pieces (also made of steel), which exert a pressure at the sides of the hole, and are 15 in. long. The points of a pair of twin wedges, 15 in. by 3 in., constituting one wedge, are then inserted in the expanding piece, and the machine is fixed in the hole. The hydraulic press, having been charged with about 3 pints of water, which may be used over and over again without loss, is then worked by a man by means of a small handle, and the ram from the cylinder is forced out, thus driving up the pair of wedges between the expanding pieces, giving a lateral extension of about 3 in. This not being in all cases sufficient to bring down the coal, the press is withdrawn, and the relief valve opened, thereby allowing the water to return to the reservoir. A second wedge is then inserted between the two twin wedges by means of a small rod, $\frac{3}{8}$ in. diameter, and, the press being again connected, this wedge is driven home in the manner before described. By this means an additional expansion of 3 in. is obtained, making a total expansion of 6 in., which in most cases is found sufficient; but a third wedge can be applied, if necessary, and the expansion thus increased to any reasonable extent. In this manner as much as 10-12 cwt. of coal have been brought down in 10 minutes.

The drilling apparatus, the principal part of the machine, consists of a screw 4 ft. by $1\frac{1}{2}$ in. diameter, to the end of which is attached the drill. The fulcrum for taking the resistance of the screw is obtained by inserting a bar of iron in the coal at the side of the place selected for the hole which the machine has to drill. This small aperture is made by punching with the ordinary instrument a hole 10 in. deep and 1 in. diameter, and the time occupied in making this preparation is usually about 4 minutes. The small bar for taking the resistance of the screw is then inserted, and it may either be fixed at the side or in the face of the coal, as the case may require. The screw is then adjusted to this bar, and the drill is driven in the coal by a man turning the handle at the end of the screw. The time occupied in drilling this hole for the machine, 3 in. diameter and 3 ft. 6 in. deep, is 10-15 minutes, according to the hardness of the strata; and if it is necessary to drill the hole in such a position that the rotary motion of the handle by which the screw is propelled cannot be obtained, a ratchet may be used, so that under any circumstances no difficulty can be felt in procuring the required motion.

R. W. Clarke thus describes the working of a Gillott and Copley machine. The wheel is 4 ft. 3 in. diameter, and contains 10 single and 10 double cutters. The former have $\frac{1}{4}$ in. lead, and the double chisels widen the cut to 3 in. The cylinders of the machines are $8\frac{1}{4}$ in. and the ratio of gearing is $5\frac{1}{4}$ to 1. The machines cut in the dirt parting, and the coal, when holed, settles down without difficulty on the nogs. To make height for the corf, the bottom coal is taken up, and partly built into packs, and partly sent out for firing the boilers. The machine draws itself along the face by means of a small flexible wire rope, which is slowly wound on a drum in front of the machine, passing round a snatch-block 100 yd. ahead, the other end being again attached to the frame of the machine. The drum is revolved by means of a ratchet-wheel and pall worked from the crank shaft. The machine runs on three pairs of single headed rails 5 yd. long, laid on 2 joint sleepers and 6 ordinary sleepers, two to each pair of rails. The sleepers are made of wrought iron, and are provided with a groove, into which the lower part of the rail fits when it is wedged tight with oak plugs as with railway chairs. The rails are fixed to the joint sleepers by bolts, one to each rail. This road, when properly laid, is very rigid, and can easily be pulled up again.

When a machine is to be started in a benk, the deputy sees that the face is left perfectly straight, and that the wood is set in a straight line. Obviously this is very essential, as the machine must move in a straight line, on account of the rope. A road layer has plenty to do laying road for the machine without finishing the collier's work for him. A hole should be cut for the wheel of the machine at the starting end of the benk, and this should be made quite deep enough, otherwise the machine will not take a full cut at first, and that end of the face will get behind. The removal of the machine into the benk is by no means an easy task. It is only by reducing this removing or "flitting" to a minimum, and making it as expeditious and simple as possible that the success of cutting coal by machinery can be ensured. The road along which the machine has to be brought when loaded up on the lorry should be as good as possible, as it is very heavy to get on the road if it gets off, especially where the roof is low.

Before the machine is brought into the face, the rails should be brought up and laid across the top of the gate to the wheel hole where the machine is to start. A flat-sheet can then be laid on the top of the rails, the machine tilted off the lorry on to this, slewed round and run into the corner, and fixed on to the wheel, which together with the wheel bracket has previously been placed in the wheel hole. Economical flitting depends a great deal on the sharpness of the men and the practice they have had in handling the different parts of the machine.

The machines should cut about 40-70 yd. per shift of 8 hours each. There are two men to every machine—driver, or man in charge, and the road-layer. The road-layer's duty is to lay a firm and rigid road for the machine, with the three pairs of rails, one of which the driver pulls up as soon as the machine clears it, and pushes over to the road-layer without stopping the machine. He has also to shift props which would be in the way of the machine, and to set additional props or bars to make the roof safe for the driver, who, owing to the noise of the machine, could not hear the roof cracking or weighting, and would run considerable risk in an unsafe benk. The road-layer should watch the road carefully to see that the machine does not thrust it out and get off the road. When he has laid his road to the proper height he should wedge it tight against the face and against the packs. If the joint sleepers are not firmly bedded on the floor, the weight of the machine has a tendency to tilt the front of the rails up, and the machine climbs up into the coal. The only remedy for this is to run the machine back and start a fresh cut, having meanwhile adjusted the road.

The driver's duty is to regulate the speed of the machine, to keep the cutter wheel clear of holing dirt, and to pull the rails up and pass them to the road-layer as soon as the machine has cleared a pair. He should also sprag the coal with thick wedges about every 2 yd. as soon as it is cut, in such a manner as will be most advantageous for the collier. He should keep the machine well oiled, and see that the cutters are firmly fixed in the wheel, otherwise they are liable to drop out and be lost. The driver is the contractor, and pays the roadman $4s.$ $8d.$ to $5s.$ per day. At Lidgett Colliery the pay is $3\frac{3}{4}d.$ per yd. cut by the machines. This includes flitting and laying pipes in the gates, which the machinemen do when the machines are not cutting.

To work machines economically, the working places should have been laid out with a view to their use. No place should be too long, as if there is any delay in filling the coals out the machine on its next journey will be stopped by them. The deputies should be men of quick observation and ready resource. They should be able to estimate how much work there is to do in every shift, in every benk, and to arrange for the regular working of the machines. If there is a fast place anywhere they should concentrate a strong staff of men there, and if possible get the benk ready for the machine, as, if it is stopped, many colliers will necessarily lose their shift, and the output will be reduced.

Some may be a little disappointed at the rate of speed named, for 12 yd. an hour is spoken of as a speed which may be taken as a basis for calculation. This may well be where the holing is very soft, but it is the average speed which tells in the end. Almost always delays will occur. Roof will want propping, coals will not be out of the way, pipes may be leaky, or air may be short from various causes, or something may go wrong with the machinery. The following is the actual performance of 4 machines on January 7, 1888:—

No. 1.—1st Shift, 8 hours, 40 yd. cut, 5 yd. per hour.
       2nd ,, 8 ,, 45 ,, 5½ ,,
       3rd ,, 7 ,, 50 ,, 7 ,,
No. 2.—1st ,, 8 ,, 50 ,, 6¼ ,,
       2nd ,, 2 ,, 15 ,, 7½ ,,
       3rd ,, 5 ,, 40 ,, 8 ,,

No. 3.—1st Shift (flitting).
       2nd ,, 7 hours, 40 yd. cut, 5¾ yd. per hour.
       3rd ,,
No. 5.—1st ,, 8 ,, 50 ,, 6¼ ,,
       2nd ,, 8 ,, 45 ,, 5½ ,,
       3rd ,, 7 ,, 45 ,, 6½ ,,

Stanley's coal-heading machines are made to cut an annular groove around the face of the heading or tunnel, leaving a core which either falls, or is got off, as the work proceeds. While removing the core, the machine is not run back, but the arms are stopped whilst the coal is being passed in large lumps to the back, there being ample room in a 5 ft. heading alongside of the machine for men to travel, and for coals to be passed as large as can be well loaded into tubs. By reference to Fig. 292, the machine will be seen to consist of a central shaft carrying radial and horizontal arms and cutters. This works in a narrow frame of angle iron, with engines and gearing attached. The front gearing causes the shaft and arms to revolve, the back gearing advances the frame and parts attached thereto. This back gearing consists of a cogwheel with threaded brass bush, fitted into its boss, which works on the threaded part of the central shaft. It is held to the frame and driven by a sliding cogwheel on the crank shaft, the cutting gear being thrown out. When it is set in motion, the frame carrying the engines and gearing, and running on central tandem wheels on the floor of the heading is caused to advance on the shaft, or the shaft is caused to rotate in the frame, according to the requirements of the case. The direction of the cut is thus kept, whilst the machine is moved forward, and all is ready for cutting when the screw pins are fixed.

When the driving gear is thrown in, and the central shaft and arms are caused to revolve, the propelling gear is thrown out by means of a sliding bolt, the wheel with the threaded bush thus being prevented from turning, causing the central shaft and arms to advance in the frame and do the cutting. Fig. 292 shows the heading machine for coal seams; it cuts an annular groove around the face of the heading. Fig. 293 is a slow-speed machine for cutting in mixed seams or shale.

The machine is light and handy (the 5 ft. size weighing about 2 tons) and running on central tandem wheels, the direction of the cut is regulated and kept in either direction without the slightest difficulty.

The machine will work to the rise or the dip, as required, and has been tested at an angle of 1 in 2, and under these circumstances did the work equally well as on the level, though of course at such an angle to the dip there is rather more difficulty in passing the lumps of coal back.

The manner of working is as follows:—The machine being up at the face of the heading, and fixed ready for work, one man takes his place on the left side with a raker, and as the cuttings are brought back by the scrapers on the back of the horizontal arms, easily rakes them towards him as the arms are revolving, passing them to the man at the back as opportunity presents itself. A floor of fine slack forms itself at the side of the machine upon which the man either stands or kneels to do this work; this floor is left behind as the machine advances, being prevented from passing under the frame by a sliding plate, working to the floor at the side of the frame. When the arms have advanced sufficiently to allow of it, he moves from the left to the front of the frame, and changing his

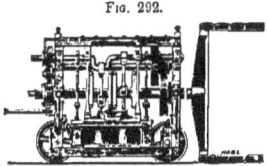

Fig. 292. Fig. 293.

Stanley's Coal-heading Machine.   Stanley's Coal-heading Machine.

raker for a shovel throws the cuttings along the left side to the back, where the man at the handle loads them into tubs, or throws them clear of the machine. Small lumps that fall from the face while the cutting is proceeding are dealt with in the same manner without stopping the arms, and for large lumps, or when the core breaks down (which it does in most seams within 1 ft. or so) the air is shut off whilst the core is disposed of. If men choose to work, this is soon done, and upon the energy displayed at this point depends almost entirely the capacity of the machine, as it does not take long to do its part of the work, the cutting being done at the rate of about 3 in. a minute, and moving and fixing the machine 5 to 10 minutes.

The arms having worked out their full length (3-4 ft.), the engines are stopped, screwpins are slacked, and by means of the propelling gear the frame is run forward and fixed for another cutting. The speed at which the entire work is done varies according to circumstances. In the slate coal at the Nuneaton Colliery, in the innumerable trials which have been made for every purpose, the yard has never failed to be done under the hour, and the distance of nearly 22 yd. has been cut in 24 hours; but in the regular way of work, in all coal headings, a speed of four times that of hand labour will be found to be the result.

On the occasion of the visit of the Chesterfield and Midland Institute of Engineers to the Nuneaton Colliery, August 30, 1888, when about 80 members were present, the time occupied in cutting the yard was $17\frac{3}{4}$ minutes, the time occupied in breaking the core and removing coal and slack, and preparing to start machine for another cut was 20 minutes = total $37\frac{3}{4}$ minutes, but this speed has often been exceeded.

The quality of the work is in every way superior to hand labour. No explosive being required, the heading remains firm and unshaken. With compressed air as motive power, the heading is well ventilated and kept cool by the exhaust, and when much gas is liberated, a tap is turned to ventilate the heading, whilst the core is being removed. Counter heads and thirls will thus in many cases not be required. The proportion of large coal made by this system is much greater than by any other, and will go far to pay for the expense of driving the heading.

The machine is already at work in many of the most important collieries in the United Kingdom and abroad.

The approximate cost of the machine is:—

Machine cutting 4 ft.	diameter		about	£ 220	Machine cutting 6 ft.	diameter, extra strong, about	£ 300
,, 4 ft. 6 in.	,,	..	,,	230	,, 6 ft. 6 in.	,, ..	,, 330
,, 5 ft.	,,	.	,,	240	,, 7 ft.	,, ..	,, 360
,, 5 ft. 6 in.	,,	..	,,	250	,, 7 ft. 6 in.	,, ..	,, 400
,, 6 ft.	,,	..	,,	270			

The 5 ft. size is most generally used. It is made by Stanley Brothers, Engineers, Nuneaton.

W. T. Goolden & Co., of Woodfield Works, Harrow Road, London, have for some time made a specialty of the application of electricity to mining and coal cutting. Fig. 294 shows an electric coal-cutting machine constructed in accordance with the patents of Bower, Blackburn, Mori, Atkinson, and Goolden, and embodying many novel features and advantages not possessed by any coal-cutting machine previously constructed. The difficulties with coal-cutting machinery that practical experience has brought forward have been, first, as to the cutters, secondly as to the motive power. Dealing with these in order, machines acting on the principle of a large horizontal circular saw, are only adapted for working in hard and strong coal, as, if the coal breaks, and the weight comes on the cutting wheel, the machine is stopped, and it is very difficult to start it again. This difficulty is

Fig. 294.

Fig. 295.

Goolden Electric Coal-cutting Machine.   Standard Goolden Dynamo.

overcome in the present machine by using a long bar revolving on its own axis, and furnished with cutters arranged in a spiral along its length. Various forms of these cutters have been used, but the form now adopted by the makers consists of straight cutters having a taper shank which fits into a taper hole in the bar where it holds tight, the head being formed to a shape adapted for dealing with the particular class of material to be cut. Coming next to the motive power: so long as only compressed air is available, the inefficiency (only 20 per cent. being transmitted if the distance exceeds about a mile), the expense of laying and maintaining air-tight long lines of pipes, and the cumbersome nature of the flexible line attachments, render the use of *power* on the coal-face almost impracticable; the electric transmission of power, properly carried out, obviates all these difficulties at once. The machine itself is made in various forms, one of which is shown in Fig. 294, where the cutting or holing is carried on at a height of about 11 in. from the floor level, where, in the particular instance for which this machine was constructed, a band of shale exists, and the cutting is carried on in the shale, thus economising coal. The coal-cutting machinery comprises a base plate mounted on trolley wheels, which can be drawn along the rails laid on the working face of the coal. The centre of this base plate carries a turn-table, in which, by means of the worm and rack, the whole machine can be turned through a right angle. The machine proper comprises an electric motor, Fig. 295, of 10 to 15 H.P., being of a special type built for mining purposes by the firm before mentioned, the revolving armature of which is geared through steel helical wheels to a lower shaft passing through part of the magnet of the motor, and projecting in front with a coupling to which the cutter bar is attached. The whole of the revolving armature and the brushes are enclosed in air-tight casings, thus protecting the armature from dirt or falling coal, and preventing the access of inflammable or explosive gases to the armature; arrangements being also made where necessary, to attach to the machine a vessel containing either pure air or carbonic acid gas under pressure, a stream of either of these gases being allowed to pass continuously into the interior of the casing, thus providing a neutral or non-inflammable atmosphere ensuring safety. The working of the machine is as follows :—The machine being turned round by the worm until the axis of the cutter is in a direction parallel to the face, the machine is started revolving by means of the switch handle provided for the purpose; while revolving, the machine is rotated in the turn-table through 90°, the cutter cutting in its own way through an arc of a circle. The cutter now being in the coal, the whole machine is drawn along the face by means of a winch, underholing the coal at a rate depending on the hardness of the coal, the power of the motor, &c.

The following results have been obtained, and will be a guide as to what the machine will do. In a trial at Briggs & Sons' Colliery at Normanton, the machine underholed to a depth of 5 ft., a depth probably exceeding that ever produced by any other coal-cutter, at an average rate of running of 10 yd. per hour. At the West Allerton Colliery belonging to T. and R. W. Bowers, near Leeds, where one of these machines has worked for a long time, an average rate of about 25-30 yd. per hour could be attained. At some trials at Lord St. Oswald's Colliery at Nostell, near Wakefield, one of these machines, underholing to a depth of 3 ft. 9 in., averaged about 30 yd. per hour while cutting. The dynamo in all these cases was at the surface of the ground, driven by a steam engine, and the distance between the dynamo and the coal-cutting machine was in some cases over 1½ miles.

The cables used are insulated with rubber, are perfectly flexible, and may be rapidly laid or taken up as may be required in the working.

MINING AND ORE-DRESSING MACHINERY.

As to the relative advantage in cheapness of getting coal by machine and by hand, the following Table by G. Blake Walker affords much information:—

COMPARISON OF THE RELATIVE COST OF COAL-GETTING BY HAND AND BY MACHINE.

	In a Seam 3 Ft. Thick.			In a Seam 2 Ft. 6 In. Thick.			In a Seam 2 Ft. Thick.			In a Seam 1 Ft. 6 In. Thick.		
	Nature of Holing.			Nature of Holing.			Nature of Holing.			Nature of Holing		
	Favourable.	Hard.	Very Hard.	Favourable.	Hard.	Very Hard.	Favourable.	Hard.	Very Hard.	Favourable.	Hard.	Very Hard.
	s. d.	s. d.	s. d.	s. d.	s. d.	s. d.	s. d.	s. d.	s. d.	s. d.	s. d.	s. d.
Price for getting coal by hand, per ton	1 6	1 7	1 8	1 10	2 0	2 2	2 6	2 9	3 0	3 3	3 7	4 0
Percentage of slack by hand	40	40	45	45	45	50	50	50	55	55	55	55
Percentage of slack by machine	20	20	20	20	20	20	25	25	25	30	30	30
Average selling price per ton:—	s. d.	s. d.	s. d.	s. d.	s. d.	s. d.	s. d.	s. d.	s. d.	s. d.	s. d.	s. d.
Round (including nuts), at 6s. per ton .. .. hand	4 7·2	4 7·2	4 5	4 5	4 5	4 3	4 3	4 3	4 1	4 1	4 1	4 1
Slack, at 2s. 6d. per ton .. machine	5 4	5 4	5 4	5 4	5 4	5 4	5 1⅛	5 1½	5 1½	4 11½	4 11½	4 11½
Price for filling out coal when coal is holed by machine	1 0	1 0	1 0	1 3	1 3	1 3	1 10	1 10	1 10	2 6	2 6	2 6
Yards holed by each machine per day of 16 hours	140	105	70	140	105	70	140	105	70	140	105	70
	tons	tons	tons	tons	tons	tons	tons	tons	tons	tons	tons	tons
Holed by 7 machines per day .. .. say	1,000	750	500	830	620	415	666	500	333	500	375	250
Holed by 7 machines in 250 days .. say	250,000	190,000	125,000	207,500	155,000	103,750	166,500	125,000	83,250	125,000	93,750	62,500
Cost per ton (based on previous estimate) for holing by machine ..	d. 3·1	d. 4·2	d. 6·3	d. 3·8	d. 5·1	d. 7·6	d. 4·8	d. 6·3	d. 9·5	d. 6·3	d. 8·4	s. d. 1 0·6
Total cost of coal getting by machine	s. d. 1 3·1	s. d. 1 4·2	s. d. 1 6·3	s. d. 1 6·8	s. d. 1 8·1	s. d. 1 10·6	s. d. 2 2·8	s. d. 2 4·3	s. d. 2 7·5	s. d. 3 0·3	s. d. 3 2·4	s. d. 3 6·6
Saving, as compared with hand labour ..	0 2·9	0 2·8	0 1·7	0 3·2	0 3·9	0 3·4	0 3·2	0 4·7	0 4·5	0 2·7	0 4·6	0 5·4
Saving in yield of coal (value)	0 8·8	0 8·8	0 11·0	0 11·0	0 11·0	1 1·0	0 10·5	0 10·5	0 5	0 10·5	0 10·5	0 10·5
Total saving	0 11·7	0 11·6	1 0·7	1 2·2	1 2·9	1 4·4	1 1·7	1 3·2	1 5	1 1·2	3 1	1 3·9

(155)

CHAPTER VIII.

PUMPING MACHINERY.

WHEN the quantity of water to be dealt with in a mine is not very great, it may often be economically raised by means of buckets or tubs drawn by the winding engine. This kind of draining machinery has the merit of being very simple in construction and action, and consequently not liable to get out of order. But when favourable conditions for its application do not exist, recourse must be had to pumps.

The tub commonly used for drawing water is of iron, barrel-shaped. The capacity varies, but frequently it is about 100 gal., when it is intended to be drawn by the engine. The tub is suspended by a bow turning on two pins, placed a little below the centre of gravity, on the outside of the tub, as shown in Fig. 296. The object of this arrangement is to facilitate the discharge of the contents on arriving at surface. Beside the larger bow which turns upon the pin forming the points of suspension, there is a smaller one fixed to the tub, and passing freely beneath the former. On one side of the tub is a spring catch, which, by laying hold of the larger bow, prevents the tub from tilting in the shaft. When the tub is raised full of water to the top of the shaft, the waiter-on seizes the smaller bow, and, releasing the spring catch, pulls the tub over, discharging the water into a shoot, by which it is conveyed away. The position of the centre of gravity above the axis upon which the tub turns, renders the operation of tipping an easy one. When the contained water has been discharged, the man pushes the tub back into

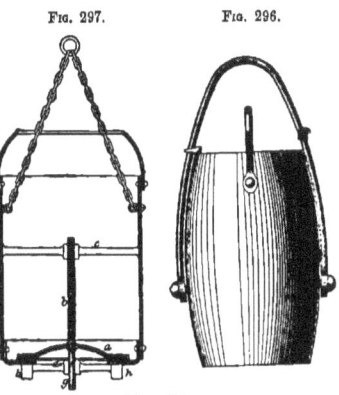

Fig. 297.　　Fig. 296.

Water Skips.

the vertical position, where it is seized by the spring catch; in this state it is ready to be again lowered into the shaft. An objection to this kind of tub is that it does not fill well, the water having to flow into it over the top.

This objection may be removed by constructing a valve at the bottom, through which the water can enter, as in Fig. 297. The bottom of the tub, which is about 5 ft. high and 3 ft. in diameter, is provided with a circular aperture 20 in. in diameter; this aperture is covered by an iron disc or valve a, mounted on a central spindle b, which moves between guides $c\,d$. The under side of the disc is faced with a ring of vulcanised rubber, to enable it to close water-tight. When this tub is

x 2

lowered into the water, the pressure of the latter forces up the valve a, and the tub fills. When the tub is full, the valve drops upon its seating, and retains the water. The mode of employing this tub, and conveying away the water, is very simple. In the end wall of the rectangle, shown in Fig. 298, is fixed a drain-pipe 2 ft. in diameter, leading to a channel provided to convey away

FIG. 298.

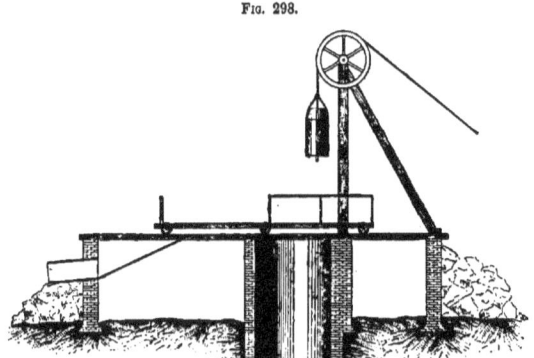

Water-barrel.

the water. A wooden shoot e leads from the flooring to this drain-pipe; and on the platform f rests a trough about 8 ft. wide and 2 ft. deep, open at the end next the shoot. When the tub is raised to a height slightly above the mouth of the shaft, as in the case of the loaded kibble, the waiters-on push the platform forward over the mouth of the shaft, until the closed end of the trough is under the tub. The latter is then lowered gently on to the trough, when the projection g of the spindle, coming in contact with the planking, the valve a is forced up till the tub comes to rest upon the stops h. The water issues through the valve aperture, and flows down the trough into the shoot, and is conveyed away through the drain-pipe in the end wall. Thus it will be seen that the apparatus is self-acting, and it has been found in practice to fulfil the purpose intended in a very satisfactory manner. A tub of this construction, and of the dimensions shown, weighs about 900 lb., and contains about 220 gal. of water.

Cylindrical water tubs have been used to a considerable extent in the French collieries. They are usually made with a capacity of about 500 gal., and weigh about 1500 lb. The water enters by a large valve at the bottom, and is discharged through a side orifice. It is necessary in order to avoid loss of time, to provide guides in the shaft, so that the tub may be drawn up and lowered rapidly. It has also been found highly advantageous to commence discharging the water as soon as the tub reaches a sufficient height above the pit. Instead of bringing the tub of water to a complete rest upon catches at the top of the shaft, it is kept slowly ascending, and strikes a movable knocker or framework, which throws open the discharge valve and lets the water escape. The motion may be stopped as soon as the valve is open; and, as soon as the tub is emptied, it may be lowered without any loss of time.

The pneumatic water-barrel, as applied by W. Galloway at Llanbradach, Figs. 299 to 303, consists of a cylindrical vessel of sheet iron, 4 ft. 2 in. diameter, and 8 ft. high, closed at the top,

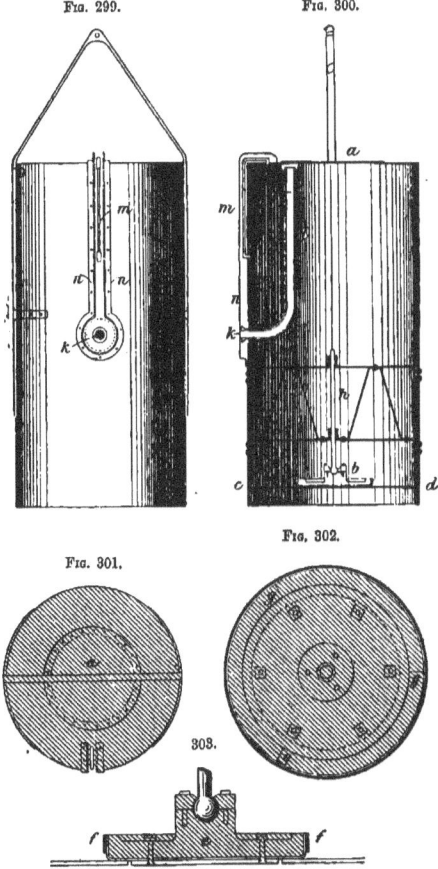

Pneumatic Water-barrel used by Galloway.

in which there is a door a, bolted to the cover, and serving as a means of access to the interior when removed. The bottom $c\,d$ is 5 in. above the bottom of the cylinder. It consists of a steel plate ¾ in. thick, with a central opening 18 in. diameter. The valve-seat is turned in a lathe, so as to secure perfect trueness. The valve b consists of a block of cast iron e (Fig. 303), having its lower face and vertical sides turned quite true. Over this turned face a sheet of leather is tightly "cupped," and held in place by a tightly fitting wrought-iron hoop f secured by three tapping bolts g. A circular plate of iron 16 in. diameter is bolted to the bottom of the valve by means of six bolts with countersunk heads. A spindle h (Fig. 300), working through two guides, and having a turned ball at its lower end, is held loosely in a socket in the valve as shown in Figs. 300 and 303. In this manner the vertical movement of the valve is secured, while the ball and socket joint enable it to accommodate itself to the seat in any position into which it may be turned. At k (Figs. 299 and 300), one half of an instantaneous coupling, identical with those used by the Vacuum Brake Company, and supplied by the same makers, constitutes the outside termination of the pipe l, which, passing through the side of the cylinder, rises to within an inch of the top in its interior. A glass gauge at m shows the height of the water in the interior when it rises to that elevation. The instantaneous coupling and the water-gauge are both protected by a strong angle-iron rib n (Figs. 299 and 300), which projects from the side of the cylinder, and serves to guard them from the blow given by any large body such as a bucket. Care must be taken not to allow the point of the leg of a drilling machine which is being let down by the winding rope to enter between the angle irons.

The vacuum is created by means of an ordinary air-pump condenser constantly working at the surface. The steam cylinder of this air-pump is 10 in. diameter by 20 in. stroke, and steam is cut off at one-fifth of the stroke by means of adjustable hand-gear. The vacuum pump is 14 in. diameter by 20 in. stroke, coupled tandem-fashion to the piston-rod of the steam cylinder. This engine produces a vacuum equivalent to a column of mercury 20–22 in. high, both in a receiver near the top of the shaft, consisting of an old egg-end boiler 24 ft. 8 in. long by 5 ft. diameter, and in a system of pipes of 3 in. diameter, communicating with it, one of which descends to the bottom of the shaft. It is there connected with a flexible hose 30 ft. long by 2½ in. diameter, provided with a stop-cock, and terminated in one-half of an instantaneous coupling corresponding to the other half, which is affixed to the pipe l of the water cylinder, in the manner already described.

When it is desired to fill the pneumatic water-barrel, it is lowered to the bottom of the shaft, and rests with its hollow end under water. One man then attaches the instantaneous coupling of the flexible hose at k, opens the stop-cock, and observes with a light at m when the water rises in the gauge-glass. As soon as he notices the water rising to the desired height, he shuts the stop-cock, detaches the instantaneous coupling, and apprises the man in charge of the signal that all is in readiness. The latter then signals to the winding-engine man, who thereupon raises the water-barrel with its contents to the surface. On its arrival there the banksman shuts the doors, draws a water-trolley under it, and signals to the engine man to lower it. When this is done, the barrel descends; its valve is arrested on the top of a conical block of wood; and, as it descends farther, the water pours out into the water-trolley, and flows thence into a wooden trough, which conveys it into a drain provided for the purpose. In this manner the water-barrel is filled in 30 seconds and emptied in 30 seconds, while the remaining manœuvres occupy about 1½–2 minutes. It has been possible, with this arrangement, to sink in the Pennant sandstone, with 5000 gal. an hour in the bottom of the shaft, at the rate of 5–5½ yd. a week; the highest rate of progress in the same ground with only 500 gal. an hour having been 6½ yd. a week. At one time, when the quantity of water in the bottom rose to 7000-7500 gal. an hour, the rate of progress attained was rather under 4 yd. a week, the rock being at the same time exceedingly hard and compact.

Although these quantities of water are, comparatively speaking, insignificant when pumps can be applied to elevate them, they are sufficient to render sinking impossible by the system of baling. The establishment of pumps in this shaft was a question that could not very well be solved at the commencement of the operations, as it was impossible to determine what quantities of water were likely to be met with; and even as the sinking progressed from day to day, the uncertainty continued the same, so long as the bottom of the Pennant sandstone had not been reached, and the Shale series entered upon. The aggregate quantity of water met with down to a depth of 135 yd. from the surface amounted to 9000 gal. an hour—namely, 5500 gal. above 95 yd., and 3500 gal. between 95 and 135 yd. Of these two quantities 5000 gal. an hour was walled out by means of brick and cement walling, leaving only 500 gal. an hour in the bottom at 95 yd. Below 95 yd. the water gradually increased up to 3500 gal. an hour at 135 yd. At the last-named point a piece of an old boiler, 5 ft. in diameter and 10 ft. 6 in. long, with one end open and the other closed, was fixed on a beam, with its centre directly below the rope of the small winding engine, and that engine was employed in raising the whole of the water running into the shaft, except some 500 gal. an hour which escaped from the sides between the bottom and the collecting curb.

As the shaft was deepened still farther, the quantity of water issuing from the rock below the collecting curb gradually increased, until at length it amounted to 5000 gal. an hour between 135 and 190 yd. But in the meantime the springs between 135 and 95 yd. had decreased to about 1800 gal. an hour. The cistern was then lengthened to 14 ft. 6 in., and fixed at a depth of 190 yd., a collecting curb having been put in, as in the former case, and the 1800 gal. an hour collected at the higher curb was also run down into it in 3-in. pipes. Another stronger engine was then put to work to raise the water from this cistern, and easily raises 5000 gal. an hour by means of the following apparatus, of which the principal details are represented in Figs. 304 and 305: a is the winding rope of the auxiliary winding engine, and e (Figs. 304 and 305) the guide ropes; the two latter ropes are

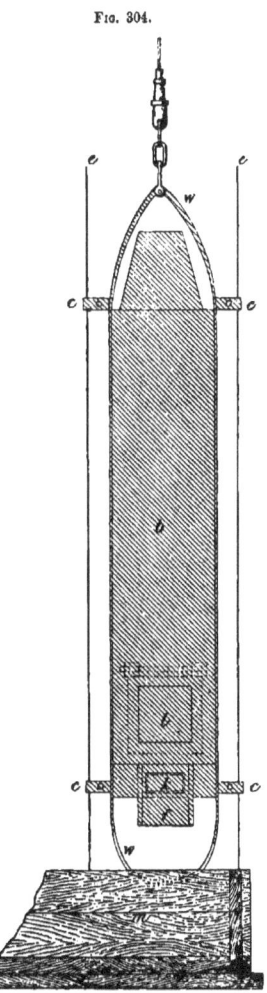

Fig. 304.

Pneumatic Water-barrel.

fixed in the bottom of the cistern at 190 yd., pass over two pulleys above the winding pulley, and are wound upon two small hand-crabs standing on the surface at some distance from the shaft, on which there is sufficient rope to reach to a depth of 500 yd. if required. The tank b is 2 ft. square inside; has parallel sides for 8 ft. of its length; terminates upwards in a pyramidical frustum-shaped top, which is bolted to its square part and can be removed when desired; and has a bottom sloping from front and back towards the centre, as shown. Four projecting studs c, one on each side, at top and bottom, clasp the guide ropes loosely. The tank is guided by these ropes in ascending and descending the shaft, but when it reaches the top, the studs on each side pass between fixed guides d, each consisting of two bars of angle iron riveted to a long plate of sheet iron and made fast to the woodwork.

A valve k in the bottom serves both as a means of filling, and as an outlet for the water which escapes into a sloping adjutage f, and precipitates itself thence into a wooden trough m, whence it runs into the drain mentioned in connection with the pneumatic water-barrel. The valve k is raised by means of the lever o, which comes in contact with a movable wooden bar p, working between two iron guiding bars, not shown. A weight u, suspended from the bar p by means of a chain s, can be regulated so as to open the valve. The bar p turns upon an iron bolt at q, which serves as a hinge to the system. The upper ends of the fixed iron guides are attached to two wooden beams t, one vertically above each side of the opening through which the tank reaches the surface, and their lower ends are secured at v in a similar manner to the sides of the opening just named. The suspending bow w of the tank passes down each side and under the bottom, the latter part being semicircular, and serving as a protection to the bottom of the tank when it is lowered to the bottom of the collecting cistern, where it comes to rest on a sheet of rubber $1\frac{1}{4}$ in. thick. The filling and emptying of the tank are purely automatic, and only 1 man is required to attend the engine. This tank can be easily filled and emptied 24 times an hour from a depth of 190 yd.; and as its capacity is about 212 gal., it brings about 5000 gal. of water to the surface during every hour it is at work.

PUMPS.—The system of pumps most frequently applied to mining purposes is that known as the Cornish. It consists in having a lifting pump at the bottom to raise the water from the sump, and a series of force pumps, set one above another, to drive it up by stages to surface, the whole of the pumps being worked simultaneously from the main rod. As the depth of the shaft in a mine is great, and as the quantity to be raised is frequently very large, it is obvious that this rod must be very strong, and therefore must possess large dimensions. Usually it is composed of balks of Memel pine, perfectly sound and straight, and without knots or faults of any kind, such as are used for the masts of ships, and of as great a length as can be obtained. The lengths are put together by scarfed joints, and secured by stout wrought-iron plates, bolted through the timber. To this main rod, the pistons of the pumps at the several levels are firmly attached by means of a set-off and strong iron straps. These piston rods work through the guides to keep them in a straight line; and for the same purpose, similar guides are placed at intervals down the shaft against the main rod. The rod where it passes through the guides is cased with hard wood, and kept well greased to lessen the friction. The rods are of enormous weight. In deep mines the main rod alone frequently weighs upwards of 70 tons. The mode of working the pumps is to make the motor raise the rods, and then to leave the weight of the latter to force up the water. As, however, the weight of the rods is usually greatly in excess of that required to raise the water, this excess is taken off by means of a loaded lever, called a balance-lever, or more commonly, a balance-bob. It consists of

Fig. 305.

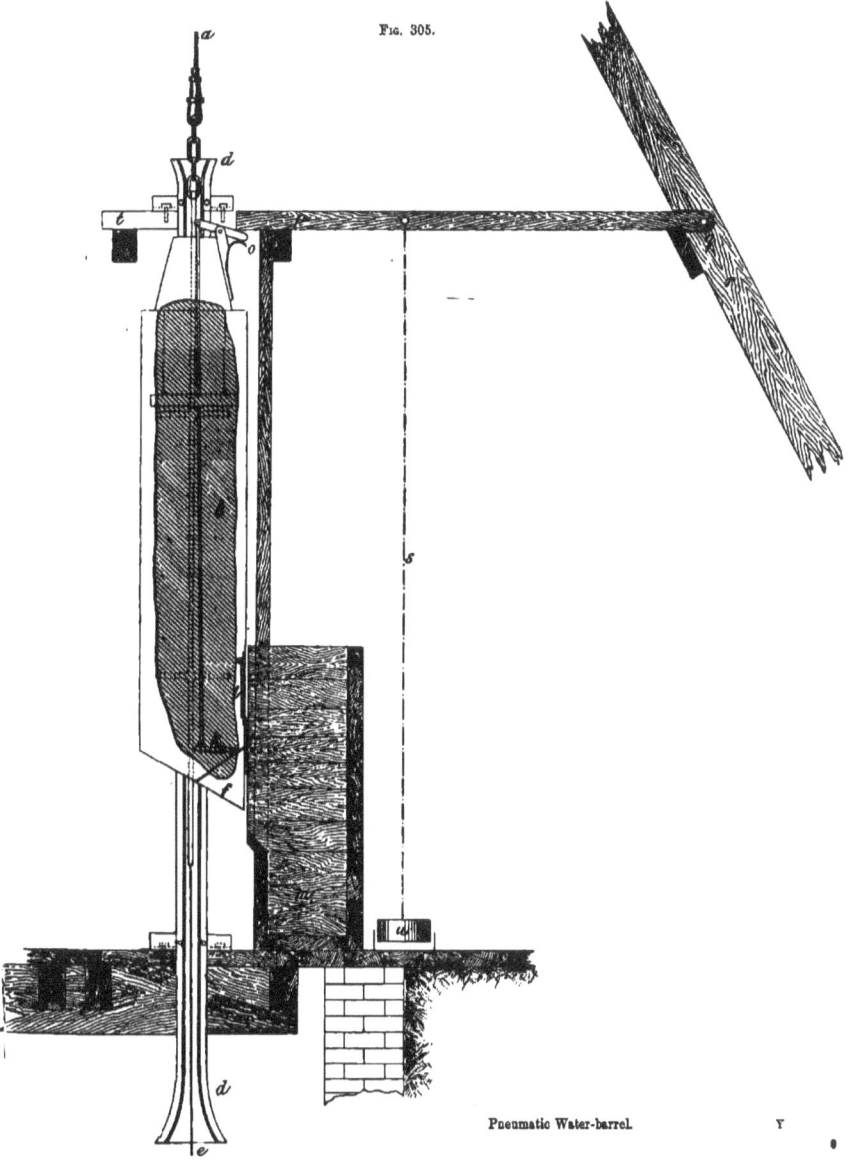

Pneumatic Water-barrel.

a stout balk of timber, often 20-33 ft. in long, turning about an axis, and loaded at the end by a box filled with stones or other heavy materials. The two ends are supported by iron ties passing over an upright support upon the axis. One of these bobs is placed at surface, and others may be set at intervals down the shaft, the unloaded end being fixed to the main rod. When the shaft is inclined, the main rod is made to rest upon friction rollers; in other respects the arrangements are unchanged by this circumstance. A vertical rod is made to communicate motion to an inclined rod, or *vice versâ*, by means of a bent lever, called a V or angle-bob. As the motor may be situate at a considerable distance from the shaft, especially when the pumps in two shafts are worked by the same motor, the rods are carried along the surface

Fig. 306.

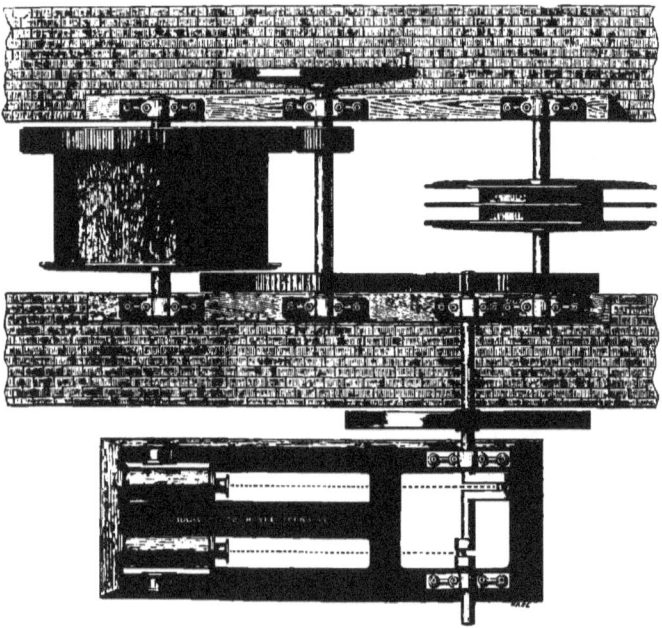

Steam Capstan for raising and lowering Pump-rods.

of the ground, and connected with the main rod in the shaft by one of these V-bobs. The horizontal, or, as they are usually termed, flat rods, are attached by means of iron straps to the arm of the lever, and the main rod in the shaft is attached in the same way to the other arm. When in this position, the V-bob is frequently double, the arms serving as counterweights.

PUMPING MACHINERY.

Flat rods are carried upon friction rollers where the surface of the ground is level, and upon vibrating rods where the surface is depressed.

The steam capstan shown in Fig. 306, which is simply a winch with a large barrel worked either by a pair of engines and gearing or by a single pinion and large wheel from the more powerful winding engine, is now generally used for raising and lowering the pump rods in the shaft, instead of a capstan worked by manual labour. As the large pumps, &c., take a considerable time to fix, it is very expensive to have a number of men employed night and day working the hand labour capstan; and consequently the work is done more quickly and at less cost by the steam winch shown.

Fig. 307 shows the manner in which pumps of the Cornish type are used, both in vertical shafts and in slant workings. At the left side of the shaft are seen two pumps at different levels, both driven by the same T-bob, the spear rods of the lower one being attached to those of the upper one. The pump in the slant working is driven in the same manner, with the intervention of a bell-crank arrangement. The general arrangement of this drawing was taken from pumps made by Hayward Tyler & Co., 84, Whitecross Street, London, for the Tapada mine. The pump body with its valve boxes and air-vessel, is shown on larger scale in Fig. 308. The rising main pipes in the shaft are omitted, to avoid complicating the illustration. The T-bobs which operate the pumps are worked by a horizontal connecting rod carried either direct from the engine or from a pumping frame driven by a strap as in Fig. 307. These connecting rods are sometimes laid for a long distance supported on rollers.

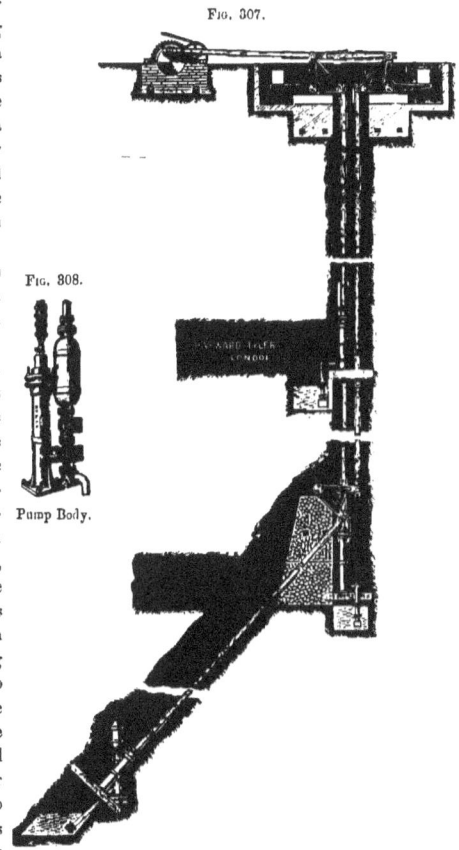

Fig. 307.

Fig. 308.

Pump Body.

Cornish Pump.

Fig. 309 illustrates an arrangement adapted for pumping where the water-level is liable to great variations. The pump barrels containing the suction valves, and the buckets, are placed at the lowest level, and above them are fixed cast-iron pipes of somewhat larger diameter than the barrels, carried to a height above the highest probable water-level; the pump covers and stuffing boxes are on the top of these stand pipes, so that the pump buckets can be drawn out for clearing, even if the water has risen many feet above the barrels themselves. This arrangement is occasionally of great value.

For general pumping purposes in mining work the direct-acting steam pump has now come into almost universal use; the great convenience and compactness of this form of pump makes it very applicable, and many thousands are at work in mines. The pump is often placed a long distance from the boiler, the steam or compressed air being conveyed with comparatively little loss. Fig. 310 shows pump and boiler on same bed-plate suitable for water supply where the water has to be raised from a valley to the mines above.

Another arrangement of great use where water is abundant in the valley and a supply is required above is to utilise the fall of the water in driving a water wheel or turbine which works a pump as

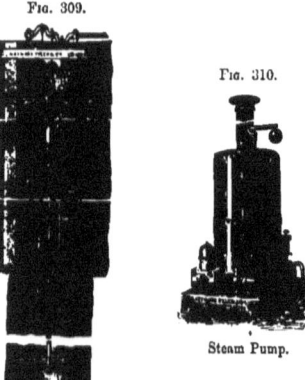

Fig. 309.
Fig. 310.
Steam Pump.
Pump for varying Level.

Fig. 311.

Water-wheel Pump.

shown in Fig. 311. These water wheels are sometimes of great size, but considerable use can be made even of a small one.

PUMPING MACHINERY.

The hydraulic system of pumping has proved successful in the Comstock and several other mines. The general plan of this system will be rendered plain by reference to the accompanying drawings. Figs. 312, 313, 314.

The mine to which this system is applied is 2400 ft. deep; the water to be pumped is to be raised from the 2400 level to a height of 800 ft. and discharged into the Sutro Tunnel, through which it is

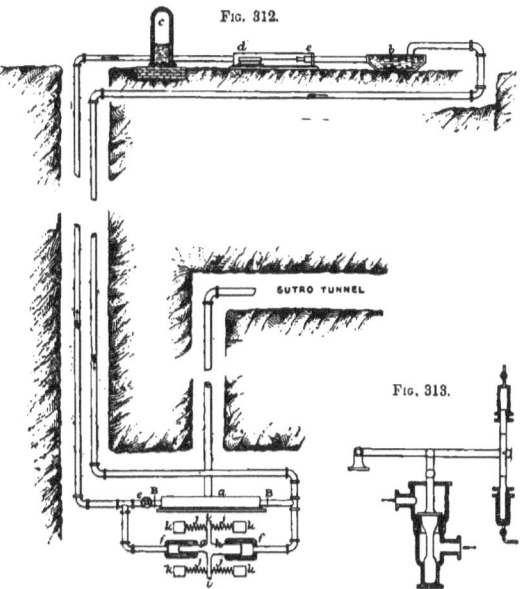

Fig. 312.

Fig. 313.

Hydraulic system of Draining Mines.

run off. Upon the 2400 level is erected a pair of hydraulic pumping engines a, which receive their pressure water through supply pipes from the surface. By these two pumps the drain water is raised through discharge column to the Sutro Tunnel, and the water used in doing the work of pumping is sent back through the return pipe to a reservoir b on the surface. Upon the surface there is a cast-iron accumulator c, which is 60 ft. high and 25 in. diameter. In this accumulator there are but 20 ft. of water, the remaining space being occupied by air.

This air is kept constantly at a certain pressure by means of plunger pumps d, which in turn are operated by a compound steam engine e. These pumps, which supply the accumulator with water and keep it at a pressure great enough to run the two underground pumps, a, take their suction from the reservoir b, into which the return water is discharged.

Now, since this pipe which supplies the pumps a is of such great length, and moreover, since the whole mass within it must be put in motion and brought to rest again during each stroke of the pumps a, it is evident that some mechanical contrivance must be introduced for the purpose of lessening the jar in the pipes caused by the sudden stoppage of such an immense mass of water. Let e be the valve which shuts off supply water from a. Close to the underground pumps there are

FIG. 314.

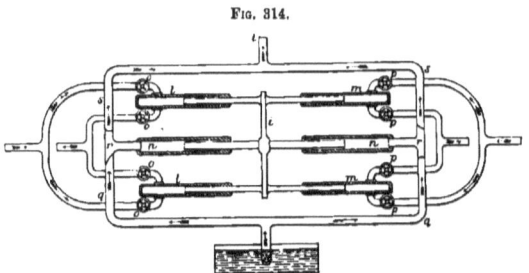

Hydraulic System of Draining Mines.

firmly fixed two cast-iron cylinders f, in which the plungers $g\ h$, carrying the crosshead i, are free to move. On either side of the arm i, are strong springs j, held in position by the abutments k. Now when water in supply and return pipes is at rest, the arm i is exactly in the middle between the cylinders f (since the plungers $g\ h$ are so proportioned that the total pressure on g is precisely equal to total pressure on plunger h). But after the underground pump a has completed its stroke, and the total mass of water in the supply pipe has to be brought from a velocity $= v$ to a velocity equal to zero, the inertia of this mass (due to velocity v) is supposed to be gradually reduced to zero by the plunger h being forced (on account of extra pressure on g) over to the right side. The distance it travels to the right depends, of course, upon the extra pressure produced by checking the velocity of supply water, and also upon the strength of the springs j, which begin to be compressed as soon as $g\ h$ begin to move. Now, after the crosshead has been forced over by this extra pressure to its maximum extent (the water in the three columns also coming to rest at this moment) the compressed springs j will immediately react to place the crosshead and plungers in their central position, ready to again take up the inertia of the mass in supply column on the return stroke of the pumps.

Thus, the object of these springs j is not only to assist the water in the return column in taking up the inertia of the water in the supply column, but chiefly to replace the plungers $g\ h$ in their central position after the masses in the three columns have come to rest.

The general arrangement of the system having been described, an account will now be given of the manner in which the work is performed by the pumps a. In Fig. 313 is shown in detail that portion of the underground pumps lying between the lines B and B on Fig. 312, leaving out, of course, that portion already explained for taking up the inertia of the supply water.

There are two of these underground pumps, situated on the 2400 level of the mine, but since they are similar to and work independently of each other, it is necessary to describe but one of them.

They both take their supply water through the valve e. Acting independently, as they do, they will not necessarily commence and finish their strokes at the same time. But, in order to discuss the system under its extreme conditions, consider the two pumps as working together, thereby making the velocity of the supply water and the inertia of the moving masses a maximum.

Therefore, in each pump, let l and m be the four pressure cylinders, and n the pump cylinders for raising the drain water to the Sutro Tunnel. Let o be the valves leading to the pressure cylinders from the supply and return pipes, and let the plungers be connected with a crosshead (as shown in drawing), so that when any movement of the plunger m takes place, the plunger n will be carried along also.

Now to follow the operation of these pumps, first assuming that the pressure of the water in the supply pipe is great enough to produce motion of the plungers l.

First, suppose the valves o to be opened (valves p remaining closed). The two plungers l will be forced over to the right, carrying with them the larger plunger n. While n travels to the right, the space left by it is immediately filled with drain water coming through the suction pipe r. Now, suppose valves o be closed and the stroke finished, the valves p are then opened, and the pressure of the supply water against the plungers m forces them to the left, and the water remaining in cylinders is forced through the valves into the return pipe, and thence into reservoir b upon the surface. At the same time, the water which was drawn into cylinder n on preceding stroke, is now forced (on account of the check valve r closing) through s into discharge column t, and thence 800 ft. into the Sutro Tunnel. Again, while this stroke to the left is being made, the space left by plunger is being filled with drain water through suction pipe r. At the completion of this stroke, valves p are closed and valves o are opened, the crossheads and plungers again move to the right, and the water remaining in the cylinders is again forced through return column into reservoir. At the same time, the water in cylinder n is forced upward (on account of check valve q being closed) through s into t, and thence to Sutro Tunnel. Thus the operation continues, the valves being worked automatically by means of tappets carried by the crosshead. The number of strokes made by these pumps evidently depends upon the pressure of water in the supply pipe, and also upon the rapidity with which the valves are opened and closed.

It will be seen that the sole object of these two underground pumps is to raise the drain water from the 2400 level of the mine to the Sutro Tunnel, a distance of 800 ft. The object of the accumulator c upon the surface, is to accumulate the water within it, under a pressure great enough to work the pumps a, and the object of the compound steam engine and pumps $d\,e$ on the surface, is to supply the accumulator with exactly the same amount of water taken from it to run the underground pumps. Furthermore it will be noticed that the same amount of water which is used to run these pumps a is returned to the reservoir on the surface, from whence it is forced into the accumulator by pumps d to be used over and over again.

Here it might be well to mention the fact, that the surface pumps d were constructed with a capacity great enough for two more pumping engines similar to those just described (a), so that when a depth of 1000 ft. more was reached, this extra set of pumps would be put in at the 3400 level, and the process go on as before.

The above description explains the action of the hydraulic pumps on the Comstock. Since it was written, the pumps have been running several years, and the springs j have been replaced by air chambers, so that the concussion of water lost in discharge, pressure, and return columns,

168   MINING AND ORE DRESSING MACHINERY.

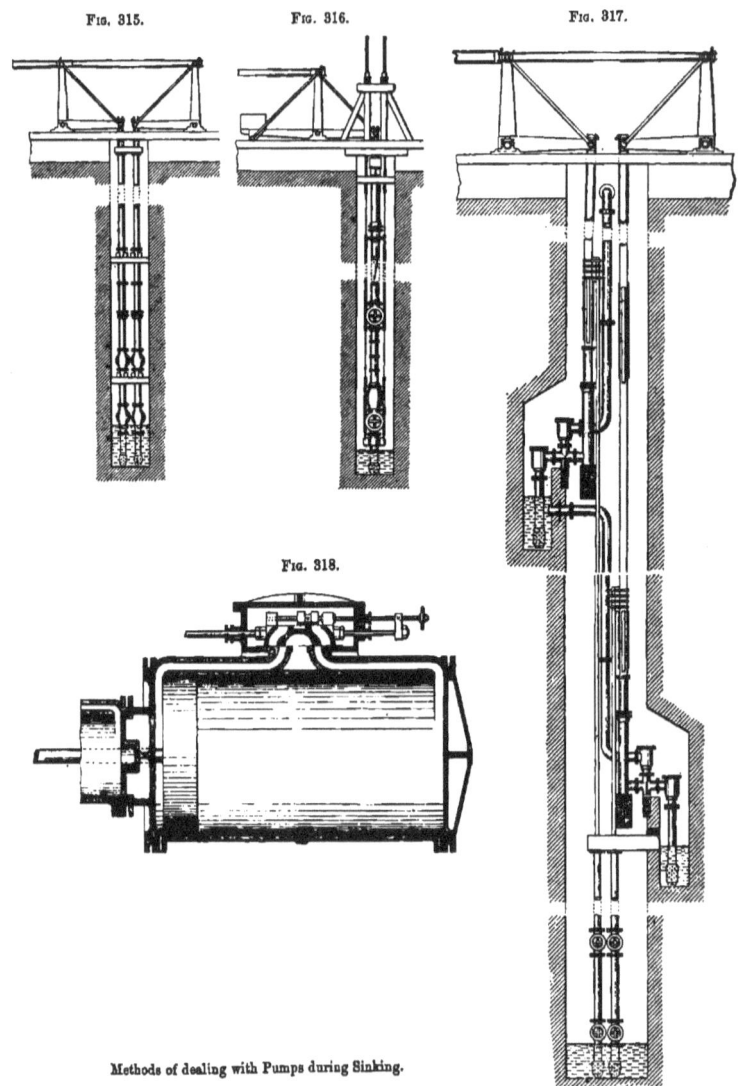

FIG. 315.   FIG. 316.   FIG. 317.

FIG. 318.

Methods of dealing with Pumps during Sinking.

PUMPING MACHINERY.

has been overcome by means of air chambers erected on the same station with the pumps. It was found impossible to properly separate the springs to work as efficiently as the air chambers have done.

There are several methods of dealing with pumps during sinking operations.

(1) To fix the rising main or pump trees as the work proceeds, and add pipes above the working barrel as the shaft is deepened. This plan involves a telescopic or flexible suction-pipe, or a telescopic pipe above the working barrel. With twin pumps, as in Fig. 315, this is a good plan, because one pump can be kept going whilst pipes are added, or whilst a bucket is changed in the other.

(2) To sling the pumps by ground spears or wire ropes, as in Fig. 316, and add pipes at the top of the lift.

(3) To use a pilot bucket-pump, and fix plungers every 40 or 45 fathoms, as is done in the Cornish mines, Fig. 317.

(4) To use a pilot bucket-pump worked by an independent engine, and fix plunger pumps every 40-45 fathoms until the bottom is reached; and then to fix plungers permanently at the bottom, for the whole height of lift.

FIG. 320. FIG. 319. FIG. 321.

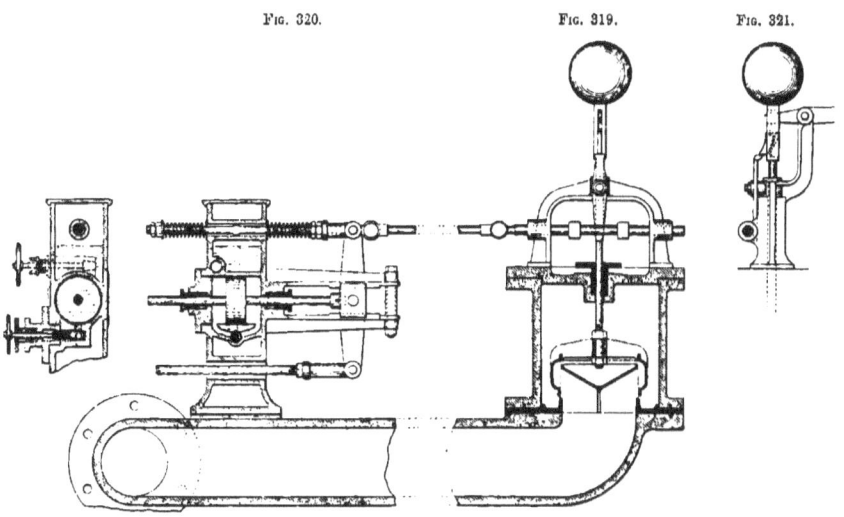

Davey's Adjustment for Pumping Engines.

All these plans involve frequent alteration in the balance of the engine, causing considerable trouble and loss of time. Davey has accordingly introduced into his own pumping engines a means of giving a different supply of steam to the two ends of the low-pressure cylinder, thus enabling the engine to be worked out of balance during sinking. The device consists simply of a shutter at the back of the low-pressure slide-valve, Fig. 318, with means of adjustment outside the valve-chest. By shifting the shutter over the forward or backward port of the slide-valve, either one or the other may be throttled to suit the want of balance in the load on the engine.

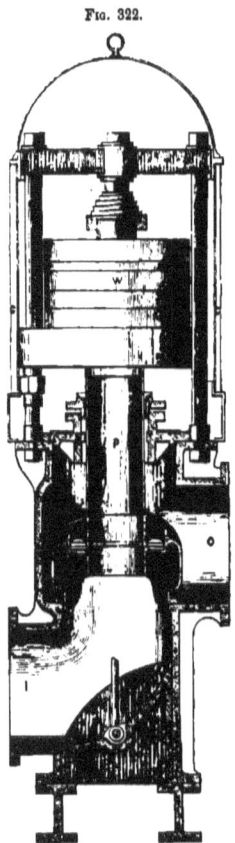

Fig. 322.

Water Supply Balance-valve.

Pumping engines employed in sinking are subject not only to loss of load from breaking of spears, &c., but also to a "riding column" when a foot-valve breaks or fails to shut, whereby the load on the engine is reversed so as to act in conjunction with the steam-pressure instead of opposing it. No amount of governing on the admission side of the piston is equal to such a contingency. The governing action of the main gear is sufficient, where there is not a great mass in motion to accumulate momentum; but it is often not equal to bringing to rest, within the limits of the stroke, a great mass moving under the exceptional condition above described.

Davey has devised a means by which, whenever the engine suddenly increases its speed during the stroke, communication between the high and low pressure cylinders is suddenly closed, thereby not only stopping the admission to the low-pressure cylinder, but also cushioning the steam in front of the high-pressure piston.

This retarding apparatus is illustrated in Figs. 319 to 321. It consists of a lever, one end of which is attached to a moving part of the engine, while the opposite end is made to actuate the trip of a double-beat valve, closing the communication between the high and low pressure cylinders. To the centre of the lever is attached a subsidiary piston, working in a cylinder filled with water. The end of the lever which actuates the trip is held stationary by means of two springs, so arranged as to oppose each other. The engine in working gives the engine end of the lever a reciprocating motion, and thereby causes it to reciprocate the subsidiary piston in the water cylinder: but the motion of this piston is resisted by means of a conical plug throttling the passage that communicates between the two ends of the water cylinder. When the engine is working at its normal speed, the conical plug is screwed up until the resistance thereby opposed to the subsidiary piston causes the trip end of the lever to partake of a slight reciprocation, and to be just on the point of tripping the valve. When working under this condition, should the engine happen from any cause to make a quicker movement, the resistance in the water cylinder would be increased, the valve would be instantly tripped, and thereby the admission would be cut off from the low-pressure cylinder, and the steam cushioned in front of the high-pressure piston.

Fig. 322 shows Husband's patent water safety balance valve, for preventing accident to engines and machinery—employed for pumping or forcing water—when the load is suddenly withdrawn by the bursting of pipes or breaking of their joints, &c., &c. It performs the office of a stand pipe, and dispenses with this costly structure. In the event of the main bursting, the load or pressure is withdrawn from the engine; this valve then substitutes another load which controls the speed of the engine, preventing racing and consequent damage to the machinery.

It consists of a plunger passing through a stuffing box and attached to a valve in the main. This plunger is loaded with a weight per unit of section equal to a unit of the head against which the engine pumps; thus if the main bursts the engine has still the same resistance to pump against. Its action will be understood by reference to Fig. 322: P the plunger, C the casing, G the guides, V V the valve, S the stuffing box, I the inlet, O the outlet, and W the weight, which is made in parts, that it may be increased or decreased at pleasure. It is made by Harvey & Co., Limited, Hayle, Cornwall, and 186, Gresham House, London.

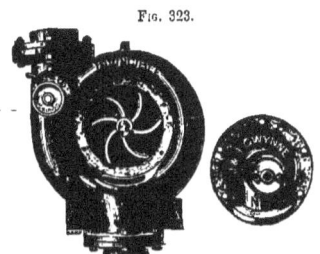

Fig. 323.

The "Invincible" patent centrifugal pumps of John and Henry Gwynne, of Hammersmith Iron Works, and Cannon Street, London, have been extensively employed in connection with gold, silver, and tin mining. The pumps are of excellent design, and, as far as workmanship is concerned, nothing better can be desired. It is claimed that they are the most efficient machines of their class. The pump volute is made in two castings, as in Fig. 323, which enables the interior to be got at, and, if necessary, the disc and spindle

Gwynne's Centrifugal Pump.

Fig. 324.

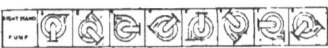

Swivelled Pump.

Fig. 326.

Fig. 325.

Gwynne's Centrifugal Pumps.

removed and replaced without disturbing a single joint in the suction or discharge pipes. In addition, handholes, with covers provided with bayonet joints, are arranged on suction branches, each side of pump, to enable any foreign matter that may have accidentally got inside to be removed quickly and without trouble. A very special feature in this machine is that the pump is arranged to swivel on the bed-plate by simply slackening a few nuts, shifting the pump to the desired angle, and retightening the nuts. For example, the branches of the pump can be swivelled to any of the positions described by the diagram, Fig. 324, or to any intermediate position.

The direct-acting pumping engine, i.e. the pump whose spindle is worked direct from the crankshaft of engine, is a very superior arrangement, and Messrs. John and Henry Gwynne have obtained far better results with it on certain lifts than have been obtained by any other description of water-raising appliance known. Hundreds of "Invincible" pumps are working in the tin mines of the Straits Settlements, where they are very suitable, on account of the moderate lift, the mines in that country being principally alluvial. In India, Africa, and elsewhere, these pumps have been extensively employed for washing purposes: they are much cheaper in first cost, the repairs are almost nil, and they last a good deal longer than reciprocating or any other description of pump, especially when raising sandy or gritty water, which soon cuts the valves of the reciprocating type to pieces.

Fig. 325 represents the arrangement of pump usually adopted for driving by strap. It has double standards, and it can be fitted either with a foot-valve or a patent steam ejector, as shown, which will charge the pump and pipes automatically. When an ejector is fitted, no foot-valve is

Gwynne's Centrifugal Pumps.

necessary. Fig. 326 represents the direct-acting arrangement having a single cylinder engine of the high pressure non-condensing type, and Fig. 327 is taken from a pair of such pumping engines; this arrangement has been supplied for mining in South Africa. Fig. 328 shows a direct-acting pump with compound condensing engine. This system is adopted in cases where economy of fuel is of importance.

W. T. Goolden & Co., of Woodfield Works, Harrow Road, London, have devoted considerable attention to electric pumping machinery, and Fig. 329 shows one form of electric dip-pump constructed by them. The pump consists of a three-throw ram pump having gun-metal plungers, the diameter of which is 3 in.; the stroke is 6 in. and the pump runs about 60 revolutions per minute. On the same base plate and geared with it by means of gear wheels is a 2 H.P. electric motor of the

"Goolden" Colliery type, possessing the features before described. This pump will deliver up to about 2000 gal. per hour against a head of 150-200 ft., and can be run night and day continuously, needing very little supervision or attention. The particular pump illustrated was made to replace

Fig. 329.

Goolden Electric Dip-pump.

a steam pump which gave great trouble, owing to the difficulty arising from the heat caused by the waste steam in the workings, so that it could only be run some 2 or 3 hours daily.

These pumps are portable, and the ease with which they can be moved about and the cables run after them, forms a striking contrast to the usual trouble with steam or hydraulic pumps and their connections. In larger sizes these pumps are made horizontal.

MINING AND ORE-DRESSING MACHINERY.

CHAPTER IX.

VENTILATING MACHINERY.

THE surface furnaces for ventilating coal mines have now almost disappeared, their place being taken by underground furnaces capable of circulating 200,000–400,000 cub. ft. of air a minute, and many kinds of mechanical ventilators have been introduced. The liability of the latter to serious injury being inflicted upon them by explosions if they are in immediate proximity to the shafts, renders it important to place them, and the engines which work them, at some distance from the upcast, and to have a second engine in reserve in case of accident. Considerable benefits have been secured by the provision of special auxiliary ventilating appliances, actuated by jets of compressed air or steam, or by water under pressure; though not adapted to extensive areas, they are very effective for local purposes, or in the event of accidents. By one large class of apparatus, ventilation is accomplished by drawing the foul air through the mine-ways to the surface; the opposite method of effecting ventilation, by forcing fresh air down the downcast shaft of a mine, has long been advocated by some, and although it has met with only limited application here and upon the Continent, it is applied very effectively in some extensive collieries in the United States.

EFFICIENCIES OF MECHANICAL VENTILATORS.

Name of Ventilator.	Dimensions of Ventilator.				Dimensions of Engines.						General Results.		
	Diameter.	Width, &c.	Theoretical Displacement per Minute.	Diameter of Inlet.	Weight.	No. of Cyls.	Dia. of Cyls.	Length of Stroke.	Direct-acting or Geared.	Volume of Air per Minute.	Mean Water Gauge at Drift Door.	Percentage of Useful Effect.
	ft. in.	ft. in.	cub. ft.	ft. in.	tons		in.	ft. in.		cub. ft.	in.	
Guibal	Fan .. 50 0	12 0	..	15 0	50	1	42	3 6	Direct	108,422	3·30	40·00
,,	,, .. 46 0	14 10	..	13 0	..	1	36	3 6	,,	246,509	1·85	52·95
,,	,, .. 40 0	12 0	..	14 0	24	1	36	3 0	,,	170,581	1·46	47·95
Waddle	,, .. 45 0	Inlet .. 6 6 Periphery 1 5	..	15 0	..	1	32	4 0	,,	163,312	3·08	52·79
Schiele	,, .. 12 0	2 1		..	..	1	25	2 0	2·57 to 1	157,176	1·91	46·12
,,	,, .. 9 6	Inlet .. 3 2 Periphery 1 8		8 0	..	1	20	1 8	2¼ to 1	106,570	2·03	49·27
Lemielle	Chamber 22 6 Drum .. 15 0	Height .. 32 0	9·9 rev. 106,900	..	..	1	55	6 0	Direct	47,307	1·37	23·40
Struvé	2 pistons 18 3	Stroke .. 7 0	6½ ,, 47,827	..	..	1	24	4 4½	4 to 1	43,793	5·11	57·80
Nixon	2 pistons, 30 ft. long, 50 ft. high	Stroke .. 7 3	7·19 ,, 120,790	..	..	1	36	6 0	Direct	72,595	2·74	45·91
Root	2 drums 25 0		13 0 16·71 ,, 96,918	..	..	2	28	4 0	,,	89,772	3·29	47·84
Cooke	2 drums 15 0 Casing .. 22 0		11 6 17·92 ,, 80,640	..	..	1	25	3 6	,,	54,190	1·12	37·33
Goffint	2 pistons 13 2	Stroke .. 10 7¾	9¼ ,, 53,020	..	..	2	15¾	10 7¾	,,	36,286	0·71	25·79

VENTILATING MACHINERY. 175

The table on p. 174 shows the results of a number of tests of the efficiency of mechanical ventilators, made by the North of England Institute of Mining Engineers in 1880-81.

The following table by Cochrane, shows the duty of ventilating furnaces at sundry collieries.

DUTY OF FURNACES AT THREE COLLIERIES IN THE NORTHUMBERLAND AND DURHAM COALFIELD.

Name of Colliery.	Downcast Shaft.		Upcast Shaft.		Area of Furnace Grate.	Temperature of Air.						Volume of Air Circulated per Minute.	Water Gauge in the Mine.	Coal Consumed in 12 Hours.	Consumption of Coal per Hour per H.P. in the Air.
	Diam.	Depth.	Diam.	Depth.		Top of Downcast.	Bottom of Downcast.	Return Air near Furnace.	Bottom of Upcast.	Half-way in Upcast.	Top of Upcast.				
	ft.	yd.	ft.	yd.	sq. ft.	Fahr.	Fahr.	Fahr.	Fahr.	Fahr.	Fahr.	cub. ft.	in.	cwt.	lb.
Rugeley	12	160	12	160	64	..	..	61	141	117	110	103,325	0·62	40	37·0
North Seaton	15¼	250	9	266	72	68	70·	65	225	206	186	99,750	1·10	91	49·2
Ryhope	15	508	10¼	460	160	62	..	76	170	..	134	126,336	1·00	120	56·3

As regards the cost of supplying air by different systems, Forster has published a table showing the capital outlay required; the size of air-pipes; the cost per cubic metre supplied, and that of a given quantity—25, 50, and 100 cubic metres per minute—for a whole year. The following are the figures obtained for 100 cubic metres per minute. The use of steam as a motive power, at a cost of 12*l*. 10*s*. per gross H.P. per annum, is assumed :—

Compressed air at 3 atmospheres excess pressure ; velocity of current 10 metres per second, escaping at full pressure from the air-pipe, 230 millimetres diameter, cost 9235*l*. per annum. The same using a Körting blower, and pipes 89 millimetres in diameter, 1510*l*. The same using a compound engine, Roots blower, and pipes of 50 millimetres, 365*l*.

Branch current taken from main fan pressure, 0·001 atmosphere, or 10 millimetres water gauge; 2 metres per second velocity of current, pipes 1030 millimetres diameter, 210*l*. This, on account of the low velocity of current and the large size of pipes, would be practically useless at the distance of 1000 metres from the fan.

Current produced by a special high speed fan at surface, $\frac{1}{30}$ atmosphere, or 250 millimetres pressure, velocity per second 6 metres, pipe 585 millimetres, cost 380*l*.

Current of 10 metres per second at $\frac{1}{10}$ atmosphere, 1033 millimetres pressure, pipes 454 millimetres, produced by a series of two or more fans, or a cylinder blowing engine. The cost in the first case is 725*l*., and in the second 870*l*. per annum ; or where the pressure is increased to ¼ atmosphere, 1510*l*., and 1935*l*. These are, however, only approximations, from very restricted data.

Electric transmission of power to mine ventilators working at low pressure ; no pipes required ; old pit ropes used as conductor.

Bells.—Where small volumes of air have to be dealt with, the simple contrivance known as the " box " or " bell " is often sufficient, and the readiness with which it may be applied leads to its frequent adoption in headings. In such situations it gives a good ventilative current, with an expenditure of a small amount of force. In Germany and Cornwall, the box is commonly employed for ventilating the ends of levels; being usually of small size, and requiring but little power, it is generally attached to the end of the pumping engine. One of these boxes is shown in Fig. 330. It consists of a wooden box of square section, open below and closed at the top, and connected by a wrought-iron rod to a cross-arm projecting at right angles from the main pump-rod, by which it is moved up and down in another box or outer case of a similar shape, partly filled with water. A pipe, in communication with the level to be ventilated, passes up through the bottom of the outer box to within a short distance of the top; it is covered with an ordinary clack-valve opening outwards; two similar valves are fixed to the top cover of the inner box. As the rod ascends, a partial vacuum is established within the box, as communication with the outer air is prevented by the water joint, and the top valves are kept closed by the pressure of the external air; the valve on the pipe inside therefore opens, and the air from the workings flows in until the change of stroke, when, by the descent of the box, the air is compressed and opens the two top valves, through which it passes freely into the atmosphere. The same principle has been applied in Belgium to the construction of large ventilating machines for collieries. At Maryhaye, near Liége, a pair of wrought-iron bells or cylinders are employed, each 144 in. diameter and about 9 ft. stroke; they are suspended by chains over guide-rollers, and are driven by a direct-acting horizontal steam-engine. There are sixteen suction and an equal number of exhaust-valves, which, owing to the small difference of pressure produced, require to be counterbalanced with weights, in order that they may open and shut freely at the change of the stroke. The amount of air drawn by this machine is about 11,500 cub. ft. per minute.

Fig. 330.

Ventilating Box.

Fig. 331 shows a so-called Cornish "duck-engine" for the same purpose. A set-off from the feed *a*, air-pump rod *b*, or a direct connection to the beam itself, as is most convenient, actuates a piston, working in a simple double-acting air cylinder *c*, the air from either end being delivered into a light air-box, which is connected by means of air-pipes *d*, to the desired points of operation underground, and as it is only necessary to deal with air slightly in excess of atmospheric pressure, the air-pipes, cylinder, valves, &c., can be of the lightest and simplest construction. A square box, constructed of wood, grooved and tongued together, provided with a wooden piston, leather geared up and down, so as to be double acting, could be easily made in any mine, and would cost but little. Such a box, say 4 ft. square, working at 3 ft. stroke, would supply when making, say, 6 double-acting strokes a minute, more than 800,000 cub. ft. per 24 hours. This quantity, though small as compared with the actual requirements of perfect ventilation, would greatly improve existing conditions in many mines.

A simple and efficient contrivance used on the Pacific coast and elsewhere is the water blast (Fig. 332), consisting of a wooden box pipe standing in a shaft some 200 ft. high, and connected at bottom with an air-pipe *b a*. The top of the box pipe *b* is open, and a shower of water being caused to fall into the box, carries with it a volume of air. The bottom of the pipe *b* dips into a box *c*, 2–3 ft. long and 15 in. deep, in which the water is allowed to stand above the bottom of the pipe, and from which the excess escapes through a sliding valve or gate d. Connected with the water pipe just above the box c is the air-pipe a, leading to the point to which the fresh air is to be forced. Sometimes, for greater depths, these box pipes are used in conjunction with a blower. They were at first made of galvanised iron, but this soon corroded in the mine waters. Californian red wood

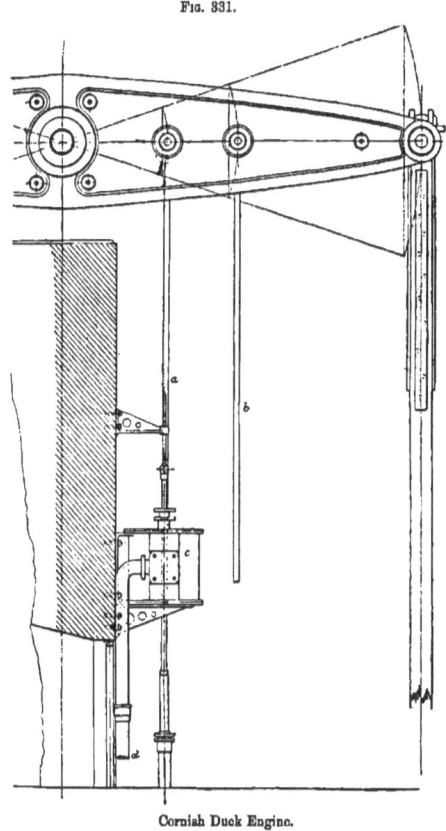

Fig. 331.

Cornish Duck Engine.

Fig. 332.

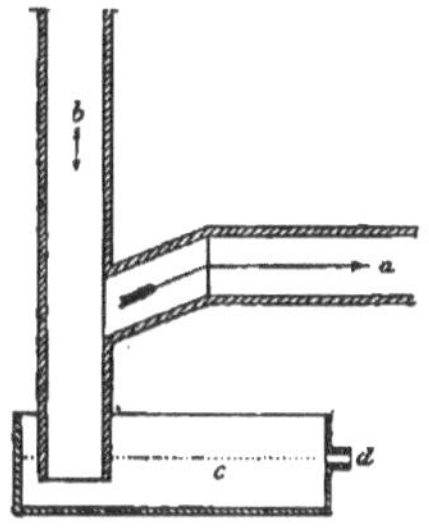

Water Blast.

superseded it, the dimensions being about 12 in. square in horizontal section, and the wood 1½ in. thick; joints tongued and grooved, well packed, painted, and bound with sheet-iron bands.

Hand Fans.—Not unfrequently, in metalliferous mines, when a small volume of air has to be put in motion, a fan driven by hand is used. This fan (Fig. 333) is of the same kind of construction as that employed for blowing iron-founders' cupolas. It has 5 radial arms with flat rectangular blades, which revolve about a horizontal axis within a cylindrical case or drum, having a circular aperture about 20 in. diameter in the centre of each of the sides; the outside diameter of the fan is about 4 ft. The air taken in at the centre is discharged through a rectangular tube 15 in. broad and 10 in. high at the bottom of the drum, and is conveyed through pipes of a similar section, made of wooden planks or sheet zinc, into the forward end of the level to be ventilated. The fan is driven by a wheel 64 in. diameter, connected by a strap with a spindle of 4 in., giving 16 revolutions of the blades for one of the driving wheel. The strap is kept at a proper tension by a friction roller attached to a board, which slides on a pair of horizontal cross timbers, an arrangement which allows the machine to be put out of work without stopping the driving wheel or disconnecting the strap in cases where it is required to be used only intermittently. By putting the central apertures in communication with the air-tubes, the fan can be used for establishing a circulation by exhausting the vitiated air. By surrounding the fan with spiral guide-plates or diffusers, the air, instead of being discharged at a useless velocity against the walls of the drum, may be led off to the discharge pipe

FIG. 333.

FIG. 334.

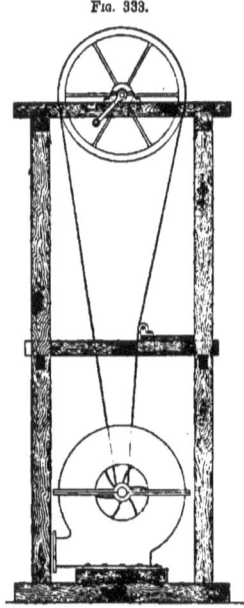

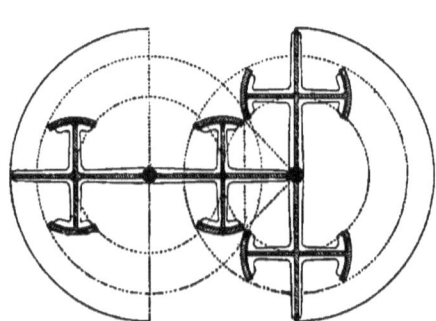

Fabry's Wheel.

Hand Fan.

more conveniently and more economically. Small ventilators on this principle are commonly used in the Saxon mines; they have 6 arms, with blades 8½ in. square and 30 in. diameter. These fans can be worked by one man at a maximum speed of 400–450 revolutions a minute, with a pipe 6 in. square; 60 cub. ft. of air can be drawn in that time from a distance not exceeding ¼ mile. The quantity of fresh air required by a man at work in the end of a level is estimated at 6 cub. ft. a minute.

Fabry's Wheel.—Fabry's pneumatic wheel (Fig. 334) is employed to a considerable extent in the Belgian collieries, and in some other localities. It consists of two fans, each of which has three broad rectangular blades, arranged radially and at equal distances apart, around a horizontal axis, connected together by spur gearing wheels, so as to revolve at equal velocities in opposite directions. The fans are hung in a·chamber of masonry, which covers about two-thirds of their circumference, the remaining parts moving in the open air. The chamber is rectangular in plan, with vertical side walls; the end walls are segments of horizontal cylinders, whose centre lines coincide with the axis of the fans. These cylindrical walls correspond to the drum in the ordinary fan-blower; they are coated with cement dressed up to a smooth face, so as to give the smallest possible interval between the ends of the blades, without actually touching. The foul air from the mine is brought in through an arched passage in one of the side walls. The space intermediate between the two axes is kept isolated from the external air by a peculiar contrivance: each of the blades has a shorter blade projecting from either face at right angles, which carries a plate curved to an epicycloidal form; these cross arms are fixed at about two-thirds of the distance from the centre of the blades towards the circumference. As the two fans turn towards each other on the inner side (between the axes), a pair of the curved heads, one on each wheel, are continually in contact, preventing any communication between the interior of the chamber and the outer atmosphere. The blades, as they rise, scoop up a quantity of air and deliver it at the outer edges of the chamber, the volume included between two contiguous blades being somewhat less than that contained in a segment of 120° of the cylinder bounded by the curved wall. A quantity of air is, however, carried in by the cross arms from without; this is, in form, an irregular five-sided prism, whose bases are enclosed by those parts of two of the blades that lie between the centre and the intersection of the cross arms, the cross pieces on one side of these blades and the cross arms on the intermediate blade of the opposite fan. The volume of this prism is, however, but little greater than that of a cylinder whose radius is equal to the length of the blade between the centre of the axis and the intersection of the cross arms with the blades of the fan. The effective volume removed by each fan, per revolution, therefore, is nearly equal to that of a hollow cylinder whose longer radius is equal to the length of the blade, the smaller one being the point of intersection of the cross arms. These machines are usually made with arms 46–48 in. long, and about 114–120 in. broad. The effective volume removed per minute is equal to rather more than 25,000 cub. ft., at a pressure of $1\frac{3}{4}$–2 in. of water, the wheels making 36–40 revolutions during that time; this requires a disposable effect of 14 steam horse-power, about one-half of which represents the useful mechanical effect.

Lemielle's Ventilator.—Lemielle's ventilator, in use at many of the Continental mines, has a vertical cylinder, within which revolves a second cylinder or drum, also vertical, the axis of which is placed eccentrically to the outer one. Two portions of the circumference of the inner drum are truncated and replaced by flat sides, to which a pair of hinged doors are articulated. The section of the inner cylinder approximates to that of a barrel, the heads representing the flat surfaces to which the doors are fixed. These doors are kept in constant contact with the inner surface of the outer cylinder by means of rods attached to an elbow or crank formed on the vertical shaft on which the drum revolves, the arrangement being similar to that of the feathering float-boards adopted in paddles wheel steamers. The central line of the aperture by which the air is introduced makes an angle of about 150° with that of the discharging orifice. The folding door as it advances pushes the air taken in at the feed aperture before it, the contact with the cylinder wall being kept up by the eccentric rod, which causes the door to open out farther, making a constant increasing angle with the side of the drum, as the distance between the inner and outer cylinders increases; this goes on until the crank has passed its centre, when the door is again gradually drawn in, as necessitated by the diminishing distance between the cylinders, until it reaches the discharging aperture, where it occupies the same angular position with respect to the side of the drum that it did at starting. The volume of the air carried through the machine by each door, as it revolves, is equal to that of a crescent-shaped solid, with truncated points, whose horizontal section is equal to that part of the base of the outer cylinder, that is truncated by a chord, joining the admission and discharging passages, diminished by half the area of the base of the drum.

Cooke's Ventilator.—This machine (Fig. 335) appears to have given good results. It is designed to deliver 180,000 cub. ft. of air a minute, with an exhaustion equal to 3 in. of water; or 50,000 cub. ft. with an exhaustion of 4 in.; or 120,000 cub. ft. if the drag of the air is increased to 5 in. It consists of two drums a, each 8 ft. diameter and 16 ft. long, mounted eccentrically on the shaft b. The amount of eccentricity of each drum is 2 ft., and each as it revolves moves almost in contact with a cylindrical casing c, of 6 ft. radius. This casing is closed at the ends by the brick walls which form the sides of the apparatus; they are coated with plaster over those portions against which the ends of the drums work, and are connected at the top of the covering. The casings are not complete cylinders, but are open throughout a portion $d\ e$ of the circumference. The air from the mine is led to the apparatus through the shaft, which is in communication with the space surrounding the casings, and it is drawn into these casings, and finally discharged at openings by the action of the revolving drums a. The portion of the casing left open is closed by a vibrating arm or "shutter" s, hung by the upper edge at j, and the lower edge of which is kept closely in contact with the surface of the revolving eccentric cylinder by means of an arm keyed upon a prolongation of the shaft j, beyond the side of the machine. Each arm is 6 ft. long between centres, this length corresponding to the distance between the centre of the shaft j and the centre m, from which the curve of the lower part of the shutter $j\ k$ is struck. In fact, the centre of each arm agrees exactly in position with the centre m, to which it corresponds. On one end of each of the main axles b is fixed a crank, each crank having a 2 ft. throw, and the centre of its crank-pin exactly corresponding in position with the centre of the eccentric drum on the same shaft. Each of these cranks is connected by a link to the end of the corresponding rocking arm, and as the length of this link is equal to the radius of the drum a added to the radius $m\ o$ of the lower part of the corresponding shutter $j\ k$, it follows that each shutter is kept in constant contact with the drum to which it belongs. The lower edge k of each shutter sweeps over a curved surface of plaster $e\ f$, this plaster, which is held in a hollow casting as shown, enabling a sufficiently tight joint to be made very readily.

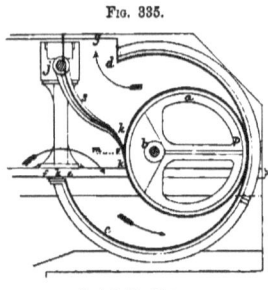

Fig. 335.

Cooke's Ventilator.

Guibal's Fan.—In this fan (Fig. 336) a sliding shutter is provided whereby the outlet may be enlarged or diminished at pleasure. The degree of opening which gives the best effect for a given case is determined by experiment. The covering enclosing the upper portion of the fan for about five-eighths of its circumference allows a clearance to the vanes of about 2 in.; from this point, the casing slopes away below the fan till it ends in the side of the chimney. This gradually enlarging outlet passage constitutes an important improvement. In consequence of the increasing sectional area of the passage, the velocity of the air is reduced, by the time it reaches the outer atmosphere into which it is discharged, to $\frac{1}{4}$–$\frac{1}{5}$, and the *vis viva* to $\frac{1}{16}$–$\frac{1}{25}$ of their original values. These conditions are obviously favourable to the utilisation of the force applied. The vanes of the Guibal fan have also been improved in some of the details of their construction. By a system of interlacing the arms, a very strong structure is obtained. Some of large dimensions have been erected. In a few cases, the diameter has been 30 ft., and the breadth about 13 ft. With these dimensions, and a velocity of 100 revolutions a minute, they discharge 100-120 cub. yd. of air a second, where a depression of the water gauge of 1¼-1½ in. is sufficient. It has been ascertained from experience that

FIG. 336.

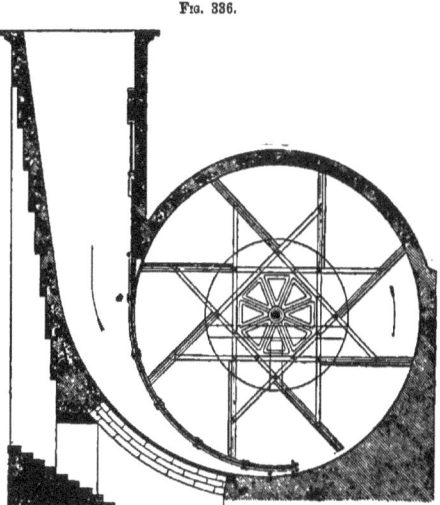

Guibal's Fan.

when the machine works under favourable conditions, the ratio between the volume generated by the vanes and that actually discharged from the apparatus varies but slightly. This ratio may be taken as having a mean value of 2·75, with a tendency to vary, within narrow limits, inversely as the speed of the fan. Not unfrequently the Guibal fan is erected in such a manner that it may be used to force air into the mine by reversing the current. When so erected, two chimneys are needed: one for delivering the air from the mine upward, the other for the reverse current when the air is taken from the surface and forced down the shaft. One of these chimneys must of course be kept closed by means of a kind of door or valve.

Schiele's Fan.—Schiele's fan is used in many places in the north of England to ventilate workings that are not very extensive. In such circumstances, it gives very good results. The air is taken in through openings at the centre around the shaft, and discharged between the partitions of the casing at the circumference. There are two fans which act successively upon the same air. The first fan drives it into an intermediate chamber at a pressure of about 6 oz.; the second compresses the air still more, so that at the delivery pipe it has a pressure of about 12 oz. to the sq. in.

Root's Blower.—A blowing machine much applied to mine ventilation is Root's Blower (Fig. 337). It occupies but a small space, is self-contained, and gives a powerful blast. In most cases, it is applied to give what is called the "positive" or force-blast, that is, it is used to force air into the downcast shaft; in some instances, however, it has been applied to exhaust air from the upcast shaft, with, it is said, better effect. The casing is usually made of cast iron, with the cylindrical parts bored out, and the head plates faced off truly upon a boring mill arranged for the purpose. The friction is limited to the journals and the toothed wheels. The wings do not touch in running, but move as closely together as possible without coming into actual contact. They are about 2 ft. in length, and they make 200–300 revolutions a minute; at a speed of 250 turns a minute, it is said to produce a pressure of about 5 lb. to the sq. in. It is inadequate to the requirements of extensive workings. But in exploring-drifts and other preliminary excavations, it will be found to be sufficient. For these cases it is very suitable, by reason of the facility it affords for erection and driving, and the lowness of its first cost. In tunnelling, it may be applied with advantage.

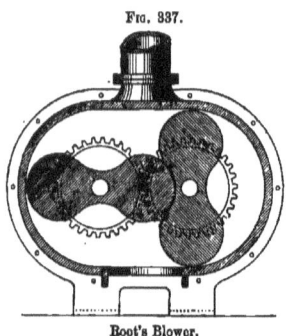

FIG. 337.

Root's Blower.

Hickie's air-cooling apparatus (Fig. 338) may be placed vertically or horizontally. It is attached to the air supply pipe *a* with the object of splitting-up the air current, and cooling it while passing

FIG. 338.

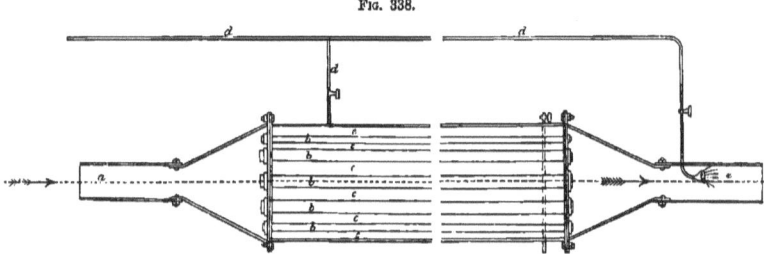

Hickie's Air-cooling Apparatus.

through the small tubes *b* surrounded by a flow of cold water *c*. The water-supply pipe *d* can be made to deliver a spray *e* after shot firing.

Anemometers.—Anemometers, or wind-measurers, are required to ascertain the quantity of air passing along a given way in a given time. In mine ventilation, the anemometer is a very important instrument, for without it there would be a good deal of uncertainty concerning the actual quantities of air circulating through the various districts of the underground workings. In collieries, the instrument becomes a necessity. Various forms of anemometer have from time to time been introduced; but Biram's has been most extensively adopted. It consists essentially of a set of vanes enclosed in a cylindrical case, and supported upon its axis in such a way as to give but little friction. The revolution of the vane-wheel gives motion, by means of endless screws, to pointers, which move over the face of suitably divided dials. The Biram anemometer is made by John Davis and Son, of Derby, and 118, Newgate Street, London.

The same firm have introduced a much improved self-timing anemometer (Davis's patent) which dispenses with the use of a watch. By holding the instrument in the current of air to be measured for a few seconds, it correctly indicates feet per second. It is exceedingly portable, being only 4 in. diameter. In general appearance it very much resembles the Biram anemometer. Every colliery manager and engineer acknowledges the difficulty and inconvenience experienced in using the Biram anemometer, which necessitates the use of a watch; and, unless he is assisted by a man to carry his lamp, he has to hold his anemometer, watch, and lamp. Davis's anemometer dispenses with the use of a watch or timer, and also of a lamp carrier; and, when held up in the current of air, without loss of time, indicates the velocity per second. Its cost is 4l. 10s. In use, it is held up with its back to the current of air to be measured, and on no account must the air enter from the face. When the vanes (Fig. 339) have revolved for a few seconds, press the spring button, the large hand then indicates feet per second. After reading, screw down the milled head until the plunger is relieved, after which unscrew the milled head as far as it will go, and the hands return to zero. Should the velocity be such that the hand travels more than one revolution, then read the inner circle of figures. The small hand shows whether the outer or inner circle should be read.

Fig. 339.

Davis's Self-timing Anemometer.

The reader desirous of knowing more about ventilation should consult 'The Theories and Practice of Centrifugal Ventilating Machines,' by Daniel Murgue, translated by A. L. Stevenson (Spon, 1883).

The general adoption of electric light and power in mining and colliery districts will open up a new industry, and a large field of business for ventilating fans of all sizes to be propelled by electricity.

It is now well known that the distribution of electric power can be more economically supplied, especially over long distances, for many purposes than steam power, more especially where power is only required intermittently. Where no current is obtainable from a central station, the smaller-sized fans up to one-man power can be worked by an ordinary primary battery for ventilating purposes. This question of electric ventilation is absorbing the attention of many engineers, amongst whom Shippey Brothers, Limited, may be said to rank as one of the foremost firms who have made this branch of the electrical industry their special study. Their new type standard electric motor of ¼ H.P., fitted with Shippey's patent xylonite 18 ft. 6 in. blade air propeller, owing to its extreme lightness, requires but little force to distribute nearly double the volume of air for same amount of energy expended to drive similar sized fans of other makers.

The motor (Fig. 340) is manufactured in 6 sizes, and wound to run on incandescent circuits of any voltage; also upon arc lighting circuits, and when required to work at a constant and regular speed, they are fitted with an automatic patent governor attached to the shaft. This governor is extremely simple and reliable in action, and does not depend upon a variation of the field magnetism by means of electrical sliding contacts, as in other constant current motors. For this reason Shippey Brothers claim perfect safety, as there are no delicate electrical adjustments to get out of order, and therefore no danger from short circuits, or fire, generally caused by defective contact points.

The motors supplied by Messrs. Shippey Brothers are arranged to be controlled by either of three methods of regulation. For constant load, for variable speed, and for constant speed, either of which is accomplished by attaching the corresponding fixture to the motor.

When the load is constant, as for example in running a fan attached directly to the shaft, the motor is set to run at the desired speed by clamping the fan to the shaft at a certain distance from

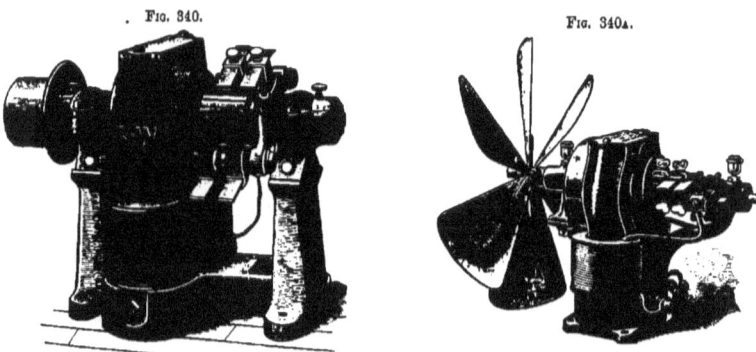

Fig. 340.   Fig. 340A.

the bearing, which allows the motor to run at a speed corresponding to the point on the shaft where the fan is attached.

Fig. 340A shows a motor in use in this way, driving a ventilating fan.

For varying the speed the hand regulator is used. This is attached to the bearing, and operates by sliding the shaft lengthwise as the variations of speed are desired.

The constant potential motors can be fitted so that they can be very easily reversed by simply inverting the brush holders upon the clamping rods, so that the upper brushes are placed below, and the lower ones above. By this device the machines can be run either right handed or left handed, without taking a part or changing any connections; and thus be used for both exhausting or propelling purposes.

The following are some special advantages claimed for the system, viz.:—

(1) They are easy to start, cannot be harmed by overloading, very simple in construction, of very few parts, and are strong and economical.

(2) Entire frame is used in magnetic circuit, hence minimum of weight.

(3) Centre of gravity is exceedingly low, and the motor has a very broad base, hence runs very steadily and noiselessly.

(4) Quick response or action of governor producing instantaneous regulation, because it does not depend upon changes of magnetisation.

(5) No switching or changing of electrical connections are employed in the governor. Therefore there is no danger of interference with the electric circuit, liability of causing fire, or wearing out of contact points, all which are the very serious objections to all other known governors at present.

(6) The automatic current governor, which not only maintains constant speed with varying currents, but actually causes motor to speed up with reduction of current, and thus makes up for the smaller power of the weaker current.

(7) No sparking at brushes when regulator acts, because neither the armature or any part of it is ever subjected to a weakened field, which is invariably the cause of sparking in motors which regulate on other principles. For this reason these machines are absolutely *non-sparking*.

(8) Governor is applied directly on and operates main shaft of machine, using no special spindle for governor, nor delicate and exposed rods to connect governor with a moving switch.

(9) No complicated multiple or differential winding of wire is used. Only a single continuous and permanently connected electric circuit in the machine.

(10) Frictional resistance to movement or action of governor is absolutely eliminated by rotary motion of shaft. This friction in regulators is ordinarily considered to be inherent, and impossible to overcome.

(11) Safety valve or automatic stop if governor breaks, because governor has to be in action and *exert* itself in order to keep the armature in the right position for power; therefore if governor fails, armature will stop automatically.

(12) Governor is free from electrical parts and entirely mechanical, therefore it can be understood, repaired, &c., by any machinist or janitor.

(13) Longitudinal motion of shaft and commutator makes them run much more smoothly, wear much more evenly, distribute heat, and, in general, work better and last longer.

(14) Governor is entirely within pulley, and therefore completely covered and protected, and occupies no extra space.

(15) Contact point of brushes is constant, since field is constant—for this reason it is very easy to set the brushes.

CHAPTER X.

LIGHTING.

OIL LAMPS.—Figs. 341, 342 illustrate two varieties of mining lamps, with naked flames, used by the miners of the Harz mountains, Germany. Fig. 341 is made of sheet iron, and is light and strong. The shape is similar to that of the old Roman lamps, the form of which has not been improved upon for general mining purposes. The body a which is closed, contains the oil and wick, the end of the wick passing out of the spout b. The screw plug c fits the hole through which the oil is fed; in this plug is a small hole, to allow air to pass in and replace the burnt oil, and another, at right angles to

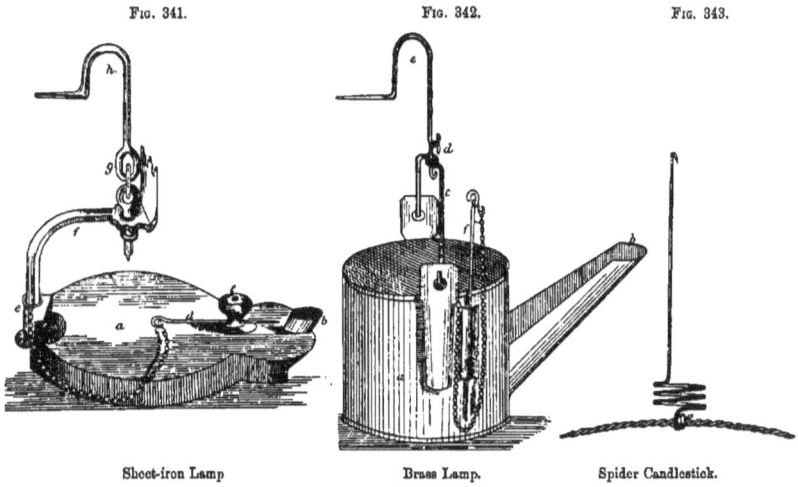

Fig. 341.   Fig. 342.   Fig. 343.

Sheet-iron Lamp   Brass Lamp.   Spider Candlestick.

it, for the reception of the pricker d, for trimming the wick, and this is connected by a light chain to a ring e, encircling the arm f, along which it can play. At the upper end of the arm, which bends over the body of the lamp so as to be above its centre of gravity, is a brass shield, on which a number can be stamped, so that each miner can be made responsible for the one placed in his charge. The arm is connected by a swivel and link g, to the hook h, by which it is carried; so that, although the hook is held firmly in the hand, the lamp may be turned round in any direction. When climbing in mines the hook h is placed over the knuckle of the thumb, on which it rests, thus giving the free use of both hands, and, if the workings are wet, the fingers can be stretched over the flame as a shelter, or a small screen of sheet iron can be rigged up. The end of the hook h is pointed, and turned at right angles to the rest; this is to enable the miner to fix it into timber when working, by striking the opposite end of the hook with his fist. When there is no timber, the point may be inserted into some crevice or the remains of an old drill hole; a very small part only of the point is necessary to stick in an object so as to support the lamp. In the rare cases where there is no suitable place to be found for fixing the lamp, a stick of timber may be placed against the wall, and the point of the hook driven into this. There is no conceivable place in which this lamp cannot be fixed, if the miner wishes, where candles are at present employed. The spike on the hook is also useful for testing the hardness of certain minerals, exposing rotten timber, &c. This lamp is made of such a size as to hold more than enough oil for the day's shift, though if it is desired to extend the time underground, an extra supply can be carried about in a cow's horn or other vessel. Oils suitable for this class of lamp are : Chinese (pea-nut), at 3s. 3d. per gal., in 10 gal. cases; Colza, at 3s. 9d. per gal., sold in 5 gal. drums; Olive, at 5s. per gal. drum. The wick used is candle cotton, sold in balls at 1s. 3d. per lb. Lamps burn 4-6 oz. of oil per 8 hours' shift, depending on the size of the wick.

Tests made with a Rumford photometer show that the ordinary light emitted by the lamp is equal to $1\frac{1}{2}$ candle power, the candle used for comparison being a stearine one of superior quality, costing 9d. per lb., six to the lb. When necessary to give a stronger light for inspection purposes, over 6 candle-power can be got from it. The oil in the above test was the ordinary colza oil.

The dimensions of such a lamp suitable for an 8 hours' shift, that will hold enough oil to burn 10 hours if necessary, are :—$5\frac{1}{2}$in. long, 4 in. broad, 1 in. deep.

From experiments made under the most favourable circumstances for the candle, as it was allowed to burn free from draught, it was found that 45·428 candles will burn for 318 hours, and that 1 gal. of colza oil serves for the same length of time; 45·4 candles, equal $7\frac{1}{2}$ lb., valued at 5s. $7\frac{1}{2}$d., while 1 gal. colza oil costs 3s. 9d. A trifle more should be added to the cost of the lamp for the quantity of wick used, which is very small, and wear and tear of the lamp itself, but this will be more than counterbalanced by the extra amount of candle wasted in reality by drippings and odds and ends.

The advantages of this lamp over the candle as usually employed are :—It is not so likely to break, and can stand plenty of knocking about. It is not affected by the heat. It is cleaner. It is more easily handled when travelling up ladders. It is not so easily upset, as it has a broad flat bottom, and when it is accidentally overturned it is not so liable to go out or waste the illuminating material. It gives a better light. It can be regulated to give a larger quantity of light when inspecting, or a smaller quantity when working in, badly ventilated places, and is designed to economise the air and oil. There is no waste from grease and oil dripping down when in draughty places, and the lamp is less likely to go out. It requires more water to fall on it to extinguish it than a candle does. There is no waste in odds and ends of candles. The cost is less. The cost of lighting is often overlooked in mines, but is nevertheless a considerable item when counted up for the year.

Fig. 342 shows a brass lamp used by surveyors, where the iron one would affect the needle of their instrument. The body of the lamp holds the oil and wick, and has a spout through which the wick passes; a cap covers the hole where the oil is fed in—it has perforation on the top to allow air to enter in and take the place of the oil as it is consumed; and a lid, connected to the body by a hinge, covers the top of the lamp, and keeps out dirt. The lamp fits into a shell, the object of this being that when the lamp is held over the instrument, should any oil dribble out of the spout, it will be caught in the gutter a, and collected in the body of the shell b, whereby, not only is the instrument kept clean, but the oil saved. An arch-shaped handle c is attached to the shell a, and on the top of this is the swivel d and pointed hook e. A pricker f is attached to the shell a, by a light chain, and rests in the sheath g, when not in use. This lamp has the same advantages for the surveyor as the former has for the miner.

Candles.—Fig. 343 shows a miner's candlestick, called the "spider," which will hang on or cling to the slightest projection in a rock-face, and can therefore be moved from place to place with the greatest facility, just as altering circumstances connected with miners' work may require in driving-levels, or crosscuts, stoping, or sinking. Lumps of clay are, and have been, employed in mining lodes, &c., from time immemorial, with which to fix candles in most suitable positions for illuminating purposes. Clay adheres to rock whilst moist, but when it dries the clay crumbles and loses its adhesive quality, and consequently candles fall, probably break, if not much burnt and over three-quarters or half in length, and at once this leads to the waste of about a quarter of the length of the candle. In falling it frequently falls into water, and leaves the miners in darkness, and when a candle is wet it is only relit with both trouble and loss of time, and even then for a while gives a poor light. The "spider" can be made by any miner in a few minutes, and at a nominal cost, as all the material required is about 2 ft. of $\frac{1}{16}$ in. iron wire, which can be got almost everywhere, but it is not always possible in mines to get suitable candle-clay. With a little care a "spider" will last for years underground (excepting accidents), and it entails no trouble, or next to none, after being first well made; whereas, clay demands a lot of kneading and moistening, and after being used with one candle or two it usually gets mixed up with candle-grease and becomes useless. In the dry hot parts of mines the "spider" will be found to possess many advantages over clay. It rarely or never falls when once fixed and left untouched, and no one unaccustomed to the "spider" would at first credit how tightly it clings, when weighted with a candle, to the smallest crevice or protrusion in a face of rock. It saves loss of time and breakage of candles, is cleaner to handle, and keeps the candle cooler when carried than clay (which absorbs the heat from the hand and softens the candle).

Safety Lamps.—The principle of the safety-lamp is founded upon the fact, first observed by Sir Humphry Davy, to whom the invention is due, that flame will not readily pass through fine wire gauze. The explanation is this :—In order to pass through the gauze, the gases in combustion must be divided into a great number of little jets, each distinct from the rest. These lose their heat by being brought into contact with the metal, and are consequently extinguished. In accordance with this fact, Davy constructed a lamp in which the wick was surrounded by a cylinder of wire gauze. This gauze was composed of 28 wires to the linear inch, giving 784 apertures or meshes to the square inch. The same principle has been acted upon in all other safety-lamps of more recent introduction. Indeed, all the safety-lamps now in use are but modifications of the Davy. The ordinary Davy lamp as at present used is almost identical in form with that constructed by the inventor. It consists (Fig. 344) of an iron wire-gauze cylinder fixed to a brass ring and screwed on to the oil vessel. The upper portion of the gauze is double for

LIGHTING.

greater protection. Externally it is guarded by three iron rods placed equidistant from one another, and attached at the top to a metal roof, above which is the loop for suspending the lamp. For the purpose of trimming the wick and extinguishing the light, a wire passes up a close-fitting tube from the bottom of the oil vessel. The average weight of one of these lamps is $1\frac{1}{4}$ lb., and the average cost 7s. A grave defect of the Davy lamp is its small lighting power. A very large proportion of the rays of light emitted by the flame are intercepted by the wire gauze. The proportion of opening to solid in the gauze adopted is about 1 to 4; that is, of the total surface of the gauze, about $\frac{4}{5}$ is solid metal. We cannot infer from this that only $\frac{1}{5}$ of the light is utilised, because some of the rays falling upon the wires are reflected; but the proportion utilised certainly does not exceed $\frac{1}{10}$. Hence the light emitted in the horizontal direction is very small. But it is evident that the proportion of light emitted through the gauze in other directions must be still less, by reason of the obliquity of the rays and the gauze, and that the proportion utilised diminishes as the point to be illuminated is situate nearer the roof of the workings. The light thrown in the upward direction is still further diminished by the double gauze and solid metal roof, so that the roof of the workings is only very feebly illuminated. This constitutes a very serious defect, inasmuch as it prevents a dangerous state of the roof from being observed, and furnishes a plausible excuse to the miner for opening his lamp.

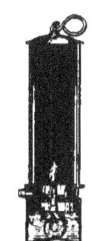

Fig. 344.

Davy Lamp.

Numerous modifications of the Davy lamp have been made for the purpose of remedying these defects. The attempts in all cases have been more or less successful, but also in all, success has been obtained by incurring defects of another kind. The chief improvement consists in employing glass in the place of a portion of the gauze. The defect of this lies in the fragility of the material, which necessitates the adoption of a great thickness. It can hardly be disputed, however, that by employing a short cylinder of thick glass of a suitable quality, properly protected on the outside by vertical iron rods, a light greatly superior to that of the Davy is obtained without incurring serious danger from the fragility of the material. It should be remarked here, that when gas fires in a lamp so constructed, there is some danger of the glass cracking if rapidly cooled.

Some modifications of the Davy lamp have been made to lessen the danger due to strong currents of air and to the heating of the gauze. It will be observed that the employment of a cylinder of glass partially accomplishes the former object; but the end in view is more or less completely attained by providing certain points of influx and efflux for the air, by means of which distinct currents are formed that are not readily affected by the agitation of the external air. To effect the second object, the air is introduced as near to the flame, and passes as directly to it, as possible, in order that an explosive mixture may burn as it reaches the flame, while the chief portion of the space inside the lamp is filled with gases that have been already burned.

Dr. Clanny's lamp consists in the substitution of a short cylinder of thick glass for the lower portion of Davy's wire gauze. The feed air enters and the products of combustion escape, through the gauze above the cylinder. This arrangement is unfavourable to combustion, and hence the gain due to the substitution of the glass for the gauze is partially lost. Indeed the light given by a Clanny lamp is but little superior to that furnished by a Davy, while it possesses the disadvantage of being much heavier and of being constructed of a fragile material. The glass cylinder, however, in the Clanny is thick, and well protected by vertical iron bars.

George Stephenson slightly increased the diameter of the Davy, and added a glass cylinder throughout the whole length of the lamp. This cylinder (Fig. 345) is placed inside the gauze, and is covered by a cap of perforated copper. The glass serves as a protection to the gauze against the heated gases inside, while the gauze serves as a protection to the glass against the blows, and also keeps the lamp safe should the glass be accidentally broken. Air is admitted to the lamp through small holes in the rim below the cylinder. The method of admitting the feed air is a very good one, inasmuch as it tends greatly to prevent overheating, and also, in a considerable degree, to preserve the lamp from injurious influence of currents of air. When the air inside becomes highly heated, the flame is extinguished. The feed air-holes must be kept free from oil and dust, and the lamp be held vertically to enable it to burn well.

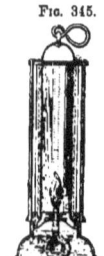

Fig. 345.

Stephenson's Lamp.

Mueseler's lamp, like the Clanny, has a short cylinder of thick glass around the flame, and draws its feed air in through the gauze above the glass; but it is provided with a central conical metal chimney placed immediately above the flame, and covered on the top with wire gauze. The products of combustion pass directly up this

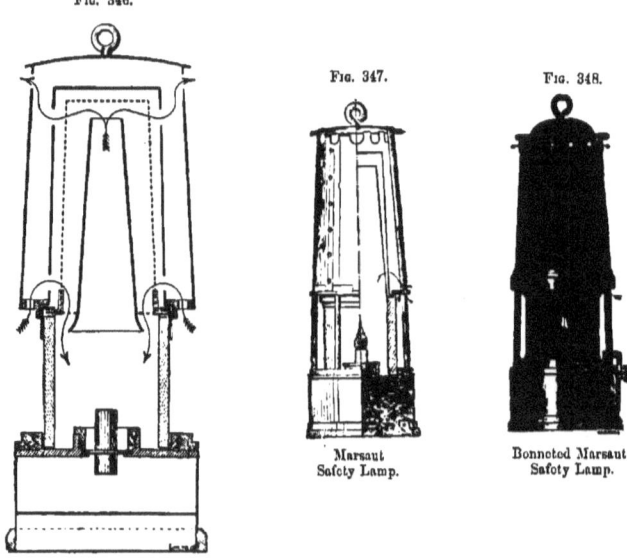

Fig. 346.

Fig. 347.

Fig. 348.

Marsaut Safety Lamp.

Bonneted Marsaut. Safety Lamp.

Davis-Ashworth Mueseler Safety Lamp.

chimney, and cause a strong upward draught. By this means, the air is drawn briskly down on the inside of the glass cylinder, thus keeping the latter cool and promoting combustion on the

LIGHTING. 191

wick. The glass cylinder is protected in the usual manner by vertical iron rods. These lamps give an excellent light, and for that reason are preferred by the miners to those already described. The fact of their being but little affected by a strong draught constitutes an important advantage over the Davy. An improvement on this is the Davis-Ashworth Mueseler (Fig. 346), made by John Davis & Sons, 118, Newgate Street, London. The main feature in the Davis-Ashworth Mueseler lamp, is in the double shield or bonnet, the intake and outlet air holes of the inner and outer bonnet, not being opposite, prevent the air reaching the flame, except at a very low velocity. This lamp has successfully baffled all attempts to explode, and is recommended for fiery mines where the velocity of air is very great. The price complete is about 7s.

Other popular modern lamps are the Marsaut and the bonneted Marsaut, costing about 7s. each, and shown in Figs. 347, 348. They are likewise made by John Davis & Son, together with many other forms of lamp, as well as the various parts that need replacing at intervals, and a host of useful mining accessories.

Electric Lighting.—This is destined to be much extended in the future. In many cases its cost is less than by other means, and it is safer. Sopwith has pointed out what facilities exist in collieries for the economical extension of electric lighting in respect of main cables. This refers to the use of old iron or steel ropes for such purpose. At Cannock Chase Colliery some 4000 yd. of old rope have thus been utilised, and in one instance a cable has been laid having only a resistance of $\frac{1}{15}$ ohm. for 1400 yds., a condition of profuse cable that would not have been thought of if copper cable had been in question. In any extensive system of lighting in bye-roads and stations, the importance of cheap cable is evident. Approximately, the relative values of old iron ropes and bare copper cable are, after allowing for difference in conductivity, as 1 to 5, and there may be conditions where the difference is considerably greater. Some of these ropes have been laid together in a trench on the surface, and only insulated with coal-dust and tar, so that little trouble is involved in insulating them in a dry mine. In fact, iron ropes lying side by side on the ground in an underground road, and extending over a distance of 140 yd. (single distance) have been found to show no appreciable leakage. The practical experience gained in laying down the cables alluded to, in trenches on the surface, in wet shafts and in roadways underground, and the economy and efficiency of the rough methods adopted in insulating, tend to prove that the problem of extensive lighting underground does not present such great difficulties as might at first be anticipated.

The following statistics show the relative cost of artificial illumination by different systems at a colliery raising 1000 tons a day —

ELECTRIC LIGHT.

1. Pit head, 2 × 200 candle power	= 14 × 16 c. p.	averaging 10 hours per day	= 40,600 lamp hours per annum.
2. Winding engine, fan, boilers, pumps, &c.	= 12 × 16 ,,	,, 10 ,,	= 34,800 ,, ,,
3. Shops, offices, &c.	= 16 × 16 ,,	,, 1½ ,,	= 6,960 ,, ,,
4. Screens, sorting, &c., 4 × 200 candle power	= 28 × 16 ,,	,, 1½ ,,	= 12,180 ,, ,,
5. Underground	= 20 × 16 ,,	,, 10 ,,	= 58,000 ,, ,,
6. ,, (continuous lighting)	= 10 × 16 ,,	,, 24 ,,	= 69,600 ,, ,,
	100 × 16 c. p.		222,140 lamp hours.

The total candle power is about 2124, though with electric light this will be more effective than with other lights, on

192   MINING AND ORE-DRESSING MACHINERY.

account of the greater facilities for reflection. To get the I.P.H. per annum we may divide the total lamp hours by 10, since each 16 c. p. lamp takes $\frac{1}{10}$ I.P. Allowing 10 lb. of slack coal per I.P.H. at 1s. 6d. per ton.

					£	s.	d.
Electricity.—Coal. 100 tons at 1s. 6d. = ...					7	10	0
Renewals of lamps at 1500 hours' burning (they usually last for 2000 hours).							
2900 hours.	1. 2 × 200 c. p. lamps renewed twice at 18s. each ..		..	..	3	12	0
2900 ,,	2. 12 × 16 ,,	,, ,, 4s. ,,	..	..	4	16	0
435 ,,	3. 16 × 16 ,,	,, ½ times ,, 4s. ,,	..	..	1	1	4
435 ,,	4. 4 × 200 ,,	,, ½ ,, ,, 18s. ,,	..	..	1	4	0
2000 ,,	5. 20 × 16 ,,	,, twice ,, 4s. ,,	..	..	8	0	0
6960 ,,	6. 10 × 16 ,,	,, four times ,, 4s. ,,	..	..	8	0	0
Interest and depreciation at 10 per cent. on capital outlay—200l. ..			..	..	20	0	0
Oil, water, waste, &c. ..	.. ..	..	..	..	5	0	0

£59 3 4

The total cost of electric lighting will therefore be about 60l. per annum, and this on 290,000 tons per annum will be ·05d. per ton raised.

Gas.—With gas, the capital outlay may be taken, for main, pipes, and fittings, at about 150l.
At 3 c. p. per cub. ft. per hour, and 5 ft. per 15 c. p. lamp.

						£	s.	d.
1.	40,600 lamp hours at 5 cub. ft.	=	203,000 cub. ft.	=	..	25	7	6
2.	34,800 ,, ,,	=	174,000 ,,	=	..	21	15	0
3.	6,960 ,, ,,	=	34,800 ,,	=	..	4	7	0
4.	12,180 ,, ,,	=	60,900 ,,	=	..	7	12	3
5.	58,000 ,, ,,	=	290,000 ,,	=	..	36	5	0
6.	69,600 ,, ,,	=	348,000 ,,	=	..	43	10	0
Taking the cost at 2s. 6d. per 1000 cub. ft. Total cost for gas				=		138	16	9
Interest and depreciation on 150l. at 10 per cent.				..	..	15	0	0

£153 16 9

On the same amount of coal per annum, viz. 290,000 tons, this will be about ·13d. per ton raised.

Paraffin.—With paraffin taking 2 × 1¼ in. wick duplex lamps, as 16 c. p., consuming $\frac{1}{10}$ pint of oil (at 8d. per gal.) per hour.

						£	s.	d.
1.	40,600 lamp hours at $\frac{1}{10}$ pint of oil per lamp hour	=	507·5 gal. at 8d. per gal.	=		16	18	4
2.	34,800 ,,	,,	,,	= 435 ,,	,, =	14	10	0
3.	6,960 ,,	,,	,,	= 87 ,,	,, =	2	18	0
4.	12,180 ,,	,,	,,	= 152·25 ,,	,, =	5	1	6
5.	58,000 ,,	,,	,,	= 725 ,,	,, =	24	3	4
6.	69,600 ,,	,,	,,	= 870 ,,	,, =	29	0	0
Total cost of oil					..	92	11	2
Wick, say					..	3	0	0
Labour, trimming, &c.					..	25	0	0
Interest on capital outlay of, say, 80l. at 5 per cent.					..	4	0	0
Depreciation, repairs, and breakages, at 20 per cent. on 80l.					..	16	0	0

£140 11 2

On 290,000 tons raised this amounts for Oil to ·116d. per ton, for Gas to ·13d. per ton, and for Electricity to ·05d. per ton. Thus it will be seen that electricity compares very well as to cost with other illuminants, since gas costs nearly three times and oil more than twice as much for the same amount of illumination.

Portable electric safety lamps are also made for use at the actual working face.

This lamp is constructed to give a light of about one candle-power for a period of about 10-15 hours, and consists of a storage battery of 4 cells enclosed in a strong teak box turned out of the solid and strengthened with metal bands, and a small burner mounted on the side of the case and protected by a strong glass cover. The case is fitted with a hinged lid secured by a cross bar fastened with a safety-nut, and having a swivel handle. The full size of the lamp is 7 in. by 4½ in., and the weight of the whole ready for use is about 7 lb. The lamp is opened by unscrewing the safety-nut by means of a special key, and lifting the lid which turns on a hinge. Then by means of a hook the battery can be withdrawn.

The lamp is fitted with a switch for turning the light on and off. Great care must be taken not to allow the lamp to continue in action after the light begins to turn red, as this will injure the battery. To charge the battery with current: the E.M.F. of the charging circuit should be 9·5-10 volts for each battery of 4 compartments, and the current ·5-·6 ampère. Ascertain the direction of the charging current by attaching a lead wire to each of the two terminal wires, holding them a little apart while immersed in dilute acid in a saucer or similar vessel. Presently one lead wire will turn dark chocolate colour: this will be the wire to be fixed to the side of the battery marked " P."

The battery should be kept on the charging current for about 9 hours. Under normal conditions this will be sufficient to fully charge it. When the battery is placed in the case, care should be taken that the lead strips on the side of the battery make contact with the lead strips inside the case. The lid is to be shut down, and the safety-nut is screwed on. The lead strips on the side of the battery and the inside of the case must be kept clean and well rubbed with graphite. The contacts must on no account be allowed to become coated with peroxide of lead. The battery plates *must* be kept covered with acid. Dilute acid of the right strength may be made by mixing one measure of acid with eight of water. The lamp should be kept in the upright position always. Should the cells become sulphated, they are best brought back to their original condition by charging with a larger current for a time, say ·75-1·0 ampère. If acid is spilt in the case the battery should be withdrawn by means of the hook, and the case rinsed out with clean water.

Fig. 349.

Gwynne's Engine and Dynamo.

Fig. 349 is taken from one of Messrs. John and Henry Gwynne's "Invincible" high-speed engines, whose crankshaft is attached to, and drives direct, the armature of a dynamo for electric illumination, very suitable for mines. This plan of working dispenses with belts and other systems of conveying power; it is more efficient, and ensures a steadier light.

CHAPTER XI.

HAULING AND HOISTING MACHINERY.

TUBS, WAGGONS, OR CARS.—The vehicle known under the several names of "tub," "waggon," or "car," is that in which the mine produce is conveyed from the working places to the shaft, and commonly from the bottom of the shaft to surface or "bank." It consists essentially of a "body," usually rectangular in form, to contain the load, and of "wheels and axles," to carry the body. These two parts of the tub will influence the system of haulage differently, the former having reference to the conditions affecting capacity, the labour of loading, and the height of the working places, and the latter relating more directly to the force of traction to be exerted upon the load.

Wheels and Axles.—The form and construction of the wheels and axles of a tub, and their arrangement relatively to each other, influence in no small degree the question of haulage considered with respect to the requisite force of traction. The principal points to be taken into account, in a consideration of this nature, are the kind of connection made between the wheels and the axles, and the diameter, and the form of the rim of the wheels adopted. There are two kinds of connection; in one the wheels are fixed upon the axles in an invariable manner, so that the latter are compelled to revolve with the former; in the other the axles are fixed, and the wheels revolve freely upon their extremities, which, in such a case, receive a particular form, and are described by the term "journals." When the connection is of the first kind, the wheels are mutually dependent, that is, the angular motion of each must be equal and take place in the same direction, or, in other words, they must turn in the same direction and with the same velocity. When the connection is of the second kind, the wheels are completely independent of each other, that is, they may revolve with different velocities and in contrary directions. Thus it is evident that the nature of the connection will, under certain conditions, operate to facilitate or to impede the work of traction.

It will have been observed that the system of fixed wheels is invariably adopted upon railways, and that the system of free wheels is as invariably applied to vehicles running upon common roads. The reasons for this are plain and easy to be understood. On a railway, the motion is in a straight line, and the surfaces over which the wheels of the vehicles roll are perfectly even. These are the conditions always sought; in practice it becomes necessary to modify them frequently, as, for example, when curves are adopted; but curves are avoided whenever possible, and when circumstances compel their adoption, they are made of the largest possible radius in order to approximate to the straight line. On common roads, on the contrary, the motion of a vehicle is continually in a curved line. It is impossible that it should be otherwise, when the wheels are not guided. But irrespective of this, road vehicles have to be very frequently directed out of their course to avoid other vehicles and obstacles of various kinds, and to be turned off at a sharp angle, or completely round in a small space. And again, the surface of a common road is very far from possessing that regularity which is characteristic of the railway. Thus the conditions of motion upon a railway and upon a common road are essentially different, and these conditions determine the kind of connection between the wheels and the axles.

When the motion of the vehicle takes place on a curve, as in the case of a common road, the arcs passed over by the two wheels are unequal, and the degree of the inequality will obviously increase with the distance of the wheels apart. One of the arcs will be reduced to nothing, if the vehicle be made to turn upon one of its wheels as a point of support; or they may be equal, but the motions contrary in direction, if it be made to turn about its centre of gravity. If the wheels were mutually dependent, as one would be required to revolve more rapidly than another, or the two be required to revolve in contrary directions, it is evident that one or both of the wheels must slide, and the same result will follow from one wheel passing over an irregularity in the road. But when the wheels are independent of each other, the requisite inequality of motion presents no difficulty whatever, since each wheel is free to move with the velocity and in the direction needed.

A consideration of the underground roads of a mine will show that both classes of conditions exist. It is altogether impracticable to construct these ways with the accuracy of direction and the solidity attained upon ordinary railways. Great irregularities have to be encountered; frequent curves of short radii occur, and often the tubs have to be turned from one road into another by hauling bodily round upon a smooth floor. Not unfrequently the tubs have to be run along roads unprovided with rails, when, of course, the conditions of the common road present themselves. Thus it will be seen that, on the underground roads of a mine, the question is greatly complicated: and hence it has happened that opinions are divided respecting the most suitable kind of connection. In some mines tubs with wheels upon revolving axles are used; in others, tubs with wheels upon fixed axles. And it will be found, when full account is taken of all the circumstances of the case, that both these systems may be justified. In this, as in all other matters relating to mining, we have to deal with conflicting requirements, and in order to effect the best attainable compromise, we must carefully consider and accurately appreciate all the determining conditions.

It would, however, appear that the system of loose wheels is generally more applicable to the conditions prevailing upon underground railways than that of fixed wheels, and that, consequently, its adoption is desirable wherever it is impracticable to lay out the roads with great regularity. It is, however, possible to combine the two systems so as to obtain some of the advantages of each, and various devices have been adopted to render the combination as advantageous as possible. The most obvious mode of combining the two systems, and one that has been extensively adopted, consists in fixing one wheel to the axle, and leaving the other loose upon a journal, the arrangement being such as to have one fixed and one loose wheel upon each side of the tub. So long as the line is straight, this system acts similarly to that in which both wheels are fixed; but as soon as a curve is entered upon, the loose wheel takes the velocity necessary to prevent slipping. In order to assimilate this system as much as possible to that of fixed wheels, the loose wheel is made to turn with a moderate friction. Another method, adopted in Silesia and some of the French collieries, consists in having as many axles as wheels. In this method each wheel is fixed upon its axle, and a pair of wheels corresponds to two revolving axles parallel to each other. Such a system solves the problem satisfactorily; but it possesses the disadvantage of complication.

Besides these combinations of the two principal systems, other expedients are resorted to for the purpose of rendering fixed wheels capable of running over curves without occasioning a great increase of resistance. Some of these expedients have reference to the road, and will therefore claim consideration in another place. One device consists in making the wheel conical towards the flanges. This form of the wheels is favourable to the stability of the tub on a straight line, and also greatly facilitates its motion over a curve. It is easy to see how these results are obtained from conical wheels invariably fixed upon their axles. If we suppose the tub moving upon a straight piece of line, and driven by some cause to one side, the radius of the wheel on that side will be increased, and that of the wheel on the other side diminished. The tendency of the larger wheel to progress more rapidly than the smaller will immediately restore the tub to its normal position upon the rails. If we suppose, again, the tub to be entering upon a curve, it will be evident that the force of inertia will throw the tub against the outer rail, and bring the flange of the wheel on that side into contact with the inside of that rail, as already pointed out. But this shifting of the tub in the

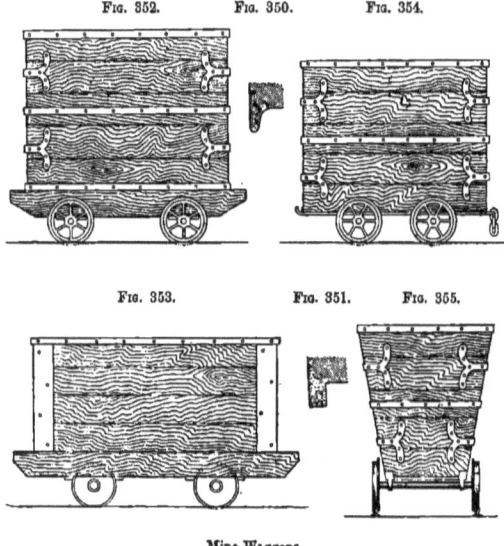

FIG. 352.   FIG. 350.   FIG. 354.

FIG. 353.   FIG. 351.   FIG. 355.

Mine Waggons.

direction of the outer rail has the effect of increasing the radius of the wheel on that side, and of diminishing, in the same proportion, that of the wheel on the other side; and hence it is clear that the outer wheel will, at each revolution, advance through a greater distance than the inner wheel, as required by the greater length of the arc on that side. On curves of a large radius, this expedient gives very satisfactory results.

In consequence of the relative obliquity of the axles and rails upon curves, it becomes necessary to allow the wheels a certain amount of play, that is, to space the rails wider apart than upon those portions of the line which are straight. The shorter the radius of the curve, the larger is the amount of play required, but generally it will be about ¾ in., or the double of that allowed in the straight. This play will necessitate the adoption of wheels of considerable breadth, as otherwise a lateral movement, such as occurs during motion round a curve, would cause derailment. A moderate breadth of wheel is also favourable to haulage along the working face, and in other situations not provided with rails. The form of the flange is a question of some importance, since the facility with which derailment takes place depends in a great measure upon it. A form commonly adopted on underground tramways is represented in Fig. 350, which, in consequence of the absence of all curvature, offers great resistance to derailment. An objection to this form, however, lies in the fact that should the flange get upon the rail it has no tendency to slip off, and to thereby restore the wheel to its proper position. A section formed of two lines, one straight, the other curved, and joined to the former by a small circular arc, has been proposed as fulfilling all the conditions required. This section, which is shown in Fig. 351, offers almost as great resistance to derailment as that represented in the last figure, and it possesses besides a tendency to return to its position, if it should from any cause get upon the rail. This quality is very important in a flange, for not only does it prevent the delays consequent on getting off the line, but it greatly facilitates the placing of a tub upon the rails. The material of coal-tub wheels is generally cast-iron, to increase their durability. Wood's observations showed that the relative durability of unchilled and chilled wheels was as 39 to 63. Lately, steel has been adopted for colliery wheels, with very good results.

Bodies.—The body or box part of a tub commonly consists of oak, ¾–1 in. thick, set upon an oak framing below, and bound with iron to give it strength. Sometimes the body is constructed wholly of iron. Such tubs are very durable, but they are not easily repaired; their weight is about the same as that of wooden tubs.

Lightness is a desirable quality in a tub, since it is important that the dead weight should be reduced as much as possible. Hence that form should be adopted which, with a given weight of material, affords the greatest carrying power. The form which best fulfils these requirements is the rectangular box, and this is therefore generally adopted.

In the design of a tub, there are two features that demand particular consideration, namely, its capacity and its height. Obviously the conditions which determine the latter will have some influence on the former, but besides these there are others which relate to capacity alone. In order to diminish the proportion of the dead weight to be moved, it is desirable that the tubs should be of large capacity, and the same quality is required by other circumstances connected with the matter of haulage. But a limit in this direction is fixed by the necessity for having tubs capable of being readily handled. It must be borne in mind that the tubs have to be dragged or pushed along the working face; that they have to be lifted and turned at the junction of lines that run in directions perpendicular to each other; and that, in consequence of the imperfections of the road, they frequently get off the line, and have to be lifted on again with little delay. Also, the onsetter and the banksman are required to drag and push the tubs over the tram-plates at the bottom and the top of the shaft, and to quickly run them on or pull them off the cage. Hence it is highly important that the weight should not be too great for one man to deal with. This condition will limit the capacity of a tub, irrespective of other considerations. Moreover, as economy often requires that the operations of haulage should be performed or conducted chiefly by boys, the weight to be dealt with should be kept within the limits of their strength. For these reasons, a capacity of 8 cwt. is not often exceeded. Another circumstance that tends to keep down the capacity of a tub is the narrowness of the roads in a mine, for as the dimensions can be increased only in one direction, the limits of convenience are soon reached.

The height of a coal tub is limited mainly by three circumstances, namely, the stability of the vehicle in transit, the difficulty of loading it at the working face, and the thickness of the seam. It has been shown that to comply with the conditions prevailing underground, tubs have to be made narrow, and that, moreover, the curves there existing are very sharp; hence it will be evident on reflection that height is inconsistent with that degree of stability which is requisite. Also, it will clearly appear that height in a tub is unfavourable to the operations of loading, since the mineral has to be lifted into it. This is a question of very considerable economical importance, and it is deserving of more attention than has hitherto been given to it. But irrespective of these limiting circumstances, that of thickness of seam may operate to compel the adoption of tubs of low height. When a seam of coal is thin, it becomes highly desirable, if not absolutely necessary, to use vehicles of such dimensions as will not require the expenditure of additional labour in ripping down the roof in order to give sufficient height; and as this circumstance limits the height, it will also influence in some degree the capacity of the tub and the diameter of the wheels. Thus numerous conditions combine to limit the dimensions of coal tubs, and it will be prudent to keep within the limits imposed in designing the rolling stock of a colliery. In some instances, tubs having a capacity of 11 cwt. have been adopted, but it must be obvious that the disadvantages incurred by the adoption of such cumbrous vehicles more than compensate the gain.

The following examples of tubs exhibit the forms actually in use, and show the various devices adopted for obtaining the greatest possible capacity for given dimensions and for satisfying the other requirements of underground haulage. Figs. 352 to 355 represent wooden tubs used in England. They are strongly built, and experience has proved that when constructed in this way the cost of maintenance is very little. Three forms are illustrated. In those shown in Figs. 352 and 353 the body is prismatic, and extends over the wheels; in Fig. 354 the body is pyramidal in form, and is brought down to the level of the axle. The latter form is very commonly adopted for coal and other mineral tubs. In different parts of the country, the design and construction of tubs vary somewhat from those illustrated, but in all essential particulars they resemble one of these types. Even when iron is employed as the material for the body, the same forms are adhered to as best fulfilling the requirements of underground carriage.

In Continental countries, the question has received greater attention than in England, in consequence of the greater irregularity of the seams and the increased difficulties of haulage, and hence we find greater variety in the form of the tubs employed. Many of these have been designed to suit special conditions, and are, therefore, not generally applicable. Others, however, have been carefully considered, and constructed, to comply with ordinary requirements, and these are deserving of the attention of all practical men. Some years ago, a commission of engineers was appointed by the proprietors of the great Anzin collieries, to examine and report on the coal tubs employed in France, Belgium, England, and Germany, for the purpose of obtaining the best possible design for the new

HAULING AND HOISTING MACHINERY.

rolling stock. The result of the labours of this commission was the adoption of the design shown in Figs. 356 and 357. The body, which is of iron, is rectangular in form, and slightly bellied. Its

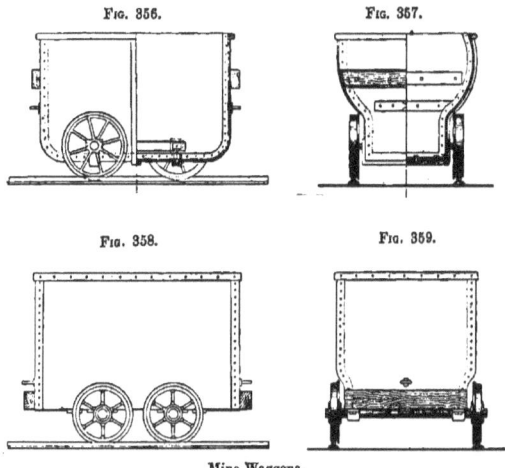

FIG. 356.   FIG. 357.

FIG. 358.   FIG. 359.

Mine Waggons.

length is about 3 ft. 7 in., its breadth 2 ft. 6 in. and its depth 1 ft. 10 in. These dimensions must, for ordinary circumstance, be considered as somewhat excessive. The axles, which are of the finest quality iron, turn upon steel bearings, which allow considerable play in all directions. One wheel is fixed upon the axle, and the other is loose, and the pairs are arranged so that there is one fixed and one loose wheel on each side. Another form of iron tub is in use at the Blanzy mines, and it appears to have satisfactorily fulfilled the requirements for which it was chosen. This type (Figs. 358 and 359) is also the outcome of a careful study of the conditions to which it must be subjected. The body is rectangular and is slightly narrowed towards the bottom, where it passes between the wheels. The wheels are loose, and, besides this arrangement, the axles themselves are allowed to rotate, so that upon curves, or in case of a defective state of lubrication, the resistance to traction due to friction cannot be great. The oval form of the journal box allows the wheels to remain upon the rails, whatever the irregularities of the road may be. The distance of the wheels apart is maintained by means of loose washers, well greased. This arrangement allows a certain degree of elasticity. Many years of experience at Blanzy has shown that, in consequences of these several devices, the tubs rarely get off the rails, though the roads are in many parts very undulated and irregular. By elbowing the axles, the bottom of the tub may be brought as low as desired. Wrought wheels stamped from a single piece of iron are much lighter and less liable to break than the cast-iron wheels.

The Pagat is a broad-faced wheel of small diameter. The oil-box appears to be the chief merit of this wheel. The end of the axle protrudes within it, and the wheel is held in place by a simple spring linchpin inserted through one of the large holes made in the hub to permit the introduction of the grease from time to time. These two openings are closed by corks only. A hard grease or tallow is used, and is inserted by means of an injector. When by movement of the wheel the axle warms a little, the grease slowly melts and runs into the bearing so gradually that it need not be renewed oftener than twice a week.

In Germany, a loose-wheeled tub is generally preferred. It is the old German *Hund*, or "dog-sledge," mounted upon wheels suitable for running upon rails. It consists of a rectangular box or body supported upon two fixed axles, the wheels being on the outside of the body. The weight of this tub is about 4 cwt., and its capacity for coal about 12 cwt. In an improved form the sides of the body are curved to increase the carrying capacity, and the wheels are set beneath the body to enable it to run in narrow ways. The upper edges are bound with iron straps, and in each of the upper angles there is a stout iron eye. The use of the latter is to allow the tub to be attached directly, by means of four short chains, to the drawing rope in narrow shafts. The wheels are loose. The weight is about 4½ cwt., and the capacity, measured for coal, about 6½ cwt. This tub is strongly built.

Figs. 360 and 361 represent a German coal waggon, used chiefly for the transport of coal at surface. The body is of the pyramidal form, and is strongly bound at the angles and the upper

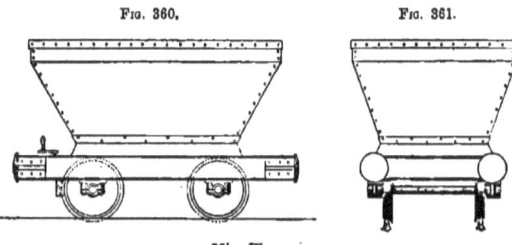

Fig. 360. Fig. 361.

Mine Waggons.

edges with iron straps. It is set upon a stout wooden framing, the sides of which are prolonged to form buffers. The wheels are set beneath the body and are fixed to the axles. Two of the wheels are provided with a brake. The weight of this waggon is about 9 cwt., and its carrying capacity about 16 cwt.

A coal waggon similar to the foregoing, but of lighter construction, is shown in Figs. 362 and 363. In this, the framing is reduced and the buffers are omitted. The same form is preserved in the body, but the wheels are made to run loose upon the axles. A modified form of brake is also applied. The weight of this tub is about 8 cwt., while its carrying capacity is about 16 cwt.

The tub in common use in the Californian mines is shown in Figs. 364 and 365. It is made of wood, and has a capacity of 16 cwt. The body is made of plank 1½–2 in. thick, lined with sheet iron, and strengthened with iron bands on the outside. The inside dimensions are 3 ft. 10 in. long, 2 ft. broad, and 2 ft. 4 in. deep. The trunk or framing upon which it is supported consists of a strong rectangular frame, the two longitudinal pieces of which have their front ends bevelled off, to allow of the body being "dumped" or tipped. A cross timber near the middle of the framing supports the body, and an iron pin attached to the bottom of the body passes through the latter, and serves as a pivot on which it may be turned to either side and tipped. Another cross timber on the

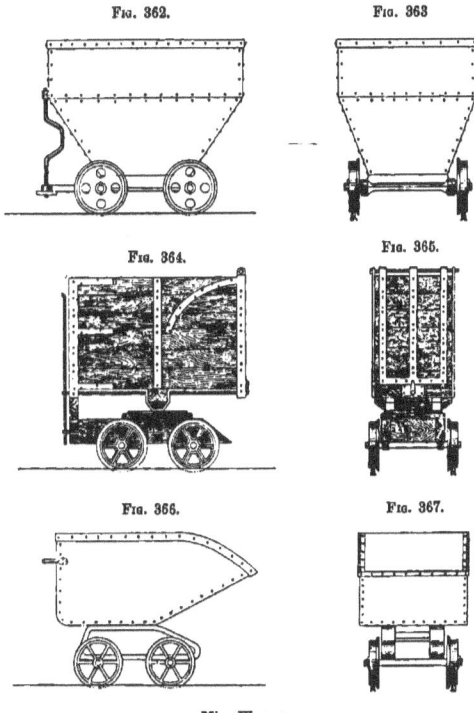

Mine Waggons.

framing supports the hinder end of the body. The wheels are of cast iron, and turn loose on the axles. The diameter of the wheels is 12 in. A little cap may be screwed on to the wheel over the end of the axle, to retain the lubricating oil and to exclude dirt. The wheels are beneath the body. The front end of the body is hinged at the top, to swing as a door for the discharge of the contents. It is closed by a button, that may be turned up to confine the door, or turned down to release it; the button is fixed on an iron rod passing under the body to the back end, and is controlled by the man who pushes the tub before him. An iron rod at the back end of the tub, which, when adjusted for that purpose, serves to prevent the body from swinging on its pivot, is so connected with the rod on which the button is fixed that the door of the tub may be opened, and body made free to swing to either side by one and the same movement on the part of the man in charge. The weight of this tub is about 4 cwt.

Tipping tubs, tipping or "teeming" waggons, or "tipplers," are extensively used in some mining operations. A simple form is shown in Figs. 366 and 367. The fore part of the body is made

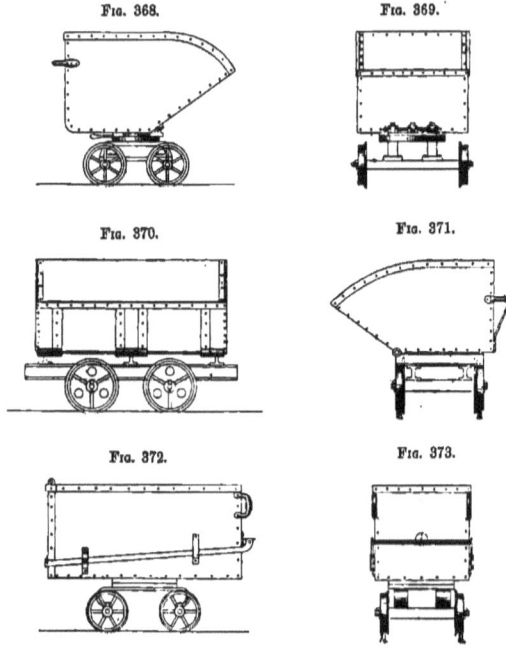

Mine Waggons.

sloping to facilitate the discharge of the contents, and means are provided whereby the body may be readily tipped forward. The weight of this waggon for a carrying capacity of 16 cwt. of coal, is about 8 cwt. ; and for a capacity of 8 cwt., about 5 cwt.

In Figs. 368 and 369, the body is supported in a manner that allows of it being tipped in any direction. Under some circumstances, this is of considerable importance. The additional parts required in the construction of this waggon increase its weight by an amount varying from 3 cwt. to 1 cwt., according to the carrying capacity of the body. The weight is about the same whether the body be constructed wholly of iron, or of wood strongly bound with iron, and supported by bolts.

Another form of tipping waggon is shown in Figs. 370 and 371. In this, one side of the body is made sloping, and means are provided for tipping in that direction. In other respects, the construction of the body is the same as in the two waggons last described. The carrying capacity of the side tippler is the same as that of the forward tippler, and the weight is about the same. This waggon may be constructed to tip to either side, as required, the additional weight of the extra parts needed in such a case being ½-1 cwt., according to the capacity.

Figs. 372 and 373 show a tipping waggon provided with a door. The body is rectangular, and constructed of wood, the angles and the upper edges being bound with iron. The door is hinged upon an iron bar, and is hung to open upwards. The framing upon which the body is supported is of iron, and the mode of support is such as will allow of the body being turned and tipped in any direction. Sometimes a wooden framing is adopted. The carrying capacity of this waggon is about 16 cwt. of coal, and its weight about 11 cwt.

Fig. 374 and 375 shows an arrangement by which waggons drawn up inclines will tip themselves. The waggon a, running upon four wheels b, is drawn up by the bow f, and the rope j. The bow is

FIG. 374.   FIG. 375.

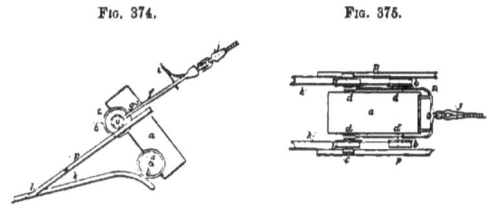

Self-tipping Waggons.

attached to the axles of the hind wheels, and in front it carries the doors of the waggon. k is the railway at the top of the incline, and p an additional outer line of rails on a steeper grade. When the waggon in its upward course reaches the point l, the rails p pick up the small outer wheels c on the hinder axles. These travel up the steeper grade, while the front wheels follow the rails k. Consequently the waggon is tilted, and, as the front end or door is attached to the bow, the contents are shot out. The stud g keeps the waggon in position, if it is drawn up too far. On lowering the rope, the waggon rights itself, and descends properly.

In the Transvaal and similar mining districts, where the cost of transport is very heavy, it is advisable to reduce the weight and measurement of plant as much as possible. Most mining waggons are more bulky than heavy, reckoning on the regulation = 40 feet to the ton, as charged for by shipowners and transport people.

With a view to reduce the transport costs, Kerr, Stuart & Co. have designed a waggon as shown in Fig. 376, in which the whole of the frame, wheels, axles, axle boxes and draw gear pack inside the box, thereby reducing the bulk to the smallest limit.

Tipping Cradles.—Figs. 377, 378, show a "tipping," "teeming," or "dumping" cradle, made to quickly empty itself on reaching the bank. The loaded tub is run into the cradle, and the latter is quickly tilted into the position necessary to allow the contents of the tub to run out. The cradle is then readily turned back into its first position, and the empty tub is run out.

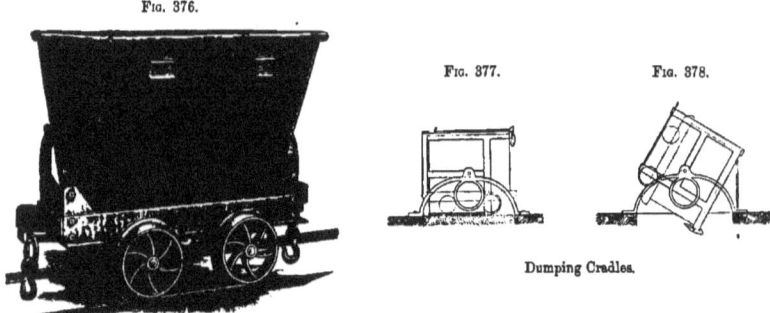

Fig. 376.

Kerr, Stuart, & Co.'s Waggon.

Fig. 377.   Fig. 378.

Dumping Cradles.

The Frongoch skip (Fig. 379) patented by Kitto, Paul, and Nancarrow, is described by Dr. C. Le Neve Foster as a valuable and welcome invention, emptying itself automatically on reaching the top of the shaft, and then righting itself, without the aid of a lander, as soon as it is lowered. The

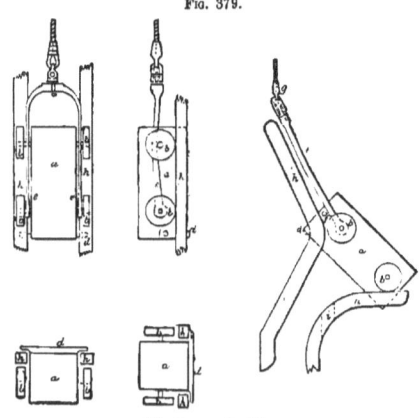

Fig. 379.

The Frongoch Skip.

time occupied in lowering the skip on to a door, knocking up a bolt so as to discharge its contents, closing it again, and raising the skip so that the door may be drawn back, is all saved, and the services of the lander are entirely dispensed with. The skip is the usual box a, made of sheet-iron or sheet-steel, with four wheels b running on the vertical wooden conductors h, and prevented from leaving them by the back (guide d), at or near the bottom. The bow or loop e, instead of being attached to the top of the skip, reaches down, and is attached to the axles of the bottom wheels. It rests against the axles of the upper wheels, and the skip is thus prevented from falling away from the guides. At the surface each perpendicular conductor terminates by a curved piece, and a front guide h is added on each side. When the skip comes up, these front guides press upon the top wheels and turn them on to the flat ends of the conductors. The partial cutting away of the conductors at i enables the back guide to pass through, and the bottom end of the skip is now raised up, and the contents are tipped or "dumped" into a large bin or pass, from which the ore can be drawn away at pleasure. If the engineman does not stop at once, the skip is simply drawn a little way up, resting upon the front guides c, the stop or stud f preventing it from assuming a wrong position.

FIG. 380.

FIG. 381.

Cornish Skip with Safety Catch.   Cornish Skip.

As soon as the engineman begins to lower, the top wheels drop upon the flat ends of the conductors, and pivoted upon these top wheels the tail end of the skip drops, the back guide passes through the slot i, and the skip, resuming its upright position, descends the shaft. One great recommendation of this system is that it can be applied to existing shafts, whether perpendicular, inclined or crooked.

Figs. 380, 381 show skips of the usual pattern employed in Cornwall, made by Harvey & Co., Hayle, who also supply many other forms. Fig. 380 is provided with a safety catch, which brings the skip to a standstill immediately the rope breaks, by imbedding itself into the wood guides in which it runs. Fig. 381 has no such safety catch run out. Several tipping cradles are required at the pit mouth of extensive collieries. The tubs are run up to them upon rails, or upon a flat surface covered with boiler plate.. In the latter case, guides are used to direct the tub upon the cradles.

Another kind of tipping cradle is shown in Fig. 382. It consists of two side discs fixed upon iron arms radiating from the axle, and connected at the top and the bottom by bars of angle iron. Upon the latter is fixed the angle iron upon which the tub is run into the cradle. A pinion is fixed upon the axle at one side of the cradle, and an endless screw turned by a winch handle gears into the pinion. The loaded tub being run into the cradle, the winch handle, which is supported by a vertical iron pillar, is turned by the man in charge until a sufficient inclination is given to the rails upon which the tub rests to cause the contents of the latter to be discharged. A few backward turns of the handle then restores the tub to the horizontal position, and it is run out to make room for the next.

In the cradle last described, the tub is tipped endwise: but sometimes it is desired to discharge the contents over the side. To allow this to be done,' the construction is modified as shown in Fig. 383. The rails are laid from end to end, and the tub is run into the cradle through the

FIG. 382.   FIG. 383.

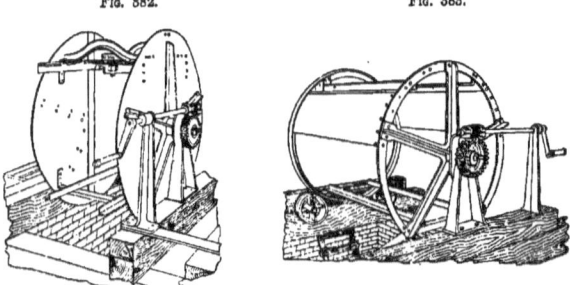

Side-tipping Arrangements.

opening left for that purpose. As this opening renders a central or axial support impossible, that end of the cradle is made to rest upon friction wheels. Over the tub is set a kind of shield to prevent the contents of the tub from being scattered. The tub may be tipped to either side. The means for tilting the cradle are the same as in Fig. 382.

PERMANENT WAY.—The importance of the chair and the sleeper for promoting the efficiency of the permanent way is sufficiently indicated by the great attention paid to the perfection of these details. The improved chair and sleeper shown in Fig. 384 is certainly nearer perfection than most of the arrangements at present in use. It will be seen from the illustration of the sleeper and rail fastening, that only one fish-plate is used at the joint, this fish-plate being provided with studs very slightly tapered, which are adopted instead of the bolts generally used. The clips for the joint and the intermediate sleepers are riveted, and for light lines, such as are used at collieries, while the vibration from the traffic will be scarcely perceptible, this form of construction will be found effective, and at the same time simple. These patent steel sleepers are made by the Chair and Sleeper Company, Limited, of Widnes, Lancashire. In Fig. 384, A is a swivel clip permanently riveted to the sleeper and free to revolve, while B is a permanent clip, which for light rails is only riveted to one side of the sleeper.

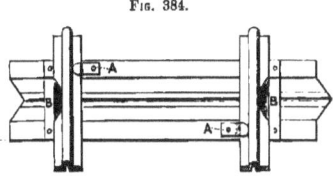

Fig. 384.

Improved Chair and Sleeper.

The sleeper packs itself in the ballast, and requires no further packing or attention of any kind after it is once laid down. The ballast is not shaken from under the sleeper, as is the case with the bulk of the steel sleepers now in use; and its great depth gives it unusual stability and strength. Another advantage is that there are absolutely no loose fastenings to get mislaid or lost in the ballast. The rail is held by a spring clip on the inside, and this has such inherent elasticity as to firmly hold the rail in its position when once driven home, without the possibility of it working loose; and further, by a most simple arrangement which cannot fail, it will take up any wear that there may be in the course of time, helping in this way to keep the fastening tight and firm. The joint sleeper has all the advantages of a "fished" joint, without the expense and trouble of the loose fish-plates and necessary fish-bolts, so that the first cost of the railway is considerably reduced, to say nothing of the absence of subsequent yearly maintenance charges, as there is nothing to work loose, and in addition, the rails can be laid entirely by unskilled labour, a great advantage for foreign countries. The sleeper is actually strengthened in the part where the rail lies, instead of being weakened, as is the case with most steel sleepers. As to practical tests, this sleeper has been tried on the permanent way of the North Staffordshire Railway with unqualified success.

Special attention should be called to the Company's improved rivetless fastenings for colliery rails to steel sleepers. We have met with nothing so efficient, so simple, and so admirably adapted to the purposes which it is intended to serve.

JUNCTIONS AND TURNTABLES.—The underground roads of a mine frequently intersect each other at a great angle, in numerous cases perpendicularly. Under such conditions, upon a common railway, the junction is effected by means of a turntable; but upon underground lines its cost and somewhat complicated construction render its use impracticable by reason of the great number that would be required, and the frequent occasion for its employment. The means adopted, however, are similar in principle, though the details have been varied to obviate the disadvantages alluded to. They may be said to consist of a fixed table, Fig. 385, upon which the tubs are turned by being lifted at one end and carried, or by being dragged round by the men or boys in charge of them. This table or platform is constructed of stout planking, carefully laid, and usually covered with iron plates, to diminish the friction and to lessen the wear and tear. The construction may be varied to suit the requirements of the case; but it will always be of a very simple character. The chief points demanding attention are to lay the floor evenly, and to give the structure sufficient stability to bear the somewhat violent strains thrown upon it. The ends of the rails are brought upon the flooring and made to curve outward, and between these curved portions, ribs, or raised guides, curved in the contrary direction and brought together at a point, are placed; the object of this arrangement is to facilitate the entrance of the tubs. The space in the centre is left clear to allow the tub to be turned round and directed as required. Sometimes a circular rib or guide is placed in the middle of the floor. The diameter of this guide is slightly less than the gauge of the line, the object being to keep the tub in the middle of the floor, and thus opposite the entrance to each of the lines, while it is being turned round. The system is applicable to the junction of the lesser roads, the pass-bys, and the points where trains of tubs are made up and distributed. It is unfavourable to the use of fixed wheels.

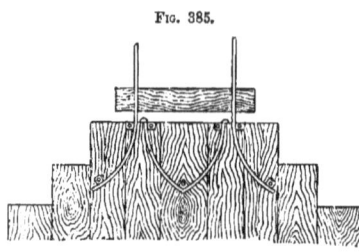

Fig. 385.

Junction.

At some of the more important intersections of the secondary with the main lines, it becomes desirable to use a turntable. At these important points, and also at some points at surface, the system just described would necessitate much labour, and would, moreover, cause great wear and tear to the tubs, and occasion delay in transit.

SHEAVES AND PULLEYS.—Besides the sheaves or winding drums, which are driven directly by the winding and hauling engines, and which will be treated of later, other sheaves, reels, and pulleys are required for the purposes of underground haulage. At the top of a self-acting plane, a reel or drum is needed to hold the rope or the chain by which the full tubs are lowered and the empty ones raised. Wherever such a system of haulage is adopted as those known respectively as the "tail-rope," the "endless-rope," and the "endless-chain" systems, several sheaves may be required, and numerous pulleys at the angles and curves in the roads. The uses of these will be pointed out as they are described. The construction of such sheaves and pulleys is very simple.

A sheave, reel, or drum, upon which is set a brake to control its motion, is the only mechanism required to work a self-acting plane. The apparatus may be made to revolve about either a vertical or a horizontal axis; the latter arrangement being the more common. In the former case, only one rope is used, which is passed round the sheave, one end being attached to a tub at the bottom of the plane, and the other end to a tub at the top of the plane. In the latter case, either two ropes coiling in contrary directions may be used, or a single rope sufficiently long to be passed several times round the reel. In some instances, an endless rope is employed, which is made to pass over a pulley at the bottom of the incline, and kept in a state of sufficient tension by means of a counterweight connected to the pulley. With the endless rope, the tubs are required to succeed each other with great regularity; the full tubs descend on one line of rails and the empty ones ascend on the other. Whatever the arrangement adopted may be, a powerful brake must always be provided. This brake may be of very simple construction, consisting merely of a segment of wood fixed to a lever, and arranged to be readily brought into contact with the periphery of the sheave. To obtain greater power, compound levers may be employed, and the same object may be attained by means of an iron band enclosing the whole of the periphery and worked by a system of jointed levers. The mode of applying a brake is a matter of some importance. It is evident that the brake may be so arranged that when left to itself it shall be in operation, or the arrangement may be such that the brake shall cease to act when left to itself. In the former case, the force is applied by means of a weight attached to the end of the lever; in the latter, the force is applied by hand. The former method of arranging the brake is generally the better, as it offers greater security against the negligence of the brakesman. It is well to adopt the principle that the apparatus of a self-acting plane should be incapable of setting itself in motion without the intervention of the brakesman. The latter, to start the tubs, releases the brake, and holds it wholly or partially released during the time of the descent of the load. In this way, he is able to regulate the motion, and to arrest it easily at the proper moment.

In order that the brake may be capable of controlling the motion of the load, as well as that of the sheave or reel, it is necessary to arrange the rope in such a way that it cannot slip. With the horizontal reel, this may easily be effected by passing the rope 3 or 4 times round it. But with a sheave turning about an axis that is perpendicular to the plane, this expedient cannot be so readily adopted. One turn round a sheave having the ordinary kind of groove would be insufficient to prevent slipping. In such a case, the groove may be made conical, so as to grip the rope, or one of Fowler's clip pulleys may be used. Another method consists in passing the rope several times round the sheave, and providing an arrangement by which the friction of the several turns of the rope against each other is avoided. The arrangement is merely the addition of one or more parallel grooves to the sheave, which is then put in relation with another sheave, of any diameter, provided with one groove less. The rope is wound and unwound upon this sheave regularly, as upon one of the ordinary kind. See Figs. 386 and 387.

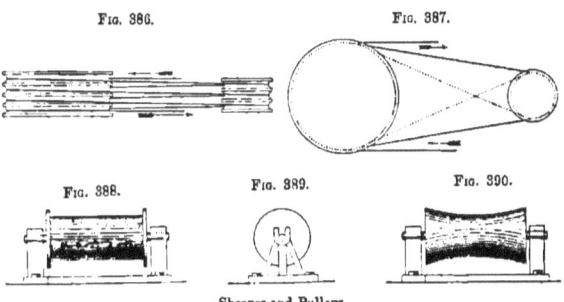

Fig. 386.   Fig. 387.

Fig. 388.   Fig. 389.   Fig. 390.

Sheaves and Pulleys.

The friction of the rope upon a self-acting plane is considerable; and as this friction not only absorbs the motive power, but causes a rapid wear of the rope, it is very important that it should be reduced as much as possible. This is accomplished by means of friction rollers, placed at sufficiently short intervals apart throughout the plane to prevent the sag of the rope from causing contact with the ground. These friction rollers should be of a considerable diameter relatively to their gudgeons, which should be kept well greased, for otherwise they will not turn, but constitute fixed points of support to the rope. Two forms of friction rollers, with the method of fixing them, are shown in Figs. 388 to 390. The latter form is used on those portions of the road where the rope has a tendency to oscillate.

The reels and sheaves used upon self-acting planes should be of a simple character, and fixed in a manner that renders them capable of being easily shifted from point to point as the workings progress. Common forms are shown in Figs. 391 to 393. In the former of these, a sheave is fixed in the middle, upon which the brake acts. The reel turns in bearings fixed upon two props securely

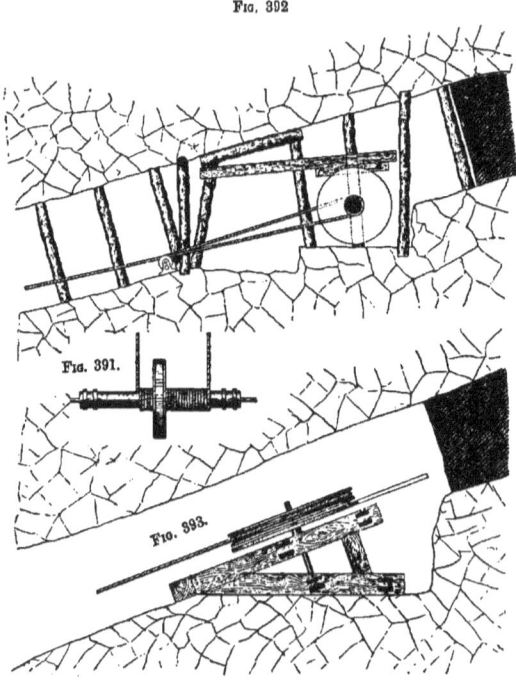

FIG. 392

FIG. 391.

FIG. 393.

Sheaves and Pulleys.

set in the roof and floor rock, and the posterior end of the brake may be fixed upon a prop, or in any other manner that may seem more suitable. The other end of the brake will be handled by the brakesman; or, by compounding the leverage, will be connected to a second smaller iron lever, an arrangement that gives greater control of the apparatus. The segment embracing a portion of the circumference of the sheave is bolted on to the lever, so as to allow of its being readily replaced by a new one when it is worn out. Soft wood should be used for these blocks. In Fig. 393 the sheave, which is provided with a conical or V groove, has its axis perpendicular to the plane. The rope in this case passes once round the sheave, upon which a brake may be made to press by any convenient arrangement. The wooden framing carrying the sheave is fixed down to the floor by means of stout iron cramps, driven into holes bored in the rock. An apparatus of this nature, like the preceding, may be quickly removed to a higher point as the workings progress: this is a quality of considerable importance in such apparatus, which has frequently—sometimes every two or three days—to be shifted higher up the plane. When the inclination becomes great, Fowler's clip pulley, with which a single turn of the rope will be sufficient, may be used. This pulley (Fig. 394) grips the rope with a force proportionate to the tension upon it.

Sheaves are required to lead the ropes and chains round curves in the several systems of haulage adopted on underground roads. In Fig. 395 is shown a sheave for carrying the rope at such points for the tail-rope system. It is usually set in walling built for the purpose. The construction is very simple.

In the endless-rope system, it becomes necessary to keep the rope very tight, as otherwise it is liable to slip. This is effected by the arrangement shown in Fig. 396. It consists of a carriage running on wheels, to the hinder end of which a chain is affixed. This chain passes over a pulley and down a pit or staple, the lower end being weighted. The weight descends as the rope stretches,

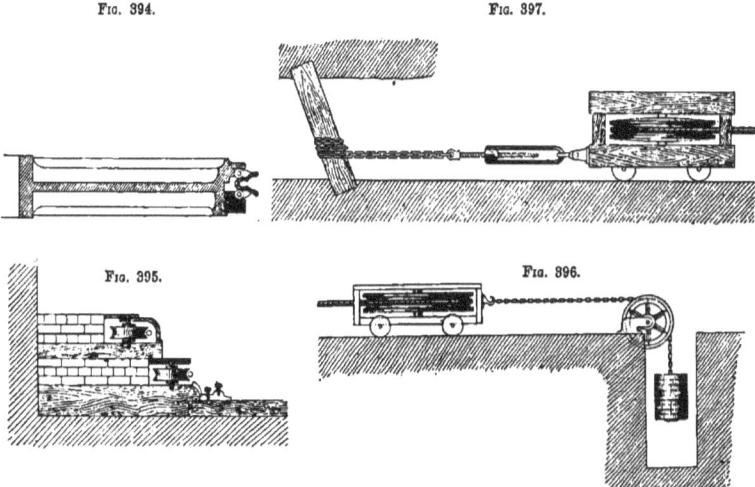

Fig. 394.   Fig. 397.

Fig. 395.   Fig. 396.

Sheaves.

and in that way keeps the latter at the same tension. The weight used is about 15 cwt. Another form of tightening pulley is shown in Fig. 397. The pulley is fixed upon a strong timber frame to which a screw is attached. The screw is secured by a chain to a balk of timber set nearly vertical. The rope is kept tight by turning the screw as occasion requires.

CONNECTIONS.—Among the means of connecting the tubs to the ropes and chains used in haulage, the following merit attention:—

Fig. 398 shows the connection used between the set of tubs and the rope in the tail-rope system. It consists of a knock-off link secured by a cottar. When the cottar is removed, the link is pushed

FIG. 398.   FIG. 400.

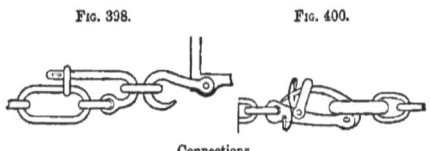

Connections.

off by the foot. Another mode of connection is shown in Fig. 399. The main rope is attached to the sets by fastening the shackle, which is on the end of the chain, to the coupling chain of the end tub, with a pin which is secured in its position by a spring cottar. The tail rope is attached by placing the end link of the chain in the centre bar, and securing it by a pin which is fixed to the end of the tub. Fig. 400 represents yet another kind of link used for these connections.

FIG. 399.

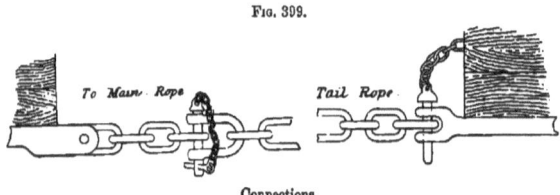

Connections.

A means of connecting the tubs to the rope in the endless-rope system of haulage is shown in Fig. 401. It consists of a chain 6 ft. long, having a hook at each end. This chain, having been connected to the coupling chain of the tub, is thrown over the rope, which is constantly in motion.

FIG. 401.   FIG. 402.

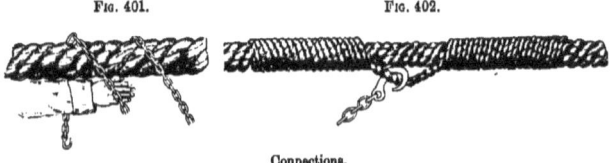

Connections.

It is passed twice over the rope, the hand being introduced under the rope to receive the coils, in order to let the chain slide loosely on the moving rope till the hook is secured. When the two coils have been passed over the hand, the latter is withdrawn, the point is brought over the hook, and the chain is pulled tight. When the weight of the tub comes upon the chain, the coils are drawn close together, and they form a very secure fastening. An expert hooker-on does not need to put his hand between the coils; but he simply passes the chain round the rope, and secures it before the rope has had time to move on. Both the fore and the hinder chains are attached in the same manner.

Another connection of a similar nature is shown in Fig. 402. Instead of the connecting chains being passed round the rope, strong loops of hemp are fastened on to the rope by a wrapping of string, at regular distances apart. One hook of the chain is first attached to the tub, and the hook at the other end is then passed through the loop. The loops are of hemp, 1 in. diameter, and are strong enough to draw 10 or 12 tubs at a time up a considerable incline. Less labour is required to make the connection in this way than in that last described.

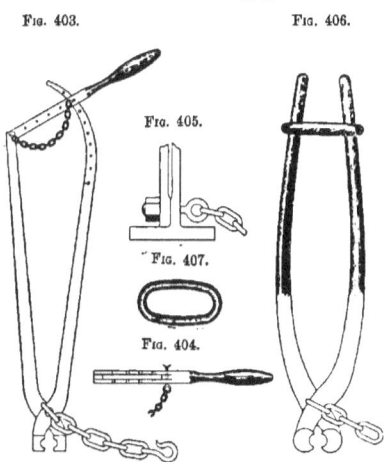

Connections.

In some cases, the rope runs along upon the floor of the waggon-way, beneath the tubs. A kind of clamp is then used to make the connection between the rope and the set of tubs. Two of these clamps are shown in Figs. 403 to 407. The clamp being closed by the lever handle and held by a pin, or by means of the link, the rope is firmly gripped. The set of tubs is connected to the clamp by a short piece of chain. Such clamps are worked by a man who rides in the first tub at the front end of set.

CAGES.—It was formerly the custom to tip the coal as it arrived at the shaft, into vessels of various forms, in which it was raised to bank. This vessel being allowed to swing loose in the shaft, rendered it impossible to wind at a high speed. Moreover, it was necessary to adopt some arrangement whereby the ascending vessel was prevented from coming into contact with the descending one, when two were used in the same shaft. This system of winding was very slow and insecure, and in consequence of the jolting occasioned by the vessel striking against the sides of the shaft, both it and the rope were speedily destroyed. Another disadvantage of this system was the delay and the injury to the coal occasioned by tipping it into the vessel at the bottom of the shaft, and by tipping it out again at surface. The necessity for raising a large quantity of mineral in a given time, for obtaining that quantity in a better condition, and for providing a system of winding more secure to life and limb, led to the adoption of cages moving between guides. These so-called cages are iron constructions, made to contain one or two or more tubs, which are in this way raised through the shaft with their contents. The tub is run on to the floor of the cage at the bottom of the shaft, and off again when the cage has arrived at surface. Thus the objections to the transfer of the load from one receptacle to another are altogether obviated. Also, as the cages are made to run between guides, they may be raised and lowered at a high speed with perfect safety. In some pits, the load is raised with a velocity of 20 ft. a second. One serious disadvantage attending this system is the great increase of the dead weight to be raised in the shaft. But this disadvantage is much more than compensated by the gain in the directions already pointed out. This additional dead weight remains, however, an important matter to be dealt with by mining engineers, the question being how to reduce this weight to a minimum.

Cages are merely receptacles for the tubs, or vehicles in which the loaded tubs are transported to surface and the empty tubs returned from surface to the workings. Their use being merely to travel up and down the shaft, they are not subject to any of the conditions which determine the construction of the ordinary rolling stock. The requirements of a drawing cage are: 1, that its form and capacity shall be such as will allow a sufficient number of tubs to be readily placed in it and removed from it; 2, that its form and mode of construction shall be such as will allow it to run easily along its guided path in the shaft; and 3, that its mode of construction and material shall be such as will allow the greatest carrying capacity with the least weight of cage.

The form of a drawing cage is determined, first, by the division in the shaft in which it has to travel; and second, by that of the tubs which it has to contain. Those divisions are always rectangular, and the tubs possess the same form. Hence it has happened that the rectangular form has been universally adopted for the drawing cage. Its capacity is determined chiefly by the requirements of the output. In many cases, it has but one floor, and is then described as "single-decked." This floor may be constructed to carry either one tub, or, what is a more frequent arrangement, two tubs standing end to end. The floor is laid with rails, to facilitate the introduction and withdrawal of the tubs. To keep the latter in their position during their transit to surface, or from surface to the shaft bottom, some kind of catch is used, often a simple latch, which, when hanging vertically of its own weight, projects over the opening into the cage. This opening is left in both of the shorter sides of the rectangle, in order that the loaded tubs may be pulled off at one side, and the empty tubs pushed on at the other. In the two, three, and four-decked cages, we have merely a repetition of this floor at different levels. The top of the cage is provided with an iron bonnet or roof, for the protection of persons riding in the cage. In the middle of the shorter sides are fixed the guide cheeks, when rigid wooden or iron conductors are used. With the flexible wire-rope conductors, rings are provided at each of the angles. The cage is suspended from the rope by four short chains at each of the upper corners, and, in the case of heavy cages, from the middle of the larger sides as well.

Drawing cages are generally constructed of wrought iron, and, as a wide margin of strength must be allowed, the parts are necessarily excessive in section, and strongly put together. These conditions make the dead weight of the cage great, and it is sought, by adopting suitable forms of sections and modes of assemblage, to reduce this weight to the lowest practicable limits. As at present constructed, wrought-iron cages weigh 5–6 cwt. when designed to carry a single tub, and 9–10 cwt. when the carrying capacity is two tubs, whether the cage be a single or a two-decker. A two-decker cage, constructed to carry 4 tubs, may weigh 1–$1\frac{1}{4}$ ton, or even more. Thus the dead weight of the drawing cage constitutes a very important item in the load to be raised. Successful attempts have been made to reduce this dead weight by substituting steel for wrought iron, and it is probable that this material will ultimately be generally adopted. The cost of drawing cages varies, of course, with the price of iron; but, taking an average, it may be said to range from about 35l. a ton for wrought iron, to about 45l. a ton for steel. The construction of these cages is shown in Figs. 408 to 414. The design shown in Figs. 408 and 409 is for a wrought-iron single-decked cage to hold two tubs; that shown in Figs. 410 and 411 is for a steel two-decked cage to contain four tubs; and that represented in Figs. 412 to 414 is for a steel two-decked cage to carry two tubs, and to be used with wire-rope conductors.

In Staffordshire, a system of winding still prevails similar in character to that of the corves generally in use in former times. Instead of using cages in which to raise the receptacles containing the coal, these receptacles are themselves suspended directly from the rope, and raised in that manner in the shaft. They differ also entirely in their construction from tubs, being composed (Fig. 415) of a platform carried upon wheels, and of two or three large iron hoops. To load these "skips," as they are called, a quantity of coal is stacked upon the platform, and the largest hoop is then placed over it to keep it in position. A second quantity is then stacked up, and a second hoop of a somewhat smaller diameter placed over it. These operations are repeated with hoops of smaller size, until the pyramid of coal has attained the limit of height allowed. The mass is further held together by the four chains by which the skip is suspended from the drawing chain. The load is then drawn by a horse to the bottom of the shaft, where it is attached to the drawing chain. On arriving at surface, it is simply drawn by the banksman from over the shaft mouth by means of a hook, and lowered upon the landing, or he pushes a platform over the mouth of the shaft beneath the load, upon which platform the load is then lowered. The loaded skip having been run off, and its place supplied by an empty one, the latter is raised sufficiently to allow the platform to be withdrawn and then lowered into the shaft. In this system, the winding is necessarily slow.

The drawing cages used in the Californian mines (Figs. 416–417) are similar in design to those already described. The bottom of the cage is a simple platform, 5–6 ft. square, according to the size of the compartment, formed of wrought-iron bars firmly joined together and covered by a floor of wood, provided with pieces of track iron on which to receive the car. The two sides of the cage, above the platform, which are next the guides in the shaft, are formed of a simple but stout framework of iron, 7–8 ft. high, joined at the top by a central cross-bar connecting them, above which is a stem or vertical rod of iron, by means of which the whole is attached to the hoisting cable. The two sides of the cage between the frames are open, for the admission or the exit of car, men, or material with which the cage is loaded.

The cage is guided in its movements in the shaft by two vertical strips of wood, or guide rods,

4 in. by 6 in. in size, attached to the lining of the shaft, one on each side of the cage, and extending from the surface to the bottom.

Attached to the cage on each side, near the top and bottom, are iron flanges, commonly called "ears," so made as to embrace the wooden guide-rods already referred to.

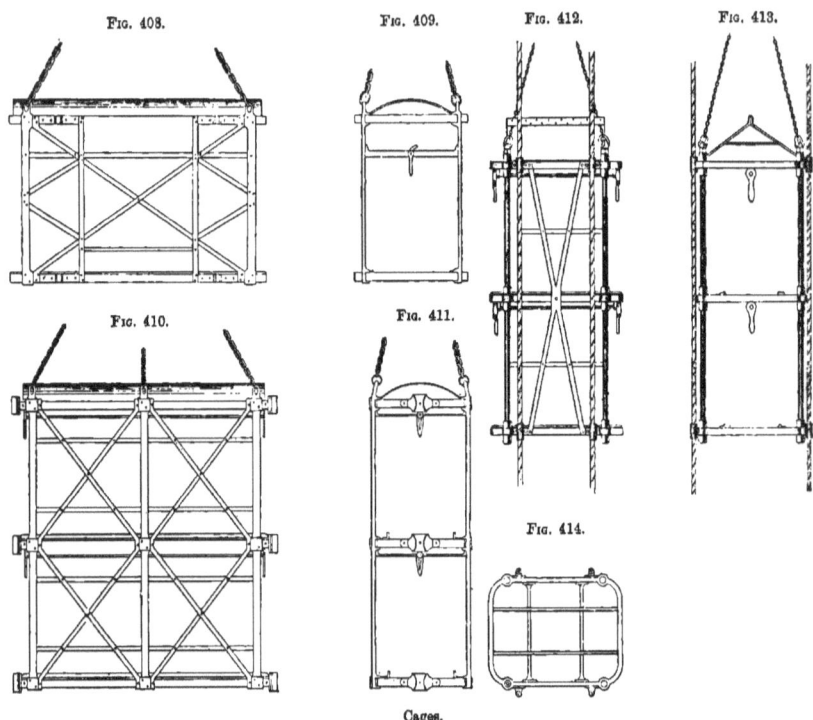

Fig. 408. Fig. 409. Fig. 412. Fig. 413.
Fig. 410. Fig. 411.
Fig. 414.

Cages.

The construction of these flanges is very simple. The wooden guide-rods are in general use, and have replaced those of iron that were formerly employed in some places. They are better adapted to the action of the "safety catches," and permit an easier movement of the cage, while allowing sufficient play to prevent the cage from binding or sticking fast, an accident which is sometimes liable to occur whenever the shaft or the guides are a little out of line, and which is likely to be followed by serious consequences.

Some of the cages in general use are constructed as simply as possible, with the only end in view of providing a suitable platform for the support and transportation of the car or other load.

HAULING AND HOISTING MACHINERY.

Others are constructed with various appliances to ensure safety, so that in case the cable or winding apparatus should break, the progress of the cage may be arrested wherever it may be at the moment of the accident, and so preserved from falling to the bottom with its load. The various devices applied for this purpose to these "safety cages" differ a good deal in detail of construction, and probably in degree of efficiency; but they generally depend on a spring so fixed, with regard to the rod by which the cage is attached to the cable, as to be compressed while the weight of the cage exerts any strain upon the cable, but if that strain is relaxed by the breaking of the cable or other parts of the winding machinery, the spring is permitted to act upon some mechanical contrivance, by means of which stout iron teeth are forcibly projected against, or caused to grasp, the guides along which the cage is moved. The teeth are so arranged, that when the spring is compressed they move along the guide without coming into contact with it; but when the spring is relieved, act with the greater force the heavier the load on the cage.

One of these contrivances may be seen in Figs. 416 and 417. A horizontal movable bar of iron crosses the cage near the top, from side to side. The lifting rod r by which the cage is attached to the cable, passes through this bar, and is so connected with it that the latter may move upward and downward between guides g, according as the rod is raised or suffered to fall. When the rod is raised

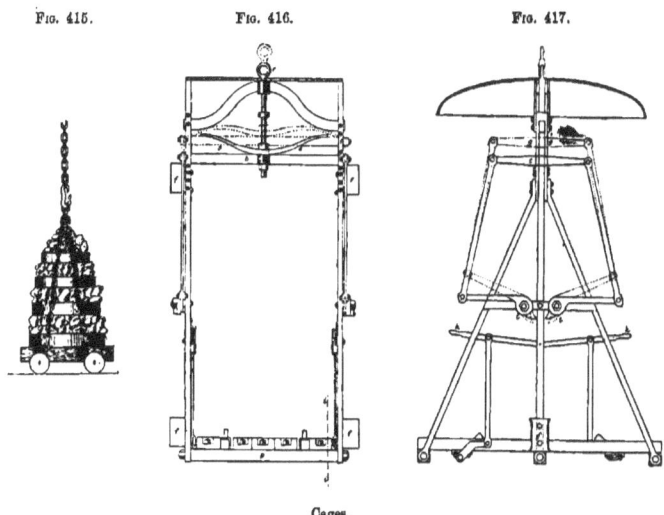

Fig. 415.   Fig. 416.   Fig. 417.

Cages.

by the strain of the cage on the cable, the bar is elevated; but if the strain on the cable is relaxed, the rod consequently falling, the bar moves downward, and a strong spring is introduced to force it down whenever this condition occurs. To each end of this cross-bar, on opposite sides of the cage, is attached at right angles a shorter horizontal bar. To each extremity of each of these last named

2 F

bars is attached one end of a system of levers, by means of which two stout iron teeth or "dogs" t, at the other end, are thrown against the guide rods in the shaft when the cross-bar is down, or drawn from the guide rods when the cross-bar is raised.

In Fig. 417 this contrivance is shown in such manner that the action of the levers can be readily traced. The cage not being suspended by the cable, the cross-bar is depressed, and the teeth are almost in contact with each other, in the position in which they would grasp the wooden guide-rods were the cage in the shaft without its usual support. The dotted lines indicate the position of the levers and teeth when the cage is hanging on the cable and the cross-bar b is raised.

Another appliance for ensuring safety is illustrated in Figs. 418 to 420. The general form of the cage may be the same as in the case already described. The contrivance for ensuring safety consists in two round shafts or rods a, which extend across the cage from side to side, parallel to the central cross-bar b of the main frame. They are supported by the main frame of the cage in such manner that they may revolve freely, and they extend beyond the sides of the cage so that their ends are opposite the wooden guide-rods c of the hoisting shaft. To each end of these two rods are attached the eccentrics a, which are circular pieces of cast iron, supported, as their name implies, in such manner that the centre of the shaft a, or axis of revolution, does not coincide with the centre of the circle. That part of the circumference of the circle which is nearest to the point of support is smooth, but that which is more remote is furnished with teeth, so that when the shafts a are in such position that the smaller diameter of the eccentrics is turned towards the guides, they may move freely, up or down, without coming into contact with the guides; but if the shafts a be turned so as to present the larger diameter of the eccentrics to the guides, the latter are grasped by the teeth just referred to. Each eccentric rod is furnished with a chain e, one end of which is fixed to the rod and, winding round it, is attached at the other end to a bolt, which passes through the cross-bar b. Between the head of the bolt and the cross-bar a strong steel spring f is interposed, the tendency of which when compressed is to cause the shaft a to revolve in such manner as to bring the teeth of the eccentrics into contact with the guides. The chains g, by which the cage is supported, are fixed at one end to the upper part of the lifting rod h, while the other end passes around the shaft a, as seen at i, and is attached to it so that the tendency of this chain, while there is any strain on the cable, is to turn the shaft a in such manner that the eccentric teeth are moved away from the guides. If, however, by the breaking of the cable or other reason this strain be relaxed, the springs f act upon the shaft a, and turn the eccentric teeth towards the guides, thus

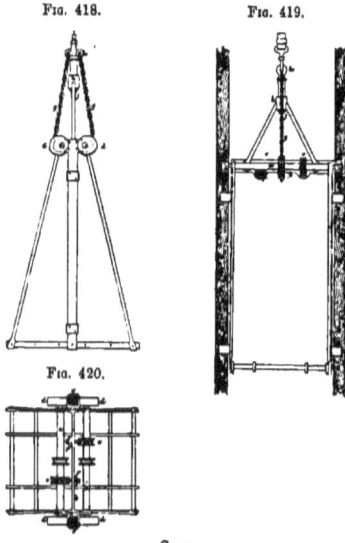

Fig. 418. Fig. 419.

Fig. 420.

Cages.

HAULING AND HOISTING MACHINERY.

preventing the fall of the cage. This movement is assisted by the spring j, which is interposed between the bottom of the lifting rod h, and the ring through which the rod passes. The cage is sometimes furnished with a hood or covering of iron, usually made of boiler-plate, for the purpose of protecting the men from the danger of the cable, if broken, or other bodies falling in the shaft. It is usually hinged in the middle, so that the two sides may be turned up when it is desired to send down long timbers on the cage. Iron hoods underneath the hook serve to keep it securely closed when so desired.

The shackles or sockets used in Cornwall, are described by Frecheville, as follows. The shackle with rivets (Fig. 421, A B); the conical socket (Fig. 421, CD); the double-pin socket (Fig. 421, EF); and the spliced shackle (Fig. 421, G.)

At East Pool (A, B), to put on the shackle, the rope is first lashed round with copper wire about 8 in. from the end, the strands are next untwisted, and the wires turned back singly; some are cut off at different lengths so as to make the requisite taper, and the whole is then bound round with copper wire. The shackle being heated to redness, is, after the tapering end of the rope has been inserted, hammered down to fit it snug. A coupling is then screwed on, and the shackle is brought as tight as possible on the rope. Finally a steel punch is driven through to make place for the rivets, which are put in and fastened in the same way as boiler rivets.

The rope end is manipulated at both South Frances (C D) and Wheal Sisters (E F) in very much the same way as described above; being made of a conical shape like the inside of the socket, it is then pulled back, and a round centre pin of steel is driven up in the middle to wedge it. With the socket used at Wheal Sisters, each chain of the runner passes over a separate heater pin: this is certainly safer. The comparative merits of these attachments have not been ascertained by testing.

The connection made by turning the rope round a thimble and splicing it, as done at Wheal Basset and Gunnislake Clitters (G), if performed by a skilful splicer, is undoubtedly the best.

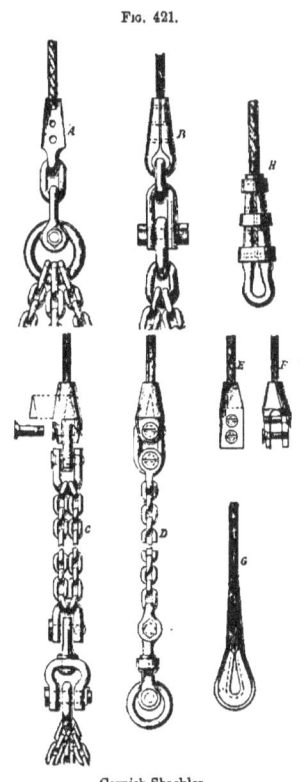

Fig. 421.

Cornish Shackles.

In many of the coal mines they use a shackle or capel (Fig. 421, H), with hoops and rivets, which is fastened to the rope as follows:—The end is untwisted for about 6 in., it is then doubled to suit the length of the capel, the loose end is twined round the main rope, and the whole is bound with hemp twine soaked in tar; rivets with countersunk heads are put through both ropes and the capel; the hoops are next put on and driven home tight. This, though doubtless a very strong connection, is not suitable for passing over pulleys and rolls, as Cornish shackles are required to do.

The screw-heater and swivel, with their pins, should be made of $1\frac{1}{2}$–$1\frac{3}{4}$ in., the runner chains of $\frac{5}{8}$–$\frac{3}{4}$ in., and the coupling chains of $\frac{1}{2}$–$\frac{5}{8}$ in., best wrought iron bar. The pins should be secured in their places by jam nuts. There should be five coupling chains, one at each corner of the cage, and one attached to the centre; the latter carries no weight, but hangs a little slack, and is provided, in case a corner one should break, to prevent the cage tipping to one side and jamming itself in the shaft. The links should be made as short as is consistent with easy play, and those at the extremities a little larger and stronger than the rest.

Chains require frequent and careful examination, as the links may wear into each other without being detected if not well looked after; also, owing to the shocks, jerks, and alternations of temperature they are subjected to when in work, the iron undergoes a change in structure, and gradually becomes hard, crystalline, and liable to snap.

Catches devised for safety-cages are very numerous; one, and perhaps the earliest class, depended upon the lateral projection of bolts intended to catch in a ratchet or ladder-way, which is fixed upon opposite sides of the pit; in another class a similar result is intended to be attained by means of bars, which hang over the cage in the form of a chevron and terminate in strong teeth; another variety operates by embracing the sides of the cage conductors, by eccentrics or by toothed clutches, which are brought into grip through the liberation of a spring, when the tension is relaxed by the breaking of the rope. Some forms are designed to bring the cage gradually to a standstill when relaxed by the fracture of the rope, instead of suddenly, as in the cases already referred to. Thus, in Cousin's contrivance, when the rope breaks, a catch is intended to clutch a second (or safety-) rope, which passes over a pulley at the shaft-top, and has attached to one end a series of heavy weights, resting upon seats. When the rope is clutched by the action of the falling cage, the weights are successively lifted, and the descent of the cage is thus checked by degrees until it is brought to rest in the shaft. Another plan, intended to accomplish the same result, has been applied in connection with two cages, having a balance-rope passing from the bottom of one cage to the bottom of the other. The tops of the two cages are connected by a pair of side-ropes; these pass over pulleys resting on spring-pedestals, which are set on the pit-head frame and act as brakes when a weight comes upon them.

The confidence in the certainty of action of many safety-clutches of the indicated types, entertained in the earlier days of their invention, does not appear to have stood the test of experience in many instances, and it seems to be even considered doubtful whether their adoption is not in some cases attended by the introduction of fresh sources of danger. Abel notes that there are many recorded examples of safety appliances having failed to come into action at the critical moment, even when automatic arrangements of the kind indicated have been supplemented by the provision of a brake-lever, under the control of an experienced operator; on the other hand, many instances are on record of clutches coming into action when not required. Casualties, due to overwinding, have been reduced in number to some extent by the employment of "safety-hooks" designed to disengage the cage, if raised too high, leaving it either to be attached to the guides by the coming into play of one or other of the devices just referred to, or to become suspended by strong catches, in the pit-head frame. A safety-hook which has been employed by Bryham of the Rosebridge Colliery, and others patented by King, Walker, Ormerod, Ramsay and Fisher, operate in the latter way. Although opinions as to the value of safety-hooks are divided, there are many in use in collieries; but more reliable means of protection have recently been provided, in connection with the more modern winding plant, in the shape of powerful steam-brakes which will bring a loaded cage to a standstill within a few feet of travel, and which are even arranged to come into action automatically. These brakes can be fitted to existing plant expeditiously and without practical difficulty.

Lupton has remarked that many mechanical engineers have set to work to devise means for preventing overwinding by disengaging hooks, so that the rope may be separated from the cage and not be drawn over the pulleys. It has unfortunately happened that just when the safety apparatus was most wanted it has generally failed. When winding up at a slow rate it acts effectually, but if it is wound up fast through some mistake of the engineman it is generally smashed to pieces. That shows how much a safety apparatus can be relied upon to prevent accidents. Many different kinds of hooks are used, and Lupton's observations apply to all of them. Safety-hooks are not regarded as of high value by Continental mining engineers, who prefer safety cages, which in their turn are despised by English engineers. Thus safety cages are adopted almost universally on the Continent, and safety hooks in England. An automatic contrivance to regulate the speed of winding, and stop the engine if it goes too far, must be better than a disengaging hook in Lupton's opinion.

Calow has invented a method of having a safety-cage so arranged, that the grip attached to the cage, which, if the rope breaks, prevents it from falling down the shaft, only comes into action when the cage actually becomes a falling body. In all other "cage-catchers," the catch itself is dependent

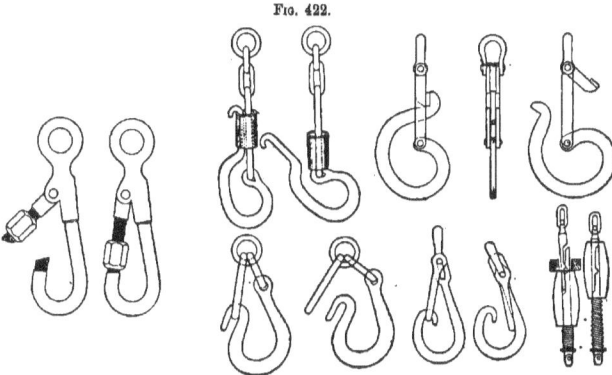

FIG. 422.

Some Safety Hooks.

upon the loosening of the winding-rope, but by means of a simple spring, Calow's is so arranged that the moment the cage becomes a falling body, the grips catch and stop it.

Fig. 422 shows a few representative forms of safety hook.

Attention has been recently invited, in Australia, to an appliance patented by Winks, Cowling, and Hosken, of Castlemaine, for preventing the overwinding of cages. The apparatus, Fig. 423, is

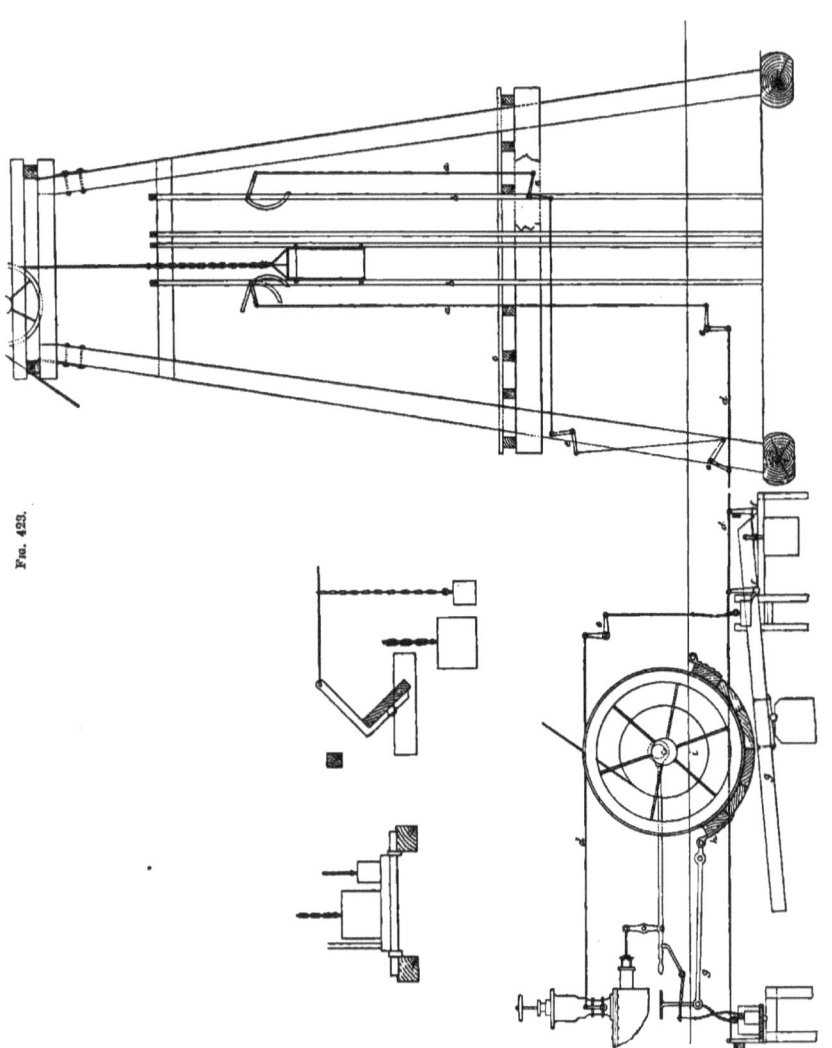

Fig. 493.

automatic in its action. The motion of the cage upwards towards the poppet-heads sets a lever affixed to the skids in action, and the gearing connected therewith then operates upon the throttle

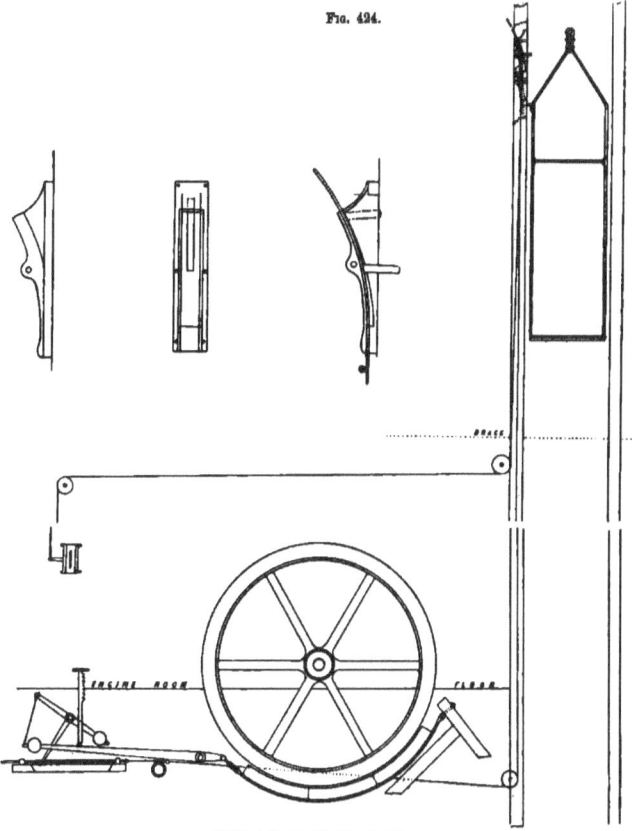

Fig. 424.

Phillips's Safety Winding Appliance.

valve of the engine and brake on the fly-wheel, which at once stops the machinery and arrests the progress of the cage. A public trial of the invention at the Ajax mine, Castlemaine, was witnessed by the Inspectors of Mines, and they report that the experiments made were entirely successful. The apparatus is said to be of cheap construction, simple in form, and easily kept in adjustment. It consists of a motive leader a, affixed to the skids b, about 7 ft. above the landing brace c; the end of this lever is connected by wire rope d, and bell cranks e, with two cams f; on these cams rest the weighted ends of two other levers g. The first of these levers acts on the brake h of the fly-wheel i; and the second closes the throttle valve and cuts off the steam from the engine. The check action to the ascent of the cage is brought into operation by the cage itself. In its upward movement above the brace, the cage compresses the motive lever a, this motion is transmitted by the connecting wire to the cams, the ends of the cams drop, and the weighted levers act on the brake of the fly-wheel and close the throttle valve. The resultant stoppage of the cage at the trial was almost instantaneous.

Fig. 424 illustrates a somewhat similar arrangement by P. E. Phillips.

KEEPS.—When the cage has been raised to the mouth of the shaft, some means are needed for supporting it in that position. These means usually consist of a system of levers, called from their use "keeps," which are raised by the cage as it ascends, and which, by being weighted, drop back into their positions as soon as the cage has passed. With this arrangement, the cage is drawn up sufficiently far above the shaft mouth to allow the keeps to fall back into their position, in which their extremities project slightly over the shaft, and then lowered upon these projecting keeps, which are incapable of further downward motion. The cage rests upon these keeps while the loaded tubs are being run off, and the empty tubs run on. When these operations are finished, the cage is again raised out of the way of the keeps, which are drawn back by the lander, and held by him clear of the shaft until the cage has descended below them. For this purpose, they are connected to a lever, and worked after the manner of a railway switch. In some instances the levers are arranged to be worked by the foot. It is obvious that a system of keeps may be contrived in a variety of ways, so that it is wholly unnecessary to describe any one in particular. Simplicity of construction and strength of parts are the only essential conditions to be satisfied in a design of this nature. It may be remarked that when the cage is two-decked, the operations of raising and lowering upon the keeps have to be repeated for the second level, and that the arrangements at the bottom of the shaft are similar to those at the top. To avoid this repetition, however, the arrangements sometimes include a staging by means of which the loading and the unloading of the cages may be carried on at the different levels at once. This is notably the case in Belgium, where four-decked cages are not uncommon.

With such an arrangement, and a two-decked cage, when the lower deck of the cage at surface is on a level with the shaft mouth, the upper deck of the cage at the bottom of the shaft is on a level with the floor of the roads entering the shaft; the lower deck is here reached by means of an inclined plane. When the cage is single-decked, the arrangements of the onsetting and the landing places, as well as the operations of loading and unloading, are greatly simplified.

One arrangement of these keeps is shown in Fig. 425. They consist of four tappets, two on each side of the shaft, just below the floor. They are fixed upon a light iron shaft which may be partly revolved, turning the tappets upward entirely out of the path of the cage when the latter is to be lowered. The cage, in ascending, striking the tappets, raising them in passing, when they fall again into place, and the cage is lowered upon them. When the cage is ready to descend again, it is first raised a few inches, the tappets are turned up out of the way by means of a lever within reach of the lander, or man who attends to the car, and held in that position until the cage has passed down. *a* is the platform of the cage, *b* are the cross-bars of the frame, to which the tappets *h* afford support. The tappets are fixed on light round shafts below the floor, and

FIG. 425.

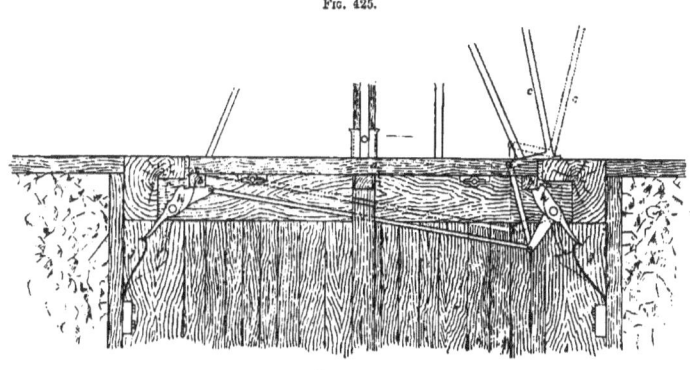

Keeps.

may be turned slightly toward or from the cage by means of levers, one end of which, the handle *c*, is within reach of the attendant. The dotted lines indicate the position of the various parts of this contrivance when the lever is drawn back, so as to turn the tappets out of the way of the descending cage. By this movement the springs *j* are forced into the position indicated by dotted lines, and cause the tappets to return to their former place as soon as the lever is released by the attendant. A similar arrangement is sometimes employed at the different stations in the shaft, though usually, when hoisting is in progress from any particular station, it is common to place a few planks across the shaft for the cage to rest upon.

The waggon, while on the cage platform, is held securely in place, sometimes by hooks fitting into staples in the body of the car, sometimes by tappets, which, being fixed under the platform, may be turned up so as to block the wheels of the car, or turned down again to admit its exit. These blocks are controlled by handles on the sides of the cage.

HEAD GEAR.—The head gear consists essentially of a pulley frame, constructed either of wood or of wrought iron, carrying a pulley, or more frequently two pulleys, over which the rope suspended in the shaft is passed, and led thence to the drum of the winding engine. These pulleys are provided with a round or a flat groove, according to the form of the rope used, and are made of a large diameter, in order to avoid giving a quick bend to the rope. The design and construction of these pulley frames, or head stocks, demand careful consideration, inasmuch as they are extremely important structures, and are required to fulfil various conditions. The two essential features which these structures must possess are height and strength. It is obviously necessary to safety that the

2 G

pulleys, over which the ropes pass, should be placed at a considerable height above the mouth of the shaft, since by this means alone can a margin of safety be allowed to the engine-man. If it be borne in mind, that with the winding drums of large diameter now in use, a single stroke of the engine is sufficient to raise the cage 50-60 ft. in the shaft, the necessity for such a margin will be apparent. For this reason the height of pulley frames is made to vary 30-60 ft. according to the speed of winding. The security of human life, however, demands that in all cases the greater rather than the lesser height should be approached. The condition of strength in the pulley frame is equally or even more important, since it is evident that a yielding of this structure must inevitably lead to disastrous consequences. The necessity for a great height renders this condition difficult of fulfilment, since height in any structure is opposed to its stability.

Hence arises the importance of carefully and fully considering the character and the directions of the strains to which the pulley frame is subjected, and of so designing and constructing it that it may possess ample strength to resist them. The essential parts of a pit-head frame are the legs or uprights, upon which the pulleys rest, and the spurs, or inclined supports, which are set on the sides of the legs next the engine. All other parts of the frame are auxiliary to these, or to some other appendage of the frame. The uprights are intended to resist the vertical strains, and the spurs the oblique strains which tend to overthrow the former in the direction of the source of power, that is, the spurs are intended to prevent the legs carrying the pulleys from being pulled over towards the engine. Thus, in designing a pit-head frame, we have to consider these two parts relatively to the strains to be thrown upon them; and in this consideration we have, first, to determine the direction of the strains; next, the value of those strains; then the best relative position of the parts of the frame; and, lastly, the dimensions necessary to enable these parts to resist the strains thrown upon them.

The condition of stability is that the line representing the strain due to the two forces shall fall within the base of the structure, and that this base is the distance comprised between the lower ends of the uprights and of the spurs. In order to ensure that the resultant shall fall well within this base, the minimum inclination of the spurs should be slightly exceeded. In many pit-head frames the minimum inclination is greatly exceeded; but inasmuch as this circumstance reduces the strength of the spur by increasing its length, the practice is to be condemned as wrong in principle. There is nothing to be gained by increasing the base of the structure beyond the limits required by the condition of stability.

The kind of wood used in the construction of pit-head frames is usually pitch or Memel pine. Though preference is generally accorded to the former, the latter will be found to be very suitable for the purpose, provided it be chosen sound and free from knots and cracks. There are various ways of arranging the several parts of a pulley frame, and also of connecting these pieces one to another.

It is essential to stability that all the chief component parts of the structure should be set upon the same wooden framing by which those parts are securely held together at their bases. This wooden framing consists of sills strongly jointed and bound together, upon which the legs and spurs are set by means of cast-iron sockets bolted down to the sill. Good workmanship is an essential requisite in the construction of pulley frames, since it is important that all the joints should be accurately fitted, and the parts made to abut evenly one upon another. The double tenon joint is generally the most suitable in such structures, and it may be rendered secure by an iron bolt passing through each tenon. Over the more important joints, wrought-iron straps will be required. After the joints have been properly fitted, they should be well covered with red lead. The legs of the frame are slightly inclined to each other towards their summits, and are braced together. The spurs are also in some instances braced to the legs. These spurs, or back-stays as they are frequently called, are sometimes made to abut against the engine-house, instead of being set upon a sill. This practice is, however, to be strongly condemned, as being inconsistent with the requisite degree of stability. In order to obtain the greatest height possible with timber of a given length, the cap or framing carrying the pulley is placed above the uprights and back-stays. As it is necessary that ready access should be had to the pulley, it is usual to provide one of the back-stays with steps, whereby the top of the framing may be reached without difficulty. For the convenience and safety of the person to whom this duty is entrusted, a hand-rail should be added.

The pulleys used on pit-head frames are of iron, and vary in diameter from 10 to 20 ft. When wire ropes are used, the pulley must be of larger diameter, to avoid straining the metal by too sharp a bend. A common diameter is 16 ft. Formerly pit-head pulleys were constructed wholly of cast iron, and this material is still used in the South Staffordshire district, where heavy chains are employed with pulleys of small diameter. But generally this system has been abandoned for the compound system, in which the central boss and the rim are of cast iron, and the arms of wrought iron. The rim of the pulley is grooved to receive the rope, and the bottom of the groove, known as the "face" of the pulley, is made either circular or flat according as round or flat ropes are to be used. It is important that the face of the pulley for flat ropes should be perfectly flat, since otherwise the rope is unduly strained. The groove in the pulley should be sufficiently broad and deep to allow the rope some degree of play. This play is desirable when flat ropes are used, to prevent any ill effects of inaccuracy in the fixing of the pulley, in consequence of which inaccuracy the vertical medial planes of the pulley and of the drum would not be perfectly coincident. But with round ropes the play is indispensable, since the rope, as it is being wound upon the drum, is constantly changing its position relatively to the vertical plane of the pulley.

Figs. 426 to 437 show some of the modes of construction that may be followed, and also some of the details of forming the joints.

Wrought iron has in some instances been substituted for wood in the construction of pit-head frames. The increasing difficulty of obtaining timber of a sufficient length to meet the requirements of the present day, has rendered the adoption of some other material than wood necessary in many cases where great height is desirable. It is evident that with an iron structure, the height is practically unlimited by the material employed; and hence we may obtain an elevation of the pulley above the mouth of the shaft of 70 or even 80 ft. without difficulty. In the construction of iron pulley-frames, the T section is generally adopted in the principal parts, and these parts are braced together by flat or by angle bars, somewhat after the manner of a lattice girder.

ROPES.—The pit rope constitutes the means through which the force developed by the engine is transmitted to the load, and is therefore an object of the first importance. The two essential requirements in a rope are flexibility and strength, and it is desirable to obtain these qualities with the least possible weight. The desirability for a light weight in the rope rests upon two different grounds. In the first place, it is important that the dead weight to be dealt with should be as little as possible; and in the second place, the strength of the rope is, in some degree, dependent upon its weight, inasmuch as the weight of the suspended portion must be subtracted from that of the useful load. Thus, if the distance between the pulley and the pit bottom be 300 yd., and the weight of the rope be 4 lb. a yard, the strain upon that portion of the rope which is upon the pulley will be equal to $300 \times 4 = 1200$ lb. when the rope is unloaded. Hence its effective strength will be reduced by that amount.

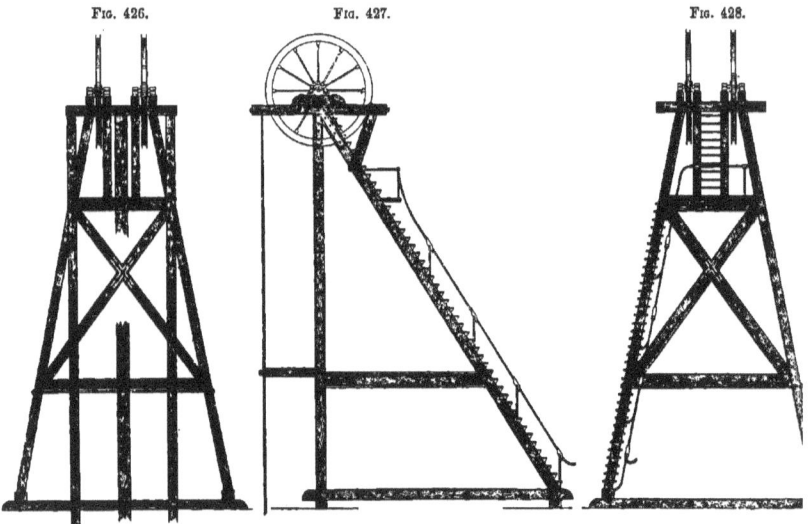

Fig. 426.   Fig. 427.   Fig. 428.

Construction of Pit-head Frames.

In order to obtain these qualities in winding ropes most fully, various materials have from time to time been chosen, and more or less extensively adopted. Hemp was a few years ago the only material employed in the manufacture of ropes; later, aloe fibre was adopted, and these two materials are still commonly used in many places. In Belgium, aloe fibre is very generally employed. The strength of ropes made of this material is slightly greater than that of hempen ropes, and their durability is notably superior. But, on the other hand, they are heavier per unit of length, so that their superiority remains on the side of durability alone. One defect in hempen and aloe fibre ropes is their liability to absorb moisture, whereby their weight per unit of length is considerably increased. The defect is probably greater in aloe fibres than in hempen rope. More recently, iron wire has been adopted as a material for ropes, and the results have proved eminently satisfactory. These ropes consist of several wires of the toughest iron, twisted together in the same manner as the strands of the vegetable ropes, but the degree of the twist is less in the former than in the latter. Theoretically, a wire rope will best resist the strains brought to bear upon it when all the wires of which it is composed are parallel to one another; but practically, by reason of the flexibility and extensibility required, the strength of a wire drawing rope is found to be greatest when the strands

HAULING AND HOISTING MACHINERY.

are arranged spirally as in the hempen rope. In the wire rope, the weight per unit of length is, for a given strength, considerably less than in the hempen and aloe fibre ropes, and the diameter is also reduced in a like degree. The flexibility, however, is less, and, for that reason, pulleys of a larger diameter have to be employed. The transition from iron to steel was an obvious step, and hence we find the most recent ropes made of this material. The greater tensile strength of steel allows the diameter of the rope to be still further reduced, so that the weight per unit of length has

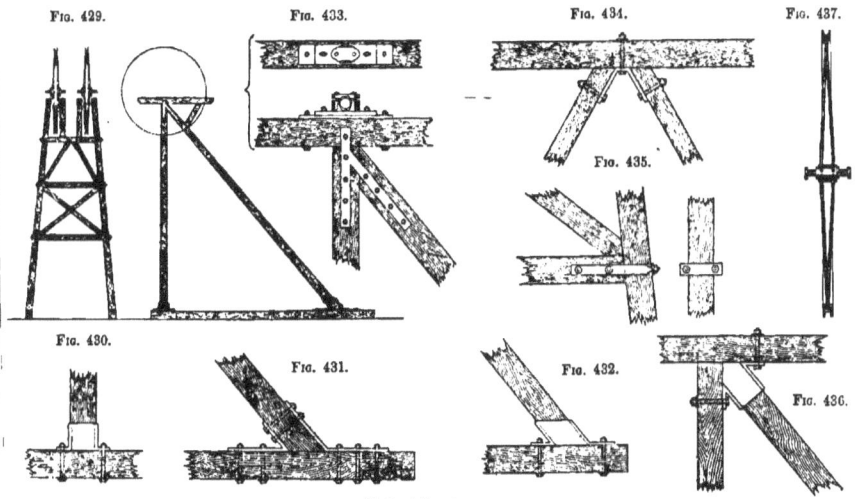

Pit-head Framing.

again to be notably lessened. The advantages obtained by the successive changes in the material employed in the manufacture of ropes are clearly set forth in the table on p. 230.

Sometimes the rope, instead of being cylindrical, is flat, and it was supposed that when arranged in this manner, the several fibres or wires of which the rope is composed would be more evenly strained than when they were all arranged spirally. This result may, however, be regarded as more than doubtful. For we have, in the first place, the fact that the fibres or wires are still arranged spirally, inasmuch as the flat rope consists merely of several small round ropes stitched together, the material forming the stitches adding to the weight, without in the smallest degree increasing the strength; and, in the second place, it does not seem probable that the separate strands are in practice more evenly loaded than they would be in the round rope. It is easy to see that even if the strain be uniformly distributed upon a new rope, that uniformity may be quickly destroyed by numerous causes. One portion of the rope may not offer the same resistance as another part, and this part, by becoming more extended than the rest, will render the strains upon the whole irregular.

Also it is evident that if the face of the pulley be not perfectly flat, the rope must be irregularly strained. To prevent, as far as possible, these accidents, each strand is made as nearly as may be identical, and they are used in even numbers. Also the direction of the twist is contrary in each pair, to counteract the tendency of the twist to come out under the action of the load. In winding, the flat rope is made to lap over itself upon the drum, so that the diameter of the latter is practically increasing or decreasing during the operation of winding. One obvious advantage of this overlap of the rope is, that the latter is kept constantly in the same vertical plane. The flat rope has not been regarded very favourably by mining engineers generally, and hence it has not been very widely adopted.

The quality of a rope of course greatly depends upon the method of its manufacture and the care bestowed upon the operations. A primary consideration is the strength of ropes; but this is a question that can be dealt with only approximately. It has been the custom to assimilate the resistance of a wire rope to that of an iron rod of the same effective section; but it is obvious that the whole section of the rope cannot be so uniformly obtained as that of the rod. Moreover, as already remarked, the operations of manufacture introduce elements of uncertainty in the rope, which either do not exist at all in the case of the rod, or exert a much less important influence. Besides, the rupture of a wire rope is due rather to the bending strains, to which it is constantly subjected, than to the tensile strains occasioned by the load suspended from it.

The following is a comparative table of the weights and strengths of hempen and of wire ropes, as given by the manufacturers:—

FLAT ROPES.

Hemp.		Iron.		Steel.		Equivalent Strength.	
Size in Inches.	Weight per Fathom.	Size in Inches.	Weight per Fathom.	Size in Inches.	Weight per Fathom.	Working Load.	Breaking Load.
	lb.		lb.		lb.	cwt.	tons
$4 + 1\frac{1}{2}$	20	$2\frac{1}{4} + \frac{1}{2}$	11	..	..	44	20
$5 + 1\frac{1}{2}$	24	$2\frac{1}{2} + \frac{1}{2}$	13	..	..	52	23
$5\frac{1}{2} + 1\frac{3}{4}$	26	$2\frac{3}{4} + \frac{5}{8}$	15	..	..	60	27
$5\frac{3}{4} + 1\frac{1}{2}$	28	$3 + \frac{5}{8}$	16	$2 + \frac{1}{2}$	10	64	28
$6 + 1\frac{1}{2}$	30	$3\frac{1}{4} + \frac{5}{8}$	18	$2\frac{1}{4} + \frac{1}{2}$	11	72	32
$7 + 1\frac{1}{2}$	36	$3\frac{1}{2} + \frac{5}{8}$	20	$2\frac{1}{4} + \frac{1}{2}$	12	80	36
$8\frac{1}{2} + 2\frac{1}{4}$	40	$3\frac{3}{4} + \frac{11}{16}$	22	$2\frac{1}{2} + \frac{1}{2}$	13	88	40
$8\frac{1}{2} + 2\frac{1}{4}$	45	$4 + \frac{11}{16}$	25	$2\frac{3}{4} + \frac{3}{4}$	15	100	45
$9 + 2\frac{1}{4}$	50	$4\frac{1}{4} + \frac{3}{4}$	28	$3 + \frac{3}{4}$	16	112	50
$9\frac{1}{4} + 2\frac{3}{4}$	55	$4\frac{1}{2} + \frac{3}{4}$	32	$3\frac{1}{4} + \frac{3}{4}$	18	128	56
$10 + 2\frac{1}{2}$	62	$4\frac{3}{4} + \frac{3}{4}$	34	$3\frac{1}{2} + \frac{3}{4}$	20	136	60

ROUND ROPES.

Hemp.		Iron Wire.		Steel Wire.		Equivalent Strength.	
Circumference.	Weight per Fathom.	Circumference.	Weight per Fathom.	Circumference.	Weight per Fathom.	Working Load.	Breaking Load.
inches	lb.	inches	lb.	inches	lb.	cwt.	tons
2¾	2	1	1	..	..	6	2
..	..	1½	1½	1	1	9	3
3¾	4	1⅝	2	..	..	12	4
..	..	1¾	2½	1½	1½	15	5
4½	5	1⅞	3	..	..	18	6
..	..	2	3½	1⅝	2	21	7
5½	7	2⅛	4	1¾	2½	24	8
..	..	2¼	4½	..	..	27	9
6	9	2⅜	5	1⅞	3	30	10
..	..	2½	5½	..	..	33	11
6½	10	2⅝	6	2	3½	36	12
..	..	2¾	6½	2⅛	4	39	13
7	12	2⅞	7	2¼	4½	42	14
..	..	3	7½	..	..	45	15
7½	14	3⅛	8	2⅜	5	48	16
..	..	3¼	8½	2½	..	51	17
8	16	3⅜	9	2½	5½	54	18
..	..	3½	10	2⅝	6	60	20
8½	18	3⅝	11	2¾	6½	66	22
..	..	3¾	12	..	6	72	24
9½	22	3⅞	13	3¼	8	78	26
10	26	4	14	..	..	84	28
..	..	4¼	15	3⅜	9	90	30
11	30	4⅜	16	..	..	96	32
..	..	4½	18	3½	10	108	36
12	34	4⅝	20	3¾	12	120	40

A German paper, quoted in *Iron*, in an article on the present methods of rope-manufacture from hemp, and the determination of the different qualities and probable strength simply from the appearance, lays down the following rule. A good hemp rope is hard, but pliant, yellowish or greenish-grey in colour, with a certain silvery or pearly lustre. A dark or blackish colour indicates that the hemp has suffered from fermentation in the process of curing, and brown spots show that the rope was spun while the fibres were damp, and is, consequently, weak and soft in those places. Again, sometimes a rope is made with inferior hemp on the insides, covered with yarns of good material—a fraud, however, which may be detected by dissecting a portion of the rope, or, in practised hands, by its behaviour in use. Other inferior ropes are made from short fibres, or with strands of unequal length or unevenly spun—the rope in the first case appearing woolly on account of the number of ends of fibre projecting, and, in the latter case, the irregularity of manufacture is evident on inspection by any good judge. As stated in *Iron*, a very simple and extremely ready means exists for ascertaining the purity or otherwise of Manila hemp rope. This consists in forming balls of loose fibre of the ropes to be tested, and burning them completely to ashes. While pure Manila hemp burns to a dull greyish-black ash, Sisal leaves a whitish-grey ash, combinations of Manila and Sisal yielding a mixed ash resembling the beard of a man turning from black to grey.

The report on mine ropes in Belgium, England, and Germany, made to the French Government by Aguillon, may be thus summarised.

With regard to the material for pit-ropes—whether hemp, or iron or steel wire—and their shape, whether round or flat—the advantage attributed to hemp ropes, of giving warning before they break, is shared equally by wire ropes when properly looked after, and the latter can be employed with as much safety as hemp ropes, when proper care is bestowed upon everything affecting their working. In wet pits, particularly where the water is at all corrosive, or where it is wound up in buckets instead of being pumped, aloes ropes are preferable. But in upcast-shafts, however slightly warm be the air-current, wire ropes should be used, in spite of the disadvantage of their hemp core. In the absence of any such special reasons, the choice of material is more a question of economy and convenience than of safety. Where it is determined that the rope shall be flat instead of round, the power of the winding engine in deep pits can be better balanced with a hemp rope; because a flat wire-rope is too thin to alter the leverage quickly enough in coiling or uncoiling on the rope roll, and would involve some kind of counterbalance, which would be a matter of difficulty. This is the practical reason while several deep pits in France have recently changed from flat wire-ropes to flat aloes-rope. With wire too there is much more difficulty in making a good flat rope than a good round one; and round ropes winding on conical or spiral drums afford a convenient means of balancing the engine-power. As to the choice between steel and iron for wire ropes, German and English practice goes to show that steel ropes, well made and of a suitable quality of steel, are capable of working better in all respects, and appear even to be safer. The exclusive use of steel-wire ropes in Germany and England, and of hemp or aloes ropes in Belgium, for all depths of pit, is attributed to the degree of excellence which has been arrived at in the two former steel-producing countries in the manufacture of steel wire, of sufficiently homogeneous quality and otherwise suitable for ropes; whereas in Belgium the manufacture of aloes or hemp ropes has always been a special industry of Flanders, where it has attained a rare degree of perfection. The whole of the winding gear should always be carefully adapted to the particular material of which the ropes are made. In France the mistake has generally been committed of ordering a rope without giving the maker any idea of the conditions under which it is to work, the very make being often specified for him in detail. Elsewhere the more sensible practice is to consult with the maker throughout, furnishing him with complete information as to the whole of the requirements to be fulfilled.

In addition to being tested, all ropes should be guaranteed by the makers. In Belgium the guarantee for aloes ropes is generally that they shall last $1\frac{1}{4}$-$2\frac{1}{4}$ years, or else for a given output; and 1-12th or 1-24th of their value is deducted for every month short of their stipulated duration. At the Royal Collieries at Saarbrücken, the ropes, of English crucible cast-steel wire, are guaranteed for 6 weeks, during which the maker is held liable to replace them if found defective.

Testing should apply, for hemp and aloes ropes, both to the raw material itself and to the spun yarn, as well as to sample lengths of the finished ropes. The twist of the rope, and the stitching of a flat rope, should be very uniform; and the rope should not contain more than 20 per cent. of tar.

Iron wire for ropes should be strong, hard, pliable, and not galvanised, and should be selected from standard makes. Steel wire should be made from crucible cast-steel, of very homogeneous and comparatively hard quality, and suitably annealed; it should have a tensile strength of 70–76 tons per sq. in., and should stretch 3–5 per cent., and be pliable. It should be tested for tensile strength, stretching, bending, and torsion; and all the wires in the same rope should be as closely alike as possible. Sample lengths of the rope itself should also be tested. The lay of the wires and strands should be regular; in flat ropes the stitching should be regular, and should be done with annealed wire. Torsion is considered an excellent test for homogeneous quality in wire: steel wires of $0 \cdot 059$ in. and $0 \cdot 118$ in. diameter should stand twisting through 40 and 20 revolutions respectively in an unloaded length of 6 in.; and the surface markings produced by the twisting should follow regular lines.

The size of the wires, and the length of their lay or pitch in the rope, should vary in accordance with the diameter of the drums and pulleys round which the rope will have to work; and particularly with the distance between the drum and the pit-head pulley, and with the angle which the inclined span winding on the drum makes with the vertical portion hanging down the pit. These are essential points for determining the stiffness requisite to prevent the rope from flapping as it runs.

Experience proves that the very material itself of every rope does certainly undergo deterioration in working, thereby diminishing the rope's strength till it becomes no longer safe. This deterioration of material is something more than mere wear by friction or rusting; in aloes ropes, the fibres lose their strength; and in wire ropes, even where testing fails to show any loss of tensile strength per sq. in. of section, there is a clear diminution of pliability and elasticity; the wires become harsh and brittle, whereby the rope is weakened. Though the deterioration is generally accompanied by unmistakable external indications, it is yet desirable to trace its progress by actual tests of the individual wires, or of the ends of the rope itself.

Large diameters for drums and pulleys are of more importance for wire ropes than for hemp, and for steel than for iron. The smallest diameter should be at least 1300–1400 times that of the iron wire in a rope, and 2000 times that of the steel wire. Its relation to the size of the rope itself matters less, because the disadvantage of too small a diameter can be obviated by selecting a suitable size of wire and by a suitable make of rope. It is well, however, for the smallest diameter of pulley or drum to be not less than 80–100 times the diameter or thickness of a wire rope, and 50 times for a hemp rope.

The rope should wind smooth on the drums or pulleys, without rubbing sideways against them, and so as to run free from jolts and flapping. For wire ropes it is desirable to line the grooves of the pulleys with wood. The larger the diameter of the head-gear pulley, the less does it matter how small be the angle which the inclined span winding on the drum makes with the rope hanging down the pit; but with smaller diameters of pulley the angle should be increased, in order thereby to diminish the bending of the rope in passing over the pulley. Opinions differ as to the minimum angle to be allowed; some assign 40° as the limit, while according to others it should never be less than 60°. In plan, the obliquity of a round rope between the overhead pulley and the drum should always be kept within the smallest possible limits.

In doubling back the rope end for attaching it to the cage, the loop should be kept as large as possible, by inserting within it an iron eye or a wooden disk; this is particularly advisable with iron wire-ropes, and still more so with steel. The attachment should also be made with springs, for easing the jerk at starting.

Iron or steel wire-ropes of large size should not work at more than one-tenth of their breaking strength; small round ropes may be worked up to one-sixth. Well-made aloes-ropes may be loaded to one-seventh or one-eighth.

Careful maintenance is indispensable to the preservation of all ropes, especially of wire ropes. Hemp ropes want tallowing regularly, and aloes ropes want keeping always damped. Wire-ropes, steel particularly, should be greased regularly, and often enough to prevent their ever beginning to rust. The grease should be soft enough to work into the strands, right through the hemp cores, but stiff enough to stick on the outside of the rope. A mixture of oil and grease, well stirred and laid on hot with a brush, answers very well; both oil and grease should be neutral.

Iron wire ropes are rapidly being replaced by steel wire, owing to the less weight needed to afford the same strength. But it must be remembered that when the mine water contains much acid, the steel will wear much faster than the iron.

The softer kinds of steel, which contain least carbon, approach wrought iron in character, having equal toughness, greater strength, and the same capacity of welding. The mildest steels contain $0\cdot15-0\cdot4$ per cent. of carbon, and the hardest $1\cdot4-1\cdot6$ per cent. The following are the breaking strains per sq. in. of wire of some of the most usual varieties employed in rope making as given by Frecheville:—

Mild Steel	from 40 to 50 tons.	Best patent steel	from 70 to 80 tons.
Best crucible steel	,, 50 ,, 60 ,,	,, plough ,,	,, 110 ,, 120 ,,

Too great stress cannot be laid upon the necessity of having ropes constructed of the best material. The selection of the material, however, somewhat depends on the conditions of working: thus with a perpendicular shaft and large drums and pulleys, a plough steel wire rope will be found the most reliable; but with small drums and pulleys and a shaft with angles in it, a rope made of best patent steel, or mild steel, will last longer, as the wires are not so apt to snap in bending. In describing a wire rope, the number of strands, the number of wires in each strand, their gauge, the quality of metal, and the material of which the centre or core is composed, should be specified.

As to the gauge. Since the ultimate strength of wire increases as its diameter decreases, and since small wires are more pliable than large ones, it would seem that the finer the wire used the better; but there is a practical limit to this, as very fine wire offers too much surface for oxidation, and is too easily injured by friction. Experience has shown that it is advisable to employ medium sized wires, between Nos. 10 and 15 of the Birmingham wire gauge.

For ordinary work, hemp cores or centres have been proved the best; they stretch with the strands, allow the wires to bed themselves solidly, and give ropes greater flexibility than could be obtained with wire centres. The latter have not given very satisfactory results in practice, although a greater breaking strain is obtained with a relatively smaller rope.

There are many modifications in the methods of laying or twisting the wires. Common laid rope often has 6 strands, with 7 wires in each, the size of the wire being altered to suit the size of the rope. Compound ropes, that is, ropes with more wires in the strands than the usual construction, in addition to other varieties, have 6 strands, with 19 equal sized wires in each, or 7 strands, with 6 wires in the middle, of about 15 gauge, and 12 round the outside, alternately 15 gauge and 12. Ropes with 6 strands, of 11, 12, and 13 wires each, are frequently manufactured. Some makers prefer the inner wires of each strand smaller, so as to be more flexible than the outer. Six strands in a rope are better than 4 or 5, as they make it more cylindrical, and consequently the friction is better distributed. Six strands, of 19 wires each, make very durable ropes; these work better than one of equal size, composed of 6 or 7 wires in a strand; as the latter, being larger and less pliable, are more likely to snap in bending round pulleys and drums. When three or four of these large wires break near together, the rope is hardly fit for work, whereas the breakage of that number of small wires would be of much less consequence. More material can be got into the same sized rope when compound instead of common laid, as the smaller wires do not leave so much space between.

On account of the many different varieties of steel wire employed in the manufacture of ropes, and the varying size of the hemp centres, and the empty spaces above referred to, it is impossible to state a formula for determining the dimensions of a steel wire rope required to bear a given strain. As the nature of a wire rope, however, is defined by the number and size of the wires, it is easy, if we know the section and weight per fathom of the gauge employed, to determine the *effective sectional area* of the rope, and its weight per fathom; given then, the quality of the metal, the breaking strain of the rope can be approximately estimated.

The following table, by Frecheville, in which the numbers of the Birmingham wire gauge most usually employed in the construction of mine ropes are compared with inches, and the weight of a cub. ft. of steel is taken as 487 lb., will be found useful in these calculations.

No. B. W. G.	Diameter in Inches.	Sectional Area in Square Inches.	Weight per Fathom in Lb.
10	·137	·01474	·2990
11	·125	·01227	·2489
12	·109	·00933	·1893
13	·095	·00708	·1436
14	·083	·00541	·1097
15	·072	·00407	·0825

Thus in the case of a steel wire rope composed of six strands, seven wires in each, of ten gauge, the *effective sectional area* will be $6 \times 7 \times ·01474 = ·61908$ sq. in., and its weight per fathom in metal $6 \times 7 \times ·2990 = 12·55$ lb.

If best plough steel wire with a breaking strain of 120 tons per sq. in. was employed in its manufacture, then ·61908, the *effective sectional area* × 120 tons = 74·28 tons, and deducting one-eighth for *lay*, we obtain 65 tons as about the breaking strain of the rope.

Again, let us suppose a compound rope made of the best patent steel wire, with breaking strain of 75 tons per sq. in., and composed of 6 strands of 19 wires each, 13 gauge. The following calculation—$6 \times 19 \times ·00708$ sectional area of each wire × 75 tons, breaking strain per sq. in. of wire, less one-eighth for *lay*, gives 52·97 tons as the approximate breaking strain of the rope. Such a rope with hemp core and fairly made would weigh about 18 lb. per fathom, and have a circumference of about 4½ in. The actual breaking strain, however, can only be found out by testing sample lengths of the finished ropes.

As the operations of manufacture introduce so many elements of uncertainty in wire ropes, it is well to allow a wide margin of safety, especially where their breakage would endanger life, and take the working-load as one-tenth of the ultimate strength or breaking strain. *The weight of the rope hanging over the pulley at the poppet heads is of course included in the working load.* In very deep mines this weight, even with steel wire ropes, becomes a matter of such serious consideration, that tapering ropes have to be used. In the case of a rope working at a very slow speed, such for instance as a capstan rope, a larger factor of safety than one-tenth may be adopted. Since any extra strain on a rope leaves it weaker than it was before, on no account should a rope used for raising men be ever worked above its fair working load.

In drawing mineral in Cornwall, the custom is to let the skip down on a "gate" put across the shaft. Probably the greatest strain the rope has to bear is when the full skip is lifted. Experiments made at some coal mines prove that when the full cage is lifted from the bottom, about double the ordinary strain due to the load is produced. This arises from the inertia of the mass to be moved. In the case of a skip resting on a "gate," the more slack chain there is, the greater will be the strain on the rope at starting.

In winding men there should be no resting place for the cage, the engine should be started gently, driven regularly, and with a speed of only about two-thirds of what is otherwise usual. The rope also should be examined every 24 hours, and this should be done by winding it slowly through the operator's hands; if he does not happen to see the broken wires, in all probability he will feel them. Occasionally the rope should be thoroughly cleaned, and its condition more minutely ascertained. When broken wires are found, the longest may be tucked underneath, and the others cut off to prevent their catching and doing further mischief. A new rope should be tested with several days' winding before men's lives are trusted to it.

It is indispensable for the preservation of steel wire ropes that they should be greased regularly. The grease used should be perfectly free from acid, and soft enough to work into the strands right through to the hemp core. It must not be of such a nature as to harden, for in that condition it allows rust to form between it and the wire, so that a rope that appears to be well greased may be corroded to a sensible depth. A mixture of Stockholm or Archangel tar, a vegetable oil, and a little lime boiled together, is often recommended. In Cornwall the tar is generally mixed with tallow. These mixtures, however, form too stiff a grease, tend to hide defects, and render the thorough examination of the rope difficult. A mixture containing gas tar is still more objectionable. Of all the lubricants for wire ropes the best is mineral oil. Some of the heavy mineral oils, such, for instance, as the Russian (their specific gravity being higher than the American), possess sufficient viscosity to be used as a lubricant for wire ropes, and will, if tried, owing to their freedom from acid and power of resisting decomposition, be found to give satisfactory results. At the Wearmouth Colliery they have a patented apparatus, consisting of a pair of wire brushes, for cleaning the ropes, and a pair of strong hair brushes fed with lubricant from feeders above for oiling them. Both sets of brushes revolve, being actuated by the travelling rope. It is claimed that this arrangement lubricates very thoroughly and effects a great saving in oil and labour.

When a rope is used for winding men, the shackle should be cut off regularly every 2-3 months, the rope thoroughly examined, and the shackle reset. This is a point of vital importance for wire ropes. In order to arrive at economical results with wire ropes, accurate accounts should be kept of their working; by this means the kind most suitable may be ascertained, and a considerable saving effected by using an article best adapted for the purpose. However well a rope may seem to be lasting, it should always be suspected as soon as its duration approaches the average that corresponds with the conditions under which it is working; it should, at any rate, cease to be used where human life depends on it. Owing to trade competition, there is great danger of inferior metal being used in the manufacture of ropes; so that when a new one is required, only the best makers should be applied to, and they should be furnished with full information as to the conditions under which it has to work. There can be no greater false economy than choosing a cheap rope. When a rope is for the purpose of winding men, it would be advisable to have a sample piece of it (say a length of 10-12 ft.) tested before use, in order to see that the quality of the metal and the breaking strain are as represented.

HORSE WHIMS.—The apparatus by which horse-power is applied to the raising of mineral in the shaft is shown in Figs. 438 and 439. In the Cornish "whim," as this structure is called, the horses are yoked to the ends of two radial arms, formed by a large horizontal beam of timber passing through a mortice in the upright axle. These arms are strengthened by two longitudinal straps or fishes applied through about two-thirds of their length. The rope barrel is a plain cylindrical cage formed by nailing straight boards to the outsides of three horizontal wooden rings, placed at different heights, and supported by arms morticed through the axle. The lower ring is carried by the top of the long beam, and another intersecting it at right angles, and the two upper ones by similar cross arms set at an angle of $22\frac{1}{2}°$ to each other. The whole cage is further supported by diagonal struts below the lower ring and resting against the sides of the axle near its lower end. The shaft is square at the intersection of the long cross bar, and is chamfered down to an octagonal section, above and below, with cylindrical ends, the cylindrical parts being tired with wrought-iron rings for securing the hold of the pivots. The framing is formed of two short inclined standards, united by a long transverse bar, to the centre of which is affixed the bearing of the top spindle. The guide pulleys over the top of the shaft are of small diameter, the framing giving a clear head room of about 8 ft. only; the axis of one is placed a little higher than that of the other, in order that the rope may lead to its proper place on the drum. The diameter of the path described by the horses is 36 ft., that of the drum being 12 ft.; the depth or height of the drum, or the receiving surface for the ropes, is 56 in. The kibbles for horse whims are estimated to carry $2\frac{1}{4}$ cwt. Round hempen ropes, of 6-7 in. (circumference), or chains of $\frac{7}{16}-\frac{1}{2}$ in. iron are employed. For depths of less than 40 fathoms, one horse is sufficient, but two are employed for drawing from any greater depth. The speed at which the load moves in the shaft is 75-100 ft. per minute, the horses during the same time passing over about three times that space, or at the rate of about $3\frac{1}{4}$-4 miles per hour.

The vertical shaft in the German horse-whim, Figs. 440 and 441, is of considerable length; the rope drums are placed near the top, and are carried on a platform formed by four arms overlapping the shaft, supported at the ends by struts, which are nailed to the axle at about one-third of its height above the bottom bearing. The lower drum is fastened to the shaft, the upper one is loose, and can be connected with the lower one by a wooden coupling pin. The brake works on a projecting part of the upper drum, and serves either to stop the whim when both drums are connected, or the top one only when the pin is taken out, which is done when altering the amount of rope out in changing the draught from a higher to a lower level, or the reverse. The upper drum in this case runs on friction rollers on the upper surface of the lower one. The bearing of the foot spindle rests upon a pair of adjusting wedges on a short pillar of masonry; to the top bearing is attached a horizontal beam, carried by two diagonal struts, which also support a conical roof covering the rope drums and a gallery projecting from the house above the shaft. The horses are attached to a

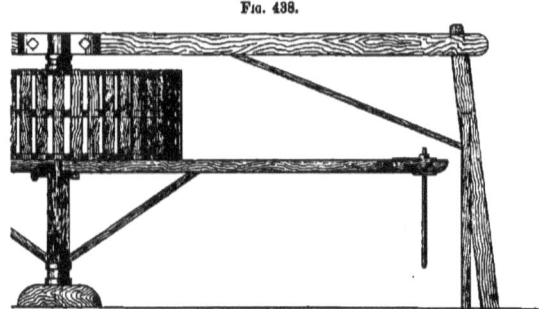

Fig. 438.

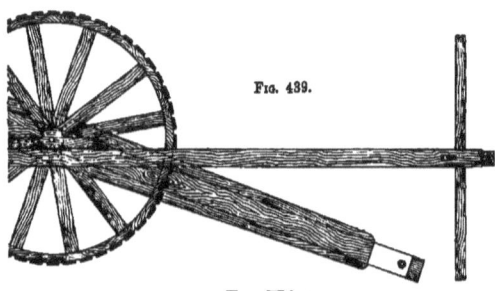

Fig. 439.

Horse Whims.

turning bar attached to the lower end of a projecting diagonal arm fixed at the upper end to the shaft and lower drum, and also supported at about half its length by a horizontal strut, which are hung by staples to the lower end of the shaft. Two poles with spiked ends are used to prevent the horses being dragged back by the weight of the loaded tub when the whim is stopped. The tubs are of a prismatic form, resting on rollers between wooden guides on the shaft, and are drawn by round wire ropes. The capacity of the tubs is about 2000 cub. in.; the radius of the path of the horses, 19 ft.; the core of the drums, 6 ft.; depth of the coil, 13 in.; the vertical shaft is 23 ft. long and 17 in. square.

The Cornish water-whim (Figs. 442 and 443) is driven by an overshot water-wheel, carrying a spur wheel of 36 teeth, which gears into a pinion with 16 teeth on an intermediate shaft, whose journals are made to slide laterally in their bearings. Two bevel wheels are fixed on to this shaft at a distance from each other somewhat greater than the diameter of the horizontal mitre wheel on the upright shaft of the whim. By means of the reversing lever, the horizontal shaft can be moved

FIG. 440.

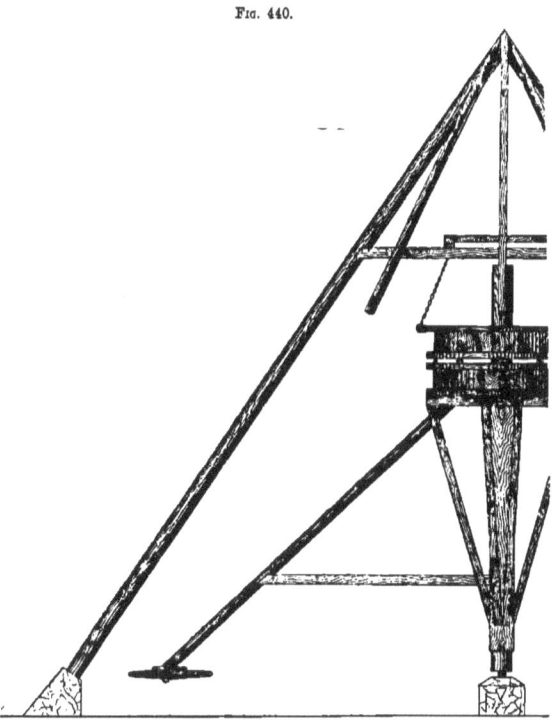

German Horse Whim.

sideways, so as to cause one or other of the two lower mitre wheels to drive that on the drum shaft, the left-hand one producing forward, and the right-hand one backward motion. The drums are intended for flat ropes, which are guided by a frame or cage with 6 wooden arms, set in cast-iron seating rings resembling those of the water wheel. The vertical shaft carries a second mitre wheel, which gears into a smaller one on a shaft carrying a small fly-wheel, serving as a brake drum. A

240   MINING AND ORE-DRESSING MACHINERY.

strap on the brake shaft drives a pulley on a small shaft close to the ground, which lifts a weight passing over a roller at the top of a signal board. The path of the weight is proportional to that of the kibble in the shaft, so that the position of the latter is constantly shown by the place of the

FIG. 441.

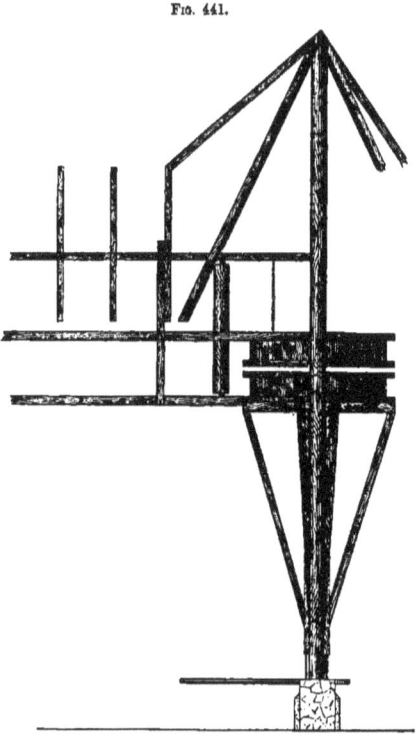

German Horse Whim.

weight on the board. The buckets of the water-wheel are made of a single board, an imperfect construction giving cells of small capacity, and now rarely employed. The load drawn at each ascent of the kibble is about $3\frac{1}{2}$ cwt.

A water-whim with underground wheel is used in the Harz mines (Fig. 444). The rope drums are on the same shaft with the overshot water-wheel, which is provided with two systems of buckets opening in opposite directions, each of which is furnished with a separate sluice. The buckets are formed of wood, in two pieces, set in wooden shroudings and backings. The framing

HAULING AND HOISTING MACHINERY. 241

of the arms is that usually adopted in Germany for large wooden wheels, known as "Dutch framing;" the ring is carried up either side by four principal arms laid in pairs at right angles to each other, and overlapping the sides of the square shaft, and eight intermediate diagonal arms or struts which are arranged in pairs forming V's, the apices of the 's resting on the centres of the

Fig. 442.

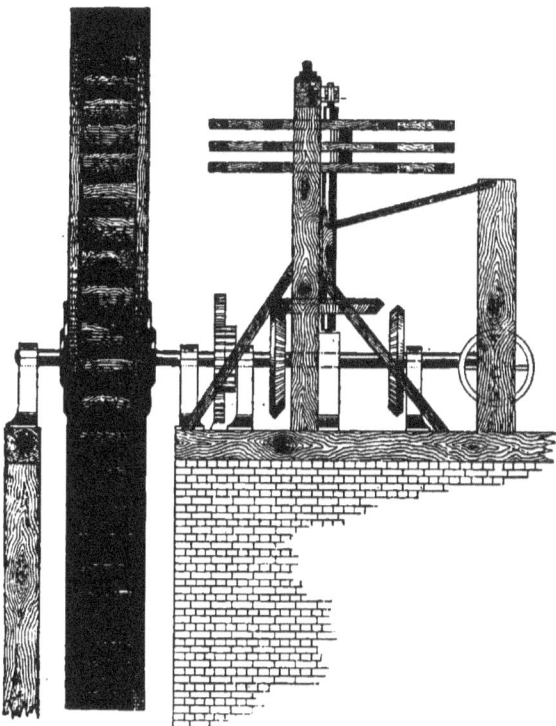

Cornish Water Whim.

sides of the shaft. The rope drum nearest to the water-wheel is fixed to the shaft; it is made entirely of wood, the framing of the arms being similar to that of the wheel. The outer rope drum is loose on the shaft; it is carried by six cast-iron arms on either side, which turn on a pair of cast-

2 I

iron rings keyed on to the shaft. Each of these rings has four square holes sunk into it to receive the points of the locking hooks, which turn on a shaft attached to the inside of the cage formed by the two sets of arms. These hooks are turned by a handle projecting from the outside of the drum, the coil of which fits into a catch. There are two brakes; one stops the loose drum when adjusting the amount of rope to be paid out; the other, which is used for stopping the wheel, works on a wooden disc placed between the rope drums and the water-wheel. The journals of the main

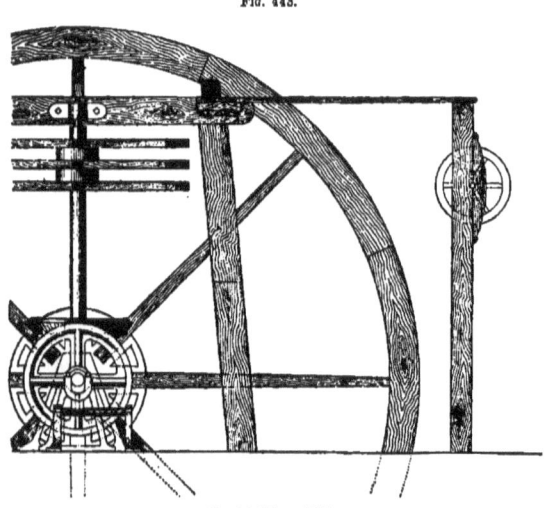

Fig. 443.

Cornish Water Whim.

shaft project from a cross formed by two plates of cast iron intersecting at right angles, and surrounded by a ring; the arms of the cross are sunk into mortices in the wood, the ring forming an outside tire.

Fig. 445 shows the arrangement commonly adopted in Cornwall in winding from shallow shafts by steam power. The engine-house is placed near the centre of the ground, the outer end of the beam, the connecting rod, and the crank, being the only exposed parts of machinery. The drums on which the drawing chains are received are fixed horizontally on a vertical shaft, which receives motion from the engine by a horizontal bevel wheel at the lower end gearing into a similar wheel of equal diameter placed vertically and at the end of the fly-wheel shaft. The load is drawn by single-link iron chains. The receiving surface of the drums is packed with wood, forming a cylinder of 4 ft. diameter.

The chains are in single lengths, one for each shaft; the ends are carried over guide rollers and hang down the shafts, having the kibbles attached to them by hooks and yokes. Each chain is carried twice round its own section of the drum, but is not made fast to it. The diameter of the path of the crank, probably equal to the length of stroke in the cylinder, is 5 ft. The wooden drum is surrounded by a skeleton cast-iron frame or cage with projecting horns, for keeping the chains in their proper places. The actual distances of the two shafts from the engine are 35 and 24 fathoms respectively.

The ores are brought up in wrought-iron buckets or kibbles, which are drawn through the shafts without the use of guide rods. The mouth of each shaft is closed by two trap doors with

FIG. 444.

Harz Water Whim.

narrow channels in the middle for the passage of the chains; they rest against inclined seats, and are lifted by the ascending kibble when it comes to the surface. The average diameter of the round iron bars, of which drawing chains employed in Cornwall at the present time are made, is about $\frac{7}{16}$ in. for deep mines with steam whims; or tapered chains are sometimes used, in which part is of $\frac{1}{2}$ in. and part of $\frac{7}{16}$ in. diameter; $\frac{3}{4}$ in. is but rarely used. The average weight of the kibble when empty is 6–10 cwt.; the load is 5–7 cwt. The working speed in shafts of varying inclination sunk on the vein is 150-170 ft. per minute; a much higher speed may be allowed in shafts where skips or boxes travelling on wooden rods are employed.

FIG. 445.

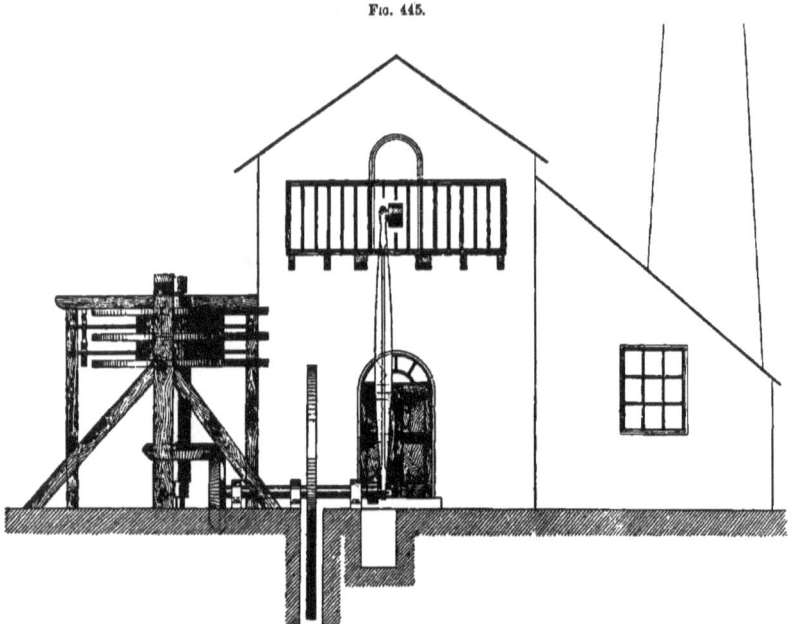

Cornish system of Steam Winding from Shallow Shafts.

HAULING ENGINES.—Where no special source of loss exists, underground hauling engines are capable of realising an efficient performance varying from 50 to 60 per cent. of the pressure of the steam upon the piston, estimated in coal conveyed. This compares very favourably with that of horse-power. The direct employment of steam in the operations of haulage is, however, not without disadvantages. If the boilers are placed underground, their position must be such that the products of combustion may pass directly to the up-cast without traversing drifts used as a travelling road, and that there may be no danger of an explosion from firedamp reaching the furnaces. These disadvantages cannot be entirely got rid of when the boilers are placed underground, and may determine in some degree the position of the engines, which ought to be chosen solely to satisfy the requirements of the traction. One means of obviating these difficulties consists in erecting the boilers at surface, and in conducting the steam down the shaft, and to the points where the engines are fixed, through iron pipes. But to prevent the condensation of the steam by the radiation of heat from the pipes, the latter have to be well coated with a suitable non-conducting material, and whatever expense may be incurred in providing this protective covering, the remedy is but partially effectual; in all cases a large amount of condensation inevitably takes place, and the evil becomes more serious as the distance of the engines from the boilers increases. Besides these drawbacks, there is in every case the disadvantage arising from the heating of the atmosphere underground by the exhaust steam.

The requirements of haulage are partially fulfilled by the system of erecting both boilers and engines at surface, and transmitting the force down the shaft by means of an endless rope. This system, however, though it gives good results in some cases, leads to complication, and is limited in its application. Perfect efficiency and completeness can only be obtained by the use of several independent engines, situate at various points in the workings, and designed and proportioned in their dimensions to the work they have to perform, and the conditions under which they are to operate. A more satisfactory solution of this problem lies in the adoption of compressed air. This may be easily and cheaply conveyed to any part of the workings through branch pipes of small diameter, laid from the primary and secondary mains in the principal roads; hence, not only may the hauling engines be placed in positions most favourable to the traction, but rock-boring and coal-cutting machines may be supplied from the same system of pipes as the hauling engines. This is a matter of no small importance.

Underground hauling may be done on any one of four systems: by the use of a tail rope, an endless chain, an endless rope, or compressed air locomotives.

In 1867 a committee of the North of England Mining Engineers made a careful investigation into the first three of these systems of hauling, as then in use. The following Table gives a summary of their results:—

System of Haulage.	Average Gradient for Full Tubs.	Cost in Pence per Ton per Mile.							
		Ropes or Chains.	Tubs.	Grease and Oil.	Coals.	Repairs to Engines and Boilers.	Maintenance of Way.	Labour.	Total.
Tail rope	Rise 1 in 213	0·276	0·114	0·186	0·558	0·098	0·064	0·583	1·879
Endless chain	„ 1 „ 59	0·083	0·173	0·155	0·256	0·072	0·068	0·572	1·379
Endless rope	„ 1 „ 36	0·252	0·309	0·138	0·323	0·196	0·083	1·692	2·993

Endless rope haulage, as adopted in the South Duffryn Colliery, at the Plymouth Works, Merthyr Tydfil, is thus described by T. H. Bailey. There are three ways in which the endless rope can be used:—

(1) By carrying it under the trams or tubs, adopted at Plymouth Works, and now to be described.

(2) By carrying it on the top of the tubs; only applicable where the coal is not loaded above the level of the tub.

(3) By carrying it on the side of the tubs; not suited to the trams of this district.

The endless-rope system has one disadvantage, that its otherwise most economical working requires two lines of railway, and consequently very wide roads to accommodate the large trams

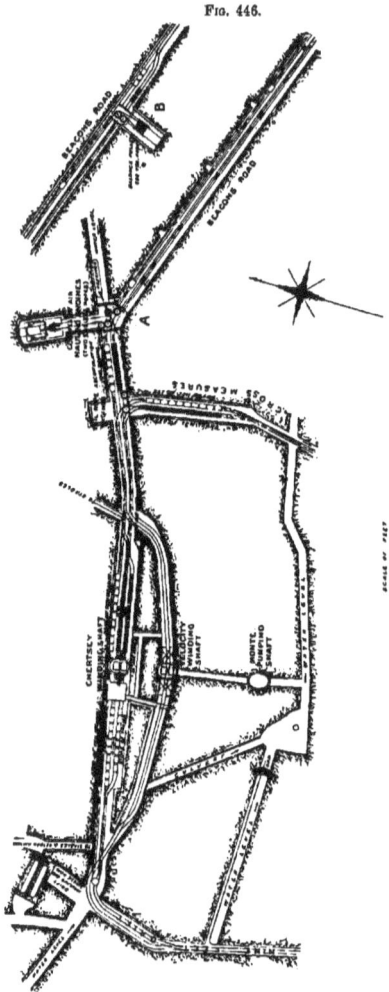

Fig. 446.

Plan of South Duffryn Colliery.

used in South Wales. Where the roof is weak, and much timbering required, the cost of making and maintaining double roadways is very great; but, notwithstanding this, Bailey is of opinion that its drawbacks are more than counterbalanced by its advantages, for the wear and tear of machinery, trams, and ropes is much less in this than in the "tail-rope" system; besides which, regular supplies of coal can be brought to the shaft, and should an accident occur, very little damage is done.

Bucknall Smith, in an article in *Engineering* on "Underground Rope Haulage," says: "The endless rope system under the tubs, driven at a moderate speed over a double way throughout, appears undoubtedly an excellent method of working underground haulage, and preferable to the intermittent and high-speed deliveries of tubs over single lines by the main and tail-system, which strain the engines, machinery and ropes spasmodically. However, on the other hand, it must not be lost sight of that certain conditions of working peculiar to different districts and mines dictate different methods of transporting the mineral produce. Where double lines can be consistently laid down, and an uninterrupted supply of coal to the shafts can be maintained from one or several workings along the route, the endless-rope system is highly advantageous and economical."

The "Beacons-road" engines, are a pair of 22-in. cylinders, 3-ft. stroke, geared 3 to 1; and two drums, 6-ft. diameter, have been constructed of cast-iron rings secured to the arms with bolts. These rings, when worn out, can be removed and replaced by new ones without any great cost. One drum has been provided to work the south-east district of the 9-ft. coal along the "Beacons road" (Fig. 446), and another drum, on the same shaft, to work the 6-ft. and 4-ft. coal districts along the Cross Measures from the 9-ft. coal, as well as the haulage on either side of the Chertsey Pit.

The rope is coiled round the drum 3 times, which gives sufficient friction for the work required; then it passes down the empty road, round the straining pulley B, and back along the full road, until both ends are brought together and well spliced. Great care is needed in straining the rope as tightly as possible, and the best way of doing this is to pull the ends together with a pair of rope blocks before splicing, and to be careful that the tightening pulley is braced up as close as possible, seeing that ropes always stretch more or less in use.

The inclination of the Beacons-road section is on an average 2-3 in. in the yard. The road is driven in a south-easterly direction, and the rope is hauled a distance of about 750 yd. The rollers for preventing the rope trailing on the ground are so fixed as not to interfere with the clip attached to the rope, nor with the trams passing over them.

The full and empty trams are attached singly, some 20 yd. apart, by means of a clip which is secured to the rope with a quick-threaded screw. It is important to attach them at equal distance from each other, say 15-20 yd., so that the weight of the rope may be carried along with the least possible friction upon the ground or the rollers—a great desideratum, as the life of the rope is thereby materially lengthened.

A variety of clips are in use in different collieries for attaching the trams or tubs to the rope. These clips have each various advantages, some being preferred by one engineer, and some by another. The screw-plate clip adopted at the Plymouth works takes, in Bailey's opinion, the best grip of the rope, with the least possible injury and this is done by holding 4-6 in. of the rope; but the clips should not be so large as to be cumbersome. By means of a hinged hook the rope is always retained in a horizontal position, whether the tram is being pushed forward or held back, thus minimising the possibility of kinking, or injury and breakage of the wires.

The next important matter is that of saving labour in hauling the trams about from the points where they are connected to or disconnected from the rope, and this is accomplished by arranging the levels of the rails as shown in section in Figs. 447, 448.

It may be mentioned that the hauling engines have been indicated with Hopkinson's instrument, giving the following results:—West engine, 8·33 H.P.; east engine, 9·75 H.P., or a total of 18·00 H.P. exerted upon the cranks. At the time the diagrams were taken, there were only 15 full trams and 15 empty trams attached to the rope, about 40 yd. apart, and the engines were making 20 revolutions per minute. The compressed air showed a pressure of 30 lb. on the gauge, and the rope was travelling about 1¼ mile per hour.

The difficulty of the double road is partly overcome in some districts by adopting the method in a partial way instead of having a double road the whole distance, by having three rails widened out to sidings or passing places, at regular distances, by which in raising coal from the "deep," the benefit of empty trams counterbalancing to some extent the loaded ones is obtained without going to the expense of widening for a double road the whole distance. Probably the adoption of the system in full would lead to the greatest economy in haulage; and where the full system may not be practicable, the adoption of three lines of rails, widened out to sidings for the passage of groups of trams (say 4 or 6) travelling along in opposite directions, would prove a considerable improvement upon the main- and tail-rope systems so largely in use. The endless rope offers a great advantage over the tail rope in that it supplies a continuous power, which can be taken off at any point, and also more readily enables extensions to be made wherever necessary, which it was not easy to do with the tail rope.

An important point arises as to whether it is more advantageous, in laying out a system of underground haulage, to carry the ropes down the pit from a steam hauling engine on the surface, or to use compressed air with hauling engines placed underground, as described. The circumstances

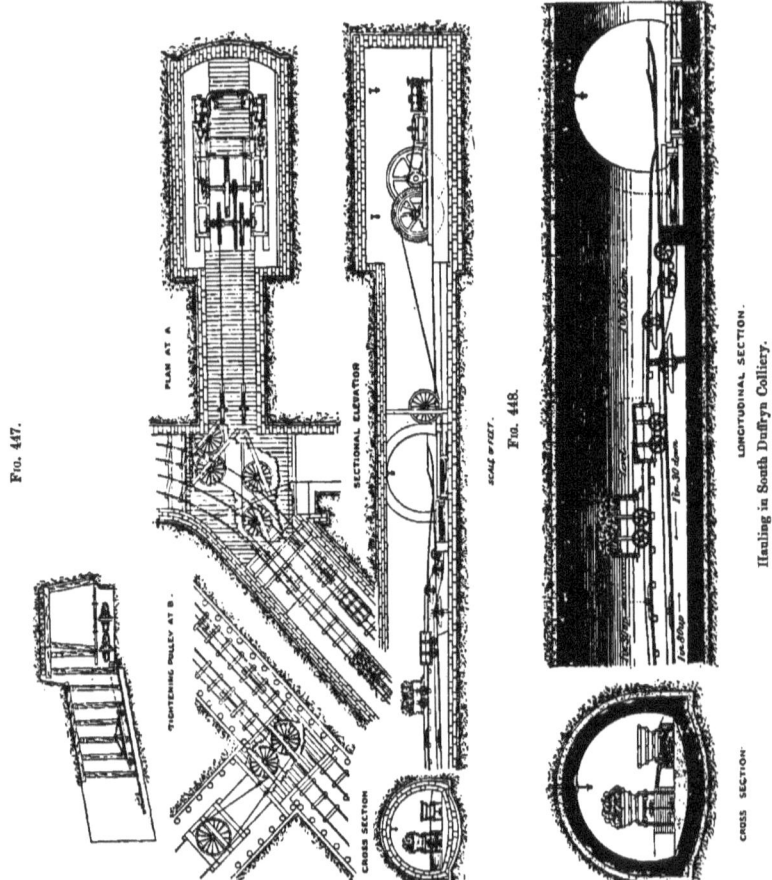

and requirements of each colliery must be fully considered in deciding this point, and probably in the case in question the number of hauling engines and pumps requiring power fully justified the adoption of compressed air. Another interesting question may arise as to the pressure at which it is most advantageous to employ compressed air; whether, for instance, at the pressure of 30 lb. as instanced, or at 60 lb. as in the endless-rope haulage at Clifton Colliery. Probably it will be generally conceded that air can be compressed and used more economically at pressures of 30 lb. than 60 lb., but the advantages of being able to adopt smaller hauling engines to develop the same power, and also of having at all times a higher power at command for underground uses, may render the higher pressure more generally serviceable.

The friction of the tram, if the latter is of good construction, assumes a pretty well ascertained proportion to the weight carried, and cannot within reasonable limits of expense be materially altered, but will remain a constant factor, quite irrespective of the system of haulage adopted. It is, moreover, insignificant, as compared to the friction that ensues in taking ropes round a number of curves and into a succession of branches, or to the increased friction that arises from ill-laid and badly kept roads, which are often allowed to become covered with dust and dirt. Besides this, the friction will be much enhanced if the length of the intervals at which the trams are attached to the rope were too great for the trams to carry the rope clear of the ground friction-rollers, and if the distance becomes greatly extended the friction will be very materially increased.

Another very important point with regard to the general efficiency and life of the ropes of any particular system of rope-haulage is the speed at which the rope is run, both in regard to friction, and freedom or otherwise from jerks and strains to the rope by the trams getting off the rope. The size of the rope employed, the proportionate size of the rope-pulleys used throughout, and also the system by which the power is conveyed to the haulage-ropes, have also most important effects on the life of the rope and the efficiency of the haulage.

Fisher's pulley consists of a rope-wheel 7 ft. diameter, with flanges 4 in. apart, the periphery being turned true, and lined with cast-steel segments 4 in. wide, having a taper in that width of $\frac{1}{4}$ in., so as to form a conical surface for the rope to coil upon, and on which it readily "fleets" (without grooving), as the full rope coils on, in the same manner as a rope does on a capstan drum. This has a very important bearing on the life of the rope, for the cone surface enables the rope to fleet with the greatest ease, and without any appreciable wear and tear, whereas on a driving-wheel with a flat surface the rope friction is excessive, and becomes greatly increased as the rope in time grinds down the periphery to a concave form.

It is most desirable in all cases of rope-haulage to remove, as far as possible, all causes of jerks and undue strains to the ropes, and to use a friction clutch, not only for the purpose of throwing the machinery in and out of gear without bringing undue strain upon the ropes, but also for the purpose of transmitting sufficient power to drive the machinery, whilst at the same time providing that if the trams get off the road, or any other sudden obstruction arises, and throws an undue strain on the rope, the slipping of the friction clutch will give immediate relief. Fisher and Walker's friction clutch is undoubtedly most effective and simple, and works exceedingly well.

Instead of the engines being so large and geared only 3 to 1, a somewhat increased air-pressure and much smaller engines might be used, running at a higher velocity and geared 7 or 9 to 1, by which it would be found that the ropes would run more steadily, and there would be very much less fear of the trams going off the road. An engine 20 in. diameter, 2 ft. 6 in. stroke, and making 20 revolutions per minute, can not well run steadily, unless it has a very large and heavy flywheel, which is impracticable underground.

Whereas Bailey employs ordinary claw clutches, it would probably have been infinitely better if

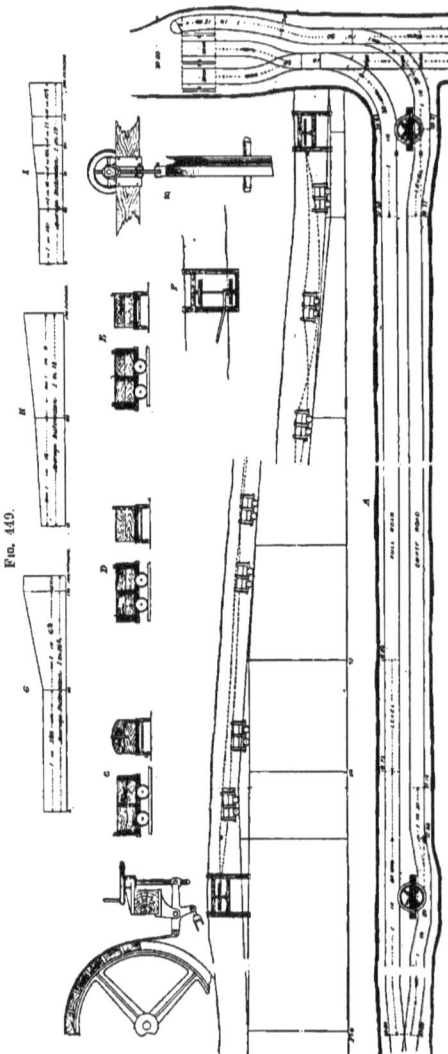

Fig. 449.

friction clutches had been employed instead. One great danger to the "life" of a rope is that, when the tubs by any chance get off the road, a tremendous strain is put upon the rope immediately. With well-appointed friction clutches the pressure can be so regulated that the friction on the drum of the clutch will be just enough to overcome the friction of the load and the weight of the trams: and if any unusual strain is put on it by trams getting off the road, the clutch immediately commences to slip, and the man in charge of the engine at once knows that something is wrong, and simply throws the clutch out of action, or stops the engine.

On the subject of haulage by self-acting endless chains, David M. Mowat has described the general arrangements required for a first-class self-acting incline. These arrangements would, of course, require some modification to suit the requirements of any particular incline to which the system might be applied.

Referring to Fig. 449, A shows the arrangements at top and bottom of the incline. These consist, at the top, of a framework supporting a wheel (B) 4 ft. diameter fitted with a screw brake, and having a groove in which the chain may get a good bearing to prevent slipping; while at the bottom there is a similar wheel without the brake. Besides these large wheels, there are generally three 6-in. bearing-up pulleys—one on the empty road at the top, and one on each road at the bottom of the incline. The incline is laid with double roads throughout its whole length, one being always used

HAULING AND HOISTING MACHINERY. 251

as a full road, and the other as an empty road. The large wheels are connected together by an endless chain, to which the hutches are attached at regular intervals. The attachment consists of a vertical plate having a fork at its upper end, which allows the vertical link of the chain to enter, but prevents the horizontal link from slipping through. This chain plate may be either on the middle of the end, or on the side of the hutch according to circumstances. When the incline is very steep, and more especially if in the course of transit the loaded hutches require to be drawn uphill by an endless chain, the chain plate should be on the middle of the end C, as in that case the tractive force is applied more directly in line with the resistance, and there is consequently less tendency to twist the hutch off the road. To this arrangement, there is one objection, that the hutch can only be loaded above the lip at each side, owing to the chain resting on the top of it. This can be partly overcome by raising the end of the hutch, as in C; and it is only an apparent objection when the inclination is as great as 1 in 4, as the hutch cannot then be filled much above the wood with any system of haulage, on account of the coals falling off.

When the incline is flat, the chain plate is better on the side of the hutch, but its position on the side should be fixed by local circumstances in each case. If, for instance, the incline dips in one direction only, so that the tendency of both full and empty hutches is to draw towards the pit-bottom throughout the entire length of the road, the proper place for the chain plate is on the start at the top end of the hutch (D), that is on the leading end of the empty hutch going uphill, and on the trailing end of the full hutch going downhill. If the chain plate is in this position, the chain drops down outside the low end of the hutch, and pressing against it keeps it from twisting, thus serving the same end as if two catches were used with the length of the hutch between them.

If, on the other hand, the road undulates, so that the chain may at one time be pulling and at another time holding back the hutch, a corner catch would not do so well, as, when the hutch came to be pushed by the corner, the leading end would separate from the chain, and a very small obstruction would suffice to put it off the road, besides which the tractive force would be very much increased owing to the twist on the hutch causing side friction between the wheels and the rails. In this case, therefore, the chain plate should be on the centre start (E), and if the chain is so tight that it might lie above the hutch instead of alongside it, the start should be extended up above the level of the hutch to a height of 4-5 in., so that the chain may be certain to drop outside of it. This arrangement serves to keep the hutch square on the rails as well as if two catches were used with a distance between them equal to half the length of the hutch. On the incline at Dykehead Colliery, a section of which is shown, the chain catch is on the end, as it was already in use for the endless chain haulage to the surface when the self-acting incline was started. On all the other inclines of this kind worked by the Summerlee Iron Company, side catches on the centre start have been adopted, as the roads were either undulating, or the hutch required to be run on two separate inclines rising in opposite directions. They have not, therefore, tried the corner catch on any of their inclines, but Mowat made experiments with a single hutch, and found that it kept its parallelism to the rails more truly than when the catch was on the side. There has been, however, little or no difficulty owing to the twisting of the hutches, even on an inclination of 1 in 4½; although, if the loaded hutches were going uphill, it would hardly be possible to work with a side catch, even if it were desirable, for the reason already stated. The speed of the chain varies from 1 to 5 miles per hour, the best speed being about 2 miles per hour, although this may be increased with advantage to 3 miles when the output is very large, and the chain is not strong enough to allow of the hutches being placed close together. The distance between the hutches varies from 10 to 25 yd., according to circumstances. If the road is very flat, they might require to be placed as close together as 10 yd. to get sufficient motive power, while, if the incline is steep, 25–30 yd. apart will give ample power. When the distance is greater than 25 yd., however, the chain is liable to trail on the pavement between the hutches, thus causing great tear and wear; so that, if possible, the distance should be shortened that the chain may never touch the ground except, perhaps, near the lower wheel, where it can hardly be avoided, and where beech planks should be laid to keep it off the pavement. The distance between the hutches is regulated by a small bell, which is placed on the incline at the prescribed distance from the hanging-on place, and which is rung by the last hung-on hutch striking it. The boy at the bottom puts on an empty hutch for every full one that comes off, thus keeping the number on both sides as nearly as possible uniform.

If the speed of the chain be taken as 2 miles per hour, and the distance between the hutches as 15 yd., the total number of hutches run off per hour, if the chain goes constant, would be $\frac{2 \times 1760}{15} = 234\frac{2}{3}$ hutches per hour; or, if the hutches hold 10 cwt. each, 117 tons per hour. These figures show clearly that a very large output can be drawn, regardless of the length of the incline, without increasing the speed beyond a creeping pace.

When a change of gradient takes place, so as to form a hollow, as in the Dykehead incline, the road must be raised in the hollow and the inclination changed gradually, so as to prevent the chain from being lifted out of the catch. In short, the road must be made to follow the natural curve of the chain. When a chain is suspended between two points of support, it forms a catenary curve, but if it is drawn nearly straight between the supports, it may be taken as a parabola, whose axis is vertical, without sensible error. In order to find the parabola which the chain will assume, it is necessary to take into account the tension of the chain at the origin or lowest point of the curve, the weight of the chain, and the distance between the hutches, all of which are known, or can easily be found.

Let T = tension at O in lb.
w = weight of chain per ft. in lb.
O H = y = $\frac{1}{2}$ span between supports in ft.
x = depression in ft.
$w y$ = approximate weight of chain O P, when O P is nearly straight and level.
Then to find the depression x in the chain between the two hutches,

$$\frac{x}{\frac{1}{2}y} = \frac{w y}{T}$$

$$x = \frac{w y^2}{2 T}, \qquad (1)$$

and the equation to a parabola, when y and q are the co-ordinates and $4\,A$ the ratio between x and y^2, is

$$y^2 = 4\,A\,x,$$

but from (1)

$$y^2 = \frac{2\,T}{w}\,x,$$

$$\therefore 4\,A = \frac{2\,T}{w}.$$

The curve is, therefore, a parabola, the equation to which is

$$Y^2 = \frac{2\,T}{w} x.$$

Again, if it is desired to find the distance from the origin O of the parabola to the tangent K, where the curve will join a straight line K L, let the inclination of the line be expressed as the tangent of the angle K L N, that is, the vertical height divided by the horizontal distance—e. g. 1 in 5 = ·20, 1 in 3 = ·3̇.

Then
$$\tan K\,O\,N = \frac{\tan K\,L\,N}{2}$$

$$\frac{x_1}{y_1} = \frac{\tan K\,L\,N}{2},$$

but
$$x_1 = \frac{y_1^2}{4\,A},$$

therefore
$$y_1 = \frac{4\,A \times \tan K\,L\,N}{2},$$

and
$$4\,A = \frac{2\,T}{w}. \tag{2}$$

$$\therefore \; O\,N = y^1 = \frac{T \times \tan K\,L\,N}{w}. \tag{3}$$

Again, to find the tension, in another place referred to as carrying tension, which must be applied in order that the chain may not trail on the ground:

$$x = \frac{w\,y^2}{2\,T} \tag{1}$$

$$\therefore \; T = \frac{w\,y^2}{2\,x}. \tag{4}$$

Of course, x must be a less distance than the height of the point of support on the hutch above the ground.

For an output of 700 hutches per day, the following oncost would be required to keep the chain going, viz.:—One man at the top, hanging on loaded hutches; one boy at the top, attending to the brake; one boy at the bottom, hanging on empty hutches; while the expense of upkeep of the incline would not be nearly so great as in an ordinary incline worked by a rope, owing to the speed being so low. To reduce the cost of attendance as much as possible, the inclinations should be arranged as in A, so that the empty hutches arriving at the top will detach themselves and run forward into the lye, for further transit by horses or drawers; while the full hutches will detach themselves at the bottom end and run forward to the pit-bottomer, or into a lye if the foot of the incline is not near the pit-bottom. When the output is small, so as to allow of the chain going at a slow speed, the brakesman may be dispensed with, the honcher being able to hang on and look after the brake as well.

The flattest inclination at which an incline of this kind will self-act depends on the comparative weights of full and empty hutches, on the weight of the chain, and on the friction of the hutches and chain wheels. As the chain should not touch the ground, there is no friction due to it, except that at the chain wheels, and the extra friction which its weight causes on the hutch wheels.

If W denote weight of full hutches and chain on full side.
,, w ,, ,, empty ,, ,, empty side.
,, F ,, friction of hutches, chain, and wheels on full side.
,, f ,, ,, ,, ,, empty side.
,, I ,, tangent of angle of inclination, or inclination expressed as a fraction—i.e. for 1 in 60—I = $\frac{1}{60}$.

Then before the incline will self-act

$$(W \times I) - F \text{ must be greater than } (w \times I) + F$$
$$(W - w) I > F + f$$
$$I > \frac{F + f}{W - w}.$$

The inclination found from the above formula would be that on which the surplus power of the full side would just balance all the resistances, so that the incline would require to be steeper than this in order that it might self-act.

The friction would require to be assumed or found by experiment, while the proper size of chain may be found from the formula,

$$D = 3 \sqrt{\text{Breaking strain in tons}}; \text{ or}$$
$$D = 3 \sqrt{\text{Working load} \times S},$$

D being the diameter of chain in sixteenths of an inch, and S being the factor of safety, which should be at least 5.

The working load on the chain is the greatest tension T, which is

$$T = (W \times 1) - F + \text{carrying tension}.$$

The chain should not be short-linked, as the chain-plate becomes fixed between the links of a short-linked chain, and tends to prevent the hutches from detaching automatically.

In an ordinary self-acting incline the road must be more or less uniform in gradient, for if steeper in some parts than in others the train must be run over the steep portion with great velocity in order that it may acquire sufficient momentum to carry it through the flatter portion; while in a great many cases it is impossible to work an incline by trains at all if the flat portion of the road happens to be at the top and the steep portion at the bottom, as a start cannot be obtained.

In working with a self-acting endless chain, if the average inclination of the road is not less than I, as found from the formula, the incline will work no matter how undulating it is, provided the average inclination is calculated from the total length of the road, and not from the horizontal distance on the section. The surplus power on a steep mine may be utilised for the purpose of drawing from a dook or level, not necessarily in the same straight line, by fitting the top wheel with a long shaft and putting on it a second driving wheel, or clip pulley, or rope drum provided

HAULING AND HOISTING MACHINERY. 255

with a clutch, as in F. In the same way water may be pumped, or almost any description of work done, if the power be sufficient.

G, H, I show three inclines on this system, and belonging to the Summerlee Iron Company. That shown at G is at Dykehead Colliery, and is used for drawing the Ell coal down a mine to the Main coal and uphill in the Main coal to the engine haulage terminus. The gradients, commencing at the top, are—

12 fathoms	..	..	..	..	level.
78 „	..	..	..	..	1 in 6¾ downhill.
98 „	..	..	..	..	1 in 588 uphill.

Or an average gradient of 1 in 16¼. There is a curve introduced at the change from the steep portion to the flat portion to keep the chain from lifting. The road is laid with edge rails, and the chain-plate, as already stated, is on the end of the hutch. This incline has been working for 7 years, and the chain was an old $\frac{9}{16}$ in. chain worn to about ½ in. The distance between the hutches is 25 yd.

The incline shown at H is at Dunsyston Colliery, and has taken the place of an ordinary self-acting incline, which did not run well, owing to the gradients at top and bottom being so unequal. The gradients, commencing at the top, are—

| 84 fathoms | .. | .. | 1 in 9 downhill. |
| 86 „ | .. | .. | 1 in 18 „ |

or an average gradient of 1 in 12. The road is laid with common tram rails, and the chain-plate is on the centre of the side of the hutch. The chain is ½ in., and the distance between the hutches is 15 yd.

I is a section of an incline in Braidhurst Colliery. This road is very undulating, the gradients, commencing at the top, being—

27 fathoms	..	..	1 in 10¼ downhill.	29 fathoms	..	1 in 16 downhill.
23 „	..	..	1 in 22 „	42 „	..	1 in 141 „
23 „	..	..	1 in 10½ „			

or an average gradient of 1 in 19. The road is laid with edge rails, and the chain-plate is on the centre of the side of the hutch. The chain is an old ½ in., and the distance between the hutches is 10 yd.

The advantages which a self-acting endless chain possesses over an ordinary incline may be summed up shortly as follows:—

(1) Small cost of upkeep of rolling stock, owing to slow speed causing few breakages; when a hutch goes off the road the chain stops.

(2) Regularity of delivery. The hutches arrive at the pit-bottom in such a manner that only very short lyes are required, and consequently the travel of the bottomers is diminished.

(3) When the output exceeds 100 tons a day, and probably before it, it can be worked much cheaper than an ordinary self-acting incline.

(4) Length makes no difference in the output or cost further than the increased upkeep of the road, whereas the difficulties in the way of drawing with a " cousie " increase with the length.

(5) Much less expenditure is required in making benches, as no long trains require to be collected on the incline as in a cousie.

(6) The cost for chains is less than for ropes, as a good chain will last 12–18 years.

On the application of electricity to underground haulage, John Fox Tallis remarks that deep shafts are being sunk, and workings are in course of development, where it is not considered practicable to take steam or ropes down the shafts, and where it would be very inconvenient to place boilers underground, to serve a number of hauling engines, even if such a course were admitted to be wise and expedient. The distances of working faces from the pits in old collieries are daily increasing, and rapidly reaching the limits of the systems of haulage at present in vogue; limits which cannot be exceeded without an enormous sacrifice in efficiency of power employed, resulting from friction on ropes, leakages of compressed-air pipes, crooked roads, uneven gradients, or other of the numerous difficulties met with in underground traction. The difficulty of extending the present systems of haulage in old collieries presents the most numerous, if not the most pressing, cases to be dealt with at the present time, and they are divided into two distinct classes.

First. Collieries that are provided with rope-haulage, deriving their power direct from steam engines fixed either on the surface or at the immediate bottom of shafts, where the coal is as a matter of necessity concentrated and collected by two or three main engine roads, extending into the heart of the properties, and depending for their supply entirely upon horses, owing to the impracticability of further extending their main roads and branches without militating against the utility of the main engine road, which must necessarily be kept waiting for its tributary branches, having only the one main rope to work the whole system.

Second. Collieries that are provided with air-compressors supplying a number of small engines, both at the immediate bottom of shafts and distributed over the workings, to feed the main engine planes; and cannot be further extended or multiplied without a complete re-arrangement of the compressed-air pipes, and probably the air-compressor, at an enormous sacrifice of efficiency and expenditure. But the most important, and perhaps the most pressing cases, are those of new and deep sinkings, where at present no haulage system has been adopted.

The transmission of power by electricity claims many advantages over compressed air or transmission by rope. The electric motor is, for equal power, smaller than any type of steam-engine used for haulage purpose, and, at the same time, the most efficient and compact machine for the transmission of power that is known.

The adoption of stationary electric motors for rope-haulage at considerable distance from the pit bottom, represents one of the cases to which the electric motor is peculiarly adapted; and for the sake of simplicity, it will be assumed that there is a main engine road 2000 yd. in from the bottom of the pit, worked by tail or other rope system. At a distance of 1500 yd. on this main engine road there is a branch heading, to the right or left, acting as a feeder to the main engine road; also at the end of the 2000 yd. main engine road there is another branch, to the right or left, and it is necessary that these two branches should be worked by rope-haulage. It is found that if these branch ropes are connected with and worked by the main rope, it will cause so much delay on the main engine road, that the output is considerably diminished; or perhaps the engine is not of sufficient power to cope with the additional friction and strain on the rope without sacrificing the output. It therefore becomes necessary to adopt some other power to work these two branches, and it is decided to put down two electric motors of 30 H.P. each at the junctions of the two branch headings with the main engine road, with a view of securing not less than 20 H.P. in ropes of each branch.

Having decided the power that is required at the distances of 1500 and 2000 yd. respectively from the bottom of the pit, the first question that arises is—What engine-power must be provided at the top of the pit to ensure 40 H.P. in the ropes underground?

This depends upon a number of circumstances, which are governed chiefly by the first cost to which the mining engineer is disposed to go, and must be decided by him rather than by the electrical engineer. The electrical engineer will guarantee a certain efficiency in the generator and motors, probably 90 per cent. in each machine if required; but if the motors were required to work at a comparatively slow speed, a considerably higher price would be demanded for high efficiency at slow speed, than for the same efficiency at a high rate of speed.

There would necessarily be a considerable loss in gearing, as in other engines, depending upon the mode of gearing adopted, which also will be governed to some extent by first cost; and there is the loss on the cable, which will depend to a great extent upon first cost. True economy will dictate the highest efficiency in each case. The various losses in an economical installation should not exceed the following figures to obtain 20 H.P. in the ropes of the two branches respectively.

	Per Cent.	On H.P.	Loss in H.P.	Electric Loss H.P.	Mechanical Loss H.P.
1. Loss between steam-engine and terminals of generator, or in belting and generator	20	80	16	6·0	10·0
2. Loss in cable to No. 1 motor	10	64	6·4	6·4	..
3. Loss in No. 1 motor	10	27·95	2·8	2·8	..
4. Loss in gearing between No. 1 motor and rope	20	25	5·0	..	5·0
5. Loss in cable from No. 1 to No. 2 motor	6	29·65	1·7	1·7	..
6. Loss in No. 2 motor	10	27·95	2·8	2·8	..
7. Loss in gearing between No. 2 motor and rope	20	25	5·0	..	5·0
Total loss	..	..	39·7	19·7	20·0

The summary of the losses being nearly 40 H.P., it would require a steam-engine of 80 H.P. on surface to provide 40 H.P. in the haulage-ropes underground, an efficiency of 50 per cent., which will compare favourably with any existing system under the same conditions. It is possible that this efficiency can be considerably increased, and probably will be; however, there should be no difficulty in getting engineers to guarantee the above efficiency.

The electrical loss represents nearly one-half of the total loss, or 20 H.P., leaving 20 H.P. for loss in mechanical gearing, allowing both the mechanical and electrical engineer equal opportunities for improvement.

Having decided the horse-power of steam-engine required, it is necessary to select a suitable engine. The first point to consider is the pressure of steam available from the colliery range of boilers, which will decide the size of cylinders.

The engine should essentially work at the highest speed compatible with continuous running, so that the generating dynamo may be worked direct off the fly-wheel of steam-engine without countershafting, and to reduce as much as possible the ratio of speed between the steam-engine and generator. Any advantage that a slow-speed engine would possess over a high-speed engine would probably be more than counterbalanced by the interposition of countershafting, and the general efficiency thereby diminished.

2 L

The engine should be provided with a sensitive governor regulating an automatic expansion-gear or throttle valve, so that the engine may be kept at a constant speed during variations of load and to prevent running wild even with no load on. The engine to meet these requirements can be either single or double; for steady running no doubt a double cylinder would be preferable, but for reasons to be explained, it may sometimes be desirable to fix a single-cylinder engine; so descriptions of both classes of engines will not be out of place. The pressure of steam in boilers is taken at 50 lb. per sq. in.

A suitable engine would be a horizontal 20-in. cylinder, stroke 36 in., with fly-wheel 16 ft. diameter and 25 in. wide on face, set so that the top part revolves from the cylinder at a speed of 60 revolutions per minute, giving a piston speed of 360 ft., fitted with governor and automatic expansion-gear and sight-feed lubricators. For a double engine, two 15-in. horizontal cylinder engines coupled, stroke 30 in., with fly-wheel 12 ft. diameter and 25 in. wide on surface, revolving at a speed of 80 revolutions per minute or a piston speed of 400 ft., fitted with governor, automatic expansion gear, and sight-feed lubricator to each cylinder.

The next point for consideration is the gearing of engine to generating dynamo. The speed of dynamo is fixed at 480 revolutions, and the engine at 60 and 80 revolutions per minute respectively; the ratio is 8 or 6 to 1.

The driving-wheel of dynamo should be not less than 2 ft. Owing to the small diameter of the dynamo wheel, steel wire-ropes would not be applicable, belting being preferable, the lead for which should not be less than 30 ft. between centres for the 16 ft. fly-wheel and 21 ft. for the 12-ft. fly-wheel.

The belt can be either a flat leather belt or patent leather chain belt arched to suit the curve of the pulley. A flat belt always retains a cushion of air between itself and the pulley, which prevents perfect grip; in the chain belt this air escapes through the spaces. However, the flat orange-tanned leather belt has the advantage of being made so that you can work another belt on top of it and running quite independently of it, so as to increase the power transmitted if desired over 50 per cent., and a double belt suitable for this installation should be 16 in. wide.

In selecting the dynamo, an efficiency of not less than 90 per cent. should be guaranteed, and for constant running with a maximum load it would be preferable to allow 25 per cent. above the required horse-power rather than work the machine continuously at its maximum power; but in the present case the maximum power would be required only when the two motors are working at the same time and with full loads. It is not probable that this would be the case excepting for occasional journeys or during parts of journeys; and as half the time of the motors would be occupied in taking in the empty journeys, it is not probable that the generating motor and steam-engine would be required to work at their maximum powers for more than 50 per cent. of the working hours. This would, of course, depend upon the nature of the engine-roads. If the roads are such as to require about an equal power in and out, the time during which the maximum power would be required would be greatly increased, and in that case it would be preferable to have the generating dynamo above the maximum power. In this instance it is assumed that the in-going journey is lighter than the out-coming journey.

The maximum power required in the generating dynamo would not exceed 65 H.P., and with an efficiency of 90 per cent. the dynamo will require to be 71 H.P., or say in round numbers 75 H.P., which would represent a dynamo giving 56,000 watts at 500 volts and 480 revolutions per minute. Approximate weight up to 6 tons according to maker.

The E.M.F. is fixed at 500 volts; this or even a higher potential has the advantage of reducing the size and cost of cables.

The maximum H.P. passing through cable should not exceed 64 H.P., which, with a voltage of 500 E.M.F., would require $\left(\frac{64 \times 746}{500}\right)$ 95·5 ampères.

The distance from dynamo-house on surface to bottom of pit is assumed to be 500 yd.; from bottom of pit to No. 1 motor, 1500; or a total of 2000 yd. lead and 2000 yd. return, making 4000 yd. of main cable, capable of carrying 64 H.P., and the loss not to exceed 6·4 H.P., or 10 per cent. of the maximum power required.

Having the loss in H.P., the resistance of cable should not exceed $\left(\frac{746 \times 6 \cdot 4}{95 \cdot 52}\right)$ 0·52 ohm in the 4000 yd.

Conductivity taken at 95 per cent. of pure copper and temperature at 60° F., it will require a cable $\left(\sqrt[2]{\frac{\cdot 003265 \times 43}{\cdot 52}}\right)$ 0·5 in. diameter weighing $\left(\frac{\cdot 19635 \times 144{,}000 \times \cdot 32}{2{,}240}\right)$ 4 tons, having an electrical resistance of 0·52 ohm, or a total resistance in cable, without allowing for rise in temperature, of $\left(\frac{\cdot 52 \times 95 \cdot 5^2}{746}\right)$ 6·4 H.P., causing an approximate rise of 8° F. in temperature of cable.

On the basis of 1000 ampères per sq. in. sectional area, this cable being 0·2 sq. in. area, would carry a current of 200 ampères, or $\left(\frac{500 \times 200}{746}\right)$ 134 H.P., with a loss of about 31 H.P. and an approximate rise in temperature of about 29° F.

A cable 0·5 in. diameter would be about equivalent to a cable of 37 strands No. 14 legal standard wire gauge.

From No. 1 motor to No. 2 motor the distance is 500 yd., or a total of 1000 yd. lead and return cable, required to transmit 30 H.P., or $\left(\frac{30 \times 746}{500}\right)$ 45 ampères, with a loss not exceeding 1·7 H.P., or $\left(\frac{746 \times 1 \cdot 7}{45^2}\right)$ 0·62 ohm resistance in cable.

To do this will require a copper cable $\left(\sqrt[2]{\frac{\cdot 03265}{\cdot 62}}\right)$ 0·23 in. diameter, or 19 strands No. 17 L.S.G. cable, weighing ·23 ton having an electrical resistance of ·57 ohm, or a total resistance in cable of $\left(\frac{\cdot 57 \times 45^2}{746}\right)$ 1·5 H.P.

On the basis of 1000 ampères per sq. in., this cable would only carry 45 ampères and allow no margin for increasing the power; therefore it is advisable to adopt a larger cable, say 19 strands No. 15 L.S.W.G., having an electrical resistance of only ·32 ohm, and equivalent to a solid wire of ·317 in. diameter, 0·0789 sq. in. area, capable of carrying 78 ampères with a rise in temperature not exceeding about 13° F. and a loss of $\left(\frac{\cdot 32 \times 78}{746}\right)$ 2·6 H.P.

The copper in this cable would weigh about 0·40 ton, and the loss in cable for 45 ampères would be only 0·8 H.P. (instead of 1·7 H.P. as allowed for in first estimate) with an approximate rise in temperature of about 6° F. An extra 20 yd. of this 0·317 in. cable would be required for connecting No. 1 motor.

All cables and wires should be covered with an insulation having a resistance of not less than 1000 megohms per statute mile at 60° F.

Having decided upon the cables and insulation, the next important question is the fixing of same down the shaft and through the workings to the motors.

In the first place, they should be laid so that they would be subject to dampness as little as possible, as very few kinds of insulation will withstand the deteriorating effect of constant changes of temperature and dry and wet conditions. They must be protected or placed out of danger of being broken or damaged by trams going over the roads, or falls from the roof or sides. They should be laid so as not to interfere with ordinary repairs, but at the same time be as accessible as possible and according to the usual rule—*leads left, returns right*; or when laid one above the other—*leads low, returns raised*. If they have from some cause been disarranged, a pocket compass will show the direction of the current. Stand with your back to the generating dynamo, place the compass *beneath* the cable; then if the N-seeking pole of the compass be deflected towards your *left* hand, the current is flowing from the dynamo to you; but if to the *right*, the current is returning to the dynamo. You only require to remember that your back must be to the dynamo and the compass placed *under* the cable, not above; the initial letters L and R will then be sufficient to guide you; to the *left* (L) for lead, and to the *right* (R) for return.

The main cable, if possible, should be laid from the dynamo to the motor without splicing.

If the cables are sheathed with steel or iron wire they can be secured to side of shaft with staples driven in gently and not too tight against cable; the weight of cable should be supported by flat iron flanges screwed on to cable and resting on the staples.

The most simple and effective protection for the cable would be to have it covered or sheathed with steel or galvanised-iron wire, finally coated with compound and two reserve tapes, and simply laid in a channel on one side of the roadway covered to a depth of about 3–4 in.

A steel or galvanised iron-wire sheathing will necessarily be expensive, but it will economise considerably in the laying down of the cable and effectually protect the insulation; so that after it has been used for several years in one district, it could be taken up and laid down elsewhere, if required, without its sustaining any damage.

Before deciding upon the motor, it is necessary to fix the diameter of drum to receive rope, the speed of rope, and the kind of gearing to be used, which will dictate the speed of motor. The speed of rope should range from 6 to 10 miles an hour, and the drum not less than 4 ft. diameter, or, say 12·5 ft. circumference.

Calculating 10 miles an hour with 12·5 ft. circumference would give $\left(\frac{880}{12\cdot 5}\right)$ 70 revolutions of drum per minute, or 7 revolutions of drum per mile. The most usual speed would probably be about 60 revolutions per minute, or not quite 9 miles an hour. This is a speed far too slow for an electric motor to work efficiently, so that some kind of gearing is compulsory.

For a high-speed motor it would be necessary to have belt or friction gearing, each of which has its drawbacks, and would no doubt meet with opposition, and it is probable what would be gained in the higher efficiency of the motor would be more than lost in the extra friction of the gearing.

The simplest and no doubt better system is the ordinary spur- and -pinion gearing, but with helical teeth to give increased strength and enable the ratio to be as great as possible, say 6 to 1, there being a 12-in. pinion on the armature spindle gearing, with a 6 ft. spur-wheel keyed on to the drum shaft. This gives a speed of 420 revolutions per minute, so that the motor should be constructed to give a maximum efficiency at about that speed, or somewhat below it.

It would be necessary to fix the motor on one side in the same line as the drum shaft, or at right angles, and work with bevel gearing; either plan would leave the drums perfectly clear and free for access, but perhaps the former would be preferable; having the engine-house the width of the drums, with an archway recess on one side in line with the drum shaft for motor, which would be 2–2½ tons in weight to give 30 H.P. at 420 revolutions per minute, absorbing about 23,000 watts. The gradual starting, stopping, and reversing of the motor would be controlled by varying the position of the brushes on the commutator by means of an ordinary switch or lever.

In the foregoing example the dynamo is supposed to be compounded and the motors series wound. It will probably be an interesting point to electricians and manufacturers as to the compounding of the dynamo; but it has no essential bearing upon the principle. Probably some manufacturers would prefer that the dynamo should be separately excited by passing a low tension current through the shunt coils and regulate the E.M.F. by the main current passing round the series coils; others may prefer simply a shunt dynamo. No. 2 motor with fittings would be simply a repetition of the above description of No. 1 motor.

It will be observed that the efficiency of the motors is fixed at only 90 per cent. and the generating dynamo at 92·5 per cent., making a total loss of 17½ per cent. in dynamo and motor; whereas Dr. Hopkinson some time back found that 87 per cent. of the energy given to the dynamo was returnable at the motor, showing an efficiency of 93·5 per cent. on the dynamo and motor respectively; so it will be seen that Tallis has underestimated the efficiency of the dynamo and motor by 4·5 per cent. Motors should be fixed in such positions as not to come into contact with gas.

There is an advantage in a system of electric haulage erected on the lines described, that no other system of haulage possesses to the same extent, viz., the facility of increasing its power at a minimum cost, should it be necessary to do so.

It is often the case that hauling-engines are erected in a colliery that are of sufficient power to cope with the quantity of coal brought out during the first 10 years, but after that period—and in many pits, probably in a much shorter period—it is found, owing to the face of the workings getting farther and farther away from the pits, necessary to extend the engine roads to even double their former length, and when that is the case the power of the engines is found deficient, and it becomes necessary, in preference to multiplying the number of engines, to take down the old engine and erect a more powerful one. This will be found especially applicable in hauling-engines fixed at immediate bottom of shafts.

In the case of the electric-haulage system here described, it is only necessary to raise the pressure of steam per sq. in. by adding a condenser, or otherwise, and work the engine with less expansion, to increase its power. The strength of driving-belt can be increased by placing another belt to work on top of present one. The dynamo can be replaced by one of the required increased power and E.M.F. The cables are ready fixed and sufficiently large to take that increased power. The motors can be removed and replaced by motors of increased power and E.M.F. with a minimum outlay in labour, and it would be especially applicable if the old motors could be utilised elsewhere. By a judicious increased E.M.F. and current, the power might be increased 50 per cent. without reducing the general efficiency to an appreciable extent.

Assuming the first dynamo and motors had been working for 8-10 years, the extra cost would be perfectly justifiable, and probably meet the exigency of the case. If required, one motor can be fixed to work two sets of drums by having two loose pinions on the motor shaft, thrown in and out of gear by friction clutches.

Approximate cost of plant for installation of two 30 H.P. electric motors, situate respectively at 2500 and 2000 yd. distant from the generating station on surface:—

	£	s.	d.
Two single horizontal 15-in. cylinder engines, coupled, stroke 30 in., with 12-ft. fly-wheel to receive driving-belt, automatic governor, expansion gear, and side-feed lubricator to each cylinder	500	0	0
Driving belt, 16 in. diameter, leather	60	0	0
Dynamo to give 56,000 watts at 500 volts and 480 revs. per minute, current and potential indicators, cut-outs, switches, &c.	500	0	0
4000 yards sheathed cable, equivalent to 0·5 in. diameter, solid copper rod, consisting of 37 strands tinned copper wire, covered with one lap of pure and two layers best vulcanising rubber-coated rubbered tape, and the whole vulcanised together, further served with jute and sheathed with 20 galvanised iron wires, 0·169 in. diameter, and finally coated with three coats compound and two reserve tapes; finished diameter 1·48 in.	1200	0	0
1020 yards sheathed cable, equivalent to ·317 in. diameter, solid copper rod, consisting of 19 strands tinned copper wire (insulation as above), sheathed with 20 galvanised iron wires 0·131 in. diameter, finally coated and taped; finished diameter 1·18 in.	180	0	0
Two 30-H.P. motors, at 420 revs. per minute, absorbing about 23,000 watts	600	0	0

Foundations, engine-houses, steam-pipes from boilers, labour, drums, gearing, and erection, extra.

Owing to the existence of farther motor in the pit from the generating dynamo fixed on surface, and the amount of power conveyed, the foregoing example is adequate to meet the most extreme cases of ordinary collieries in the South Wales district, and the cost per H.P. useful effect represents about the outside limit to which it would be required to go. The general efficiency of the whole system would of course depend upon the quantity of coal that is brought out of the two districts. If the motors are kept constantly going, a high efficiency will be attained; if not, the efficiency will decrease in proportion to the reduced output. But it must be remembered that if the motors are only required to work intermittently, there would be no difficulty in running branch cables from bottom of pit, or from any point on main cable, to motors situate in other parts of the colliery, providing motors running at the same time are not required to develop more than 64 H.P., including loss in cables, just as is done at present with compressed air.

For instance, one 60 H.P. motor, or two 30 H.P. motors, or three 20 H.P. motors, or four 15 H.P. motors, or any other combination of motors, not exceeding about 60 H.P. situate at bottom of pit, or distributed over the colliery, could be worked together and at the same time; so there is practically no limit, excepting that of expediency, to the number of motors that the engine and dynamo at generating-station could work intermittently in any part of the colliery, providing the number working at the same time are not developing more than 64 H.P., including loss in cables.

It can be seen how very important it is that the cables in the first instance should be highly insulated with materials not subject to deterioration owing to changes of temperature or dampness, are well protected, and of a generally useful section; as a small cable will only transmit little power and a short distance, whereas a large cable will transmit either little or great power, and for either a short or long distance.

To arrive at the power required in an electric motor, the better and safer way is to calculate the work in foot-pounds that it is required to do. For instance, assume a drift of 1000 yd. long, with an average dip. of 3 in. per yd., and it is wanted to take up a journey of 20 tons (including rope) in 5 minutes of time, which represents a little over 6 miles an hour. By reducing the above to foot-pounds, and dividing by 33,000, you have 67 H.P., and to this must be added 20 per cent. loss in friction of gearing, and 10 per cent. loss in motor, to get the actual H.P. of motor required, which would be 88 H.P. It is also necessary to take into consideration, the expediency and economy of working a motor at its maximum power, and perhaps in all cases, especially for constant running, it is advisable to add 25 per cent. to power of motor, bringing the total H.P. of motor up to 110 H.P.

In erecting an installation for a new colliery only partly developed, it is assumed that the distance from generating station to bottom of pit is 700 yd. At the present time, or in the course of a year or so, they require one 100 H.P. and three 50-H.P. motors, or a total of 250 H.P. and will ultimately require in the course of 10 years, say, double that power, or 500 H.P. It is necessary to erect plant at the present time to give 250 H.P., such plant being designed and erected so that at any time the power can be increased to 500 H.P., and without sacrificing in any way the plant first installed.

Allowing 5 per cent. loss in cable,
,, 10 ,, ,, dynamo,
,, 10 ,, ,, gearing,

the size of steam-engine required would be 650 H.P.

The pressure of steam is taken at 70 lb. per sq. in. in boilers.

An engine capable of doing this would be two single horizontal 28 in. cylinder engines, coupled 4-ft. stroke, with two 20-ft fly-wheels for driving-belts, fixed between the two cylinders, revolving at 50 rev. per minute.

Only one of these engines to be erected at present, but so designed and arranged that the other one can be coupled to it when required, driving a 300-H.P. dynamo at 400 rev. per minute.

If one pair of large cables were put in the shaft at the outset to take the full power that would be ultimately required, i. e. 527 H.P., to give 500 H.P. at bottom of pit, which is equivalent to 655 ampères at 600 volts, it would require a cable 1 in. diameter, or ·7854 sq. in. sectional area, weighing 5·6 tons of copper, and giving a loss of about 26 H.P., or 5 per cent.; which for a cable insulated and sheathed as per previous estimate, finished diameter 2·37 in., would cost 17s. per yd.

But if two pairs of cables are used, the cable will require to be exactly half the sectional area and weight of copper per yard to take the same current, with a like loss of only 5 per cent., and they have the advantage of doing so at a considerably less rise in temperature in the cables, which is better for the insulation and resistance of the cable. The price of two pairs of small cables would be 20s. per yd., or 18 per cent. more than the larger cable. Therefore, as the labour cost of laying another pair of cables down the shaft is very trifling, it is a decided advantage to put down a pair of cables only sufficient to convey the power that is at present required, i. e. 263 H.P., to give 250 H.P. at bottom of shaft, which is equivalent to 327 ampères at 600 volts; and a cable to meet these requirements would be 0·707 in. diameter, with a sectional area of ·3925 sq. in., weighing about 2·8 tons of copper, giving a loss of about 13 H.P., or 5 per cent., with an approximate rise in temperature in cable of 24° F., as against about 45° F., for the large cable conveying 655 ampères. The horse-power of this installation could be doubled at any future time, without sacrificing any of the plant already installed. The engine can be coupled with one of similar design and make, with another fly-wheel driving a second dynamo; by this means increased steadiness of running would be gained.

Another pair of cables would be required in the shaft, where they could be joined at the bottom to the existing system of cables; or, if thought preferable, the whole of the newly-installed plant could be kept perfectly separate from engine to motors, and so secure two independent systems; but this would depend upon different circumstances and requirements existing at the time of erecting the new plant.

If the two installations and systems are kept separate, it would perhaps be convenient to have a switch at bottom of shaft, so that one engine and dynamo can work either of the two groups of motors; in this way a great safeguard would be secured in case one of the engines or dynamos on surface from any reason failed to work, as the other engine and dynamo could work each group of motors alternately, and so prevent a complete stoppage of the haulage system.

Systems of electric lighting should be kept perfectly independent of the haulage system, as great fluctuations would be caused in the haulage-cables by starting and stopping of motors, which would be most injurious, and probably fatal to lamps. There are also several other reasons why the systems should be perfectly independent.

Tallis has not gone into the question of electric locomotives, as he does not consider them suitable either from a practical or technical view of the question. It is impracticable to have an electric locomotive to do a reasonable amount of work without having great weight to maintain adhesion or grip to the rails. They would be of no use excepting on tolerably level roads. Where used to greatest advantage on surface, they are supplied with current by means of overhead loose wires, or cables carried in a channel under the road, either of which is impracticable on colliery roads underground.

Locomotives worked by secondary batteries would be more feasible, but as they are easily subjected to derangement and breakage, they are never likely to gain any degree of popularity for underground work. Imagine a locomotive, 6 tons weight, carrying a number of secondary batteries, travelling at 6-8 miles an hour, going over the road, and coming in contact with the sides. It is not impossible to keep good and suitable roads for locomotives of this weight underground, but you cannot always secure against accidents occasioned by small falls of top, or lumps of coal, or stones on the rails, which would cause the locomotive to leave the rails; and in addition to the difficulty of putting such a locomotive on the rails again, if the secondary batteries or the machinery was broken, it would be impossible to get a 6-ton locomotive out of the way without considerable trouble and stoppage of the work.

Lebreton has recently discussed the application of electricity to haulage in the underground workings of mines, as exemplified by the installations at Zaukeroda near Dresden, at Beuthen in Upper Silesia, and at Neu-Stassfurt; all of which have been carried out by Siemens and Halske, of Berlin.

(1) Installation at Zaukeroda, opened September 1st, 1882. The electrical railway is fixed in a level 240 yd. deep, and is 785 yd. long, of which 676 yd. are used for transport, the remainder being employed for the formation of the trains. The line is double, and the rails are of steel, $13\frac{1}{4}$ lb. per yd. It is practically level throughout. The generating dynamo is series-wound, and runs 750-850 rev. per minute, being driven by a steam-engine of about 15 H.P. It is fixed at a distance

of about 60 yd. from the pit-mouth, and the current is conveyed above ground by naked copper conductors, and in the shaft by a cable with gutta-percha insulation with lead covering. The outward lead has a further sheathing of galvanised iron-wire and tarred canvas, while the return lead has no iron sheathing. Experience has shown that the latter stands equally well with the former. In the level the conductor is formed by two rails of inverted T iron, bolted to insulators and carried from the roof. The T iron has a section of 10 lb. per yd.; and the ends of consecutive lengths are soldered together. The collector consists of a carriage, which slides along the webs of the inverted T iron, and which is provided with contact-springs, pressing against the under surface of the web, and connected to the locomotive by cables, which also serve to drag the collector carriage along the rails. The trains are formed of 10–15 waggons, each weighing about 5 cwt., and carrying 10 cwt. of coal. The average speed is about 6 miles per hour. The total quantity of coal drawn in two shifts of 8 hours each is about 380 tons. The cost of the whole installation delivered and erected, including steam engine, conductors, and locomotive, was 810l.

(2) Installation of Hohenzollern-Grube, at Beuthen, opened September 24th, 1883. The line is along a level at the depth of 200 yd., and is about ½ mile long. There are no gradients, and the velocity attained is about 7 miles per hour. The weight of the locomotive is over 2 tons. A second somewhat heavier and more powerful locomotive has since been added. This can be developed up to 6·8 H.P. The general arrangement of the line and of the conductors is the same as at Zaukeroda. The tare of the waggons is a little over 7 cwt; and they can carry 11 cwt. of coal. The line is worked at an electromotive force of 320 volts, and a current of 25 ampères. The resistance of the conductor is 0·8 ohm. The total cost of the installation, including one locomotive, but exclusive of the steam-engine, was 1000l.

(3) Installation of Neu-Stassfurt, opened December 21st, 1883. The line is along a level 360 yd. below the surface, and is about 1100 yd. long, and practically horizontal. The locomotive can work up to about 6·8 H.P., and weighs 39 cwt. The line, in its general arrangement, is similar to the two previously described, and the trains consist of 10 waggons, the gross weight being about 12 tons.

At Zaukeroda experiments have been made to determine the efficiency of the system. As measured between the indicated power of the steam-engine driving the generator, and the useful work done by the locomotive, it is 30 per cent.; as measured between the power absorbed by the generator and the power absorbed by the motor, it is 46·6 per cent. The cost of working is given as: 11s. 9d. for 660 waggon-loads in 16 hours, to which must be added 8s. 1d. for interest and depreciation at the rate of 15 per cent., making a total of 19s. 10d., or 0·36d. per waggon.

The following table shows the cost as compared with previous results for traction by horses and men:—

	Cost of Electrical Traction, including Depreciation.	Cost of Horse Traction.	Cost of Traction by Men.
	d.	d.	d.
For 660 waggons	0·36	0·45	0·74
	Reduced to ton-miles.		
,, ,, ...	2·0	2·5	4·1

	Rope on Drum at Aniche.	Trailing Cable at Searbrucken.	High-speed Endless Cable at Shire Oaks, Nottingham.	Low-speed Endless Cable at Bridge Pit Wigan.	Low-speed Endless Cable at Meadow Colliery, Wigan.	Endless Chain at Saarbrucken.	Endless Chain at Mine, Hessnent, Belgium.	Steam Locomotive at Doman.	Steam Locomotive at Cessous.	Compressed Air, Petsu System.	Compressed Air, Mekarski System.	Electricity at Zankerode.
Cost of installation .. .. .. £	1600	2300	750	2300	750	4000	5600	1600	1600	650	1000	800
Interest and depreciation per ton-mile .. .. } d.	1·38	0·26	0·28	0·37	0·40	0·79	0·23	0·42	0·15	0·80	0·82	0·45
Cost of working per ton-mile d.	0·53	1·02	1·82	3·71	2·02	0·74	0·22	0·60	0·52	1·20	1·35	1·11
Total cost per ton-mile .. .. d.	1·91	1·28	2·10	4·08	2·42	1·53	0·45	1·02	0·67	2·00	2·17	1·56
Daily ton-miles	93	708	200	491	145	429	1950	298	867	65	97	143
Distance traversed .. .. .. yd.	600	4110	920	2163	638	1920	3500	2530	5041	676	676	676
Speed .. .. yd. per minute	330	220	220	40	33	123	100	151	220	100	100	172
Weight of locomotive .. .. tons	..	..	..	..	..	..	..	4·4	8	2·7	2·3	1·6

Professor W. Schutz has published the preceding table, showing the comparative costs of the various systems. The increased cost of ventilation of the mine, when a steam locomotive is used, is not taken into account.

At Hohenzollern-Grube, Lebreton calculates the cost of haulage, exclusive of interest and depreciation, to be 0·9d. per ton-mile. The comparison of this, with the results obtained by other systems of haulage in some English and Scotch mines, is interesting :—

System.	Tons Drawn per Shift.	Length of Line.	Cost per Ton-mile.
		yards	d.
Tail rope	480	2130	1·15
Endless rope	429	1050	1·25
Floating cable	403	850	0·83
Trailing „	325	1400	2·01
Cable in Cadzow Colliery, Hamilton ..	842	1310	1·33

Vogel has arrived at the following, as the comparative cost of the various systems, all reduced to a tonnage of 400 tons over a distance of 2200 yd., interest and depreciation, at the rate of 15 per cent. on the first cost, being allowed for—

                        d.
1. Self-acting chain ..   ..   0·45 per ton-mile.
2. Electrical locomotive ..   ..   0·88 „
3. Endless chain   ..   .. 1·31 to 1·46 „
4. Various systems of traction by cables ..   .. 1·66 „ 2·00 „
5. Horses   ..   ..   3·24 „

WINDING DRUM.—The drawing rope, after passing over the pulley at the top of the headstock, is led to the winding drum, upon which it is coiled. This drum may be either cylindrical or conical in form, and it may be made to revolve either upon a horizontal or upon a vertical axis. The latter arrangement is now, however, rarely adopted, and we shall therefore consider only the case of horizontal drums. A drum consists of a barrel, upon which the rope is wound, and two sidepieces or flanges, to prevent the rope from slipping off the barrel. These two portions are carried upon arms connected to a central boss, through which the shaft passes. The material used in the construction of winding drums is most frequently iron, a combination of both cast and wrought iron being usually adopted. The barrel is cast in segments, and put together by being bolted through flanges provided for that purpose. The arms are also of cast-iron, and are bolted to the flanges of the barrel, a portion of the rim being cast upon each arm, in some cases. The inner ends of the arms are fitted into a cast-iron boss, and secured in position by turned bolts in bored holes. The shaft is of wrought-iron, and should be forged from the best scrap; to secure the bosses, which should be bored out to the exact diameter of the shaft, the latter is turned and divided with key-beds cut into it. A similar mode of construction is adopted when the drum is conical in form, in so far as its essential component parts are concerned. With this form the drum presents the appearance of a double cone, or two cones, or frustra of cones placed base to base, and the rope is fixed so as to be ascending upon one cone while it is descending upon the other. The principal object of this arrangement of the drum and the rope is to ensure the regular coiling of the latter; but it contributes to equalise the resistance to be overcome by the engine.

The diameter of a winding drum is determined mainly by the nature of the rope to be used, a much larger diameter being required for wire ropes than hempen ropes; but it should also bear some proportion to the diameter of the rope of a given material, since it is obvious that the thicker the less readily it will coil upon a cylinder of a given diameter. A suitable diameter of the drum may be obtained in the following manner: Assuming 10 ft. to be the minimum diameter for a wire rope 1 in. in circumference, add 6 in. to the diameter of the drum for every increase of $\frac{1}{4}$ in. in the circumference of the rope. Thus a rope $2\frac{1}{4}$ in. in circumference will require a drum of $10 + 4\cdot 5 = 14$ ft. 6 in. diameter, and a rope of $3\frac{1}{4}$ in. will require a drum of $10 + 7\cdot 5 = 17$ ft. 6 in. As the diameter of the pulley and drum is increased the life of the rope is lengthened, and it is obvious that, determined by the conditions of wear in the rope, the diameters of the pulley and of the drum should be equal.

Round rope is wound upon the drum in parallel coils, and in some instances it is made to rise and return upon itself on cylindrical drums for the purpose of diminishing the length of the latter; the arrangement is, however, unfavourable to the durability of the rope. When the drums are conical the overlap is, of course, impossible, and the same necessity for it does not exist. A flat rope is always wound upon itself, so that its coils are all in the same vertical plane; hence, practically, the diameter of the drum is constantly increasing or decreasing, and the velocity of the load consequently accelerated or retarded. This variation tends, of itself, to render the work of the engine unequal during the raising of the load. But it will be observed that this tendency is counteracted by a variation in the value of the load during the same time, and that, consequently, this overlap of the rope results in an equalisation of the work of the engine. When the load starts from the bottom of the shaft it has its maximum value, for at that moment the weight of the whole length of rope is added to that of the cage with its contained load; and the resistance due to the inertia of the mass must also be overcome at the moment of starting. But when the load has thus its maximum value, the diameter of the drum is at its minimum value since the rope is then wholly uncoiled, and hence the leverage in favour of the load will also have reached its lowest limit. Moreover, as the other portion of the rope will, at the same moment, be wholly coiled upon the drum, the latter will, relatively to this portion, have attained its greatest diameter, and consequently the leverage in favour of the descending load, consisting of the empty cage, its highest value. These circumstances are evidently favourable to the equalisation of the work of the engine, and continue throughout the time of winding. For, as the one portion of the rope ascends and diminishes in weight, the leverage in favour of it increases in a like degree; and as the other portion descends and increases in weight the leverage in favour of it is diminished in like manner. The same advantages are obtained with round ropes, though under less favourable conditions, by making the drum conical. When the drum has this form, there is a liability of the rope slipping, if any hitch should occur to slacken it, and such a slipping would probably cause rupture of the rope. The length, or as it is sometimes described, the breadth of the drum is obviously least with the rope.

When both portions of a round rope are wound upon the same drum, the length of the latter will be that required by a single rope, since one portion is being unwound while the other is being coiled upon the drum, so that the sum of the lengths coiled at any given moment is equal to the length of one portion of the rope. In such a case one portion of the rope is wound over the drum, and the other portion under the drum. As both portions are wound over the pulley, one is thus wound in contrary directions, a circumstance unfavourable to its durability. The evil is removed by the use of two drums revolving in contrary directions, an arrangement which allows both portions of the rope to be passed over the drum. The details of fixing the rope to the drum are very simple. Usually a notch or a groove is provided on the drum to receive the end of the rope, which is held in by wedging. To avoid bringing the strain of the load upon the fastened end of the rope, the length is always regulated to leave two or three coils upon the drum when the cage is at the bottom of the shaft.

The position of the drums is a matter of importance. Relatively to the engine, they may be placed with their axes in the horizontal plane passing through the piston rod, or they may be placed above the cylinders with their axes in the vertical plane passing through the piston rod. Each of these positions possesses some advantages: the former appears, however, to be preferable, and it is more commonly adopted. Relatively to the pulleys, the level of the drums should, where easily practicable, be so adjusted that the inclined portion of the rope shall not make a very acute angle with the vertical portion; hence the higher the pulleys the greater should be the interval between the drums and the pit mouth. Too great a distance is, however, objectionable, by reason of the sagging and swaying of the rope. The best arrangement, where it can be adopted without difficulty, consists in erecting the drums at a higher level than the pit mouth. This is one of the advantages obtained by placing the drums over the steam cylinders. An essential condition to be observed is to place the drum and its corresponding pulley in the same vertical plane, and strictly perpendicular to their axes of rotation. A slight irregularity in this respect, by forcing the rope to deviate from one side to the other, gives rise to considerable lateral friction, which tends to rapidly destroy the rope.

The question of regulating the load to be lifted is one of the most important relating to the operations of winding. The variation in the value of the load is due to the constantly diminishing length of the ascending rope, and the constantly increasing length of the descending rope. As the weight of the rope is great relatively to that of the useful load, it is obvious that this variation must be great also. To take an example. Suppose a depth of shaft equal to 340 yd., a useful load of 16 cwt. of coal and wire rope weighing 10 lb. a fathom, or 5 lb. a yd. As the cages are equal in weight they may be left out of the question. At starting the load to be lifted is 16 × 112 = 1792 lb. of coals × 340 + 5 = 1700 lb. of rope, = 3492 lb. We are not now considering the strain upon the engine, which question would involve the taking into account of the inertia of the mass, but only directing attention to the alteration which takes place in the value of the load during the time of ascent. Now it will be observed that as the length of the ascending rope is constantly diminishing, its weight is constantly decreasing from 1700 lb. at starting to zero at the landing place at the mouth of the shaft. And as the length of the descending rope is constantly increasing, its weight is being constantly augmented, from zero at starting to 1700 lb. at the moment of stopping at the bottom of the shaft. Moreover, as this weight acts as a counterbalance to the ascending load, the latter, on arriving at surface, will be reduced to 3492 − (1700 × 2) = 92 lb. Thus during the time of ascent the value of the load has been diminishing from 3492 lb. to 92 lb. It is easy to see that this value may become negative. Suppose the depth of the shaft to be 366 yd. instead of 340 yd. In such a case the weight of the load on arriving at surface will be 3592 = 3600 = − 8 lb.; that is, the descending load will have overrun the ascending load, and the engine will have to oppose a retarding force of 8 lb.

This great variation in the load to be raised is manifestly very unfavourable to the work of a steam engine, and hence it becomes necessary to provide means for regulating the load. They have been found in the counterweight and the conical drum. It has been already pointed out that the regulating effect of the conical drum is more or less fully obtained, when a flat rope is used, by coiling the rope upon itself, whereby the virtual diameter of the drum is made to vary. We shall, therefore, consider the counterbalancing of the load and the coning of the drum relatively to the case of round rope. These means solve the problem in a satisfactory manner; and it may be remarked that the former is more common in England, where it was first employed, and the latter on the Continent, where it has received the most attention.

The counterweight usually consists of a number of excessively heavy iron links, suspended in a pit or well 30–50 yd. deep, provided for that purpose. To these links is attached a rope, which is fixed to the drum-shaft. The length of the balance-chain is equal to the depth of the pit in which it hangs, and it is connected to the drum-shaft in such a manner relatively to its length, that when the drawing ropes are at the starting point, that is, when one cage is at surface and the other at the bottom of the shaft, its whole length is hanging in the pit. The rope by which it is wound up is also arranged so that the whole of the balance-chain may rest upon the bottom of the pit when the ascending and the descending cages arrive at the same point in the shaft. This rope is made to pass over the drum-shaft in a direction contrary to that of the drawing rope which it is intended to counterbalance. The action of the counterbalance will now be readily understood. At the moment of starting the engine, the whole of the links are suspended, and these, by their great weight, hold the drawing rope in equilibrium. As the latter ascends and is diminished in weight, both by reason of the reduction going on in its own length, and of the increase taking place at the same time in that of the descending rope, the links are being deposited at the bottom of the pit, and, as previously pointed out, the whole of the links will be resting upon the bottom when the cages meet in the shaft, at which moment the ascending and the descending ropes balance each other. From the time when the cages pass each other, the weight of the descending rope preponderates, and this preponderance goes on increasing until the bottom of the shaft is reached. But from the moment when the descending cage passed the ascending one, the counterbalance chain is again being wound up, this time in the contrary direction; and as it is raised link by link, its weight counteracts the preponderating weight of the descending rope. This system of counterbalancing solves the problem of regulating the load with sufficient completeness for practical purposes. The weight of the balance links must, of course, be proportioned to that of the rope, account being taken in the calculation of the diameter of the pulley or drum upon which it is wound. This diameter is related to the depth of the pit or well in which the chain hangs. The pit is generally situate on the side of the drum farthest from the shaft. Sometimes, instead of the chain, a heavily loaded tub, or truck, is used as a counterweight. In this case, the tub is made to run upon rails suitably inclined. The inclination of the road is made to vary so as to be sharp near the upper end and flat at the lower end, for the purpose of obtaining a constantly increasing or diminishing resistance. During the time of drawing a load, the tub runs twice over the road, first descending and then ascending. Thus the force of traction exerted by the tub upon the rope to which it is attached is greatest at the moment of starting, null at the end of its course when the cages are at the same point in the shaft, and greatest again when the cages have reached the landing place; whence it will be seen that the action of the tub is precisely that of the balance-chain in the pit. By carefully determining the curve required, the counterbalancing of the rope may be in this way very completely accomplished, and often more easily, and at a less cost than by means of the chain.

The other means of regulating the load by means of a conical drum solves the problem less completely than the counterweight, but it possesses the advantage of leading to less complication; for every additional piece of machinery needing constant inspection increases the risk of failure. The question to be determined relatively to the conical drum is, what, under the given conditions, shall be the value of its mean diameter? This question, however, practically resolves itself into another, namely, what, under these conditions, can be its initial or least diameter? Here we have to deal with considerations of a conflicting character. The initial diameter most favourable to the durability of the ropes is the largest possible. But the initial diameter most favourable to an equalising of the moments of resistance in a deep shaft is the smallest possible, the number of coils upon the drum increasing as the diameter diminishes. It is evident that when wire ropes are used, the wear of the ropes will require a large initial diameter, since that wear will be determined by the least, and not by the mean diameter. The initial diameter should be proportioned to the thickness of the rope, in the manner already described for cylindrical drums, and the mean determined according to the conditions of the case. The limits of variation are very narrow, and hence it results that the regulating effect is more or less imperfect. In practice, a common size of conical drum is 16 ft. at the smaller end, and 20 ft. at the other.

Large conical drums are sometimes provided with a spiral channel for the reception of the rope, the object of this arrangement being to prevent the rope from slipping. The slipping of the rope is a danger to be feared with conical drums; but if due care be taken to wind the rope on very tightly at first, this danger is not great upon drums having the inclination usually adopted. Of course, cheeks or side rims are required, as in the case of cylindrical drums, to guide the rope from slipping off the drum altogether. This matter has been made the subject of legislative control, and it is enacted that there shall be on the drum of every machine used for lowering or raising persons, such flanges or horns, and also, when the drum is conical, such other appliances as may be sufficient to prevent the rope from slipping. It is hardly necessary to add that the component parts of a winding drum should possess ample dimensions and be strongly connected together, and that the foundations upon which the bearings of the shaft rest should be massive and securely placed, so as to render the drum capable of resisting not only the ordinary but accidental shocks, and of serving as a protective medium interposed between the force and the engine.

Cornish winding engines do not meet the ideas of modern practice. Their first cost, and the expense of erection, are both excessive. They are besides clumsy to handle. The type that has found favour in other districts is a double-cylinder high-pressure engine, fitted with variable expansion and reversing gear, the pistons connected directly to the fly-wheel shaft, on which also are the drums and a powerful brake, worked by means of a counterpoise, or, better still, by steam, and capable of stopping the machinery instantly.

When the tooth-wheel gearing for reducing speed intervenes between the fly-wheel shaft and the drum-shaft, as is common in Cornwall, the brake should be arranged to act on the drum-shaft, and not on the fly-wheel shaft, so that the consequences which would ensue from the breakage of the cog-wheels may be avoided.

Coal mining, as requiring very large outputs, involves the use of direct-acting engines for quick winding; whilst for metalliferous mining geared engines give sufficient speed, and are more commonly used. With geared winding engines, the load in the shaft travels so slowly, that its inertia, momentum, and speed do not practically interfere with the application of ordinary methods of expansive working. Few direct-acting winding engines work expansively, although several expansion gears have been brought out from time to time. Of these, all which require special attention from the engine-driver, who is sufficiently occupied with the simple reversing lever, have fallen into disuse; whilst those requiring no more manipulation than that given by the reversing handle are gaining in favour.

An excellent expansion gear, illustrated in Fig. 450, came under Davey's notice in Germany. The valves are lifted by cams, the expansion cams a being shaped after the manner of the old

FIG. 450.

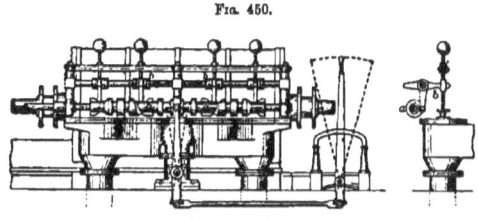

Expansion Gear.

stepped cam, but with the steps made so numerous as to form a continuous curve, thereby rendering it possible to shift the cam longitudinally under the lifter b. The cams are so made and arranged as to perform the functions of a reversing and expansion gear, by the simple movement of the reversing handle. Other forms of expansion gear are in use, some having a trip on the spindle of the steam valve.

The chief modern improvements and changes in direct-acting winding engines are as follows:—

(1) Expansive working;
(2) The counterbalancing of the rope by means of a conical drum, and also by means of a tail rope suspended under the cages;
(3) The short-rope system, in which the rope makes rather more than half a turn round a single large driving pulley, instead of a number of coils round a drum;
(4) The application of separate condensing engines.

Direct winding is done at enormous speeds, as will be seen from the following example. At the Bestwood Colliery, near Nottingham, a pair of direct-acting winding engines, with cylinders 36 in. diameter and 6 ft. stroke, are employed in raising coal from a depth of 1300 ft. One complete run, including changing, is made in 55 seconds. The weight of coal raised each time is 2 tons 2 cwt. Therefore this engine is capable of raising 1150 tons in 8¼ hours from the depth of 1300 ft. The average speed of the cages while running is 22 miles per hour, and the maximum about 35 miles per hour.

Of the two methods of counterbalancing—one by a conical drum, and the other by tail ropes suspended under the cages—each has its disadvantages and difficulties. In counterbalancing a load by a conical drum, there is the difficulty that one cage is moving through a greater distance than the other cage at top and bottom of the pit, which is a great inconvenience for changing the tubs in cages of more than a single deck. There is indeed a mode of getting over this difficulty; but the difficulty itself is one which increases with the depth of the pit and with the greater amount of coning in the drum. In counterbalancing by the tail rope suspended under the cages, there is the disadvantage that a taper winding rope can not be employed. The weight is constant on the winding rope at any stage of the winding, and consequently a parallel rope is needed. But the use of a taper rope has advantages for great depths. For example, a parallel steel rope 806 yd. long, which is the depth of the Rosebridge Colliery, will carry only twice its own weight; at 580 yd., which is the depth of the Monkwearmouth Colliery, it will carry three times its own weight; at 433 yd. (the Bestwood Colliery) four times its own weight. By a taper rope a much greater weight can be lifted in proportion to the weight of the rope used. The economical limit of depth, for the method of counterbalancing by a tail rope, will probably be found to be about 500 yd. A convenient unit is adopted by Daglish for comparing the work done at different pits by different winding engines: it is the number of tons raised per hour per hundred yards depth, or multiplied by the depth in hundreds of yards. This is not an absolutely correct unit.

The most important points in quick winding are to get away from the pit bottom as quickly as possible, and to spend as short a time as possible in changing the cages. In order to get away quickly from the bottom, light cages are desirable, which could be best obtained by using steel. For the Sandwell Park Colliery, Hall supplied some steel cages which were only half the weight of the preceding iron cages; and the total saving in the weight to be shifted each time amounted to 17 per cent., which would of course quicken the winding.

For the changing of cages quickly, George Fowler's plan of hydraulic loading and unloading is excellent. On this system, when a two-decked or three-decked cage is raised a certain height above the pit mouth, two dummy cages are raised alongside it by hydraulic power. One of these is ready charged with empty tubs, which are then, also by hydraulic power, pushed on to the winding cage, driving the loaded tubs before them on to the second dummy cage. Then away goes the winding cage with the empty tubs, and while it is proceeding on its journey down the pit, the second dummy cage is unloaded, and the first one charged with a relay of empty tubs; the same process being simultaneously carried on at the bottom of the pit shaft. This mode of working is shown in Fig. 451. *a* are the 3 empty-tub platforms, lifted up into position by the press ram *b*; *c* are the 3 unloading platforms, each on a level with one deck of the cage; they are lowered by the hydraulic press *d*, so as to withdraw the tubs one by one. The actual work of shoving the tubs off and on the cages is done by horizontal rams *e* above the bank level, and by the ordinary banksman at the bank level. Thus the men usually required to change the upper tubs are saved. There is a further economy in the wear and tear of ropes, as more damage is done to the ropes by the repeated lifting of cages than by running in the shafts. At the Cinderhill Colliery of the Babbington Coal Company, near Nottingham, this system is now in operation; and whereas about 30 seconds used to be occupied in the run and 30 seconds in changing the cages, now only 12–13 seconds are required for changing the cages, making that amount of saving in the total time. The same result might be obtained by means of a balance, without the hydraulic apparatus; but in that case the benefit would not be obtained of hydraulic rams to shove the tubs off the cages, and a man would be wanted for that purpose on each deck.

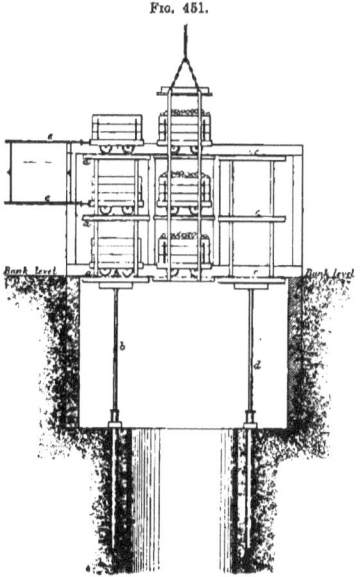

FIG. 451.

Fowler's Hydraulic Loading and Unloading.

Ransomes, Sims, & Jefferies, Limited, of Ipswich, who have devoted considerable attention to the manufacture of mining machinery for various parts of the world, have recently designed a series of standard winding and hauling gears, embodying all the improvements suggested by a long and practical experience.

Messrs. Ransomes divide these gears into 4 types, the arrangement and special qualifications of which are as follows:—

Type A.—This gear (Fig. 452) consists of a single drum, which is fitted with a powerful brake and clutch disconnecting gear.

It is suitable for winding where only one rope is required, and for hauling on single line inclines where the load descends by its own weight. The engine need not be fitted with reversing gear, for when the lift has been effected the drum may be disconnected without stopping the engine, and the empty bucket or truck lowered by the brake. The engine running always in one direction, single cylinder engines may be used for light loads.

Type B.—In this gear the drum is double the width of that in type A. It is usually keyed fast on the shaft. It is fitted with a brake, but unless specially ordered no clutch gear is provided.

This gear is intended for winding with two ropes. These start from either end of the drum and are led off in opposite directions. When one rope is paid out the other is coiled on the drum, and the length of lift for which this gear is suitable is therefore limited by the length of the number of coils the drum will hold without the rope overlapping; this gear is therefore only recommended where the length of lift is a short one in comparison with the other gears. The engine must be

Fig. 452.

Ransomes' Winding Machinery.

fitted with reversing gear, and double cylinder engines should be used in preference. If required for sinking, or for cases in which the length of the lift is constantly varying, the gear type D is preferable to this one. This gear is not recommended for hauling purposes.

Type C.—In this gear the drum is divided by a central flange. Each division is of the same width as the drum in type A. The drum is usually keyed fast on the shaft. It is provided with a brake, but unless specially ordered no clutch gear is fitted.

This gear is intended for winding with two ropes. These are led off from each division of the drum in opposite directions. The length of rope the drum will hold being only limited by the depth of the flanges, this gear is well suited for cases where a long length of lift is required. The engine must be fitted with reversing gear, and double cylinder engines should be used in preference. If required for sinking, or for cases in which the length of the lift is constantly varying, the gear type D is preferable to this one. This gear is not recommended for hauling purposes.

Type D.—This gear (Fig. 453) consists of two drums mounted upon a single shaft; each drum is of the same width as in Type A, and each is fitted with disconnecting clutch and brake gear.

This arrangement is adapted for hauling on single lines where a tail rope is necessary, one drum carrying the main, and the other the tail rope. It is also adapted for working two adjacent and opposite inclines on a double line of railway where the trucks descend by their own weight, the drum working the shorter of the inclines being thrown out of gear while the other completes its lift. When used for this purpose, the drums are usually placed below the level of the roadway so as to allow the trucks, after they have been hauled up one side of the hill, to be passed over the winding gear, when they are lowered down the opposite side by the brake. This arrangement of gear is also adapted for working two parallel inclines, raising the loads either independently of each other or alternately. It is also adapted for winding from shafts where the length of the lift is constantly varying, and for all descriptions of hauling and winding work where two ropes are employed.

Fig. 453.

Ransomes' Winding Machinery.

When used for hauling under the tail rope system, when both ropes are led off either from the top or bottom of the drums, it is advisable that the gear should be fitted with one double clutch instead of a separate clutch to each drum; by this arrangement all possibility of straining the rope by having both drums in gear at the same time is avoided.

PARTICULARS OF STANDARD WINDING GEARS, TYPES A, B, C, AND D.

Engine, H.P. nominal	4	5	6	8	10	12	14	16	20	25	30
Dimensions of drums—	ft. in.	ft. in.	ft. in.	ft. in.	ft. in.	ft. in.	ft. in.	ft. in.	ft. in.	ft. in.	ft. in.
Diameter of barrel	3 0	3 0	3 6	3 6	4 0	4 0	4 6	4 6	5 0	5 0	5 6
Width of face, type A	1 3	1 3	1 4	1 4	1 6	1 6	1 9	1 9	2 0	2 0	2 3
Depth of flange	0 5	0 5	0 6	0 6	0 6	0 6	0 6	0 6	0 7	0 7	0 8
Approximate gross load—											
Raised at a speed of about 400 ft. per minute	cwt. 7	cwt. 8	cwt. 10	cwt. 14	cwt. 17	cwt. 20	cwt. 24	cwt. 27	cwt. 35	cwt. 43	cwt. 50
Rope—											
Circumference if of steel	in. 1¾	in. 1¾	in. 1⅞	in. 1⅞	in. 2	in. 2¼	in. 2¼	in. 2¼	in. 2¾	in. 3¼	in. 3¾
Maximum length of lift—											
Types A, C, and D	yd. 1225	yd. 1225	yd. 1270	yd. 850	yd. 1040	yd. 810	yd. 980	yd. 940	yd. 1010	yd. 800	yd. 1120
Type B	..	..	..	195	240	210	260	250	280	245	290

MINING AND ORE-DRESSING MACHINERY.

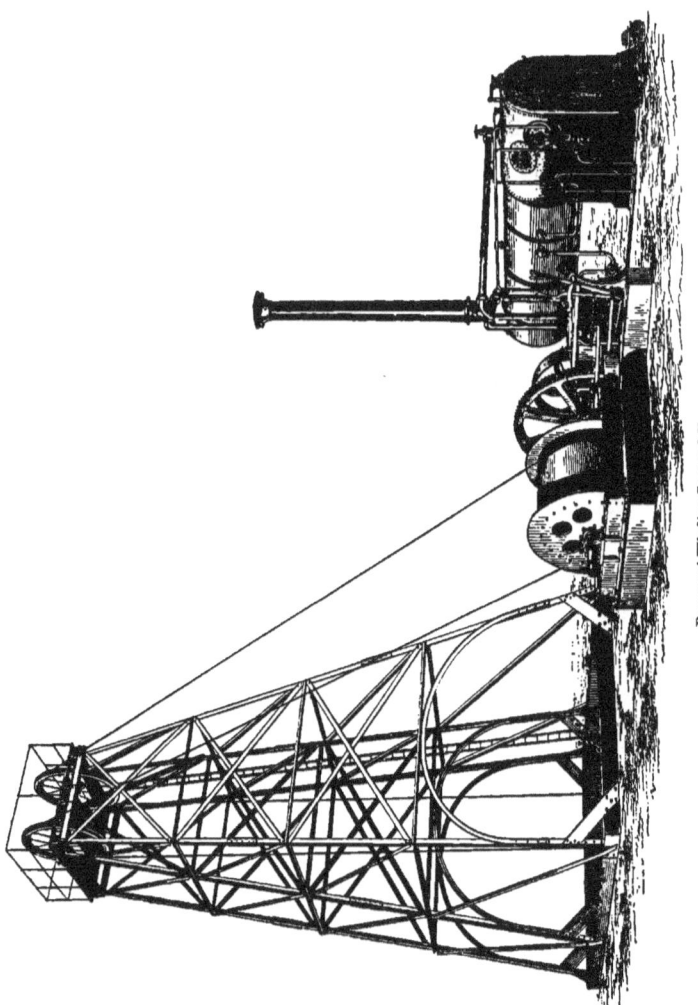

Ransomes' Winding Gear and Frame.

HAULING AND HOISTING MACHINERY. 277

Fig. 455.

Hornsby's arrangement of Pumping and Winding Engines separate.

These standard gears, as will be gathered from the illustrations, can be combined with engines of the portable, semi-portable, and "underneath" type, or they can be driven by stationary girder frame engines, with independent boilers. The engines, when combined with gears fitted with disconnecting clutches, may, when not required for winding, conveniently be employed in driving other machinery by means of a belt from the fly-wheel. When the gears are not provided with clutches, the driving pinion upon the engine crank-shaft can be arranged to slide in and out of gear at a slight extra charge.

A pumping crank arranged for different throws can be fitted at the end of the drum shaft if required, but if the winding gear adopted be either that described as Type B or C, it is important that a disconnecting clutch, which for these two types of winding gear is charged extra, should be fitted to the drums, otherwise, if at any time it is required to pump without winding, it will be necessary to disconnect the ropes, which cannot be done without considerable inconvenience.

An improved dial indicator, with bell adjustable to ring at top and bottom of lift, may be fitted to any of these winding gears. In this apparatus a dial is mounted on a neat cast-iron pillar of such a height that the face can be easily seen by the driver when standing at the starting handle of the

FIG. 456.

Hornsby's Vertical Winding and Pumping Engine.

engine. This indicator will show at any time the exact position of the cages or trucks, and the bell can be set to ring automatically when the cage is at any desired distance from the top or bottom of the shaft.

Fig. 454 shows one of Messrs. Ransomes' gears, type B, in connection with a stationary engine on a wrought-iron girder frame with independent boiler, and with an improved wrought-iron pit-head gear, which has been lately supplied for the Transvaal, and will be found to present many advantages both in regard to transport and facility of erection.

Messrs. Ransomes, in addition to the gears already described, have manufactured many other forms of winding machinery, some being specially designed for flat ropes; they are also makers of pit-head pulleys, head stocks in wood or iron, and the general accessories in connection with mining machinery.

R. Hornsby & Sons, Limited, of Grantham, who are now well known as makers of mining machinery, have arranged some very compact and complete sets of pumping and winding engines, as shown in Fig. 455. In most cases it is found advantageous to have the pumping engine separate

FIG. 457.

Hornsby's Geared Winding Engine.

from the winding engine, and arranged as shown, and one or more boilers are used. The plan illustrated consists of a 16 H.P. pumping engine complete with gearing, T-bob, and the necessary pit pumps, which are made in various sizes. The winding engine is a 16 H.P. double cylinder with a drum either made in one piece with a division in the middle, or, as is generally more convenient, with two separate drums which can be worked independently. Two 16 H.P. locomotive type boilers are shown driving the two engines. The pit-head gear shown is made of wood; light steel pit-head gears are, however, made if preferred, and either of them (wood or steel) can readily be taken to pieces and packed for shipment. The cages shown are also made so as to be taken to pieces and packed in a small space.

The small vertical winding and pumping engine (Hornsby's) shown in Fig. 456, is mounted on iron frame and wheels, and commonly used in the earlier stages of mining. It is readily moved from place to place, and can be started at work without requiring any foundation. The engine is also available at any time for driving other machines such as air compressors, &c.

Fig. 457 shows one of Hornsby's 20 H.P. geared winding engines, similar to the one used in the general arrangement shown in Fig. 455.

These engines are made from 10–50 H.P., are mounted on steel frames, and have also steel drums, and are generally of high-class construction.

Fig. 458 represents a side elevation of a whim engine, with winding cage attached. This is the ordinary engine employed for raising the productions of the mine, and is sometimes adapted for the threefold purposes of winding, crushing, and pumping. These engines are made in the best modern style, and of any power required. For winding, it is only necessary to state the height and daily tonnage of the lift. If for crushing, state the number of tons and description of material required to be crushed in a given time; and if for pumping, the depth of mine and quantity of water (in gallons) to be raised per hour. The makers are Harvey & Co., Limited, Hayle, Cornwall, and 186, Gresham House, London.

FIG. 458.

Whim Engine and Winding Cage.

Fig. 459 shows a portable hauling and winding engine introduced by John Wild & Co., Limited, Falcon Iron Works, Oldham, specially designed to meet the demand for engines for colliery use, which can be fixed and handled by unskilled men, and which can be readily moved about from one part of the mine to another like an ordinary pit tub, re-fixed, and set to work in a few minutes, by simply coupling the flexible tube to the air mains.

Underneath the bed-plate or truck, there are a number of screws fixed for tightening down to

HAULING AND HOISTING MACHINERY.

the floor, so as to take the weight off the wheels, and so secure greater rigidity whilst at work. Engines of this type are made with cylinders up to 10 in. diameter, but the general range runs from 4 in. to 8 in., as will be seen by reference to the list below. They are designed for either steam or compressed air.

Nominal Horse-power.	Diameter of Cylinders.	Length of Stroke.	Winding Drum.			Revolutions of Engine per Minute.	Space Occupied.				Approximate Prices.	
			Width Outside.	Diameter.	Depth of Cheeks.		Length.	Width of Wide Type.	Width of Narrow Type.	Height.	With Two Drums, keyed on Shaft, and One Brake.	With Two Loose Drums. Clutch Gear, and Two Brakes.
	in.	in.	in.	in.	in.		ft. in.	ft. in.	ft. in	ft. in.	£ s. d.	£ s. d.
4	4	8	6	18	4	200	4 6	3 6	2 9	2 9	70 0 0	75 0 0
6	5	10	8	21	5	180	5 6	4 0	3 0	3 4	87 10 0	90 0 0
8	6	12	9	24	6	160	6 6	4 6	3 3	3 9	105 0 0	112 10 0
10	7	14	10	27	7	150	7 6	5 3	4 0	4 6	125 0 0	132 10 0
12	8	16	12	30	8	140	8 6	6 0	4 6	5 0	150 0 0	160 0 0

Fig. 459

Wild's Portable Hauling and Winding Engine.

Figs. 460, 461 illustrate a pair of semi-portable hauling engines, made by John Wild & Co., Falcon Ironworks, Oldham. These engines have been specially designed for underground work, and are the outcome of a long practical experience with mechanical engineering work in collieries. The larger sizes of this type of engine are intended for main roads and inclines, whilst the smaller sizes are more especially intended for hauling from the workings to the main roads. The engines are mounted on a strong framework composed of wrought-iron girders and channels, whilst the drums are

FIG. 460.

Pair of Wild's Semi-portable Hauling Engines.

fitted with wrought-iron cheeks which are secured to the body of the drum by bolts. The brake consists of a cast-iron ring turned and fitted into a prepared recess in the drum. When pressure is applied to the brake lever, the cast-iron ring is expanded so as to come in close contact with the drum.

In the engine illustrated, the cylinders are 12 in. diameter, and have an 18-in. stroke. The pistons are of the "Mather & Platt" type, 4¼ in. wide. The piston-rods are of the best mild steel, 2 in. diameter. The stroke of the slide-valve is 3¼ in. The slide-valve spindles are of the best mild steel, and the outer end is carried by an adjustable slide to receive the strain due to the angularity of eccentric-rod. The connecting-rods are of the best hammered iron, 3 ft. 9 in. long, 2 in. diameter at their small ends, $2\frac{3}{4}$ in. diameter at their large ends, and $2\frac{7}{8}$ in. diameter in the centre. The crank shaft is of the best mild steel, $4\frac{1}{2}$ in. diameter; and the cranks are of good cast iron, and of the balanced type. The crank pins are of the best mild steel, $2\frac{7}{8}$ in. diameter, and $4\frac{1}{4}$ in. long between collars. The drum shaft is of mild steel, $6\frac{1}{2}$ in. diameter, and the bottom step of its

FIG. 461.

Pair of Wild's Semi-portable Hauling Engines.

bearings is of gun metal. The drums are 4 ft. diameter, and 12 in. wide between the cheeks. The cheeks are 9 in. deep, and are made of wrought-iron plates $\frac{1}{2}$ in. thick. The body of each drum consists of one casting. The brakes consist of cast-iron expansion rings fitting inside the drum, and they can be operated without the engineman leaving his place.

CHAPTER XII.

TRANSPORT.

For the conveyance of minerals from the pit mouth to the dressing floors or the shipping port, animal power is rapidly going out of favour in most districts, though there are some in which it will long survive.

In some localities self-acting jigs or inclines are used, where there is a falling gradient all the way. This is the case at the Somorrostro ironstone beds, Bilbao, where the following method is adopted.

The Mac Lean inclined plane is about 330 yd. long, with a gradient of 1 in 2. The full and the empty waggons are attached to either end of a single rope passing round two horizontal pulleys at the head of the incline, and controlled by a brake. The useful load for each trip is about 6 tons contained in two waggons.

The Orconera plane is about 1300 yd. long, with an average gradient of 1 in 7. It has two parallel lines of rail, and about one-half the length is on a curve, necessitating inclined guide-sheaves for the ropes, which are coiled in reverse directions on two drums, about 16 ft. 6 in. diameter, keyed on the same axle. Each drum is furnished with 2 brake-sheaves, the whole controlled by 4 strap-brakes shod with cast-iron brake-blocks, and operated simultaneously by the brakesman. A train consists of 7 or 8 4-ton waggons, or a net load of 30–32 tons, and about 2000 tons of ore a day can be dealt with.

The Cadegal plane is about 660 yd. long, with a total fall of about 175 yd., the gradients varying from 1 in 2·9, 1 in 3·3, and 1 in 4, on the upper, middle, and lower sections respectively. It is laid with a double track of 3 ft. $3\frac{3}{8}$ in. gauge. The drums, about 16 ft. 6 in. diameter, are of slightly conical outline, and are formed of wrought-iron plates $\frac{3}{4}$ in. thick, carried on 3 cast-iron frames, the two outer ones being formed to receive brake-straps, while the centre one is cogged, and gears into a pinion in the ratio of 8 to 1. The shaft of this pinion carries a large "fly," with 4 straight wings, about 6 ft. 6 in. wide, and 16 ft. 6 in. outside diameter, formed of wooden planks on iron frames. By adding or removing one or more planks the speed can be regulated to a nicety, and with 90 rev. per minute of the fly a train-speed of 200 yd. per minute is permitted and never exceeded. The run of 660 yd. takes about $3\frac{1}{3}$ minutes, and, as 6–7 minutes are occupied in making up the trains at each end, they can be despatched at intervals of 10 minutes. A train consists of 8 2-ton waggons, so that about 1000 tons can be dealt with in a day of 10 hours, and by increasing the number of waggons in each train this might easily be brought up to 1500 tons.

The ropes are of steel $1\frac{1}{2}$ in. diameter. The drums are mounted at a sufficient height above the rails to allow the waggons to pass beneath them, and by means of two short inclines in opposite directions, between the drums and the head of the plane, the trains are made up with a minimum of labour.

TRANSPORT. 285

Wire tramways for the carriage of ore over a difficult section of ground have been largely employed, and have proved themselves most efficient. The situation of many mines is such that a road can only be made to them at great expense, while the output, especially of those producing the valuable metals, is generally too small to admit of any large expenditure in the form of railways, inclined planes, &c. Under such circumstances wire tramways present themselves as a most efficient means of transport. Such wire tramways are made by Messrs. Bullivant & Co., of 72, Mark Lane, E.C., from designs prepared by Mr. W. Carrington, M. Inst. C.E., the tramways being constructed on various principles, each being suited to a special situation.

In one case a continuous running rope from which the loads are hung is used. Such lines are capable of being worked to a length of 3 to 4 miles in each section, and by multiplying these sections, distances of 20 to 30 miles can be traversed. Quantities of from 50 to 500 tons per day can thus be carried. Ravines and obstacles can be passed over by single spans, having a length of 500 ft. or more. Water power, where available, can be used for driving such tramways, thus reducing the cost of their working to the wear and tear, which is very small, and that of the labour required for supplying and delivering the material brought for transport. In many cases, where

FIG. 462.

Wire Tramway by Bullivant & Co.

the gradients are favourable, such tramways will self-act. In other cases, where the gradients are not sufficient to cause them to self-act, the amount of power required for driving is greatly reduced. Such tramways are similar to that in Fig. 462, and are now in use at mines in England, Russia,

Italy, Spain, Norway, South America, Australia, New Zealand, China, Japan, India, Cape of Good Hope, and many other places.

Another system of wire tramway also largely used in connection with mines is a species of self-acting incline, where ropes are used as rails, and the carrier runs suspended from them, the loaded carrier descending on an incline, drawing up an empty carrier on the adjoining rope. Such a system of wire tramway is especially suitable for mountainous situations where long leaps may be taken from ridge to ridge. Spans up to 3000 ft. and 3500 ft. without support can thus be made, and quantities of material up to 100 to 200 tons can be carried easily. The labour required is that which is needed to fill and empty the buckets, with one man appointed to control the speed by means of a brake. With such tramways a distance of two miles may be traversed in four to five leaps, the difference in level between the loading and unloading points amounting in some cases to as much as 3000 to 5000 ft.

The speed with which such tramways can be erected, and the small amount of permanent work required for their installation, renders such a system exceedingly suitable for mines, which having

FIG. 463.   FIG. 464.   FIG. 465.

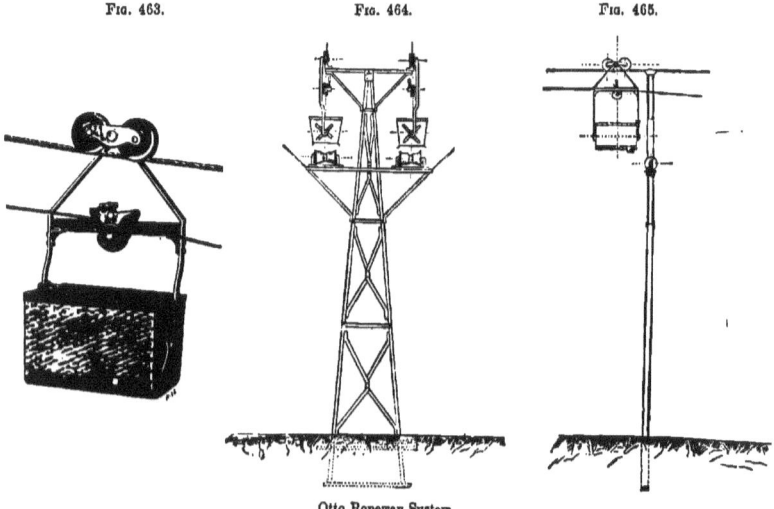

Otto Ropeway System.

been opened, have not yet been thoroughly proved. Such tramways are now employed in France at the mines of The Pierrefitte Mining Co., The Castillon Mining Co., Société des Mines d'Arre, also in Italy, Spain, Norway, Sweden, Cape of Good Hope, India, China, &c., &c.; in fact, in all parts of the world.

Many other special applications of wire rope transport can be furnished for special situations.

The Otto system of wire ropeways, of which there are now over 400 lines at work in various

TRANSPORT. 287

Fig. 166.

Otto Ropeway System.

parts of the world, is now very largely adopted. Commans & Co., of 52, Gracechurch Street, London, E.C., are the sole licensees for the sale and manufacture of these ropeways for England and the Colonies. These ropeways consist of two fixed carrying ropes resting on supports, and an endless light hauling rope, to which the buckets or skips are attached by means of special grips. At either terminus the buckets are switched off on to fixed rails for loading and unloading. Fig. 463 shows the style of bucket generally used for mineral transport, and Figs. 464 and 465 the standards or supports for the carrying ropes. Fig. 466 gives a very good general view of one of these lines passing over comparatively level country, and Fig. 467 a line up a very steep hill

Fig. 467.

Otto Ropeway System.

with gradients, in places, of 1 in 3. The first line is transporting 700 tons, and the second, 500 tons of material per day. There is now one of these lines working at Garrucha, in the South of Spain, transporting 400 tons of iron ore per day, over no less a distance than 9¾ miles. These lines are now being introduced into the De Kaap district of the Transvaal, and have already been adopted by the Sheba, the Edwin Bray, and other leading Companies. The Section Fig. 468 gives some idea of the mountainous nature of this district, and how the introduction of these aërial ropeways enables the ore from the mines to be delivered direct to the stamps situated alongside the

TRANSPORT. 289

De Kaap River, from which the power to drive the same is derived. This line is about three miles in length, and has some very large spans, the longest being nearly 1500 ft. The saving in many instances effected by the adoption of these ropeways, has enabled mines previously idle, owing to the heavy cost of transport, to be now worked at a profit.

Fig. 468.

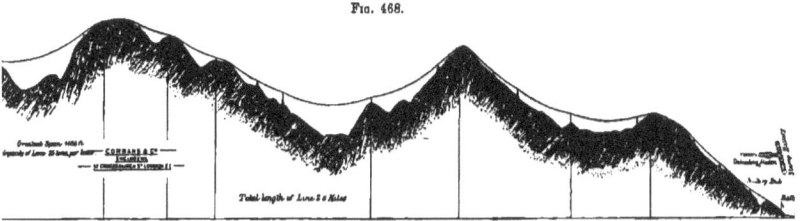

Section of Otto's Aerial Ropeway for the Sheba Gold Mining Company Limited (Transvaal.)

Otto Ropeway System.

CHAPTER XIII.

REDUCING MACHINERY.

THE first step necessary in the separation of ores from their associated gangues, is their reduction to a more or less fine state of division. The machines used for this purpose are first "breakers," which reduce the ore to fragments not larger than hens' eggs; and then "stamps," "rolls," or other forms of mill to complete the operation.

BREAKERS.—These are of many patterns, each possessing some peculiar advantage. Fig. 469 illustrates a very simple form, made by Robey & Company, Lincoln. To render its transport easy, over bad or mountainous roads, steel is largely used in its construction, replacing the usual heavy cast-iron portions which add so much to the weight. The parts are few and simple, and afford access for oiling, cleaning, and adjusting, thus minimising the risks of breakage.

Fig. 470 shows an approved form of breaker, made by Calvert, Cornes, & Harris, London.

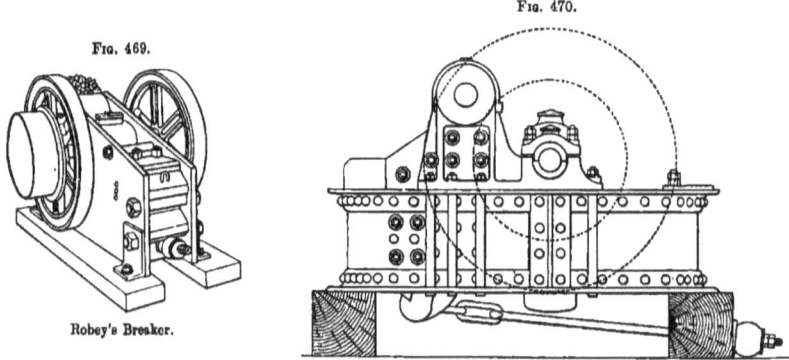

Fig. 469.
Robey's Breaker.

Fig. 470.
Calvert & Co.'s Ore Breaker.

Fig. 471 shows a breaker made by George Green, Aberystwith, South Wales. It is a machine of great power, strength and durability, made expressly for mule transit through rough and mountainous country, where it is impossible for wheel carriages to travel. The machine is made in parts not exceeding 330 lb. in weight, whilst the 12 in. machine can be made in parts not exceeding 120 lb. The whole strain of the work is taken up by the wrought iron longitudinal bolts, which, being well secured at the ends, bind the whole machine together, so that notwithstanding the lightness a breakage is almost impossible. The ends are made of wrought iron, when specially ordered, at a proportionate extra charge. The jaw faces are made of best hematite chilled iron, and are composed of reversible sections; those on the swing jaw being held by strong wrought iron arms, giving at once strength combined with lightness, whilst the sections themselves may be changed and

Fig. 471.

Green's Ore Breaker.

re-adjusted in a few minutes, bringing the unused parts into the former position of the worn ends, which is a great advantage. Machines of heavier construction are also made, having cast-iron ends made in one piece, held in place by the longitudinal bolts; with cast-iron swing jaw, and connecting-rod of the same, having similar reversible jaw faces as the former machine.

Dimensions at Mouth of Machine.	Approximate Total Weight.		Capacity per Hour.		Approximate Horse Power Required.	Approximate Cost.
in. in.	tons	cwts.	tons	cwts.		£
16 × 9	4	0	8	0	8	..
14 × 8	3	10	6	0	6	..
12 × 7	2	10	3	0	4	..
9 × 6	1	5	2	0	3	..
6 × 5	0	15	1	0	1¼	..

STAMPS.—Stamps consist of a series of heavy pestles of iron, which are lifted to a varying height and allowed to fall upon the ore that is to be reduced. They work in a mortar or trough, also of iron, into which a constant supply of ore is introduced, and from which the crushed material escapes through openings furnished with closely fitting screens, as soon as it is reduced to the desired degree of fineness. The mortar is usually rectangular in form, and contains any number up to 6, but

2 P 2

commonly 5, stamps, forming what is usually called a "battery," or set. The mortars rest on a solid foundation, and are established in a substantial framework of timber. The stamps are lifted by means of revolving cams or arms of iron, keyed to a cam-shaft, which is placed directly in front of the batteries, and which receives its motion from the driving power of the mill. The stamps move vertically between guides that form a part of the battery frame.

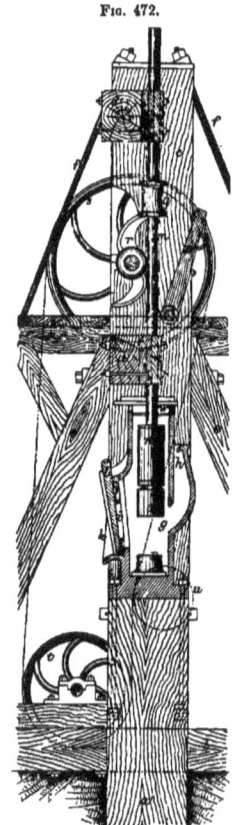

Fig. 472.

The general construction of the several parts of the battery is shown in Fig. 472 (scale ⅜ in. = 1 ft.):—a, foundation timber or mortar-block; b, transverse sill; c, battery-posts; d, tie-timbers; e, braces; f, tie-rods; g, mortar; h, feed-aperture; i, screen or grating; j, screen-frame; k, lugs to secure frame; l, wedge or key; m, stamp-stem or lifter; n, stamp-head; o, shoe; p, die; q, tappet; r, cam; s, pulley on cam-shaft; t, driving-pulley; u, tightener; v, guides; w, battery-covers; x, prop for supporting stamp when not at work.

Foundations.—The foundation-timber or mortar-block for batteries of this character often consists of heavy vertical timbers, placed close together, and firmly connected by cross-timbers and iron bolts. The timbers may be 6–12 ft. long, according to the nature of the ground and the proposed height of discharge from the mortar. Sometimes the timbers are laid horizontally, so as to serve as the base of two or more batteries. When the foundation timbers are in place, the space about them is packed and stamped as firmly as possible with clay or earth. When the ground on which the batteries are to be built is a hard compact gravel, or a firm clayey material, the surface is sometimes levelled off so as to admit of laying the transverse sill-timber b of the battery-frame, and a narrow pit is then excavated, some 6–14 ft. deep, and long and wide enough to receive the ends of the mortar-blocks; the posts or blocks are introduced into the pit in a vertical position, their bottom ends resting directly on the ground without any intervening horizontal timber. The remaining space in the pit is then compactly filled with clay, which is pounded or stamped firmly into place. The sill-timbers b and battery-posts c are securely bolted to the foundation-timbers. The posts c are braced by the timbers e and rods f, and are connected by the tie-timbers d, which also support guides v. Foundation-timbers should be well tarred or kyanised.

Battery.

It must ever be borne in mind that the foundations are of prime consequence. When improperly constructed, the battery cannot be run at its full speed and capacity, without shaking itself to pieces, whereby great delay, expense, and actual loss of metal are sure to arise. Extra care in securing complete solidity for the battery in the first instance will be amply repaid, while nothing will compensate for a rickety structure. Often it is advisable to excavate the foundations down to the solid rock, where that is not more than

14 ft. below the surface. At some mills, the trench itself is cut in the solid bed-rock, leaving about 2 ft. all round for packing.

Fig. 473 shows the details of the foundations and framing more minutely. The mortar-blocks *a* are 30 in. square, and 12 to 14 ft. long. They are made quite true, and thoroughly coated with

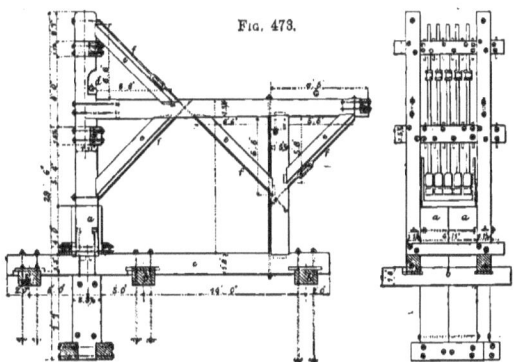

Fig. 473.

Details of Foundations and Frames.

Stockholm tar, applied hot, then bolted together by six 1½-in. pins and nuts. The transverse sills or foot-timbers are 18 in. square, 6 ft. long, and are let 6 in. into the mortar-block, freely tarred, and bolted together by six 1½-in. pins after being squared. At 5 ft. from the top, the mortar-blocks are cut to 59 in. by 29 in. The prepared blocks are let down upon the floor, and levelled up by putting sand beneath. When in place, the height is accurately determined, and a level is run across the whole set. The tops are planed smooth and dished about $\frac{1}{10}$ in., to prevent the surface becoming rounded; they are kept covered till the mortar is fixed on the top.

Frames.—Battery-frames in America are usually made of the best red spruce (*Abies rubra*) or sugar-pine (*Pinus Lambertiana*). First, three battery-sills *b* (Fig. 473), 18 in. by 24 in., and 28 ft. long, are placed parallel to the direction of the cam-shaft, one being 5 ft. from centre to centre behind the mortar-block, a second 5 ft. in front, and the third 14 ft. from the second. They are secured by bolts 8 ft. long, keyed into the masonry or to the bed-rock. In the latter case, holes 3 ft. deep and 1½ in. diameter are bored in the rock; the bolts are slotted at 6 in. from the lower end, and wrought-iron wedges, ⅜ in. by 1 in., 5 in. long, with a head 1 in. square, are made to fit the slots; the bolts are inserted in the holes, and driven so that the wedges enter up to their heads, when the holes are filled up with molten brimstone. Cast-iron washers and nuts retain the bolts in the sills. Next, the outside-line timbers *c*, measuring 20 in. by 14 in., and 28 ft. long, are wedged into the battery-sills, and secured by bolts. The top of the sill should be 4 ft. above the top of the mortar-block. The centre-line timbers measure 20 in. square and 28 ft. long. The intermediate line timbers measure 20 in. by 14 in. by 28 ft., and are dressed on the upper side and reduced to 13¼ in. and 19¼ in. where they pass the battery-blocks; they are let 3 in. into the sills, and are secured by keys driven both ways and by two iron bolts 33 in. by 1¼ in. The outside battery-post measures 23 in. by

13½ in., and is tarred and let into the sills. The posts for as many as four batteries may be raised simultaneously. The middle one is usually of somewhat larger scantling, say 23 in. by 19½ in. The posts are secured to the line-timbers by two 1-in. joint-bolts, 44 in. long. In the upper part of the posts is cut the cam-shaft journal-seat d. The posts are held together by the tie-timbers d (Fig. 472), carrying the stamp-guides v. The bracing and tie-bars ef are all arranged to ensure the greatest possible steadiness during work. A travelling block and tackle suspended over the battery will be found very useful for inserting and removing the stamps. Where white ants exist, or good timber is scarce, iron framework must be resorted to.

Mortars.—Mortars are often fixed directly upon vertical mortar-blocks, without any horizontal sill intervening. When the frame is ready, the temporary covering is removed from the mortar-block, and the holes for the mortar-rods are bored from the template taken from the bottom of the mortar. All cracks in the block are filled up with molten brimstone, and its surface is again planed and tarred. In the Western States of America, it is a favourite custom, before fixing the mortar, to cover the top of the block with a triple thickness of common domestic blanket, thoroughly tarred on both sides. Upon this cushion, the mortar is bolted with 1½-in. pins d (Fig. 474). This plan

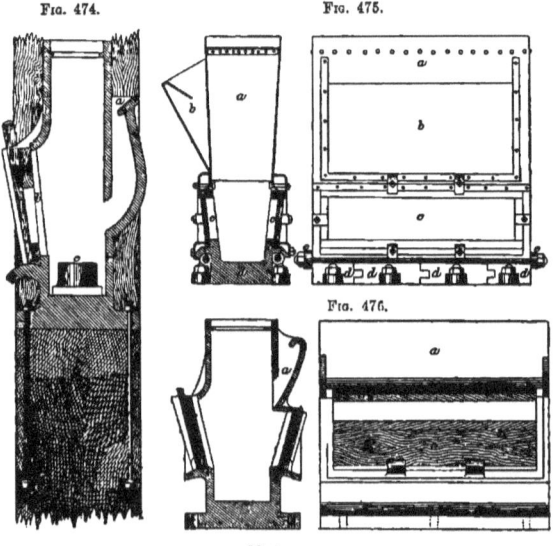

Fig. 474. Fig. 475.

Fig. 476.

Mortars.

reduces the "jar" to a minimum, and prevents the gradual loosening of the mortar from the block, and consequent introduction of material into the space, whereby the perfect level of the mortar is destroyed. The rate of discharge is partly governed by the width of the mortar, increasing as the mortar narrows; but if the ore is very hard, it will cause frequent renewals of screens should the mortar be very narrow.

Mortars may be constructed partly of wood and partly of iron—the sides and ends being of wood, and the bed-plate of solid iron—or they may be entirely of iron. The latter plan is now general, as with the compound method there is great trouble in keeping tight joints. The form commonly adopted is an iron box or trough, 4 or 5 ft. long and deep, and 12 in. wide inside, preferably cast in one piece, but sometimes made in sections bolting together, where transport is difficult. The bottom is always made very thick, as it has to bear the chief strain; but in positions' remote from iron-foundries, it is an advantage to have the sides cast thin, and to attach a lining which can be renewed at will. This form of mortar is shown in Fig. 475. The feed-opening a is an aperture 3 or 4 in. wide, and nearly as long as the mortar, by means of which the ore, suitably sized, is fed into the stamps. On the opposite side is the discharge, furnished with a screen b, by which the "pulped" material escapes; this opening is almost as long as the mortar, and 12 to 18 in. deep, the lower edge being 2 or 3 in. above the top of the die c. The bolts d hold the mortar on the block e. A cover is useful to exclude foreign matters.

The Californian high mortar varies in weight from 3000 to 6000 lb.; it is usually about 4 ft. 7 in. long, 4 ft. 2 in. to 4 ft. 4 in. high, 12 in. wide inside where the dies are set, and 3 to 6 in. thick in the bottom. Mortars which are made in sections are termed "section-mortars." The one illustrated in Fig. 475 is 4 ft. long, and will take five stamps. The upper portions a are of boiler-plate, strengthened with angle-iron; the feed-opening is at b; there is a double discharge with screens at c, the screens being attached by movable lugs or clamps; the bottom is cast in four sections d, which are accurately fitted together with tongued and grooved joints, planed, and held by heavy iron bolts e running through them from end to end, and secured by strong nuts on the outside.

Donnell's mortar, much used in gold-ore crushing, is shown in Fig. 476. The ore is fed in at a, and the discharge is at back and front. The screen b is narrow, placed high above the dies, and occupies only a part of the opening in front. The lower portion of this opening, and the opening at the back, is closed by a wooden door c, covered on the inside by a sheet of amalgamated copper.

Screens.—The action of the stamps, when properly supplied with ore and water, results in the reduction of the solid matters to such a degree of fineness as will enable them to flow off with the water, which wells and splashes up, at each blow of the stamps, through the screens or gratings placed at the exit from the mortar. The position of the screens in the mortar has been already shown in Fig. 472. The screen should incline outwards at the top, to facilitate the passage of the pulp through its meshes. The length of the screens will vary with the length of the mortar, and the width is usually 10 to 15 in.

The shape, disposition, and size of the orifices in the screens are subject to the greatest possible variety. A few of the many patterns in use are shown in Fig. 477. The "gauge," or number of holes per sq. in., adopted in Victoria, ranges between 60 and 800; in America, it runs from 900 to 10,000. When the holes are round, their size is graduated by the numbers of sewing-machine needles, from 0 to 10: thus No. 5 is about $\frac{1}{18}$ in. diameter, and No. 8 about $\frac{1}{14}$ in. When the holes are slots, they are usually $\frac{3}{8}$ in. long and of the same diameter as a No. 6 needle. As regards material used in the construction of screens, Americans are universally in favour of Russia sheet iron or sheet steel, $\frac{1}{32}$ in. thick, weighing about 1 lb. per sq. ft., very soft and tough, with a

MINING AND ORE-DRESSING MACHINERY.

Fig. 477.

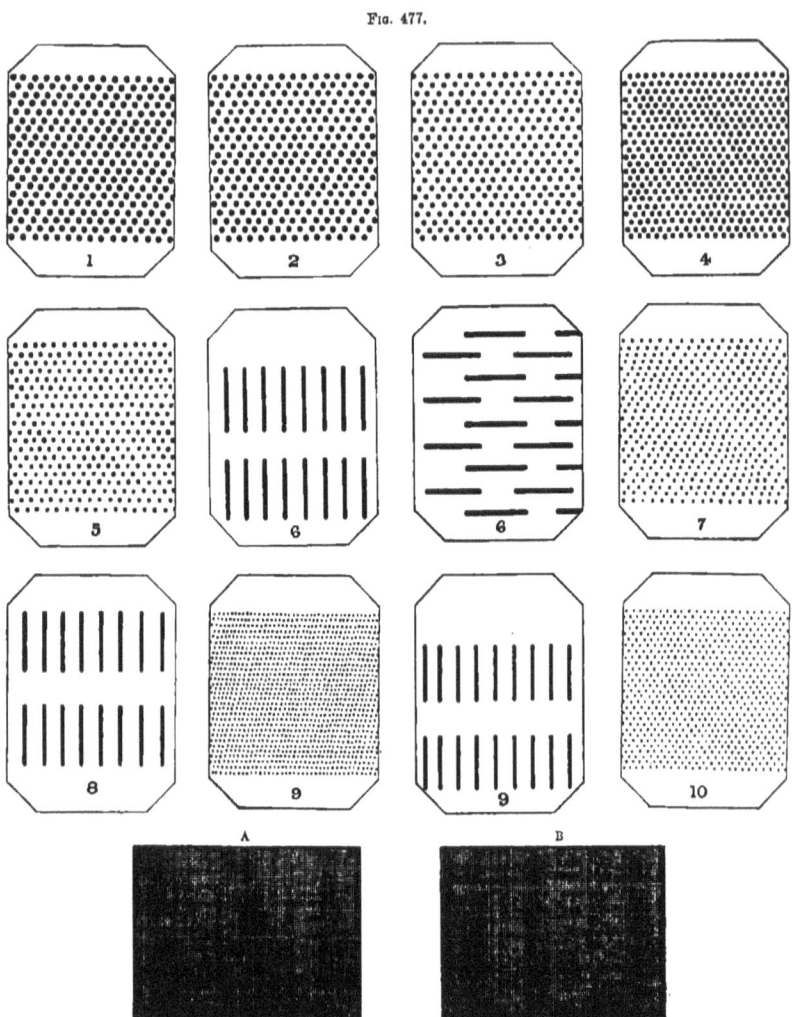

Screens.

REDUCING MACHINERY.   297

clean smooth surface, and perfect freedom from rust or flaws; in Australia, sheet copper is often employed, that at the Port Phillip works being $\frac{1}{16}$ in. thick, with 84 holes per in. The holes should always be punched. This operation leaves one side in a rough state, like the outside of a nutmeg-grater. The rough side is turned inwards, towards the wear of the issuing pulp; and, as the inner end of the orifice is smaller than the outer, there can be no fear of the meshes becoming clogged, as everything which enters from the inside can more readily escape on the outside. A 5-stamp battery usually requires 13 sets of screens per annum, a set consisting of 5 sheets of 1 to $1\frac{1}{2}$ sq. ft.

The "gauges," or number of holes per sq. in., in the samples of screens shown in Figs. 161 and 162, are approximately as follows :—

No. 1, 150 holes per sq. in.; Nos. 2 and 3, 140; No. 4, 200; No. 5, 160; No. 7, 220; No. 9, 575; No. 10, 320; A, 3600; B, 3250.

It is very essential that the material of which the screens are composed, whether stamped sheets or woven wire, should be of the best quality attainable, as the wear in any case is very rapid, and with inferior material the loss of much valuable mineral may result through the escape of particles not sufficiently reduced in size to liberate the metal. In this connection mention may be made of the well-known "agate steel," stamp battery cloth, specially manufactured for metal mining by N. Greening & Sons, Limited, Warrington, England. Nothing can surpass their screens for regularity of mesh, and the heavy qualities are remarkably substantial and enduring.

Some mills adopt tinned iron, first burning off the tin; being thinner than Russia iron they allow of more rapid discharge, but do not last so long. Brass wire screens endure 10-14 days; Russia iron, 15-40 days. Common steel wire is liable to rust.

The "stamp duty" or effective capacity of a battery is governed entirely by the facility with which the pulp can escape. It is therefore obvious that proper adjustment and gauge of the screens are most important matters. The object sought is to reduce the mineral just sufficiently to enable the gold to free itself from the gangue, and to thus reduce the greatest possible quantity in a given time. Microscopical examination of the ore and of the pulp will afford some guide in this respect, as will also the same test applied to the tailings escaping from the mill. It is hardly necessary to point out that the rate of discharge will also be proportionate to the area of screen presented to the pulp, whence it follows that the screens should be as deep and wide as possible. The most rapid delivery would be attained by having a single stamp surrounded by screen on all sides; and every battery should at any rate have a back and front delivery. Where the feed is on one side and the discharge on the other, the working power of the battery is lessened by one-half. The importance of this is apparently not yet recognised by some engineers. Experiments have shown that pulp fed back into the mortar takes nearly as long to escape through the screens as rock which has first to be reduced to pulp. Fine stamping does not necessitate the use of fine-gauged screens, as the same result may be gained by elevating the screen, though with a certain loss of effective capacity.

Dies.—In order to save the mortars from wear and tear, "dies" or "false bottoms" are placed in them, to receive the blows from the stamps. In America, the die is a cylindrical piece of cast iron, corresponding in form to the shoe of the stamp that falls upon it, and 4 to 6 in. high. Some mortars are made with circular recesses in the bottom, for the dies to fit into. In others, to prevent the material from working in under the die and displacing it, the circular recess in the bed-plate is cast with a flange, and the die with a small projection or lug. A groove is also made in the bottom of the mortar, so that the die may be introduced, with its lugs dropping into the groove; the die

2 Q

being then turned about 90°, the lugs come under the flanges of the recess, and the die is thus held in place. A simpler and more general plan is to cast the die with an upper circular "boss" or die proper a, on a square flat "foot-plate" b (Fig. 478). The bottom of the mortar is then also made flat, and the dies are dropped in, resting on their foot-plates, which just fill up the space in the floor of the mortar. The corners of the foot-plates of the dies are bevelled off, so as to allow of the insertion of a pick for effecting their removal when necessary. The foot-plate of American dies is usually $1\frac{1}{2}$ to 2 in. thick and 10 to 12 in. square; the boss is 3 to $5\frac{1}{2}$ in. high and 8 to 10 in. in diameter. It is of hard tough cast iron, and is chilled down to the foot-plate. In Victoria, it is found to be a good plan to allow the dies to rest immediately upon a layer of finely-broken quartz, at least 3 in. deep, by which means an opportunity is provided for the liberated particles to get into the gravel, out of reach of the stamps, and whence they can readily be recovered. In some localities, the die consists of a simple slab of iron filling the whole mortar, and which is turned over when one side is much worn; but the wear is liable to be uneven, and the die often breaks before it is worn out.

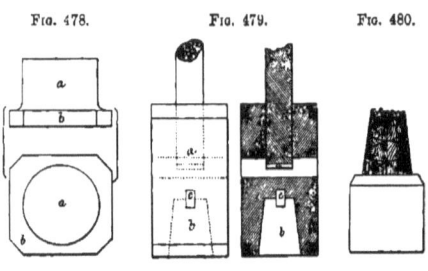

Fig. 478. Fig. 479. Fig. 480.

Die. Stamp Head. Shoe Fastening.

Stamps.—The stamp consists of the stem or lifter; the head or socket attached to the lower end of the stem, and furnished with a shoe; a movable part which sustains the force of the blows and the wear of the operation; and the collar or tappet, by means of which the revolving cam lifts the stamp for its fall. The stem is a round bar of wrought iron, about 3 in. in diameter, usually turned in a lathe. Its length is 10 or 12 ft. Its lower end is slightly tapered, and corresponds in form to a socket or conical hole in the upper part of the stamp-head. The rest of the stem is usually made round throughout its entire length, the method, now in general use, of attaching the tappets to the stems not requiring any modification in the form of the latter, as was formerly the case.

The stamp-head, illustrated in Fig. 479, is a cylindrical piece of tough cast iron, about 8 in. in diameter and 15 in. high. In its upper end is a socket, shown by dotted lines, corresponding with the axis of the cylinder, and conical in form, designed to receive the slightly tapering end of the stem, to the dimensions of which it must be adapted. This conical hole or socket is about 7 in. deep. At its bottom is a hole or key-way a, passing through the head at right angles to the cylindrical axis, by which passage a key may be driven in to force the head from the stem when necessary.

To attach the stamp-head to the stem, the latter is placed in its position between its guides, the head standing immediately under it. The stem, being dropped, enters the socket, and a few blows of the hammer drive it in with sufficient force to cause the head to be raised when the stem is lifted. The stem and the head, being suffered to drop together a few times, become firmly connected. In the lower end of the head is a similar hole or socket b, but larger than the upper one, likewise tapering or conical in form, made to receive the stem or shank of the shoe, which is thus connected with the head in a similar manner; a rectangular hole or passage c through the head at the end of this lower socket, permits the removal of the shoe in the same way as the stamp-stem is forced out from the upper socket. A stout wrought-iron hoop encircles each end of the stamp-head, being fitted and driven on when hot, and allowed to shrink in place.

The shoe in common use is a cylindrical piece of cast iron about 8 in. in diameter and 6 in. high, above which is a shank or stem, the base of which is 4 or 5 in. in diameter, tapering in form, and about 5 in. high. It is made of the hardest white iron. It is attached to the head in manner somewhat similar to that just described for connecting the head and the stem, but is wedged on by means of strips of pine-wood. These strips, which are cut about as long as the stem of the shoe, ¼ in. thick and ½ in. wide, are placed around the stem of the shoe, and tied with twine, as shown in Fig. 480. They must be thick enough to wedge the stem of the shoe firmly in its socket, without allowing the head to come into contact with the body of the shoe. When the shoe is ready to be fixed to the head, it is placed in proper position, with the stem of the shoe directly under the socket of the head, and the stamp and head are then allowed to drop upon it. If necessary, a few blows of a hammer are struck upon the top of the stamp-stem. The whole may then be raised, the shoe keeping its place, and suffered to fall repeatedly, until the shoe is firmly established in its socket. During this operation, a piece of plank is interposed between the die and the shoe, for the latter to strike upon. When a shoe is worn out, it can be removed from the socket by driving the key into the key-way c, and forcing it off. Care must be taken that the shoe does not become so thin as to permit the head to sustain undue wear, and so become weakened. Shoes should be renewed when worn down to a thickness of 1 in.

It was at one time objected that the effective capacity of round stamps was less than that of square ones; but it has been proved, by careful experiments under uniform conditions, that they are equal in this respect. At the same time, the circular stamp possesses a great advantage in that it can be caused to revolve on its own vertical axis while at work, making a partial revolution at each blow, the rotary motion being continued during the free fall of the stamp, which produces a grinding effect upon the material between the shoe and the die, increasing the effective duty of the stamp, and equalising and reducing the wear of the shoe. Bland declares himself satisfied that the square heads used at the Port Phillip works are more efficient and economical, the cam-barrel being simpler and more easily kept in order. Round revolving stamps are almost a necessity when the feeding is done by hand from the back only; but with self-feeding all round the stamps, he considers the square head the better.

Weight of stamps.—This is subject to great variation. In Victoria, the figures usually range between 224 lb. and 1232 lb. per head; in America, 700 to 950 lb. Generally a medium weight of 560 to 672 lb. best suits the character of the ore, but some ores are met with requiring the higher figures.

Height of drop.—The height of the drop or fall of the stamps varies from 2 to 12 in. in Victoria, and from 7 to 11 in. in America. It should not be less than 7 in., and may be increased with advantage if the stamps are light.

Speed.—The number of drops made by each stamp per minute is the "speed" of the battery. In Victoria, it varies from 45 to 85 blows, 70 to 80 being generally considered most effective; in America, 70 to 100. The greater speed adopted in America is suited to the system of amalgamating in the battery.

Order of drop.—The order in which the stamps drop varies in different mills, but the desired conditions are (1) that the work of raising the stamps shall be uniformly distributed on the camshaft, so that the weight lifted shall be, as nearly as possible, the same at any period of the revolution; and (2) that each stamp shall fall effectively upon the material to be crushed, and maintain its proper distribution in the mortar. If all the stamps fell at the same time, the structure would soon be knocked to pieces; and if they fell in regular succession, from one end of the battery to the other, the material would accumulate at one end, and the effective duty of all the stamps would be greatly diminished. In a 5-stamp battery, the common sequence is 3, 5, 2, 4, 1, or, in other words: (1) the middle stamp, (2) the end one on the right, (3) the second on the left, (4) the second on the right, (5) the end one on the left. Another sequence, which makes a backward and forward wave, and thus keeps the mortar very evenly filled, is 3, 4, 5, 2, 1. Others which find favour are— 3, 4, 2, 1, 5; 2, 4, 5, 3, 1; 3, 5, 1, 4, 2; 1, 5, 2, 4, 3; 1, 5, 4, 2, 3. It is thought that the middle stamp dropping first secures the greatest discharge, and that the end stamps dropping first effect the maximum of work done.

Character of blow.—The character of the blow delivered by the stamp upon the ore demands attention. The hardness of the mineral containing gold is almost always so much greater than that of the gold itself, or even of auriferous pyrites, that the same amount of stamping on the three substances will render the two latter much finer than the former. But it is of the utmost importance to prevent the gold being smashed too fine, or beaten flat, for in those conditions it is very difficult to save effectively. The tendency of slow heavy blows is to flatten the gold-particles, while that of smart light blows is to effect disintegration without materially altering the shape of the particles.

Tappets.—The collar or tappet is a projecting piece firmly secured to the upper part of the stem of the stamp, by means of which the revolving cam may lift the stamp and let it fall upon the substance to be crushed. Tappets vary in form and method of attachment to the stem, but that which seems to combine the greatest number of advantages, and to have been most generally adopted on the Pacific coast, is that which is known as Wheeler's " gib-tappet." Fig. 481 shows an elevation and vertical section of this contrivance. It is a piece of cast iron, cylindrical in form, about 8 in. in height and diameter, and hollow at the centre, so as to receive the stamp-stem. To secure the tappet to the stem, there is a gib *g*, about 2 in. wide, and nearly as long as the tappet, having its inside face curved so as to correspond in form to the circular hole through which the stem passes. The gib being fixed in its place in the tappet, and the latter being upon the stem, it is pressed against the stem by means of two keys *k*, driven into the key-ways with force sufficient to hold the tappet and stem firmly together, and prevent slipping between them. This is found to be a very effective method of securing the tappet, while permitting it to be fixed at any desired point on the stem, according to the wear of the shoe. The stem is uniform in size, and the work of cutting facings, screw-threads, and key-seats on the stem, required by other methods, is thus avoided. The revolving cam, meeting the tappet, and raising the stamp, causes it, while being lifted, to make a partial revolution about its vertical axis.

Guides.—The stamp is held vertically in its movements by guides, between which the stem passes. These were formerly made of iron, but such have been almost entirely replaced by wooden ones in Nevada and California. One set of guides is placed below the tappet, about 1 ft. above the top of the mortar; the other set is placed near the top of the stem, so that 6 in. or 1 ft. of the latter may project above the guides. They are supported by the cross-timbers or ties *d* (Fig. 472), which form a part of the battery-frame, connecting the two uprights or posts. They are usually made of

REDUCING MACHINERY.

pine, though hard wood is preferred, and are 10 to 16 in. wide. One part of the guide is made in a single piece for the whole battery, and bolted to the cross-timber; the other may be in one piece, like the first, or cut into as many pieces as there are stamps in the battery, as in Fig. 482, which are then secured to the corresponding part by bolts. In each part are cut semicircular recesses, which form, when the two parts are put together so that the recesses correspond, the holes or stemways for the reception of the stamp-stems. When the guides are so much worn by friction as to permit too much motion of the stems, they may be dressed down on their adjacent faces, by which mean the recesses are reduced to nearly the proper dimensions.

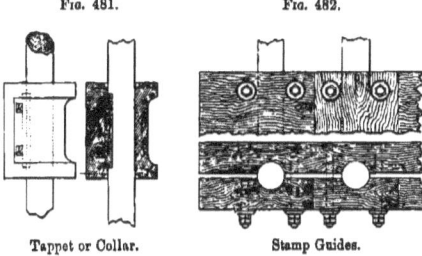

Fig. 481.   Fig. 482.

Tappet or Collar.   Stamp Guides.

Cams.—The cam is a curved arm fixed to a shaft, which is so placed in front of the battery that, by the revolution of the shaft, the cam is brought into contact with the tappet of the stamp-stem, causing the tappet to rise to a height determined by the length of the cam, and to fall at the moment of its release from such contact.

In Nevada, the cams are made of tough cast iron, and are usually "double-armed," that is, have two arms attached to one central hub. Fig. 483 shows the form of cam generally in use: a is the hub; b, the arms; c, the face; d, a strengthening-rib. Where the speed of the stamps is great and the fall slight, single-armed cams are necessary, or the constantly recurring impact of the cam would virtually keep the stamp suspended, or at least so lessen its fall that it would not reach the material in the battery.

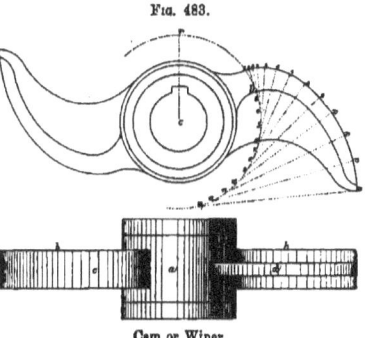

Fig. 483.

Cam or Wiper.

The proper curve of the face of the cam, in order that it may perform the required duty with the least friction, is the involute of a circle the radius of which is equal to the distance between the centre of the cam-shaft and the centre of the stamp-stem. This produces a line for the face of the cam, which meets, better than any other, the various requirements. The bottom of the tappet is constantly perpendicular to the radius of the curve of the cam; the tappet, and with it the stamp, is lifted vertically and uniformly, so that the lift of the stamp is always regularly proportioned to the revolution of the cam-shaft.

The cam-curve may be constructed on paper by means of tangents, as shown in Fig. 483. If c represents the centre of the cam-shaft, and cr the distance from the centre of the cam-shaft to the centre of the stamp stem, the circle described about c, with cr as a radius, is the developing circle of the involute. The distance representing the height to which the stamp is to be lifted, is laid off upon the circumference of this circle, as from the point 1, which distance is subdivided into a convenient number of equal parts, determining, as in Fig. 483, the points 2, 3, 4, to 13. From each one of these points in a circle, a tangent is drawn, on which is laid off a distance equal to the length of arc between the point 1 and the point from which the tangent is drawn. All the points thus determined in the tangent lines are points in the cam-curve, and may be connected as shown in the figure, thus producing the line for the face of the cam.

In practice the line of curvature is produced by cutting from a thin board a circular piece, the radius of which is equal to the horizontal distance from the centre of the cam-shaft to the centre of the stamp-stem. At a given point on the periphery of the circular piece is fixed one end of a thread, which must have the length of the greatest desired lift of the stamp, and to the other end of which is attached a pencil-point. The circular piece, with the attached thread wound on the periphery of the circle, is laid on a smooth board, on which the line is to be traced, and the thread being constantly stretched to its farthest reach, is unwound until it forms a tangent to the circle at the point where the other end is attached. The line described by the pencil-point is the desired curve.

Some builders slightly modify this curve, giving to the cam-arm a greater curvature near each of its ends, in order that the cam in its revolution may come into contact with the tappet at the least practicable distance from the cam-shaft, where the concussion is less than at a greater distance, and to diminish the friction between the extreme end of the cam and the face of the tappet. The face of the cam is 2 or $2\frac{1}{2}$ in. wide. Its extreme end is fashioned so as to correspond to the outer edge of the tappet, which is circular. The cam is placed as near the stamp-stem as practicable without coming into contact with it. The cams are caused to revolve by means of the cam-shaft, to which they are secured by one or sometimes two keys or wedges.

McNeill's patent cam is designed to facilitate the replacing of a broken cam without removing any of the other cams in the same shaft. It is shown in Fig. 484. The boss of the cam is of wrought iron or steel, while the arms or wings may be wholly of steel, or of wrought iron with steel-faced edges. The connection between the boss and the arms, or wings, may take various forms. That which is considered the best is shown, consisting of two flat parallel faces, one at each side of the boss a, cutting or forming a dovetail recess b in the boss. These dovetails are slightly tapered, the larger end being at the working face c, of the cam. The cam arms are arranged centrally on the boss, or flush with one side. A second form of connection is shown, and consists in having the dovetail projection c upon the boss a in the ends of the wings e, while a third form f consists in having the connection take a partially circular sectional form, instead of a dovetail section. By constructing lifter cams in three separable parts it enables the shaping and steel facing of the arms or wings to be easily accomplished, as the work is lighter and capable of being more readily handled.

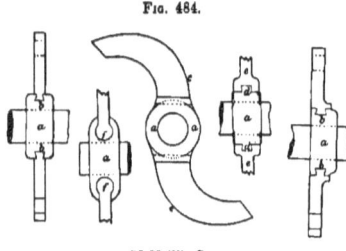

Fig. 484.

McNeill's Cam.

Cam-shaft.—The cam-shaft is a round shaft of iron, which is smoothly turned and finished, having one or two key-seats or grooves cut in it lengthwise, for the purpose of securing the cams in their places. The shaft rests in boxes, which are usually supported by shoulders cut on the upright posts of the battery-frame. Cam-shafts vary in diameter from 4 to 6 or 7 in., according to the number of cams to be fixed upon them, and the weight of the stamps to be raised. In some mills, a single cam-shaft is made long enough to carry all the cams for as many batteries as there may be. In Nevada and California, however, short cam-shafts are in general use, a separate shaft being employed for each battery, or in many cases one shaft for two batteries. Separate cam-shafts are preferred, on account of the independence of each battery, so that if one be stopped by any accident to the cams or stamps, or for repairs of any kind, the operation of the others is uninterrupted. Each shaft in such case is driven by its proper pulley, which receives its motion by means of belting from a counter-shaft. The order in which the stamps are to fall being determined, it is carried into effect by fixing the cams on the shaft in such position that each cam, by the revolution of the shaft, will lift its respective stamp at the desired moment. For this purpose, the key-seats cut in the hub of the cam must be determined with care; one common key-seat being cut on the cam-shaft, when the desired position of any given cam has been ascertained, the key-seat in the hub is cut to correspond with that of the shaft.

Props.—When it becomes necessary to hang up a stamp so that the cam may revolve without reaching the tappet, it is supported by a prop, stud, or "finger" *x* (Fig. 472). The lower end of the stud, of which there is one for each stamp, is pivoted on a small shaft fixed across the battery from end to end, resting in boxes, which are secured to the uprights. Each stud is just long enough to support the stamp, when placed under the tappet, at a height which is about 1 in. above the highest lift given by the cam. To bring the end of the stud into this position when desired, the workman lays a smooth stick on the face of the cam as it is rising to the tappet, and holds it there while the stamp is lifted. The stick is as wide as the face of the cam, and long enough to be held conveniently, and 1¼ in. thick at the end which comes between the cam and tappet. By this means the stamp is raised high enough for the stud to be put in place, which being done, the stamp is supported above the reach of the cam. To set it again in motion, the operation is repeated, the stud being withdrawn at the moment when the stick on the face of the cam has lifted the stamp clear of its support. To lift the stamp-stem, a chain or rope pulley is provided above the battery and should be arranged to travel on rails so that it can be readily shifted over any particular stamp.

Feeding.—Much of the effectiveness of the stamps depends on the degree of care devoted to keeping the working parts in good condition, and on the regularity with which they are supplied with ore. This is commonly done by hand labour, the rock being shovelled in at such a rate as it is crushed and discharged. In some mills, however, automatic feeders are employed, which give satisfaction. These consist of a hopper filled with ore, from which a trough or chute leads to the feed-opening of the battery, so inclined that the ore will slide down from the hopper to the battery if the chute, which is hung on a pivot, be agitated. A rod is attached to the chute, and so placed that the tappet of the stamp, when the latter gets so low as to require an additional supply of rock, will strike its upper end, thus giving a shock which causes the ore to move down and fall into the battery. But a great objection to most automatic feeders is that they are arranged to supply one constant quantity, and make no difference in the work for the several parts of the mortar under varying conditions. With hand-feeding, the ore is received into a large bin or pocket, the floor of which is made in such a way that the ore will run easily towards the mortar. Here, in a space of 6 to 12 ft., stand the feeders, whose duty consists in keeping a constant amount (generally 2 in. deep) of ore between the shoes and dies, whereby the drop of the stamps is maintained at an equable height.

Automatic ore-feeders are broadly of one type, as just described. Stanford's feeder, as made by Joshua Hendy, 49, Fremont Street, San Francisco, is shown in Fig. 485. It consists of a hopper

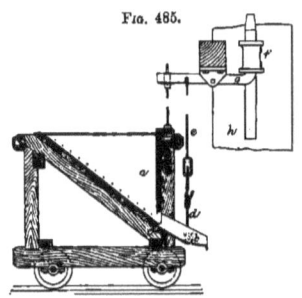

Stanford's Ore Feeder.

Tulloch's Ore Feeder.

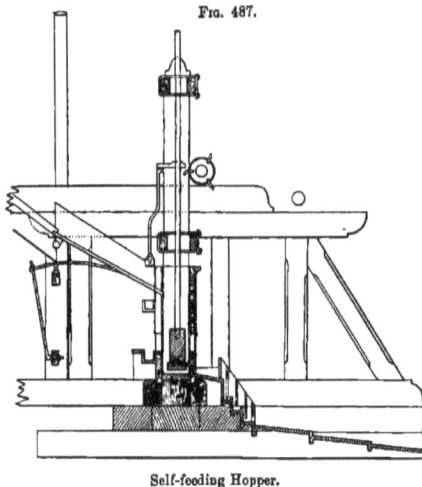

Self-feeding Hopper.

a, with adjustable spout b, swung on trunnions c, and attached to a cross-bar d, suspended from an adjustable rod e. A feeding-tappet f is keyed upon the battery-post h, and a lever g rotating on pivots is fixed to the rod e, so as to be struck by the cam-tappet, the lever g being forked that it may span the stem. While the battery is supplied with sufficient ore, the tappet does not descend far enough to strike the end of the feeding-rod; when the ore gets low, the tappet does strike the rod, and the effect is an oscillation of the front spout on its trunnions, whereby the ore is thrown forward. The apparatus being on wheels can be readily removed. It is simple in construction and operation, maintains the feed at any desired degree, and materially reduces the wear and tear of the stamps, while increasing their duty, it is said, 25 per cent. It seems best adapted for dry-crushing.

The Tulloch feeder, also made by Hendy, is likewise largely used. It is illustrated in Fig. 486. In Fig. 487 is shown the self-feeding hopper employed by the Port Phillip Gold Mining Company; and in Fig. 488, the roller ore-feeder made by the Golden State and Miners Ironworks, San Francisco.

Hendy's "Challenge" feeder is perhaps the most popular for wet, sticky ores. The mode of attaching it (to the second stamp in the battery) is shown in Fig. 489: a, tappet; b, lever; c, lower guide; d, hopper; e, carrier-table; f, shoot; g, bumper; h, stamp-stem. This feeder is said to effect a reduction of the wear and tear by 15 per cent., and an increase of duty by 20 per cent. It is by many considered the best feeder for all ores, but is more costly than some other forms. Recently an improvement has been introduced, whereby the feed side of the mortar is rendered more accessible.

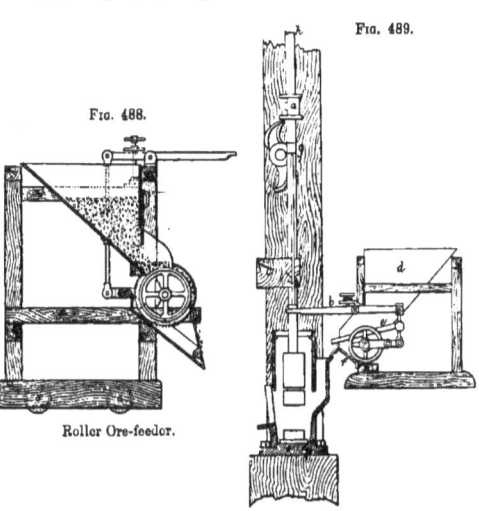

Fig. 488.
Fig. 489.
Roller Ore-feeder.
Hendy's Ore-feeder.

Water.— The quantity of water used in wet crushing depends in a great measure upon the character of the ore and the degree of fineness to which it is crushed. In Victoria, the proportion varies from 30 to 1200 gal. per stamp-head per hour, though 300 to 500 gal. would appear to satisfy all requirements. In America, about 93 gal. is commonly thought sufficient. Egleston says the consumption in Nevada and California is 200 to 300 cub. ft.* per ton of rock stamped, or $\frac{1}{8}$ to $\frac{1}{4}$ cub. ft. per stamp per minute, for all purposes, leaving probably about $\frac{1}{4}$ cub. ft. for the batteries alone (or nearly 94 gal. per hour). In Colorado, the amount is 28 cub. ft. per ton of rich ore, and 33 cub. ft. per ton. of poor ore. The cub. ft. of ore averaging 108 to 125 lb., this will give $\frac{1}{4}$ cub. ft. per stamp per minute, or about the same as Nevada and California. When the water used is purchased from a "ditch company," it is measured by the miners' "inch."

The water is fed to the stamps by horizontal piping just above the feed-slot of the mortar, with orifices opposite each stamp, capable of being closed if necessary; or the main feed may be brought higher up, and vertical supplies be carried from it down to each stamp, with valves at the ends. The main is often a 3-in. gas-pipe. A second main of half the size is placed in front, to help carry the pulp. In America, arrangements are made to warm the water supplied to the stamps, by waste steam or otherwise, during winter. The water should always be strained before use, to exclude leaves, chips, grass, &c.

Power.—To find the horse-power required to drive a battery, multiply the weight of one stamp by the number of stamps in the battery by the height of lift in feet by the number of lifts per minute, add one-third of the product to the result for friction, which will be the number of foot-pounds per minute; divide this by 33,000, which is the number of foot-pounds per minute equal to 1 H.P. and the

* 1 gal. = 277¼ cub. in.; 1728 cub. in. = 1 cub. ft.

result will be the H.P. required. For example, supposing a stamp weighs 800 lb., that there are five in a set, that each stamp has a lift of 9 in. = 0·75 ft, and gives 80 blows per minute, then :—800 × 5 × 0·75 × 80 = 240,000; one-third of 240,000 = 80,000; this added to 240,000 = 320,000; 320,000 divided by 33,000 = 9·7 H.P. or 1·9 H.P. per stamp. The total weight of a battery, including mortar-box, stampers, &c., may be roughly estimated at 1 ton per stamp. Medium sized stampers, *i.e.*, stem, tappet, head, and shoe, weigh 600 to 700 lb., and require about ½ H.P. Heavy stampers weigh 800 lb. and over.

Tables of Dimensions and Duty.—The annexed tables reveal at a glance the dimensions and working results of a number of mills in various parts of the world, including some that may be considered representative.

Mill or District.	Weight of Stamps.	Drops per Minute.	Depth of Drop.	Crushed per 24 Hours.	No. of Stamps.	Duty per Stamp.	H.P. per Stamp.	Holes per Sq. In. of Grating.	Water per Stamp per 24 Hours.	Mercury used per Stamp.	Mercury lost per Stamp.
	lb.		in.	tons		tons			gal.	lb.	oz.
Grass Valley	850	61	10	40	20	2			..	..	..
"	700	68	10	32	20	1·12			..	..	..
Eureka	950	80	9	..	60	2½–3			..	..	..
Brunswick	..	..	..	160	56	3			..	..	..
Keystone	750	75–80	..	75–80	40	2			..	..	..
Idaho	950	80	9	..	..	..				..	..
Metacom	900	90	10	..	..	..	..				..
Port Phillip	672	75	..	..	56	2·2	1			..	..
"	896	75	..	..	24	3	1¼			..	..
Nova Scotia	650	55	6 –9	..	..	1 –1½	..			..	..
	cwt.										
Ballarat	4 –8½	50–85	7 –10	..	..	1 –4	1 –2	40–200	950– 8,640	5–75	1 – 8
Beechworth	4¼–7¾	40–90	5 –14	..	..	⅓–4	¾–1½	60–140	720–11,520	5–70	½– 8
Sandhurst	5 –8	25–75	6 –18	..	..	1 –3¾	⅔–2	64–140	4,000– 8,640	10–40	½– 5½
Maryborough ..	4½–8	50–75	6 –22	..	..	1 –3	½–2½	70–144	900– 8,640	3–30	1¼– 8
Castlemaine	4½–8	35–75	6 –15	..	..	1 –3¼	½–2	40–144	4,800–12,960	6–40	¼–24
Ararat	5 –6¾	60–72	7½–10	..	..	1¼–1½	¾	90–120	4,320–12,960	6–47	½– 7
Gippsland	6 –7½	60–80	7 –10	..	..	1½–2	¾–1½	70–250	1,600–25,000	10–37	⅛–32

Name of Mill.	Length of Stem.	Diam. of Stem.	Weight of Stem.	Height of Shoe.	Diam. of Shoe.	Weight of Shoe.	Height of Boss.	Diam. of Boss.	Height of Tappet.	Diam. of Tappet.	Weight of Tappet.	Height of Die.	Diam. of Die.	Weight of Die.
	ft.	in.	lb.	in.	in.	lb.	in.	in.	in.	in.	lb.	in.	in.	lb.
Douglass	12½	2⅞	290	9	8	115	18	8	12	8	120	..	..	..
Cons. Virginia ..	13	3⅛	320	7	8	110	16	8	10	7½	95	..	..	..
Lincoln	13	3⅛	320	7	8½	119	18	8½	10	7¾	93	5½	8½	99
Brunswick	15	3⅛	375	10	9	125	18	8	12	9	125	..	..	..
Electric	11⅔	3	258	8	8⅛	123	16	8⅛	8½	7½	83	6	8⅛	100
Eureka	14	3½	450	..	..	160	..	..	..	..	120	..	..	120
Keystone	..	..	..	..	..	100	..	..	..	..	100	..	..	113
Stanford	..	..	..	..	..	120	..	..	..	..	114	..	..	..
Walhalla	10¾	3⅜	..	9	10	..	14	10	..	..	..	..	..	..

The wide differences in the figures given in the preceding tables are a sufficient warning that the weight and speed of stamps most suitable to a particular ore can only be ascertained by actual trial of that ore. It is vain to expect that the best attainable working results can be got by applying the figures found successful in one case to another case apparently similar. Ores vary in the most marked degree, especially in resistance to disintegration, and such variances cannot be judged in an out-of-hand fashion. Therefore careful experiment with large and representative samples of the ore should precede the choice of dimensions for the stamp battery.

In general terms, it may be said that there is a tendency towards the adoption of heavy stamps, with slight fall and rapid stroke, in preference to the opposite pattern which ruled at one time. A series of experiments made at Metacom in America, clearly showed that the effective capacity of the battery was very largely increased by augmenting the speed of the stamps. In a dry stamping silver mill the increase of work was 244 per cent. for an increase of speed amounting to 85 per cent.; or in other words, by quickening from 60 to 102 drops a minute the outturn rose from less than 1 to over 3 tons per stamp. These figures will not be attained in wet crushing gold ore, because in dry crushing it is the compression of air by the fall of the stamp which alone forces the pulp out of the mortar, while in wet crushing the movement of the water greatly assists in freeing the battery, and it must be borne in mind that the rate of escape from the battery is the rate of effective work done by the stamps.

Against the extra horse-power required to move the heavier stamp, there must be set the gain in efficiency, which is always in advance. There is also a saving in labour. But certain considerations control the limits to which these features may be economically developed. For instance, excessive weight or speed would soon injure the structure of the battery, and excessive speed alone would be abortive, as it would not allow the stamp to fall at all, the rapidly revolving cam keeping it suspended.

Life of Battery.—The duration of the several parts of a stamp battery is very unequal. An ordinary die should last about twice as long as the shoe, or say six weeks to two months; the rate of wear is commonly $\frac{1}{4}$ lb. to 1 lb. per ton of ore treated. The conditions which govern the rate of wear are the speed of the battery, regularity of feeding, hardness of ore, and quality of iron. Good cast iron is as enduring as steel, but much of the cast iron employed for this purpose in remote localities is far from good. Shoes generally last four to six weeks, wearing at the rate of $\frac{1}{2}$ to $1\frac{1}{4}$ lb. per ton of ore put through.

Stamp stems, especially when running in wooden guides, should last three or four years, and, with regular feeding, very much longer; the jar occasioned by a short supply of ore in the mortar is apt to cause them to break off at the ends. When this happens, the injury is remedied by welding on a piece, which may be done repeatedly. Stamp heads sometimes split when not bound with iron rings, and the sockets become enlarged by wear; but the latter drawback may be overcome by inserting wooden or iron wedges. Tappets should endure 4 or 5 years, but are sometimes split through wedging up too tightly.

In order to change a stamp-stem, an operation occupying about half an hour, the whole battery must be hung up on the props already described (p. 303). The die and shoe are placed on the foot-plate, and three or four strips of wood, $1\frac{1}{4}$ in. wide, are put into the stem socket exactly as if the battery were new. The stem is then let down to within about 3 in. of the die, and the tappet is keyed on in this position; the stem and tappet are then hung up upon the prop. While the whole of the rest of the battery remains hung up, a strip of wood 2½ ft. long and 2 in. wide, covered with leather, is placed between the cam and the tappet and held in one hand, while with the other the workman seizes the handle of the prop and pulls it out from under the tappet. The stem falls into the socket of the die. Keeping the stick in its place, the workman allows this single stem to work until it is quite fast in the die; he then hangs it up again and going below raises the shoe to its proper level, supporting it there, and then sets the tappet in its proper place. This is a delicate operation, requiring considerable skill, as for a time the stem is held up by the pulley alone, and just the right moment must be chosen to set the stick under the cam. If the stick is not held exactly in the proper position, it may result in breaking the stem.

Dry Stamping.—The idea of stamping the ore dry is by no means new; in fact it was probably the original method, but gave way to wet stamping because it was found that the water assisted in clearing the mortar of fine matters, and served as an aid to separation of the valuable and valueless matters afterwards. Whether it might not be better to stamp dry and not add the water till the ore is in a fine state is, however, exercising the minds of some mill managers, since means have been perfected for drawing away the fine particles from the mortar as soon as they attain the desired degree of comminution.

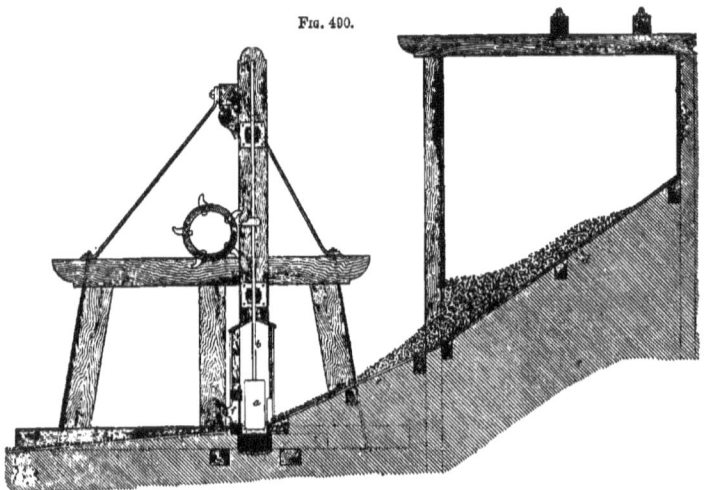

FIG. 490.

Cornish Tin-stamps.

Recently, Mr. W. H. Nash, of Sydney, has brought under the notice of the Victorian Mining Department a dry-stamping process invented by him for extracting gold from quartz. The apparatus consists of a tall chimney of sheet-iron or suitable material attached to an air-tight enclosure in front of the gratings of the stamping-boxes, and communicating with the external air through the gratings and stamping-boxes only. Water not being used, the stamps, stamping-boxes, and quartz quickly become warm in working, and the heat thus generated causes a strong draught of warm air to pass up the chimney, which being supplied only through the gratings, draws the fine powder with it. This arrangement, it is said, has been found in practice to discharge the stamping-boxes as rapidly

Fig. 491.

Husband's Pneumatic Stamp.

and effectually as if water were used. Combined with this apparatus is an enclosed elevator, into which the pulverised quartz falls as soon as it has passed through the gratings, and which, continually working, raises it to a high level for subsequent treatment. Nash claims that the process gives a much higher average yield of gold per ton, and that the cost of treatment is less, there being no handling. The particular process invented by Nash has not yet been introduced on the Victorian gold-fields; but some years ago a machine for dry-stamping quartz was erected at Ballarat by the late Mr. Costin, and it pulverised finer than an ordinary battery, at a cost of only 6½d. per ton.

Tin Stamps.—The ordinary Cornish tin-stamps (Fig. 490) are arranged in sets of 4 heads each in a wooden coffer. The head *a* is of cast iron, the stem *b* of wrought iron. The height of lift is about 10 in., and each stamp makes 50–70 blows a minute, and weighs 6–7 cwt., stamping 15 to 20 cwt. of stuff per 24 hours. The stuff is shot from waggons on to the pass *d*, inclined at 1 in 1½, and goes thence into the half pass *e*, inclined at 1 in 2¼. The overflow from the coffer is through the grating *f*, which is a thin copper plate containing usually about 144 holes per sq. in., though this varies somewhat according to the ore.

Pneumatic stamps.—The application of an air-cushion to the delivery of the blow of the stamp very greatly increases its efficiency. This has been most successfully demonstrated in Husband's stamp, as made by Harvey & Co., Hayle, Cornwall, and 186, Gresham House, London. The stamp is shown in Fig. 491. It is an exceedingly compact and simple machine. The driving-power being applied by means of a belt pulley on the end of the crank shaft, is communicated by a piston-rod to the air-piston; the air-cylinder therefore oscillates, and a portion of the air contained in the cylinder having no means of escape, becomes compressed between the piston and the cylinder-ends. The effect of this compression is to raise and lower the cylinder alternately with the rise and fall of the crank. The stamps head, with its shoe, is attached to the cylinder, and the blow falls on a die or anvil; when the machine is working at its normal velocity, the elasticity of the compressed air forces the head with rapid, violent blows into the coffer, where the ore is pulverised with water, and passed through the screens in the usual manner. As the screens entirely surround the battery-box, the outlet for fine stuff is greatly in excess of that provided in ordinary stamps, and hence the effective work is proportionately increased in that direction also.

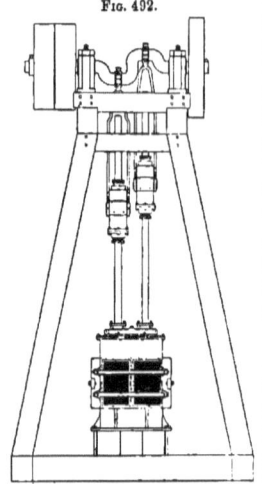
Fig. 492.

Pneumatic Battery.

Another pneumatic battery is shown in Fig. 492. The capacity of a battery of two stamps is 24 tons hard stone per day of 24 hours, crushed to go through a No. 40 screen, which is equal to 1600 holes per sq. inch. There are four outlets, which gives a large area of discharge in comparison to the size of the box. The shoes and dies or bottoms are made of the best Bessemer steel. The number of blows given by each stamp is 180–200 per minute. Twelve-horse power required to drive this battery. The price is 250*l*. The improvement claimed for this battery is the compressing of the air in the cylinder by the piston, which is fixed on the stamp-shank. The method by which this is done is as follows:—The connecting rod lifts the cylinder without lifting the shank until the air is compressed underneath the piston to such a degree that it overcomes the weight of the shank and shoe; so soon as this compression is attained, the shank rises with the cylinder until the crank passes the top centre, and then begins to descend. Great economy, however, is claimed to be obtained by the strongly-compressed air in the cylinder, which is utilised by expansion on the return of the crank, causing an elastic spring of air to be released at each stroke. No extra power is required to compress the air. Ore which can pass through a 6 in. mesh can be fed in the machine. Its weight, complete, is not quite 3 tons.

Batteries.—Messrs. Harvey & Co. (Limited), of Hayle, Cornwall, and 186 and 187, Gresham House, Old Broad Street, E.C., have introduced the high-speed principle into some of the stamp batteries, which they have recently supplied to mines in Cornwall, with most encouraging results. This departure enables an almost unlimited speed to be attained, and it would seem to follow that

Fig. 493.

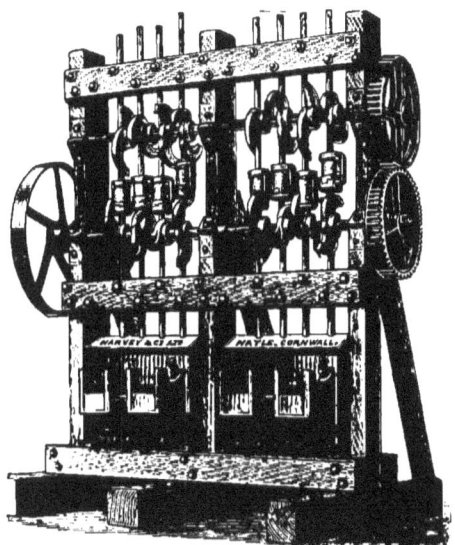

Harvey's High Speed Revolving Stamps.

the logical result must needs be an increased output per head. The exact increase cannot yet be stated definitely, but the working of batteries in several of the leading mines in Cornwall has been very satisfactory.

Opinion is not invariably favourable, amongst mining engineers, to revolving stamps, but this mainly arises doubtless from the fact that, as such stamps are usually made, more than, say, 90 blows per minute cannot well be made in cases of a drop of 8 to 10 in., because of the head and appendages not being able to fall sufficiently quickly before the cam comes round again to lift them. In Harvey & Co.'s new arrangement, however, this difficulty is overcome by the provision of a duplicate cam-shaft and set of cams independent of and above the ordinary ones. These are so set that as soon as the upward movement of the heads ceases the downward cam movement commences. This supplementary active power is calculated to increase the rapidity of stamping to almost any extent, and to be equally effective at a high or a low lift.

This new system points to securing a maximum of active stamping capacity from a battery, and this signifies so much economy in working. When applied to a low lift of head the system should be especially suitable for the treatment of ores of small size, and of a softish character. Fig. 493 gives a general indication of the arrangement of the cams and cam shaft in Harvey's high-speed revolving stamps.

Fig. 494 represents one of R. Hornsby & Sons' arrangement of gold mill, with wood frame battery, launders, and improved Wheeler pans and settlers, &c., which are now being sent to all parts of the world.

FIG. 494.

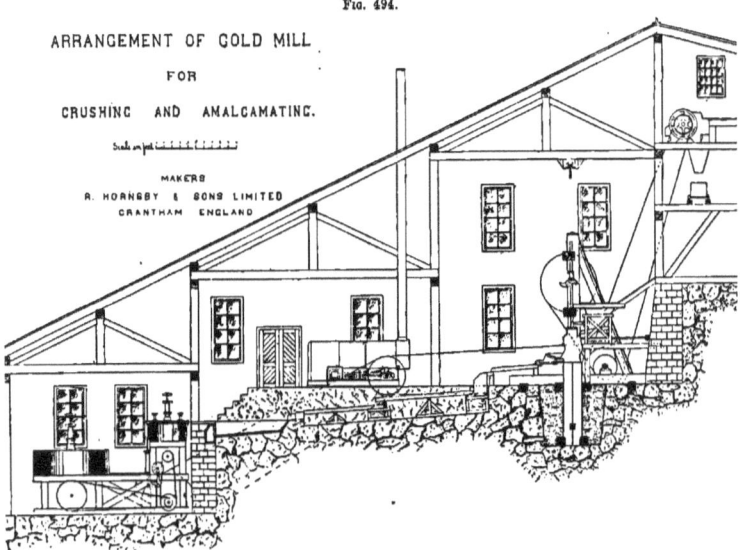

Fig. 495 shows one of R. Hornsby & Sons' type of Californian gravitation stamp mills, which have been largely constructed for the South African and other gold fields. This mill is constructed with timber framework, having diagonal stays placed at the back, so that the front of the battery is quite clear of all stays, thereby allowing ample room for wide launder plates. These mills are of very massive proportions, and are made so as to stand the hardest work. The shoes and dies supplied with them are made of special steel, which in practice has been shown to have very great endurance. Blanket strakes are shown at the end of the launders.

Messrs. Hornsby also construct stamp batteries, with the frames of wrought iron and steel and channel iron put together with turned bolts. Friction clutches or bolt-tightening gear are used as

Fig. 495.

Hornsby's Gravitation Stamp Mill.

preferred. The counter shaft is shown and arranged at back, so as to be quite clear of the front of the battery. These batteries are supplied with automatic feeders, and they are on an improved plan of the well-known Tulloch or Hendy type, which are proved to be the most efficient feeders in use.

Hornsby & Ogle's patent all-round discharge mortar box is shown in front and back views in Figs. 496, 497. It will be seen that the screening surface embraces the complete outside surface of the mortar box, which is the greatest surface possible. The ends of the screens are circular, so as to be equidistant from the shoes and dies. The advantage of such a mortar box, especially where the material is easily crushed, or where a large quantity is required to be discharged, is at once seen.

The screens are made in four pieces, viz. front and back frames, and two end ones; these are readily removable by taking out the wedges and the hoop-iron straps shown. The sole makers are R. Hornsby & Sons, Limited, Grantham.

FIG. 496.   FIG. 497.

Hornsby and Ogle's Mortar.   Hornsby and Ogle's Mortar.

CRUSHING ROLLS.—These consist of a pair of cast-iron cylinders placed horizontally and nearly in contact, and connected together by spur-wheels of equal diameter, so that their surfaces revolve towards each other with equal velocities. Motion is given to one of these rollers either by steam or by water power, and the stuff to be broken is dropped from a hopper between the rollers. It was formerly the custom not to gear the two rollers together, but to allow one to be carried round by the friction of the stuff against it. Experience has, however, shown that the product is greater when the rollers are geared. The diameter varies between 14 and 34 in., a common diameter being 27 in. The length, or breadth of face, varies from 12 to 24 in.

When the whole of the crushing is done by rolls, it is a common practice to have three or more of them. An upper pair, with fluted surfaces, is set so as to take in large masses; the fragments falling from this upper pair are divided between two pairs set below and pressed close together.

A pair of Cornish rolls are shown in Figs. 498 and 499. The rollers of this machine are of unequal length, the driver being 24 in., and the follower only 18 in.; both are 27 in. diameter. These rollers are formed of thick cast-iron shells, keyed on to cylindrical bosses on a pair of plain shafts, which have couplings outside their bearings connecting them with a pair of lighter shafts, carrying the gearing wheels. The bearings of the shorter roller slide between parallel guides, and are kept in position by round bars passing through holes in the frame, which are pressed against by the shorter arm of an unequal armed bent lever, whose longer arm carries a loaded box; the relation of the two arms to each other is as 1 to 9. The object of this arrangement is to save the rollers from fracture in case any unyielding substance should get between them; for when the resistance of any fragment is greater than the horizontal thrust exerted by the loaded arms on the bearings, the shorter roller slides apart from its fellow and opens a passage for the unbroken substance to pass through. The broken material passes into a tubular or drum sieve, 40 in. long, and 24 in. diameter, whose axis is set at an angle of 25° to that of the driving roller, with which it is connected by a pair of bevel wheels; the smaller wheel on the sieve shaft has 11 teeth, and receives motion from a larger one of 40 teeth. The gauze has 6 apertures to the sq. inch; the particles passing through are received in a box, closed by a door, through which they are loaded into a truck on the railway below; the coarse fragments are thrown out into the buckets of the raff-wheel or lifting wheel, which resembles a reversed water-wheel, being closed on its outer circumference, but provided with a ring of buckets opening inwards; it is 15 ft. diameter, and, being

FIG. 498.

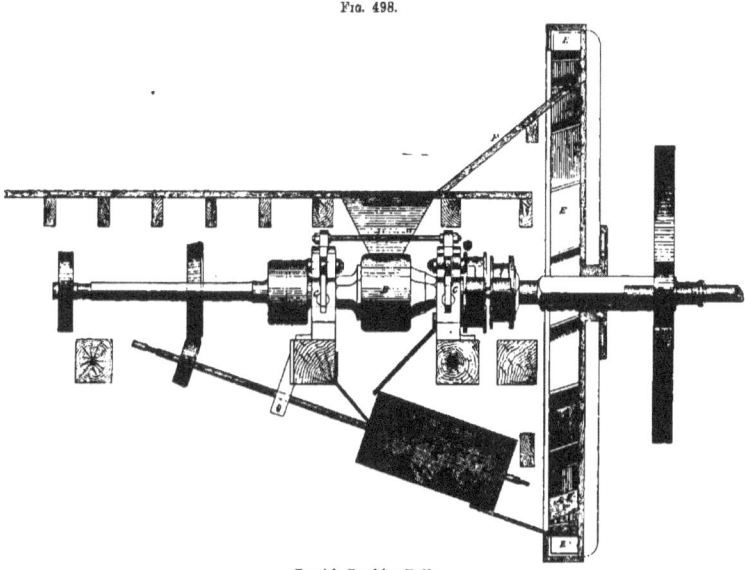

Cornish Crushing Rolls.

coupled to the driving roller, makes the same number of revolutions; it discharges its contents on to a floor immediately adjacent to the feed hopper, to which they are returned to pass a second time through the rollers.

A crusher of this size makes 30-50 rev. per minute, requiring an effective driving power of 12-20 H.P.; it will break down a quantity of ore varying from 40 to 60 tons a day, according to the hardness of the associated gangues.

The surfaces of the rollers soon become much worn, and when made of chilled iron, the irregularity of the chilling is soon made manifest by the unequal wearing away, the softer parts being hollowed out, while the harder are left in ridges and irregular bulges. It is preferred, therefore, to use ordinary hard pig-iron, or a mixture of hard white iron, similar to that used for the dies and shoes of stamps. The rollers are also made with an outer casing or shell, as in the foregoing example, that can be slipped upon the axis or core of the roller, and removed from it when too much worn. This hollow cylinder is usually cast so as to make a firm lock-joint upon the core, or it is keyed by means of two or three keys driven into recesses extending throughout the length of the cylinders, one half of the key being in the central core, and the other half in the shell or casing.

Fig. 499.

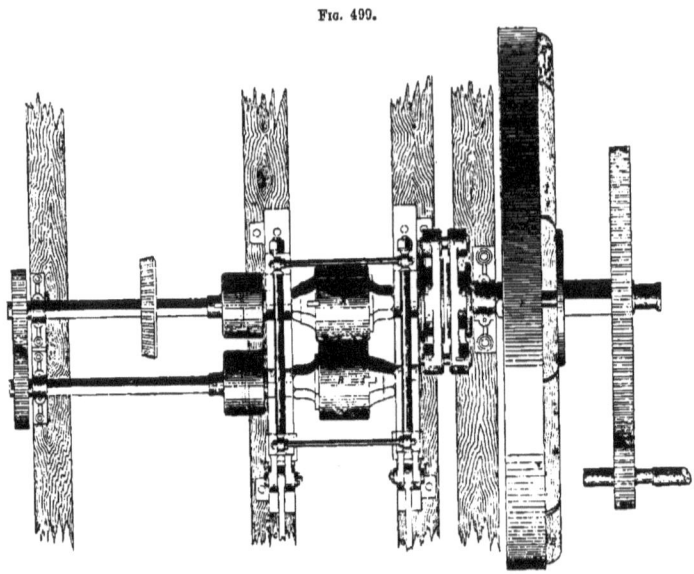

Cornish Crushing Rolls.

Fig. 500.

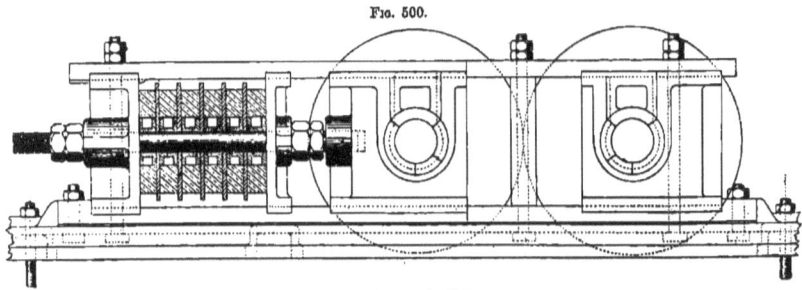

Rubber Springs for Rolls.

The means commonly used to keep the rollers in contact is a weighted lever; but rubber springs are also employed for that purpose. These springs, one of which is shown in Fig. 500, are placed one on each side of the frame. They consist of alternate discs of rubber and iron, the former being 1 in., the latter $\frac{1}{4}$ in. thick. The number of rubber discs is generally 6. The requisite initial pressure is obtained by means of strong screws passing through the axes of the springs; by tightening up or loosening the nuts on those screws, the pressure may be increased or diminished as occasion requires. A great advantage of this mode of applying pressure to the rollers is that it allows the latter to be driven at a high speed. The system has been very generally adopted in Germany.

The following table gives some important details concerning a number of crushing rolls in use:—

DIMENSIONS AND PRODUCT OF CORNISH ROLLS AT VARIOUS MINES.

Name of Mine.	Rollers.				Total Pressure on Rolls.	Sifter.				Diameter of Raff-Wheel.	Horse-power.	Quantity Crushed in Ten Hours.	Cost of Crushing per Ton.
	Diameter.	Length.	Revolutions per Minute.	Crushing Area per Minute.		Diameter.	Length.	No. of Holes.	Revolutions per Minute.				
	in.	in.		sq. in.	cwts.	in.	in.	sq. in.		feet.		tons	pence
Grassington Mines ..	27	12	$5\frac{1}{2}$	5,593	91	21	48	$6\frac{1}{4}$	37	14	..	80	..
Minera	14	14	8	4,920	$73\frac{1}{3}$	24	42	9	48	10 6–10	6	20	$2\frac{1}{2}$
Cwmystwith, No. 1 ..	27	14	4	4,748	78	20	33	9	24	16	..	32	$2\frac{1}{2}$
Cwmystwith, No. 2 ..	27	14	$4\frac{1}{2}$	5,341	85	24	36	9	24	16		35	$2\frac{1}{2}$
Goginan	30	14	$5\frac{1}{2}$	7,254	39	20	39	9	36	16		30	$2\frac{3}{4}$
Cwm Erfin	27	14	$7\frac{1}{2}$	8,902	293	26	32	9	30	16	..	20	3
Lisburne, No. 1 ..	27	15	6	7,632	180	22	36	$12\frac{1}{4}$	30	16	..	42	$2\frac{1}{2}$
Lisburne, No. 2 ..	27	15	6	7,632	224	22	36	$12\frac{1}{4}$	30	16	..	42	$2\frac{1}{2}$
Derwent	27	14	7	8,309	227	22	60	16	..	15	..	60	$2\frac{3}{4}$
Goldscope	14	18	14	11,060	6	..	..	..	..	..	..	25	$2\frac{1}{2}$
East Darren	30	18	6	9,996	207	24	36	16	45	16	..	25	$2\frac{3}{4}$
Cefn Cwm Brwyno ..	20	13	5	4,080	84	20	48	16	$27\frac{1}{2}$	14	..	20	$2\frac{1}{2}$
Lisburne, No. 3 ..	18	16	8	6,432	169	22	36	25	30	16	..	42	$2\frac{1}{2}$
Llandudno	18	15	15	12,705	61	..	..	..	..	..	..	30	..
Wheal Friendship ..	23	12	10	8,670	123	24	36	36	30	13	13	20	$11\frac{1}{2}$
Pontgibaud	25	12	$12\frac{1}{2}$	12,075	36	22	44	36	60	15	15	17	$2\frac{1}{2}$
Devon Great Consols	34	22	7	16,443	458	24	84	64	21	..	..	65	$3\frac{3}{4}$

Fig. 501 represents a pair of 30 in. diameter, by 16 in. wide crushing rolls, mounted in cast-iron frames, made by Green, Aberystwith. This is a very powerful machine. The shafts are made of 6 in. square scrap hammered wrought iron, on which the cores are keyed on by 8 well-fitted keys at each end, the beds for which are truly planed and slotted. The shells are cast from specially prepared iron—to ensure a maximum hardness and toughness; these are wedged on the cores with wood and iron, which experience has proved to be the best, simplest, and only safe method, thus making the changing of the shells an operation that can be performed by an ordinary mine carpenter. The bearings of the roller shafts are of gun metal, and are made as long and as large in diameter as possible to give large wearing surface. Rubber cushions are used for giving the pressure to the rolls; they are of ample size and power to crush the hardest mineral; the pressure is regulated at will by the screws provided. The pressure may be applied by levers and weights instead of rubber cushion when preferred. The whole has been modified and improved by long and extensive experience, and every care has been taken in the construction of the entire machine to ensure the utmost durability and efficiency in its work. They are adapted to crush all kinds of metallic ores.

FIG. 501.

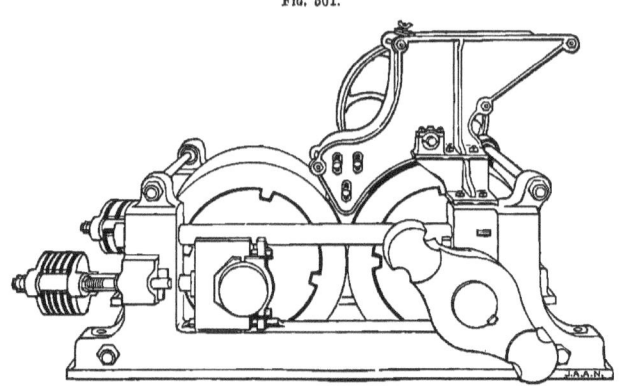

Green's Crushing Rolls.

Steel shafts and crucible cast steel roller shells are supplied when required at an extra charge. These crushers are constructed in five sizes, as under, and may be fitted with self-acting feed regulating hopper as shown. The cost of connecting gear to motor varies according to kind of motor used. With a high speed engine or turbine the crusher is driven by a belt.

Diameter and Width of Rolls.	Approximate Weight.	Capacity per Day of 10 Hours.	Horse-power required.	Prices of Machinery delivered in Aberystwith.	Extra for Wrought-iron Trough and Shoot to Convey the Crushed Ore to Classifiers and Dressing Machinery.
in. in.	tons cwt. qr.	tons		£ s. d.	£ s. d.
30 × 16	8 10 0	45	14	135 0 0	8 10 0
27 × 16	6 10 0	35	12	124 0 0	7 15 0
24 × 14	4 10 0	26	10	110 0 0	7 10 0
22 × 14	3 15 0	20	8	95 0 0	6 15 0
20 × 14	2 5 0	16	6	68 0 0	6 10 0

Fig. 502 shows a roller ore crusher with rolls 24 in. diameter by 14 in. wide, also by Green, Aberystwith. No part is made more than 300 lb. in weight except roller shells; these by putting a temporary axle in them, can be taken up the steepest mountains. The machine has been constructed with a view to securing the utmost strength and durability combined with efficient effect in working, and has, where used, given the utmost satisfaction. The best materials are used in the construction and the whole is carefully fitted together. The pressure may be applied by levers and weights instead of rubber cushions, when preferred. Steel shaft and crucible cast steel roller shells are supplied when required at an extra charge. The machines are made principally of wrought iron

Fig. 502.

Green's Roller Ore Crusher.

and steel, rolls up to 18 in. diameter, in parts not exceeding 300 lb. in weight except roller shells. These crushers are made in three sizes, as under, and fitted with self-acting feed regulating hopper.

Diameter and Width of Rolls.	Capacity per Day of 10 Hours.	Horse-power required.	Price of Machine Delivered in Aberystwith.	Extra for Wrought-iron Trough and Shoot to Convey the Crushed Ore to Classifiers and Dressing Machinery.
in. in.	tons		£ s. d.	£ s. d.
24 × 14	26	10	120 0 0	7 10 0
22 × 13	20	8	108 0 0	6 15 0
20 × 12	16	6	98 10 0	6 10 0

Since the introduction, in 1882, of steel crushing rolls in the place of stamps, in the Bertrand Mill, Nevada, attention has been concentrated on this method of reducing ore to a fine state. This looked somewhat like a return to obsolete methods, for the modern stamp battery was currently supposed to be immensely superior to the old Cornish rolls. But when mechanical genius was directed to improving upon the Cornish rolls, it soon became evident that there was very much to be said in their favour. The recognised leading pattern of modern rolls is that introduced by Stephen R. Krom, of New York, and made in England by Bowes Scott & Western, Broadway Chambers, Westminster.

The machine, as constructed by them, consists of two sets of cast-iron rolls a on steel axes b carrying steel tyres c (see Figs. 503, 504). The rolls vary from 26 in. to 30 in. in diameter, including the steel tyre. The tyres are made of the best open-hearth steel, and are $2\frac{1}{4}$ in. thick on the 26-in. rolls, and $2\frac{3}{4}$ in. thick on the 30 in. rolls, and can be worn down to $\frac{1}{2}$ in. with safety. They have been worn as low as ¼ in., but there is danger that they will then spring and become loose. The pillar block d of one of these rolls is firmly bolted to bedplate e by nuts f. The second roll is set in a swinging pillar block, fixed in two strong reinforced cranks g, which rotate in a journal h set

FIG. 503.

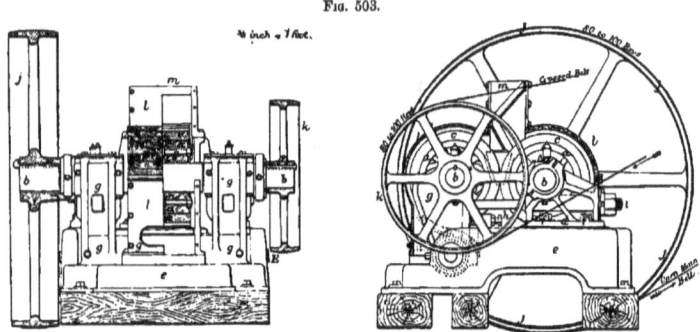

Krom's System of Crushing Rolls.

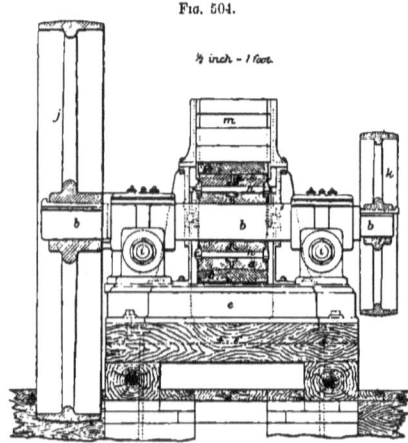

FIG. 504.

Krom's System of Crushing Rolls.

in the cast-iron frame e. These double cranks are of exactly the same width as the roll. The shaft which supports them is 11 in. in diameter. Their journals are therefore pivoted with those of the rolls, and the two are thus forced to move together, so that this roll and shaft are always parallel to the other roll. The distance between the two rolls is regulated by screws h, one on each side, with jam nuts to prevent any motion after they are once set. The rolls are held to their position by two heavy bolts i, also with jam nuts, so that the position being once fixed, the distance of the surfaces cannot be changed by any action of the machine but the wear of the rolls. Formerly the distance was fixed by having the pillar boxes slide on the bedplate, but this was found not to work well, as they constantly became loose and thus pounded on the bed. It was also found almost impossible to get the surfaces of the two rolls parallel, and they wore unevenly. To the axis of the fixed roll a large driving wheel j, 7 ft. in diameter and 15 in. wide, is keyed, and to the movable one a smaller wheel k, 50 in. in diameter and 8 in. wide, is keyed. The rolls are covered with housing l, to which an exhaust fan is attached, so that no dust escapes into the air; this protects the journals of the machine from abrasion. This housing carries a feed box m, arranged with a series of inclines so as to spread the ore in a continuous and even sheet between the surfaces of the rolls. Magnets are provided in the feed box to catch any pieces of steel which may be in the ore from broken tools, and which might indent the surface of the rolls.

The whole machine is self-contained and very compact, a pair of 26-in. rolls occupying a ground space of only 7 ft. by 7½ ft. The tyres, which for the 26-in. roll weigh 816 lb. each, are held in place by two cast-iron heads, which are slightly conical in shape. One of these is shrunk on to the shaft and permanently fixed, the other is slit on one side and slips on to it. Both the heads are so placed on the shaft that the smaller diameter will be towards the centre. The steel tyre is turned out on the inside to correspond to this, so that it can be easily slipped over the permanent head and the loose one brought up to it. The two are securely fastened together by bolts n, so that when the movable head is drawn up to the permanent one, the slit in it closes, and makes it perfectly tight on the axle. It has been found expensive, on account of turning the inside of the tyre, to have it all made in one piece, so that it is sometimes cast in two pieces, which are turned to fit together in the centre. The bolts n draw them together perfectly tight. It was thought at first that the steel would wear unevenly on the two parts of the tyres thus joined, but it does not, and the economy and convenience realised by the construction is great. When the tyres are worn so thin as to be in danger of springing, they can be very easily removed and others substituted for them. To prevent delay, it is a good plan in every mill to have extra hubs with the tyres on, so that there may be no delay when the tyres are worn out. The worn-out tyres can then be replaced at leisure. In order to provide for abrasion on the sides of the machine, cheek pieces of composition metal are provided, which can be easily taken out and replaced when too much worn. The putting on of a new set of tyres requires but a short time, so that the delay when the rolls are to be replaced need be only very short, more especially if extra sets of rolls are kept on hand with the tyres already shrunk on the hubs, so that they may be adapted at once. It has been theoretically objected against the use of rolls, not only that they might wear unevenly, but that any stoppage would involve suspension of the whole work, while the stopping of the stamps would involve the suspension only of that particular head or battery, which could be repaired at leisure. This objection, however, in practice falls to the ground, for it has been found that repairs to the rolls can be very quickly made, and that stoppages are much less frequent.

Krom's rolls are generally arranged in double sets, one of which receives the coarse ore from the crusher and delivers it on to a screen of a determined size. That portion which does not pass through the screen is carried back to the first rolls. What passes the screen falls upon the second pair and from there to another screen. Part passes this last screen and goes on for treatment; part remains and goes back to the second rolls again. Whenever the screens show any wear in the rolls, the latter are stopped and reset. When very fine crushing is required, three pairs of rolls can be used, but generally two only are required. For quantities of 10 to 20 tons a day, a single roll can be used.

Krom's rolls are made of three sizes, with varying lengths and diameters. These are 22 in., 26 in., and 30 in. in diameter, and 14 in., 15 in., and 16 in. long. It has been suggested that they should be made longer, but there is no practical advantage in making them so, while the difficulty of fitting and making the tyres would be very much increased. In considering their efficiency and capacity, it must be borne in mind that the whole of the surface of the faces of the rolls is fitted to act on the ore, since it is even and parallel, and wears so. The ore escapes from the rolls by gravity as soon as it is crushed fine enough to fall through the space between them. If the feeding is automatic, or even when it is not, no clogging of the ore is possible. In order to form a comparison between ores and stamp rolls, the surfaces actually in contact at any given moment must be compared. A 30-stamp mill, with stamps weighing 750 lb., dropping 90 times a minute, with shoes and dies 8 in. in diameter, have, in round numbers, 50 sq. in. of surface on each stamp, or 50 × 90 × 30 = 135,000 sq. in. of surface acting on the ore every minute. It may be assumed that two sets of 22-in. rolls will have an average diameter of 21 in. If these rolls make 80 revolutions per minute, and are 14 in. on the face, they will have a contact surface of 141,120 sq. in. per minute, or a little more than a 30-stamp mill. If the number of revolutions is increased to 100, the capacity is brought up to 171,000 sq. in. of surface, a little less than a 38-stamp mill, whose capacity is 176,400 sq. in. of surface. If the diameter of the rolls is increased to 26 in. with 15 in. of face, the average diameter may be considered as 24 in. If they make 80 revolutions a minute, the face surface will be 162,800 sq. in., equal to the surface of a 36-stamp mill. If the number of revolutions of this same roll are increased to 100, the surface capacity is the same as that of a 48-stamp mill. If the diameter is increased to 30 in., taking the average as 28 in. and the length of face 16 in., and the number of revolutions 80, the surface capacity will be equal to 47 stamps, and at 100 revolutions to 60 stamps. If, in each of these cases, where two sets of rolls have been considered, a third set is added, the capacity will be increased 50 per cent., and if four sets are used, doubled. These calculations are made on the supposition that the surfaces acting are of equal efficiency in both rolls and stamps, and are, for that reason, more favourable to the stamps than actual practice shows. With rolls properly constructed the pressure is constant at each instant of time, while with stamps, on account of the varying height of the ore in the mortar, the cushioning of the stamp, and the fact that the stamp must not only crush, but also force the ore through the screens, it never can be constant. Actual practice in mills has shown that the output of rolls is always larger than that given by these calculations. That the work can be done as well with small rolls in a compact space, as with stamps which occupy a much larger one, is shown by the experience of mills where they are used; that they can crush as fine, is shown by the fact that rolls are now being used to crush after the stamps.

Power was formerly applied to rolls by gearing, which was constantly liable to get out of order, and limited the speed at which it was possible to run. Pulleys are now used, a belt passing over both large and small wheel. This allows of attaining a speed of 80 to 100 revolutions a minute, or higher if desirable, which was entirely impracticable with gear. The capacity of the rolls is thus increased, while the wear is confined almost entirely to the crushing faces. It does away with the noise of the gearing, and reduces the risk of breakage almost to nothing. Both rolls, when the machine is in operation, travel with the same speed of surface, but the small pulley is so speeded that when there is no ore between the rolls, it travels one or two revolutions per minute faster than the large one. The reasons why this double system was adopted is that one of the rolls being movable and the other stationary, it would not be a good construction to place a large pulley on a movable pillar block. When ore is between the rolls, a single driven roll will cause the other to revolve, so that if most of the power to do the work is applied to the large wheel, only a small part of it will be necessary to ensure that the other roll will always bite, when fed with ore, and be kept in motion when the ore at any instant fails. It will be seen also that most of the strain of the rolls is taken upon the bolts i, and that, by the use of the two bolts i and h with the nuts f, all of which are provided with double nuts, no pounding action is possible. The defects of almost all the improved rolls has been that they were not provided for this pounding movement, and that they have been too light in weight to do the duty required of them. Both these defects have been remedied in this machine. The surprisingly small number of parts, all of which are easily accessible, their simplicity, as well as the small space occupied for such a very large output, and the possibility of adjusting the distance between the rolls at any time and in a very few minutes, makes this machine, which is entirely self-contained, not only most efficient but very economical.

It has recently been shown by Professor Egleston that two sets of 26-in. rolls at the Bertrand Mill easily crush 150 tons of hard ore in 24 hours, so as to pass a 16 screen. The Mount Cory Mill has crushed 50 tons in the same time through a 30-mesh screen. The usual capacity for the best stamps on the same kind of ore is 2 tons per stamp, which make the rolls equal to a 50-stamp mill. At the Bertrand Mill, 9000 tons of ore were crushed without costing a dollar for repairs, with the expenditure of less than one-half the power required to do the same work in a stamp mill. At this mill 15,000 tons of ore passed the fine crushing rolls before new tyres were necessary. After crushing 20,000 tons, the coarse crushing tyres were still good for two months' wear. It is considered that each set of rolls with a set of repair linings of composition metal, will be capable of crushing 20,000 tons; the only expense for repairs being the tyres and the cheek pieces. There is no absolute limit to the fineness of the crushing.

The quantity of material which can be pulverised in any given time will depend on the amount of actual crushing surface which is in contact at each moment of time. It is evident that that machine will be the most efficient which presents the greatest amount of crushing surface at every instant of time. The cost depends on the amount of power required to do the work, and on the expense of repairing the wear and tear of the machine, and keeping it up to its greatest efficiency. The cost will be greatest when any part of the power is badly expended or wasted, and when the number of parts to the machine is so large as to make the construction or repairs expensive, and when for any reason skilled labour is difficult to obtain. For comparison, suppose the rolls have been used and slightly worn to a diameter which will represent the average wear of a 26-in. roll, and that the rolls are 15 in. long, running 100 revolutions per minute. The crushing surfaces of two such sets of rolls in actual contact at every instant of time will be equal to that of a 50-stamp mill with shoes and dies 8 in. in diameter, making 90 drops per minute. The action of the rolls is continuous. They have nothing to do and do nothing but crush the ore which falls between them. If during any instant of time there is no ore, the power expended by the machine is stored up in the flywheels. The power of the stamp is spent in first crushing the ore and then forcing it out of the screens. As the whole of it cannot be forced out at every blow, a considerable quantity of it will be acted on several times without passing the screen, and without being reduced any finer, and this ore not only consumes the force of the blow of the stamp, but cushions it, and thus still further reduces its effectiveness. The weight of the stamp and the height to which it must be lifted are constant quantities, hence the power required to move it will be constant, while the effectiveness of the blow will depend on the size of the ore from the crusher, the quantity of it in the mortar, the size of the screen, and the effectiveness of each blow to force the crushed material out, which takes as much time as it does to crush it. No careful experiments have as yet been made to find the actual amount of power exerted in a stamp mill to crush the ore only, but the stamp has so much else to do that it is safe to say that it does not exceed 50 per cent. of the power actually expended. With the rolls held rigidly together the action is constant at any given time. The power, if not effective, is stored, and no piece of ore, not reduced, remains between the rolls; it falls by gravity, and is returned to the roll, if too large, by an elevator, so that no cushioning is possible. Stamps have a maximum of velocity which is very soon reached, beyond which the fall is reduced. It may be carried so far that the stamp head may be kept suspended in the air, and consequently do no duty whatever; while the increased speed of rolls always gives increased output, which beyond a certain limit is gained at the cost of the expenditure of power, but there is no cessation of action possible, as in the stamp. If we make a comparison of parts required in the construction of each machine, we shall find that the two machines of the same capacity will have the following working parts. For the rolls—3 shafts; 4 journals; 4 pillar blocks for journals; 2 crushing rolls; 2 side plates as repair pieces; 12 wearing surfaces for each two sets of rolls; total, 27 parts.

All of these parts are large, made so strong that they are not liable to break, and the wear is not rapid. They can all be easily replaced, and all are easily accessible at any time. The stamp mill has 15 journals; 5 shafts, 10 stamps to a shaft; 5 cam shafts; 15 bearing boxes for the journals; 50 stems; 100 guide-boxes for stems; 50 stamp heads (bosses); 50 shoes; 50 dies; 50 cams with keys; 50 tappets; total, 440 parts.

All these parts are subjected to rapid wear, many of them are liable to break, most of them can be easily replaced, but the greater part of them are small and require much time to watch them, and if left to take care of themselves may cause much damage and stoppage in the machine. While most of them are accessible to make repairs, one or more of the stamps, or even batteries, must be hung up, and while any single repair does not consume much time, the multiplicity of them in the course of a year, does. The screens have not been considered, as they are common to both machines, although as they receive part of the blow of the stamp which thrusts the crushed ore against them with considerable force, they must wear more rapidly than in the rolls, where the ore falls on them only by gravity and for a very short distance. This simple comparison of the parts speaks for itself.

It has been found by experience that the condition of crushed ore from rolls makes the pulp much better suited for lixiviation than that from stamps, as the particles of ore coming from rolls are not only more uniform in size, but contain much less dust, which interferes with the rapid and effective passage of the solutions used. It has been found, too, that in chloridising, great fineness of the ore is entirely unnecessary, except in the rare case of the precious metal being very finely distributed through the gangue, so that with the exception of this single case, rolls seem to give the most favourable results. This is of importance in view of the favour which is being extended to chloridising processes.

The wear of rolls is almost exclusively confined to the steel tyres and cheek pieces; that of stamps to the great number of parts already enumerated. Something over 80 per cent. of the steel of rolls can be safely used in crushing before it is necessary to put in a new roll. Generally it is not safe to run more than 50 to 60 per cent. of the shoes and dies before replacing them. On account of the number and complicated nature of the parts of the battery, skilled labour has to be used to a considerable extent, while only a nominal amount of it has to be used with the rolls. Every one who has ever seen a mill, knows how often the stamps have to be hung up, while rolls, if they are made strong enough at the outset, require little or no repair except the change of tyres, which is rapidly done and occurs at long intervals.

REDUCING MACHINERY. 325

Centrifugal Roller Mill.—The centrifugal roller quartz mill, designed by F. A. Huntington, of San Francisco, and sold by Calvert, Cornes & Harris, 76, Cannon Street, London, is shown in Fig. 505. It is designed on an entirely different principle from its predecessor, the stamp battery. The illustration represents a 5-ft. mill. At the upper part of the figure will be noticed a circular horizontal frame of cast iron, keyed at its centre to a vertical spindle G. This frame or "disc

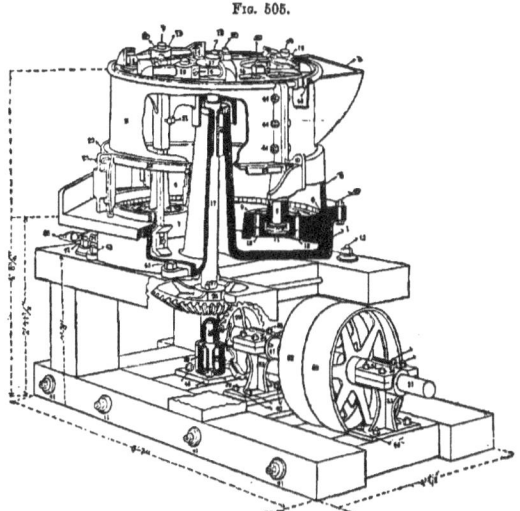

Fig. 505.

Huntington's Centrifugal Roller Mill.

driver" carries the rollers and scrapers, and has shrunk on its periphery a wrought-iron ring B. The method of suspending the rollers and scrapers will be best understood by the plan of the disc in Fig. 506. Round the edge of the ring four pairs of pockets P, are cast with it and project above it. Into each pair of pockets are set journals to form bearings for the ends of the yokes, one of which is shown at Y. In Fig. 507, one of the rollers and yoke is shown in sectional elevation. It will be seen that the journals of the bearings at each end of the yoke are fixed into the pockets. This allows the rollers to swing in a radial direction with reference to the vertical spindle G, and, when in action, centrifugal force carries them hard against the steel ring die fitted on the inside of the rim. Each roller consists of a steel ring R (Fig. 506), 13 in. bore, and 1¼ in. thick. An internal flange on the ring forms a seat on which the hub C rests. There is ½ in. space all round between the hub and the steel ring, into which are driven the wood wedges W for securely fixing the ring to the hub. In addition to this, the ring is supported at the flange by two claw bolts passed through the hub. A hollow lubricating chamber H is left in the hub C, and a slot in the side of the roller shaft S forms a channel of communication from the chamber to the top, where oil may be added. This hole is kept plugged to prevent the entrance of grit. The lubricating chamber is hermetically sealed by a faced plate J, and rubber joint, rendering impossible any ingress of oil into the pan or amongst the amalgam. The roller shaft S is of steel and firmly keyed to the yoke at the top: it does not therefore rotate on its own axis, but is the centre round which the hub and rollers rotate. The weight of the roller is supported by the circular steel head at the lower end of the shaft. The rotation of the rollers on their own axes is caused by the centrifugal pressure

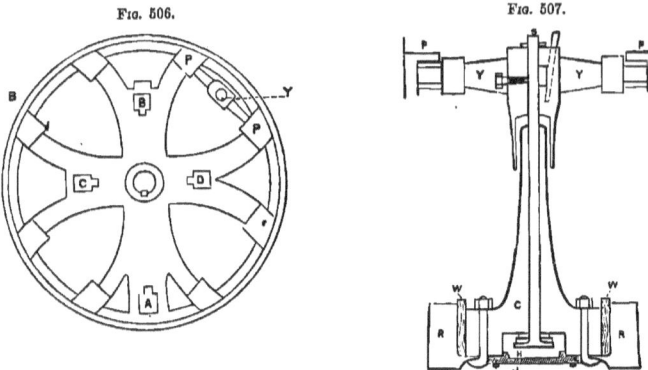

Details of Huntington Mill.

against the ring die when the mill is in action. A clearance of 1 in. is left between the lower edge of the rollers and the bottom of the pan, so that the former pass freely over the amalgam without coming in contact with it, or in any way flouring it, whilst at the same time there is sufficient agitation to render the amalgamation perfect. Referring again to Fig. 507, the wooden supports for the scrapers are driven and keyed into the holes A B C D, which are at different distances from the centre. The iron scraper fixed at the lower end of the wood carrier can be seen in Fig. 506 at F. By fitting these at different distances from the centre, the action of the scrapers tends to keep the quartz against the steel ring die. The outside shell enclosing the mill is a casting, the upper part being in two pieces, bolted together by vertical flanges. The whole of this shell is again bolted by a horizontal flange at its lower end to another casting of ring shape. This latter contains, wedged inside it, the steel ring die, against which the rollers bear. To render this part of the machine in every respect equal to the wear and strain, and prevent fracture of the casting when the ring is wedged in, a wrought-iron band or ring is shrunk on externally. The action of the mill will now be readily understood. Power being applied to the driving pulley, the countershaft is driven and the horizontal motion transmitted to a vertical motion by bevel wheels under the framework of the mill. The central vertical spindle is therefore rotated, carrying with it the disc driver, rollers, and scrapers. The quartz or material to be treated, being fed into the pan with water, at the hopper A, is immediately thrown against the steel ring, where the rollers come in contact with it and pulverise it. The precious metal in a free state at once amalgamates with the mercury run in the bottom of the pan (about 40 to 60 lb. of mercury are required in a 5-ft. mill), and the pulverised quartz, as soon as fine enough, is discharged through gratings extending along half the circumference of the mill. These gratings or perforated plates are made in various degrees of fineness to suit any required fineness of discharge. The delivery is both rapid and perfect, producing the fine grit without slime, which is then conducted to the distributing boxes, and passes over the concentrators in the usual manner. About 75 per cent. of the precious metal is retained in the amalgam. For the purpose of comparing the first cost, saving in transit, and labour of installation at the mine, let us take a ten stamp battery, whose output per hour is somewhat less than a 5-ft. Huntington mill. The ten stamp battery of 850 lb. heads, together with its boxes, frames, foundations, driving gear, &c., when erected, will cost 650*l*. and weigh from 12 to 14 tons, in addition to which the cost of transit alone, in such districts as the Transvaal for instance, will cost nearly as much as the mill itself. Now a 5-ft. Huntington mill weighs only $5\frac{1}{4}$ tons, including frames, gear, &c., and the cost, including foundations and fixing, is but 350*l*. A most important matter is the time taken in preparing the foundations and fixing the mill at the mine. A Huntington mill can be completely fixed by two men in 15 working hours, whereas under the most favourable circumstances the ten stamp battery would require at least four men for 30 days, and frequently take many months before it is got into action. Even when started, it is often found that the foundations are not satisfactory, causing further delay. Again, as regards the commercial efficiency of the two mills, the Huntington stands a long way ahead, for while working at a higher rate of output than the stamp battery (in some cases from 10 to 50 per cent. higher) it absorbs considerably less than half the driving power. The cost for renewals of rings and rollers in the Huntington mill is about equal to that for dies and shoes in the stamp battery, but the latter has frequently against it heavy items of expense for breakages and renewal of wearing parts in the stamps, which are entirely avoided in the centrifugal mill.

Fig. 508 illustrates a mill arranged on the Huntington system by Calvert, Cornes & Harris, 76, Cannon Street, London.

A new form of edge-runner mill is shown in Fig. 509. It has the shape of a Chilian mill of medium size, with heavy rollers; but three conical tooth wheels are fixed to it; one, in horizontal position, is fixed to the inner part of the basin, through which the spindle goes, teeth upward; the other two are in vertical position, one fixed to the inner side of each roller. The three wheels are in gear, and the number of their teeth are so enumerated that when the spindle turns, and thereby the rollers revolve, the latter (forced to remain in gear) must obtain a higher speed than in case they were rolling round without being influenced by the said set of wheels. The motion of the rollers is that of a carriage wheel in a hole, the progress of which is supported by turning the wheel by the spokes round its axle. If it is considered that the roller transmits the pressure resulting from its heavier weight to a line representing the breadth of the roller (where it touches the false bottom of the basin), and that in consequence of its higher rotating speed, squeezes the grains between bottom and roller-periphery, it is not surprising that no mineral can resist the pulverising action of the mill, and that it is not necessary to have the same grain passing more than once between roller and bottom, whilst this is unavoidable with common Chilian mills. One rotation of the new mill may in its effect be considered equivalent to hundreds of rotations of some other grinders, and, although the new mill requires (compared with others) a large amount of driving power, it is more economic in this respect, because almost the whole power imparted on it is

328   MINING AND ORE-DRESSING MACHINERY.

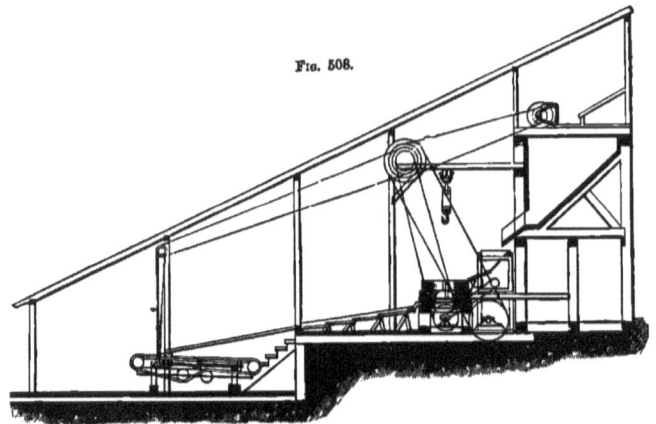

Fig. 508.

Arrangement of Mill on Huntington System.

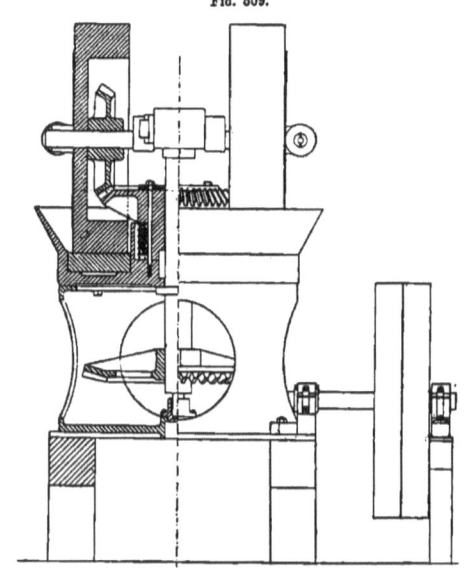

Fig. 509.

Edge-runner Mill.

consumed in useful work. In consequence of the acceleration of the rollers, they have an excellent grip; in fact, they suck the ground between them and the bottom. The mill can be used with or without water, but the latter is preferable, as the water carries away the ground sludge, and the mill can work without interruption. If the stuff is not too clayey, the water can be used over and over again. The difficulty of keeping the three cog-wheels in gear has been successfully overcome. The horizontal cog-wheel is not absolutely fixed to the inner part of the basin, but allowed to glide on it vertically; a set of rubber rings separated by flat iron rings presses the said cog-wheel upwards against the two vertical ones. When the roller rings wear out, the horizontal cog-wheel can be lowered by three screws. The common linch-washer is replaced by a horizontal strong cast-iron arm, which carries on both ends friction-rollers pressing against the roller near its periphery. Thereby the tendency of the rollers to run off in the direction of the tangent, and to oscillate, is remedied.

The Globe Mill Company, Limited, of Blomfield House, Old Broad Street, London, E.C., have introduced a crushing mill which is possessed of valuable qualities for the treatment of ores, gold or silver, the guaranteed saving being 30 to 50 per cent., as compared with the older system. The mill will pulverise to such a degree of fineness, by wet or dry process, that 99 per cent. of the material will pass a 55-mesh, and 75 per cent. a 100-mesh sieve, equal to 10,000 holes to the square inch, and it can be supplied to turn out any given quantity of material of uniform fineness from 10 to 100 tons per day of 24 hours. It is shown in Fig. 510.

Fig. 510.

Globe Mill.

These mills, which are known in the market as "Globe" mills, have very much in their favour besides the rapidity with which they grind, and the evenness and fineness of the grinding. The simplicity of their construction, and the small number of wearing parts, will go a long way in bringing them into general use. The mill proper really only consists of a spherical ball revolving in a vertical circular path, which ball is driven or thrown round the path by the centrifugal force imparted to it by two flexible discs fixed on a horizontal shaft running through the centre of the mill. These parts of the mill, namely, the path, ball, and discs, are made of a special hard steel, much harder than the best tool steel.

The mill requires 20 H.P. to pulverise a quantity equal to the work of a 30-stamp battery. The amount of water required, for wet process, per ton of ore, for the pulveriser, is also claimed to be less than one-half that required for a Californian stamp mill, a very important consideration, when we understand the great difficulty in obtaining water which frequently exists in mining regions. We may add, in conclusion, that the "Globe" series of mills is capable of reducing practically any non-fibrous material to any required degree of fineness.

2 U

330   MINING AND ORE-DRESSING MACHINERY.

Jordan's Centrifugal Mill.—One of the newest, and certainly by far the most important, of recent innovations in reducing machinery, is the centrifugal mill introduced by Rowland Jordan, Esq., C.E., and made by T. B. Jordan & Son, 15, George Street, Mansion House, London. The plant is

Fig. 511.

Jordan's Reducer.

Fig. 512.

Jordan's Amalgamator.

especially designed for dealing with gold ores, and consists of two distinct portions, the reducer and the amalgamator. The reducer is shown in Fig. 511, the amalgamator in Fig. 512, and the general arrangement of the whole in Fig. 513.

The following reports, made on trial runs of machinery, by the present author, and other practical gold-miners, are reproduced below, with permission.

The author's report says:—

"The plant employed is remarkable for its simplicity, and but little description is necessary. The usual stone-breaker begins the process. This is followed by a revolving pan, set at an angle, and carrying three massive balls of white iron, which work in a suitably shaped bed, also of white iron, round the greatest circumference of the pan. The ore and water are fed automatically into the bed of the pan, and, by the rotary motion of the latter, are conveyed under the rapidly revolving balls, whereby the comminution of the ore is effected. The inner half of the floor of the pan rises as a shallow dome surrounding the central shaft, and is fitted with movable frames carrying wire screens of any required mesh. The feeds of ore and water, and the inclination of the screens, are so adjusted that as the ore is reduced to a sufficient degree of fineness, it is washed over the screens, and passed away into a launder for conveyance to the amalgamator.

"A comparison of this machine with the most approved form of stamp battery reveals some highly important facts, which may be summarised thus:—

"Cost.—A patent pan, equal in efficiency to a ten-stamp battery, is considerably less in first cost. To this saving must be added the greatly reduced cost of transport to the mine, as the pan weighs less than half a battery. A third economy effected is in the erection—eminently simple and expeditious in the case of the pan, but a long and expensive operation with stamps.

Fig. 513.

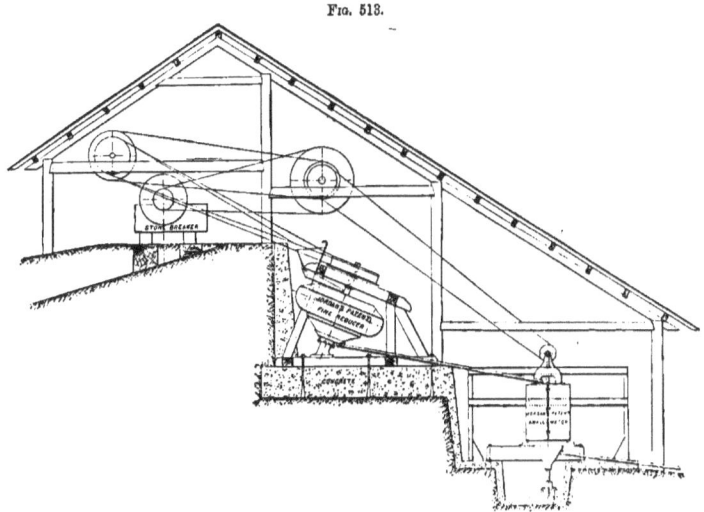

Arrangement of Jordan's Reducer and Amalgamator.

"Generally, it may be stated that the total cost of the patent plant erected at the mines will not exceed one half that of a stamp battery.

"Efficiency.—The simple principle of the stamp battery (that of a falling hammer), which proves such an attraction to its advocates, carries with it several disadvantages. Not the least is its intermittent action. The time during which actual work is being done bears but a small proportion to the time the battery is in motion—each hammer must be lifted to a greater or less height, and though the fall produces an effective blow, it is only at the moment of impact between stamp and ore that work is really done, and then the work is not of the most useful kind, as part of the blow always falls upon material which is already in a sufficiently fine state, but has been unable to escape.

"The patent pan, on the other hand, has a *continuous* action on an *evenly distributed* and *constantly changing* layer of ore, and thus accomplishes a greater amount of work. In fact, it disposes of a rapid feed at an astonishing rate.

"Trials with various ores showed a power of reducing 20 to 25 tons per 24 hours to a size that would pass an 80 *mesh screen*. Now, few 10 stamp batteries can do more than 15 to 20 tons, even when the mesh of the screens is as low as 30. This is one of the great faults of the stamp battery—fine grinding is impossible at anything like a reasonable rate and cost. The outlet for the stamped stuff is so limited, and so ill-adapted, that it takes almost as long to pass pulp through a battery as to crush coarse stuff and pass it through.

"With a patent pan this drawback is obviated. The screen area is much greater in proportion, and the screens are set at a very low angle (almost horizontal). These features, combined with the wash produced by the rotation of the mill and inclination of the jet of feed water, increase the facilities of outlet in a most remarkable degree.

"Another advantage which, in many cases, would be of primary importance, is that the *consumption of water is only about half* the amount required by a battery. In crushing of "Edwin Bray" ore, which I witnessed, the water-feed was only a one-inch pipe (where a 10-stamp battery would have had a three-inch), and the ore, fed from a stone-breaker, was passed *through an 80-mesh screen* at a surprising rate, not less *than 20 tons per* 24 *hours*. Such a result speaks for itself; moreover, the power used to effect this is about half that needed with stamps; while the labour is virtually *nil*—with automatic feeds one man could watch a dozen pans; and the prodigal use of mercury, so common with stamp batteries, is dispensed with.

"Wear and Repairs.—Here again the patent pan has advantages over the stamp battery. There are only *two wearing parts* in the pan, *i. e.*, the balls and their bed. An examination of the latter, after six months' intermittent use, shows that the two surfaces wear in such a way as to maintain their proper relation to each other. When worn too thin, the bed can be removed and replaced by a new one with very little trouble. The screens are much less exposed to injury and wear, and are readily detached and renewed, any desired mesh being adopted. The frequent minor stoppages inherent to stamps, with their numerous working parts, are quite avoided.

"The only other apparatus employed in the process is a novel form of amalgamator. The stream of pulp, without any addition of water, flows into the hopper of this amalgamator.

"This machine consists of a series of shallow dishes, attached one below another to a central revolving shaft, and enclosed in a fixed circular casing, which is kept under lock and key. Secured to the inner side of the casing, and alternating with the dishes, are slightly inclined shelves, also amalgamated. The pulp fed into the amalgamator enters the first dish, in which it is revolved, until impelled by the centrifugal motion over the edge of a dish. It then falls on one of the shelves and is thus conveyed to the centre of the second dish, there to undergo similar treatment. This is repeated to the end of the series, where the tailings escape. The free gold and silver contained in the pulp are completely arrested by the amalgamated dishes and shelves.

"The very high efficiency of this patent amalgamator is apparently due to several causes.

"In the first place the ore is reduced in the pan to such a degree of fineness that all precious metal not actually in chemical combination is set free. Then the shape of the dishes, and the manner and speed of rotation, all tend to ensure intimate fractional contact between the atoms of precious metal and the amalgamated plates, so that even the finest particles of float gold are retained and the gentle attrition of this flowing pulp maintains the amalgamated surfaces in a constantly bright and favourable condition. Any amalgam which may become detached is caught in a well at the bottom of the machine, together with such mercury as may have escaped from the dishes, when it is thought desirable to employ it.

"To such a high state of perfection is amalgamation carried in this apparatus, that it is capable of extracting nearly all the gold even from pyritic ores, without calcination or any other treatment. In making this statement I expect to incur the condemnation of theorists : but a study of the table of actual results achieved, recorded below, will substantiate my opinion. And a more startling piece of evidence, which may be deduced from this same table of results, is that in all cases, the patent process is far cheaper, and generally more advantageous than chlorination itself; for whereas chlorination alone costs between 2l. and 4l. per ton, and seldom less than 3l. for a return of 90 to 94 per per cent. of the assay value, the total cost by the patent process is only about 5s. to 10s. per ton, and the yield is often equal and sometimes superior to that from chlorination.

"The subjoined figures, though quoted on the authority of the Inventor of the process, are in my opinion reliable. To verify them, I made a trial with 'Edwin Bray' ore, its refractory character affording a crucial test of the process. Three samples before treatment gave an average of 1-oz. 16-dwt. 2-grs. of gold, and the average gold contents of three samples of tailings, taken as they left the amalgamator, was as nearly as possible 3-dwts., or 91·686 per cent. extracted.

No.	Description of Ore Treated.	Gold contained in Raw Ore before Treatment.			Gold contained in Tailings after Treatment.		
		oz.	dwt.	gr.	oz.	dwt.	gr.
1	Welsh ore : sulphides of iron, copper, lead, &c. ..	0	1	12	0	0	13
2	Tailings from Transvaal	0	2	7	0	0	23
3	"Indian Consolidated" ore : refractory sulphides	0	1	23	0	0	12
4	Australian "Black Jack" : arsenical sulphides, very complex ..	1	9	9	0	10	16
5	Tailings from "Johnson and Matthey"	1	6	3	0	1	23
6	"Devala Moyar" ore : complex arsenical pyrites	0	19	22	0	6	11
7	"Edwin Bray" ore : pyritic	2	0	22	0	3	0

"From the foregoing results of my examination of the patent process, I am of opinion that, notwithstanding the enormous capital represented by existing stamp batteries, the patent pan possesses so many advantages that it will force its way into use and displace them, whilst for new mills, extensions and renewals, it will be adopted without hesitation.

"As to the patent amalgamator, there is not a mill in existence where it will not be found advantageous, and where it will not surely supplant the blanket tables and miscellaneous appliances now in use for saving gold that would otherwise be lost.

"Another direction in which it will find highly profitable employment will be in re-working tailings containing much gold that has been passed over by the less efficient processes hitherto employed.

"C. G. WARNFORD LOCK."

"LONDON, November 1889."

Report by Mr. Michael Thomas, for thirty-two years miner, manager, director, and proprietor of gold mines in Victoria, upon the above-mentioned mill and amalgamator :—

"The subject of reducing and amalgamating auriferous ores receives, perhaps, more attention at the present time, than at any previous period of the world's history; and although these operations are carried on only to a limited extent in England itself, yet the development of the African and other gold fields seems to have brought the subject nearer home to the minds of the people; so that it is only natural, in common with many other things, that important innovations should be proposed upon the old methods of carrying out these processes. Progress in mechanics and chemistry being a marked feature of this age, why should it be thought presumptuous to attempt to supersede the old time-honoured stamps, and to dispense with the almost equally well-known ripple table.

"Improvements hitherto have been so few and far between that it is rather astonishing to find a Company attempting at once to introduce two fundamental changes in these matters, and yet this is what the above-named corporation are aiming at in their new machines, and, if I am not very much mistaken, with an excellent prospect of success.

"My experience in reference to these matters has been gained in Australia, and in that part of it known as the Bendigo Goldfield, where stamping is almost the only method of crushing ores, and the ripple table, amalgamating barrel, and Berdan, the usual methods afterwards adopted of extracting the gold, resort having to be made to calcining, &c., for the difficult arsenical pyrites. It was therefore with much interest I visited the premises of this Company, and saw their three processes in full operation.

"The first calls for but few remarks, being that of breaking down the rough ores by the stone breaker, which appeared to be of the ordinary kind, and to which, I understand, the corporation make no claim; but when we come to the second, viz. that of reducing the ore to the necessary minuteness for the extraction of the gold, our curiosity is at once aroused, for it is seen at a glance that the new method is essentially different from the old. The milling machine consists of a large cast-iron basin, about seven feet in diameter, having the upper edge curved inwards, thus forming a trough or path for three white iron balls, about five hundred-weight each, to travel in. The basin being tilted at an angle of 30°–40° from the horizon, the balls naturally fall to the lowest part; and when the basin is made to revolve on a central shaft, about fifty times a minute, the balls also revolve, but in an opposite direction, and thus the ore, coming between the bed of the basin and the under surface of the balls, the operation of pulverising goes on. The discharge of the pulp is in the centre of the basin where the grating, in a conical form, surrounds the shaft. This grating or netting is very fine, having eighty apertures to the lineal or six thousand four hundred to the square inch. The particles passing through this are necessarily very minute, sufficiently so to render the work of amalgamation easy and complete. The grating, however, can be of any degree of fineness or coarseness, and can be changed at any time with facility. The centre discharge I deem to be of great importance, as by this means the netting is protected from the violent action of the crushing operation. In stamps it is quite different, for the grating gets the full force of the splash from the falling heads, often resulting in their speedy destruction.

"Assuming that one of these machines is equal in capacity to a ten-head battery, which I understand is the claim of the inventor, and which, given the same mesh, seems to be a just one;

then I do not hesitate to say it takes much less power to drive it than such a battery. This conclusion is derived not from actual competition, but from the slimness of the shafting employed. Had shafting of the same strength been employed to a ten-head stamp battery, it would collapse at the first revolution, whilst in this case it bore the strain without injury of any kind.

"With respect to the weight of the plant in comparison with that of stamps or other crushing appliances, this has a decided advantage; and as weight is generally a criterion of expense both as to first cost as well as that of transportation, I firmly believe this machine will distance all competitors.

"It only remains to dwell on the cost of erection.

"Again, making comparisons with the stamps (for notwithstanding all attempts hitherto made to supersede them, they remain the foremost pulveriser of the present day) I am safe in saying the advantage is altogether on the side of this mill; for whilst the erection of the former is a prolonged, elaborate, and very costly affair, requiring deep and heavy foundations, and ponderous framework, the latter may be erected cheaply, speedily, and securely on ordinary foundations, and by persons of moderate mechanical skill.

"The amalgamator is even more original than the mill, and I believe it is based on the soundest principles. Although all the mechanical and other appliances in and about a gold mine are interesting and more or less important, the amalgamator is the vital part. It is in vain that capital is raised, shafts sunk, and quartz brought to the surface, if, after all, the precious metal, through imperfect treatment, is allowed to escape. It is admitted, however, on all hands that great waste does take place, and therefore any well-devised method to reduce the loss to a minimum is worthy of attention. It has been said that he who makes two blades of grass to grow where only one grew before is a benefactor, and with equal truth it may be said of him who saves two particles of gold where previously only one was retained.

"The question is does the amalgamator, under review, accomplish this? It must be judged by its conditions and results. The form of this machine is a model of compactness, it being so arranged and worked as to present a vast surface of silvered copper sheet to the auriferous pulp in the smallest amount of space. The pulp, with the requisite quantity of water, is distributed over this bright surface by centrifugal force, but the dishes being circular in form and inclining at an angle from the horizon, gravity gives it a tendency towards the centre, it being the deepest part. Between the two forces operating at the same time, the pulp has a beautiful spiral motion until it is eventually thrown over the edge into the next dish below, where the same operation is repeated. There are six of these dishes, over each of which in succession the material has to pass, besides as many corresponding fixed shelves silvered in the same manner, and which in their turn conduct the pulp ejected from one dish into the centre of the next below it; hence it will be seen that every particle of matter is kept in contact with the mercury for a long period, and in such a way as to give the gold every possible chance to adhere thereto.

"The dishes are attached to a perpendicular shaft, and revolve at considerable speed. The whole being enclosed by a strong iron case, giving it the strength and security of a burglar-proof safe.

"After seeing the machines at work and carefully comparing them in the light of a long and varied experience with all other machines with which I am acquainted, I have deliberately come to the following conclusions, viz.:—

"That this mill or pulveriser is the simplest in the world, and that it does its work more effectively and cheaply than any other I have met with in my extensive experience. At the same time it takes less water and requires less power to drive it and less supervision, and I have no doubt, when in practical use at the mine, the results achieved will in every respect justify my conclusions. Then as to wear and tear, I do not see how it could be possibly less than here, for all the work is done on smooth surfaces, and the wearing parts are renewable in the simplest manner.

"Again, it is a safe machine, as none of its parts are likely to meet with sudden damage, as it is very strong, and its action equal and uniform. This is important, especially when used in parts distant from supplies.

"With respect to the amalgamator, I can only repeat what I have already stated. It is the most compact and best arranged I have ever seen, and I must add that the gentle friction of the revolving pulp on the silvered surface ever keeps it bright and in the very best possible condition for arresting the particles of gold.

"Finally, as I am perhaps not well known in England, having been absent for many years, I ought to mention that I have had great experience in gold mining in all its branches, having been during the past 32 years miner, manager and director, as well as proprietor, and at the time I left Victoria in November, 1888, I was chairman of several gold mines. I also hold a certificate of competency for the office of Inspector of Mines and Mining Machinery under the Victorian Government. "M. THOMAS, J.P.

"LONDON, *November* 8, 1889."

Report of Josiah Thomas, Esq., Manager of the Great Dolcoath Mine, Cornwall :—

"I have inspected the mill owned by the above company, and witnessed the pulverising of some hard tin stone and tin sand which I sent to the works for that purpose. I may observe that I have had very large experience in the reduction of tin ores, and in the various mines under my management we are at present reducing to a fine powder about 400 tons of tin stuff per day. This work is chiefly being done by stone breakers and stamps; a small portion, which contains very fine particles of tin (known as tin sand), being further reduced to an almost impalpable powder by pulverisers of various kinds, but none of them at all resembling the above-named mill, which, so far as my experience has gone, is on quite a new principle. An inclined circular cast-iron pan, about 6 feet in diameter, in which are three loose iron balls of about 5 cwt. each, is made to revolve about fifty times per minute, so that the balls are constantly kept rolling in the lowest part of the pan, thus pulverising the ores with which they are brought in contact. The grating through which the pulverised ores are discharged, is in the centre of the pan surrounding the shaft, so that in all probability the wear and tear of the grating would not be nearly so great as in the stamps. In my short visit to the works I had not the means of accurately testing the actual power employed in working the mill, nor the exact amount of stuff that could be pulverised in a given time. I may say, however, that the pulverising of the tin stone and sand were done in a satisfactory manner, and apparently with a very moderate power. I am favourably impressed with the general construction of the mill, and think it will probably be found to be very effective and economical, especially for pulverising sand to a very fine powder. In order to expeditiously reduce very hard tin stone to the required fineness by this means instead of stamps, the pan should, I think, be made larger, and the balls heavier than in the mill I saw at work. "JOSIAH THOMAS,

"*Manager of Dolcoath and other Mines.*

"CAMBORNE, *November* 22, 1889."

REDUCING MACHINERY. 337

The following table is considered a fair comparison between a Stamp Battery installation capable of dealing with a moderately difficult ore and Jordan's Patent Plant. The figures include all necessary machines and accessories in both cases :—

	Cost of Plant.	Weight of Plant.	Freight at 15l. per Ton.	Cost of Erection on a Moderate Site.	Time Occupied in Erection.	Power Required to Drive Plant.	Cost for Treatment per Ton of Ore.	Cost of Complete Plant Erected, say, in Transvaal, without Power.	Labour for Double Shift.	Cost of Wear and Tear and Renewals per Annum.
	£	tons	£	£		h.p.	s.	£		£
Jordan's plant	750	10	150	50	2 to 3 weeks	8 to 10	5 to 10	950	2 skilled mechanics 2 labourers	50
Stamp battery	1399	51½	772	say 275	2 to 3 months	15 to 20	12 to 18	2446	4 skilled mechanics 6 labourers	150 to 200

CHAPTER XIV.

DRESSING MACHINERY.

THE object of dressing ores is to separate the useful from the useless portions, and to sort the valuable minerals from each other. It should be carried out as near the mine as possible, to avoid carriage of worthless material. Water is essential, and the floors should be arranged so that the matters can be moved forward in a measure by their own specific gravity.

The specific gravity of minerals commonly met with in wet dressing floors as follows:—

Gold	12·7–19·3	Proustite	5·5	Chalybite	3·6–3·9
Mercury	13·6	Iron pyrites	4·8–5·2	Calamine	3·3–3·6
Silver	10·5	Fahlerz	5·0–5·1	Fluorspar	3·1
Copper	8·4–9	Purple copper ore	4·9–5·1	Calcite	2·6–3·0
Iron	7·5	Magnetite	4·8–5·2	Felspar	2·5–2·9
Galena	7·5	Stibnite	4·6	Quartz	2·5–2·7
Cassiterite	6·4–7·1	Barytes	4·3–4·7	Gypsum	2·2–2·4
Cinnabar	6·7–8·2	Copper pyrites	4·1–4·3	Coal	1·2–1·5
Mispickel	6·0–6·2	Zincblende	4·1	Lignite	1·2–1·4

Decomposed silver and other ores are difficult to dress, especially if easily powdered, e.g., malachite, argentiferous cerussite, cinnabar, and spangles of native silver. It is difficult to separate zincblende, copper pyrites, iron pyrites, mispickel and barytes from silver ores; wolfram from tin ores; chlorite and epidote from copper ores; and chalybite from copper pyrites and galena.

The following minerals are injurious to:—

Hæmatite, limonite, and chalybite (iron and copper pyrites and apatite).

Cassiterite (iron pyrites, copper pyrites, mispickel, and zincblende; bismuth makes the colour dull; copper makes it brittle).

Lead (arsenic makes it brittle; antimony makes it hard; fluorspar promotes its fusibility; chalybite and barytes are also advantageous).

Zinc (lead spoils it).

Copper (lead must be separated from it if it is to be treated by precipitation process).

Silver (lead and antimony are injurious for the amalgamation process; also talc for chlorination).

Cobalt (for blue paint calcspar, brownspar, manganese spar, hornstone, ferruginous quartz, and galena are injurious; nickel, when predominant, imparts a red tinge; arsenic intensifies the blue colour, and renders it more agreeable).

Magnetite (mica, lime, garnet, augite, and hornblende promote its fusibility).

Auriferous pyrites (for chlorination process, talc is injurious).

When pieces of different minerals of the same size are allowed to settle in water, the heavier particles fall first. When pieces of different minerals of the same weight are subjected to a flow of water sufficient to move them along, the specific lighter mineral having a larger surface exposed is washed away quickest.

It will be convenient to deal with the generalities of dressing machinery first, and then to supplement that with details of the systems adopted for lead, zinc, tin, &c. Metalliferous mining is practically the same in the case of all metals or ores till the treatment of the mineral for market is reached. Then each requires its own special method.

Jiggers.—The process of jigging or hutching is resorted to chiefly in the dressing of minerals in fragments of a comparatively large grain, such as we obtain from the crusher. The charge or ore is placed in a sieve, or in a frame having a bottom of wire gauze or perforated metal plate, where it is subjected to a series of small lifts or jerks in rapid succession from a column of water forced through the perforations; by these means, the specifically lighter earthy fragments are gradually brought to the top, the clean ore being found immediately above the plate. To produce

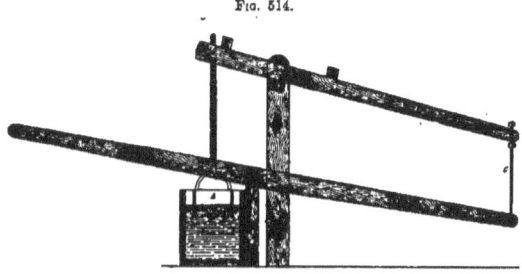

Fig. 514.

Fig. 515.

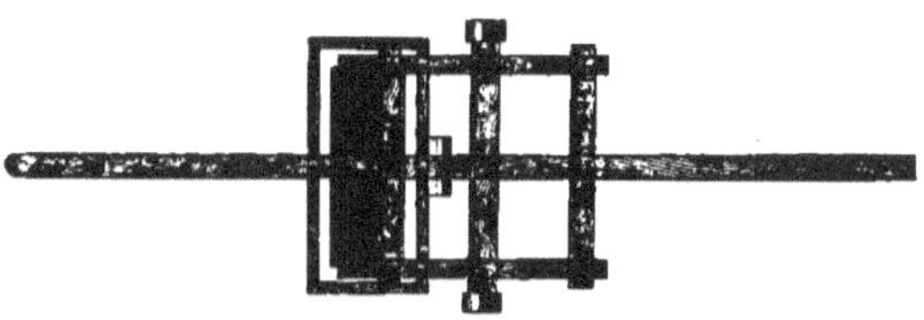

Hand-lever Jigger.

the jigging motion, either the sieve is jerked up and down in a cistern of water, or the water is forced up through the sieve by means of a piston. The process is often an intermittent one, the jigging motion being suspended during the time of charging and emptying the sieves. But it may be made continuous by the addition of feeding and discharging apparatus.

The common hand-lever jigging machine used in Cornwall is shown in Figs. 514 and 515. The

2 x 2

sieve is rectangular; to each of the short sides a vertical iron bar is attached. These bars are perforated at their upper ends with three long holes, through which a pair of bolts are passed linking them to the two parallel arms of an oscillating frame. By altering the holes through which the suspension bolts pass the sieve is made to hang at a greater or less depth in the rectangular water cistern or hutch in which it is worked. The suspension frame is an unequal-armed lever, the sieve is attached to the shorter side, while the longer arm is terminated by a slotted part, in which works a T-headed fixed link or connecting rod attached to the shorter arm of a second lever placed below it. The motive power is supplied by a boy, who jerks the longer arm of the second lever, moving it through a height of 48 in., while the sieve is only moved through 8 in. Clean water is introduced through a square pipe on one side of the cistern to replace the muddy waste which is carried off through a similar pipe fitted with discharging apertures at different depths on the opposite side. The sieve is emptied by scraping out the contents with an iron scraper or limp; they are usually classified into three parts, the uppermost being thrown away; the middle, containing mixed ore with earthy matter, requires a further treatment, while the bottom is clean ore fit for sale. The hutch work or fine stuff passing through the sieve collects in the cistern, and is subsequently treated on the round buddle or some other slime-washing machine. The sieve shown is 57 in. long and 24 in. broad, and 10 in. deep, the hutch measuring 90 in. long, 43 in. broad, and 45 in. deep. The power arm of the jigging frame is four times as long as that to which the sieve is attached, while the relation of the same parts in the second lever is as 1½ to 1.

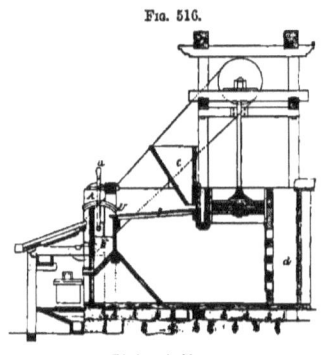

Fig. 516.

Rittinger's Jigger.

Rittinger's jigger is represented in Fig. 516, and is characterised by the inclination of the grates and the lowness of the front partition, over which the poor and lighter stuff falls continuously, and with very little water, while the heavier and rich portions fall through the opening or slit, at the base of the partition. This partition is the segment of a cylinder, and is supported upon the lever or arm a, so as to be movable backward and forward in such a manner that the slit may be increased or diminished at pleasure. The heavy stuff, passing through the opening, falls into the box b, from which it is removed as required. The inclination of the grate in this machine is 5°–8°. It is fed through the hopper c, which plunges below the surface of the stuff accumulated on the grate. The loss of water which occurs at each stroke of the piston is replaced from a reservoir d at the back of the apparatus. According to Rittinger, experience has shown that the duty of self-acting machines of this kind is generally three times as great as that from the ordinary intermittent working apparatus.

Huet & Geyler have constructed an excellent form of jig, Fig. 517, constructed of cast iron, and very compact. Most self-acting jigs require a large quantity of water, and this in many localities is a great objection to their use; but this jig is designed to work with but little loss of water, and, at the same time, by the aid of an automatic scraper, to increase the product. The tub is shaped like the letter U, and is divided into two compartments, one for the piston and the other for the working grate. Water is supplied through the valve a, at the side, and the fine stuff or slime which falls through the sieve settles upon the bottom, and is discharged through an opening b, controlled by a lever reaching out to the front of the apparatus. The piston is operated by means of a shaft and crank, which works in an inclined slide c, connected with a lever carrying the piston, so as to give a rapid descending stroke with a period of rest at the bottom, and then a slow upward movement; thus giving the most favourable conditions for the rapid and perfect separation of the stuff upon the grate.

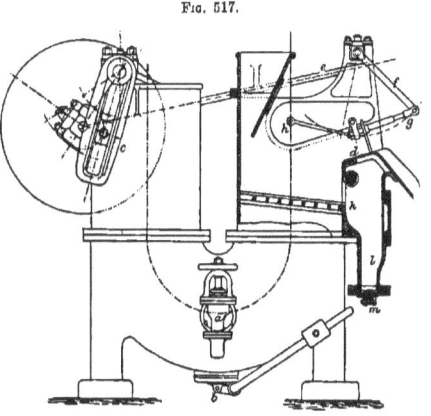

Fig. 517.

Huet and Geyler Jig.

The motion of the piston may be varied at will, in order to secure the best flow or motion of the water for different grades of ore. This adjustment is effected by shifting the position of the head of the piston along the lever or arm, and by this means increasing or diminishing the amplitude of its motion. The construction of this slide is shown in the figure. By turning fixed screws, the head of the piston may be moved forward or backward.

The machine is provided with a scraper d, actuated by the long rod e, which is attached to an eccentric on the main shaft and moves the levers f g, giving to the scraper a forward and backward motion over the top of the stuff upon the grate, and throwing out a portion of it at each movement. The path of the scraper is determined by the guides h, attached to each side of the tub. It can be varied by means of screws upon the lever or arm g. In passing backward, the roller or projection on the scraper, which follows the guides, rises upon the movable inclined plane, and on its return passes below this plane, following the double-dotted line in the figure. The poor stuff from the top, which is constantly thrown forward and off by this scraper, falls over the front of the tub at d, along the chute i. The grate is inclined as in Rittinger's machine, and the opening for the escape of the heavier and rich portion is similarly placed at the foot of the incline and just below the bridge over which the poor stuff is scraped. The opening is shown at k. It is closed by a valve which extends along the whole front edge of the sieve, and can be opened and closed at pleasure by a lever. The stuff passing through this valve falls into a receptacle d, from which it may be removed at pleasure through the opening m. The scraper is so made of perforated sheet-iron that it does not throw the water out together with the waste. These jigs are made with great care and accuracy, and work in a satisfactory manner.

A two-sieved continuous jigging machine, used at Clausthal, is shown in Figs. 518 to 520. It works with 1½-in. stroke, 100 strokes a minute being made for the coarser, and 120 strokes a minute for the less coarse ore. The ore is led on to the first sieve through the opening a at the bottom of the hopper. The stuff, which during the jigging forms the lower layer next to the sieve, passes underneath the cap b, which by means of two thumbscrews can be raised or lowered to suit the requirements of the ore to be jigged, into the slit c, from which it falls out through the small shoot d,

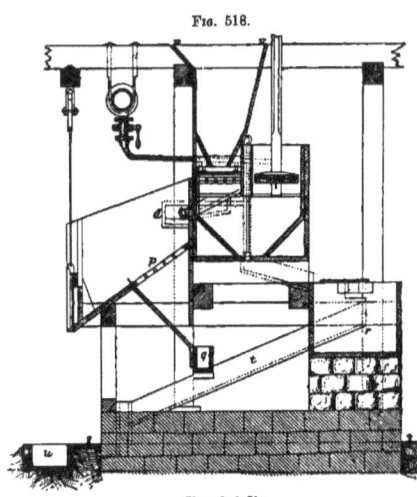

Fig. 518. Fig. 519.

Clausthal Jig.   Clausthal Jig.

Fig. 520.

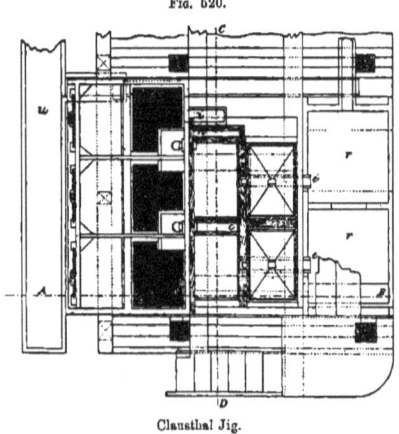

Clausthal Jig.

into the receiver m. The remaining portion of the ore passes from the first sieve over the cap on to the second sieve, and from here the raggings pass off beneath the cap (likewise adjustable), in doing which they pass over sheet iron projections into the slit f, and from which they fall out through the small shoot g into the receiver n, whilst the remainder, as attle or skimping, passes over the cap e into the dam box h, and from thence along the small shoot i into the receiver o. The very fine ore (hutch work) which may happen to be present, or which is formed during the jigging process, goes through the meshes of the sieves, and is let off through the conical openings k, and passes along the small channels e into the receivers r. The meshes of the sieves are 0·79–0·118 in. wide, according to the size of ore they are intended to jig.

DRESSING MACHINERY.

The water which passes over with the jigged ore and attle falls into the receivers m, n, and o, and escapes through the sieve p, into the channel q, from whence, in conjunction with the water passing off from the receivers r, it flows through the channel t into the chief collecting channel u. This channel conducts it to a settling tank, whence, after clearance, it is led away to the lower-lying works to be used as clear water. The hutch work, which falls through the sieve, and is consequently under $0 \cdot 079 - 0 \cdot 118$ in., is emptied from time to time out of the receivers r, in which it is collected.

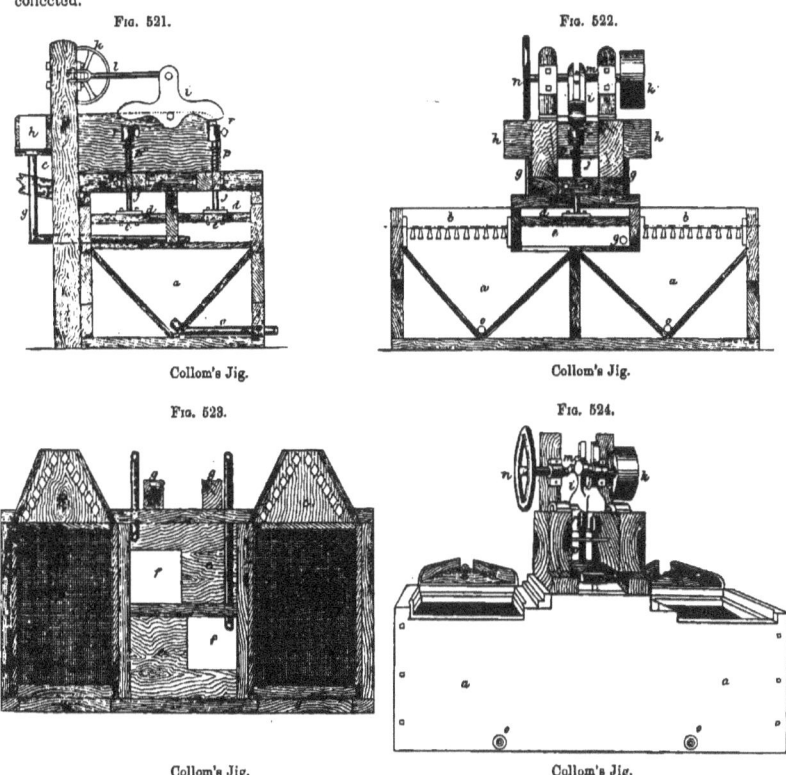

Fig. 521.   Fig. 522.

Collom's Jig.   Collom's Jig.

Fig. 523.   Fig. 524.

Collom's Jig.   Collom's Jig.

Collom's jigging machine (Figs. 521 to 524) consists of a box or tank about 7 ft. long and 3–4 ft. wide, divided by a middle partition into two parts. Each of these parts is fitted on the inside with inclined partitions sloping from the four sides toward the centre of the box, and thus forming two cisterns a, above each of which is placed a sieve b. The sieve frame may be furnished with a wire-cloth sieve of any desired degree of fineness, according to the character of the ore to be dressed. Between the two sieves are the piston or plunger compartments e, separated from each other, and each connecting by an aperture f with one of the cisterns a. Each aperture f affords communication to the cistern nearest to it, but without any connection with the other cistern. The plungers d move up and down in the compartment e, being forced rapidly downwards by the rockers i and lifted again by the action of springs p. The rockers are set in motion by pulleys k, with which they are connected by eccentric rods l. The cisterns and plunger compartments are supplied with water by pipes g, and when the outlets o are closed, the machines are filled with water, the overflow being at q, in front of the sieves. The movements, therefore, of the plungers, which follow each other in rapid succession, produce an agitation in the water, which rises through the sieves with a constantly throbbing motion. The crushed ores, consisting of heavy mineral and gangue, are brought upon the sieves b by a stream of water that enters through the distributing boards c, and, being subjected to the agitaton caused by the plungers d, are held in a state of partial suspension, during which the heavier metallic particles sink, while the earthy matters rise to the top, and are carried off by the water at the overflow q. That portion of the metallic substance which is fine enough to pass the meshes of the sieve falls through into the hutch or cistern a, and may be withdrawn thence at stated intervals by the outlet pipe o; while the coarse part remains upon the sieve, and is cleaned up from time to time, leaving a stratum on the sieve for continued operations. The thimbles r, on the plunger-rods p, serve to adjust the length of the stroke. The action of these machines is excellent. They effect the separation of the galena in a very thorough manner, not only from the earthy gangue, but from the lighter metallic minerals, such as the zincblende and grey copper. The last two are obtained together, owing to the similarity of their specific gravities, and they are also mingled with heavy spar and some quartz.

FIG. 525.

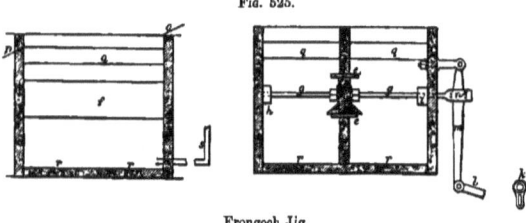

Frongoch Jig.

The peculiarity of the Frongoch jig (Fig. 525), as described by Dr. C. Le Neve Foster, is that the piston is vertical, and works in the partition between two tubs or hutches. Its construction is plain from the illustration. $a\,b$ are the two hutches, c is the middle partition, and d the piston working between two plates of iron e. The piston occupies the whole length of the jig, as shown by f. The piston is worked by the rod g, guided at h, and passing through a stuffing-box i. The reciprocating motion is given by a crank k through the connecting rod l and lever m, which traverses the head of the piston-rod n. The crank has a long loop, which enables the stroke to be varied. The same end can be attained by an eccentric with a slot, which allows the eccentricity to be altered at pleasure. o shows where the ore is fed on, and p is the place of discharge of the waste or impoverished ore. q is the sieve, and r are holes with plugs manipulated by handles (not shown) by which the concentrates that pass through the sieve are drawn off. s is the pipe bringing in fresh water. These patent jiggers of Messrs. Kitto and Paul are doing good work with ores containing blende and galena at Frongoch, and have been favourably spoken of for the treatment of tin ore. They are made by Williams & Metcalfe, of Aberystwith.

Settlers.—These differ somewhat in details of construction, but they usually are round tubs of iron, or of wood with cast-iron bottoms. A hollow pillar or cone a, Fig. 526, is cast in the centre of the bottom, within which is an upright shaft b. This shaft is caused to revolve by gearing below the pan. To its upper end is attached a yoke or driver c, that gives revolving motion to arms d, extending from the centre to the circumference of the vessel. The arms carry a number of flows or stirrers, of various devices, usually terminating in blocks of hard wood e, that rest lightly on the bottom. No grinding is required in the operation, but a gentle stirring or agitation of the pulp is desired in order to facilitate the settling of the valuable portions. The stirring apparatus, or muller, makes about 15 revolutions a minute.

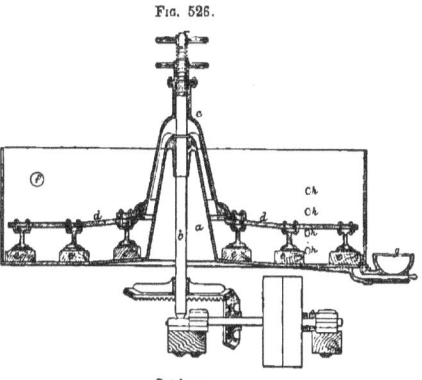

Fig. 526.

Settler.

The settler is usually placed directly in front of the pan, and on a lower level, so that the pan is readily discharged into it. In some mills, two pans are discharged into one settler, the operation of settling occupying 4 hours, or the time required by the pan to grind and amalgamate another charge. In other mills, the settling is allowed only 2 hours, and the two pans connected with any one settler are discharged alternately.

The consistency of the pulp in the settler is considerably diluted by the water used in discharging the pan, and by a further supply, which in many mills is kept up during the settling operation. In other mills, however, the pulp is brought from the pan into the settler with the addition of as little water as possible, and allowed to settle for a time by the gentle agitation of the slowly revolving muller, after which cold water is added in a constant stream. The quantity of water used affecting the consistency of the pulp, and the speed of the stirring apparatus, are important matters in the operation of settling. Since the object of the process is to allow the valuable portions to separate themselves from the pulp and settle to the bottom of the vessel, it is desirable that the consistency should be such that the lighter particles may be kept in suspension by a gentle movement, while the heavier particles fall to the bottom. If the pulp be too thick, the metal will remain suspended : if it be too thin, the sand will settle with it. Too rapid or too slow motion may produce similar results.

A discharge-hole f, near the top of the settler permits the water carrying the lighter portion of

2 Y

the pulp to run off; and, at successive intervals, the point of discharge is lowered by withdrawing the plugs from a series of similar holes h, in the side of the settler, one below the other, so that finally the entire mass is drawn off, leaving nothing in the settler but the valuable portions. There are various devices for discharging these. Usually there is a groove or canal in the bottom of the vessel leading to a bowl g, from which the fluid amalgam may be dipped or allowed to run out by withdrawing the plug from the outlet pipe.

The agitators through which the pulp passes after leaving the settlers are, in general, wooden tubs, that vary in size from 6 to 12 ft. in diameter and 2 to 6 ft. in depth. The main object in letting the stream of pulp pass through them is to retain and collect as much as possible of the valuable portions that are carried out with the pulp discharged from the settler. A simple stirring apparatus, somewhat resembling that of the settler, keeps the material in a state of gentle agitation, the revolving shaft carrying 4 arms, to which a number of staves are attached. In some mills there are several agitators, in most cases only one, and by some they are not used at all. The stuff that accumulates on the bottom is shovelled out from time to time, usually at intervals of 3 or 4 days, and worked over again.

Sizers.—Labyrinths.—The slime labyrinth is a German apparatus. It consists of a number of contiguous connected settling-pits, which, if, for instance, in connection with 3 batteries of 3 light stampers, increase in size from 1 to $1\frac{1}{4}$, $1\frac{1}{2}$, and $1\frac{3}{4}$ ft. square in transverse sections, and respectively from 15 to 21, 24 and 36 ft. in length, with inclinations in the same sequence of $\frac{1}{2}$, $\frac{1}{4}$, and $\frac{1}{8}$ in. per foot, the largest being horizontal. Such a labyrinth classifies the stuff into 4 portions, differing in size of grain, which form deposits during the passage from the smaller to the larger pits, the coarser grains depositing in the former. This classification is, however, far from perfect, and is besides attended with expenses on account of transport and re-puddling of the settled stuff, previous to further treatment. It also entails a large loss through escape and waste of fine material, that generally amounts to 10 or 12 per cent., but may, in unfavourable cases, rise to 15 or 20 per cent. The labyrinths are therefore now in operation only in some of the older and smaller establishments, where want of space and other circumstances prevent the use of either of the other two following classifying apparatus.

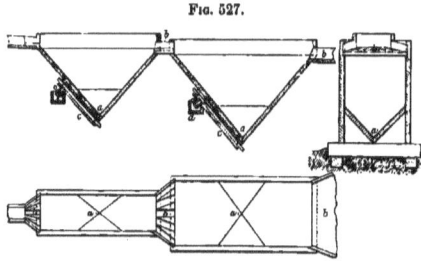

Fig. 527.

German Pyramidal Boxes, or Spitzkästen.

Pyramidal boxes. — The pyramidal boxes or *Spitzkästen* of the Germans, Fig. 527, are, as their name implies, hollow, generally rectangular, pyramids. They are constructed of strong boards, well joined together (strong sheet-iron may be employed also). The sides are inclined at angles of not less than 50°, and there is a small hole in one side close to the apex. They are fixed horizontally, in an inverted position, and the crushed material is introduced at one of the narrow sides, a few inches below the top by means of a launder. The result is that, as soon as the box is filled, a certain portion of the crushed matter—i. e. the coarsest and heaviest, which the water, on account of its diminished velocity, is not able to carry farther—sinks and slides down the inclined sides of the pyramid, and escapes through the small hole a near the apex, whilst the finer and lighter matter passes off at the top by an outlet b in the centre of the side, opposite to the point of entrance. If now a second larger box be attached to the first, a third still larger to the second, and so on—each succeeding box at a slightly lower level, in order to prevent any settlement of stuff in the passage-ways—it follows not only that the same process of settling and escaping of the particles from the apex will take place in every box, but also that their size will decrease nearly in inverse proportion as the surface of a succeeding box is larger than that of the preceding one, or directly as the velocity of the water is diminished in it. According to this principle of the boxes—if they were made of only very gradually increasing size, and the apex holes proportionately small—it would be possible to classify the stuff into a great number of portions, different in size of grain, before it had entirely settled—i. e. till clear water passed off from the last box. Experience has, however, shown, that for fine ore-dressing in general, classification into 4 different sizes by an apparatus of 4 boxes is quite sufficient. The sizes of the different boxes, in order to ensure the most perfect classification, depend both on the amount of material which has to pass through them per second, and the size and character of the grains; and by theory and practice it has been found, that for the supply of every cub. ft. of material, the width of the first or smallest box must be $\frac{1}{10}$ ft.—i. e. for instance, for 20 cub. ft., 2 ft.—and for every succeeding box it ought to be about double that of the preceding one, or, generally, the widths of the boxes must increase nearly in geometrical progression, 2 : 4 : 8, &c., and their lengths in an arithmetical one, 3, 6, 9, &c.

For the stuff under notice, their dimensions are thus in different large establishments as follows :—

The first box is 6 ft. long and 1½ to 1¾ ft. wide. | The third box is 12 ft. long and 4 to 5 ft. wide.
„ second „ 9 „ 2¼ to 3 „ | „ fourth „ 15 to 16 ft. „ 8 to 10 „

Their depths depend on the angle of inclination of the sides, which, as already stated, is generally 50°, because if less, the stuff would be liable to settle firmly and choke the central orifice, and if larger, unnecessarily great height of the boxes would be required. The form of the two smaller boxes is commonly such that the two short sides are inclined at the above angle, and the two long ones, which would become far steeper, are broken—i. e. are for a certain depth from the top vertical, and afterwards inclined at the normal angle. This modification has, however, no influence upon the action of the boxes, but simply facilitates somewhat their construction and firm fixing. The sides of the larger boxes are generally even throughout. The way in which the outlet-holes a at the apexes are constructed has an important bearing on the operation of the boxes. At these points, the hydrostatic pressure is considerable, and the holes should naturally be kept small, in order to prevent too much water passing with the particles of stuff; such small outlets are, however, especially in the treatment of coarser material, very liable to become choked. This difficulty has been met by the holes being made of conveniently large size, but connected with pipes c, ¾ in. in diameter, which rise up the side of the boxes—i. e. of the smallest box to within 3 or 3½ ft., and of others to within 2 to 2½ ft. from the top—and are there furnished with small mouthpieces d, supplied with taps for regulating the outflow. This arrangement, on account of the outlets being so much higher, has the further advantage that a considerable amount of fall is gained (especially as regards the large boxes), which, for the subsequent treatment of the material, is in some cases of special value. There are two more points that require attention, in order to ensure good action of the apparatus, namely, the introduction of the material into the different boxes equally and without splashing, and prevention of the entrance of chips of wood, gravel, or other impurities that are likely to stop or obstruct the outlets. The first point is met either by having the supply-launders expanded fan-like and furnished with dividing-ledges b, or by the interposition of small troughs, the sides of which nearest the box to be supplied are perforated near the bottom by equidistant small holes. The cleaning of the material previous to its entering the first box, is generally effected by the main supply-launder being made a little wider near the point of entrance, and the insertion at this place of a fine wire-sieve across the launder and somewhat inclined against the stream. This sieve must be occasionally looked after, to remove any impurities collected in front; and this, in fact, is the chief attention the whole apparatus requires, for otherwise it needs hardly any supervision. If once in proper working order, its action is constant and uniform, provided the material introduced does not change in amount and quality; and it has this further advantage, as compared with the slime labyrinths, that the classified stuff can, from the outlets, be directly conveyed in small launders to the concentration-machines for treatment, without any previous preparation. One point, however, not in favour of the apparatus, is that, having to be placed between the mills and the concentration machines, a great fall of ground is required, to permit the direct introduction of the material and allow sufficient fall for the tailings; and thus, where local circumstances are unfavourable, it has to be erected at a higher level, and necessitates the use of elevators or pumps for lifting the stuff. The action of the different boxes on the material under notice, with regard to the percentage of fluid matter and the quantity and character of its solid contents, which they respectively separate, is according to experiment as follows :—

The small box separates 38 to 40 per cent., containing per cub. ft. 16 to 18 lb. coarse sand.
The second box separates 20 to 22 per cent., containing per cub. ft. 13 to 14 lb. fine sand.
The third box separates 18 to 20 per cent., containing per cub. ft. 15 to 16 lb. coarse slime.
The largest box separates 10 to 12 per cent., containing per cub. ft. 10 to 12 lb. fine slime.

These results are satisfactory for the further concentration of the ore. As regards the loss caused through the final escape of impalpable ore from the last box, it amounts for rich galena-ores to about 6 per cent., and for quartziferous silver-ores to about $2\frac{1}{2}$ per cent.

According to J. M. Adams (' School of Mines Quarterly '), " several forms of pointed box are in use. Their dimensions vary according to the duty required. In some cases, it is desired to settle all the pulp, including the slimes, when there is too much water present for subsequent concentration. In such event the pointed box should be about 6 ft. deep, and 3 ft. by 7 ft. at the top, the longest sides sloping till they meet at the bottom. Such a box will settle and save about 6 tons of ore in 24 hours, discharging it automatically and continuously from the bottom by a siphon hose, with the proper amount of water for subsequent concentration. This form is used when the tailings from pan amalgamation are to be concentrated, after leaving the settlers and agitators, for they contain a large excess of water, which must be got rid of, so that the tailings are of the proper consistency for concentration. Fig. 528 shows a form of pointed box used in cases where the slimes are to be separated from the battery pulp and saved. Each box is 40 in. square at the top, and 40 in. deep, coming to a point at the bottom; and one box will handle 6–10 tons of pulp in 24 hours, making a good separation. The pulp from the battery, entering the box a at the top, is confined by partition b, until it passes into the box proper c, near its bottom. Clear water is conveyed from a launder d above, through a $\frac{1}{2}$-in. pipe e, which delivers it into the box at the bottom. Care must be taken that this pipe is kept full, so that no air bubbles are carried through it, as they create agitation, and cause sand, &c., to pass off with the slimes. The amount of clear water needed varies, so it is a good plan to have a cock in the pipe just below the clear water box d, or else to partially close, with a wooden plug, the opening of the pipe in the clear water box. At f is a hollow plug, and to it is attached a piece of hose g, which is used as a siphon, so that the pressure is lessened, and too violent discharge of the pulp is prevented. Without the siphon hose, ⅛-in. opening would not be too small, while with it ⅜-in. opening is about right, and the end of the hose is plugged accordingly. Inasmuch as foreign coarse material occasionally gets into the box (prevented as much as possible by a screen over the top), it is advisable to use in place of the hollow wooden plug shown, a 1½-in. iron T with one end plugged, and with ⅜-in. side outlet, attaching the siphon hose by nipple." The launder h carries off slimes, and the launder i conveys rich matters to the concentrators.

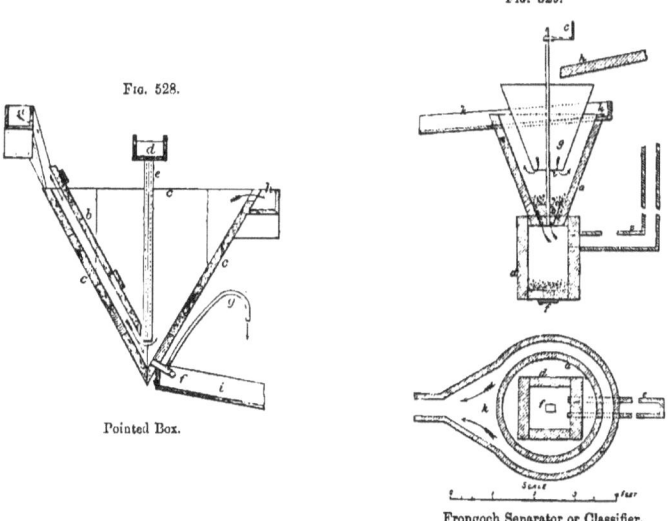

Fig. 528.

Pointed Box.

Fig. 529.

Frongoch Separator or Classifier.

The classifier introduced by Messrs. Kitto & Paul at the Frongoch mine (Fig. 529) consists of an inverted wooden cone a, which can be more or less completely closed at the bottom by a plug b, controlled by a handle and screw-nut c. The cone a stands upon a wooden box d, which receives water under pressure from a pipe e, and is provided with a discharge-valve f, a mere flat plate of iron working on a pin, which can be pushed sideways so as to close the orifice more or less entirely. Inside the wooden cone a is a sheet iron funnel g, which receives the stream of ore and water from a launder h, and causes it to descend to the level i. There it meets with the upward current of clear water, and a separation is effected. The coarse and heavy particles which can overcome the stream pass into the box d below, and flow out continuously at f, while the fine and light particles are mastered by the current and carried over the top edge of the wooden cone a, which is surrounded by a circular launder k. By altering the flow of the upward current of clear water and the size of the discharge orifice, the separator can be adjusted to the requirements of any particular case. At Frongoch, this separator is used for classifying blende containing galena, just as the stuff comes from the crusher after passing through a sieve with 12 holes per sq. in. (3 holes by 4). The coarse goes to the jigs, the fine to the buddles. The foregoing description is taken from Dr. C. Le Neve Foster's 'Mining Notes in 1887,' (Transactions Mining Association and Institute of Cornwall.)

Triangular Double Troughs.—Classification in the triangular double troughs or *Spitzlutten*, an invention by Rittinger, is based upon the principle that, if material composed of particles differing in size and density is exposed to a rising stream of water, the velocity of this stream may be so regulated that particles of certain size and character sink and may be conveyed off, whilst the remainder is carried upward by it; and that, consequently, by repeating this operation a certain number of times with a gradually decreased velocity of the rising stream each time, the material can thereby be separated in as many different classes of grains. The *Spitzlutten* (Fig. 530), by which this action is now very simply produced, are constructed as follows:— Within a triangular trough a, of certain length and width, with two opposite sides vertical and two inclined at angles of 60°, is a similar smaller one b, having the vertical sides in common with the larger trough, but its inclined sides fixed at certain equal distances from, and parallel to, those of the latter. There is thus an open V-like space c left between the inclined sides of the two troughs, representing, as it were a rectangular pipe, sharply bent in the centre; and it is through this that the stream of material has to pass—i. e. to fall and rise. The velocity of the stream depends on the size of this space, and consequently so does the

FIG. 530.

Triangular Double Trough, or Spitzlutten.

size of the particles that will rise or sink in it. The cross-section and respective velocity stand in inverse relation to each other, and their determination for each double trough of a complete apparatus is a matter of mathematical calculation, in which the size of the largest particles and the specific weight of the material to be classified form the main figures. For galena-ores, such as those under notice, and which are crushed so fine that the largest grains are not more than 0·6 millimetre in diameter, the most satisfactory classification into 4 different kinds of grains is, according to Rittinger's calculation, arrived at by a series of four double troughs, with the velocity of the stream decreasing from the first to the succeeding troughs in the progression of 2·3, 0·94, 0·15 in. per second; and if the width of the channel for the first trough is 1·1 in., and its

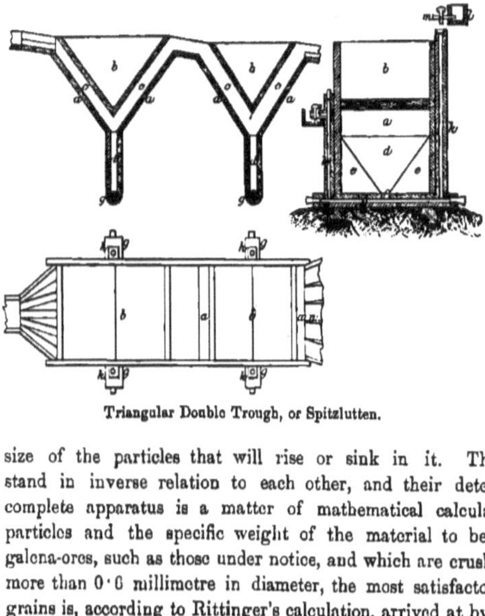

DRESSING MACHINERY. 351

length 2 ft., the dimensions of that of the second trough follow as 2·75 in. : 2 ft. And as it is not advisable to increase the width of the channels beyond 3 in., the channels of the third and fourth troughs are each 3 in. wide, and respectively about 54·5 in. and 135 in. long. The mean depth of the channels, measured from the line of inflow of the material to the lowest part of the inside trough, is, for the two smaller double troughs about 3 ft. ; for the two larger ones, 4-6 ft. In order to carry off the coarse particles that sink in the channels, the inclined sides of the outside troughs do not meet below, but are continued downward, forming a long and narrow pyramidal opening d, about $1\frac{1}{2}$ in. wide at top. The short sides e slope inwards at an angle of not less than 50°, contracting the opening to a small hole f of about 1 in. sq. at bottom, through which the material is discharged into a horizontal pipe g, that extends both ways a small distance beyond the sides of the apparatus, and is connected at the ends with vertical 1-in. pipes. One of these, h, serves for the outlet of the classified material, and is carried up to within 36 to 21 in. of the water-level in the channel c, according to the degree of fineness of the particles that have to pass through it (the same as in the pyramidal boxes). At the top it is supplied with a tap for the regulation of the outflow. The other pipe k conveys a supply of clear water, furnished from a launder l supplied with a tap m, and as the water in the pipe stands 6 to 8 in. above the water-level in the trough, a small uniform pressure is produced, causing a forced influx of water at the point f, which is essential for good classification. This water—opposing itself to the downward current, charged with sediment in the pyramidal channel d—prevents all but the coarser particles and pure water passing into the pipe h, and thus only grains of the desired size are carried to the outlet i. With regard to the relative positions of the different double troughs of the series, they are fixed exactly horizontal, and sufficiently below each other to prevent any settlement of material in the communication launders n, which are necessarily very broad. Other particulars regarding proper working, supervision, &c., are the same as those given for the pyramidal boxes. According to present experience, a series of 4 of these double troughs classifies as well as, and, for the two coarser kinds of grains, even better and cleaner than, a set of 4 pyramidal boxes, though for the fine slimes these latter are generally preferred, as they effect the desired settlement of the stuff more completely. A complete apparatus of troughs requires also less fall and space than one of pyramidal boxes, and is more easily regulated in cases of increased or diminished influx of material. The necessary additional supply of clear water might, however, form a drawback to its application in cases where this medium is scarce. As regards the results of classification by the different troughs of the series, they are stated to be as follows :—The first or smallest trough separates about 30 per cent. of coarse sand ; the second, about 25 per cent. of fine sand ; the third, 20 per cent. of coarse slime ; the fourth, 15 per cent. of fine slime.

Concentration.—Having classified the material according to size, the next step is to submit each separate size to a process of concentration with the object of eliminating the valuable portion. For this purpose many apparatus are in use, all working upon the principle of taking advantage of the greater specific gravity of the part sought to be saved.

Percussion-tables.—The most highly perfected of the various percussion-tables or shaking-tables is Rittinger's continuously-acting sidethrow percussion-table, shown in Fig. 531.

To simplify the construction and movement of these tables, they are generally made so that they represent one large table, divided by a cheek b in the centre into two (a^1 and a^2), for the movement of which consequently only one arrangement is required, rendering the percussion simultaneous for both. The floor or platform of each table (a^1 and a^2), measured inside the head-board and cheeks c, which are about 4 in. high and 1¼ in. thick, is 8 ft. long and 50 in. wide. It is generally double boarded, the upper surface being made of tongued-and-grooved 1¼-in. boards of some even, close-grained wood (generally sycamore), planed as exactly as possible, and slightly blackened by weak

FIG. 531.

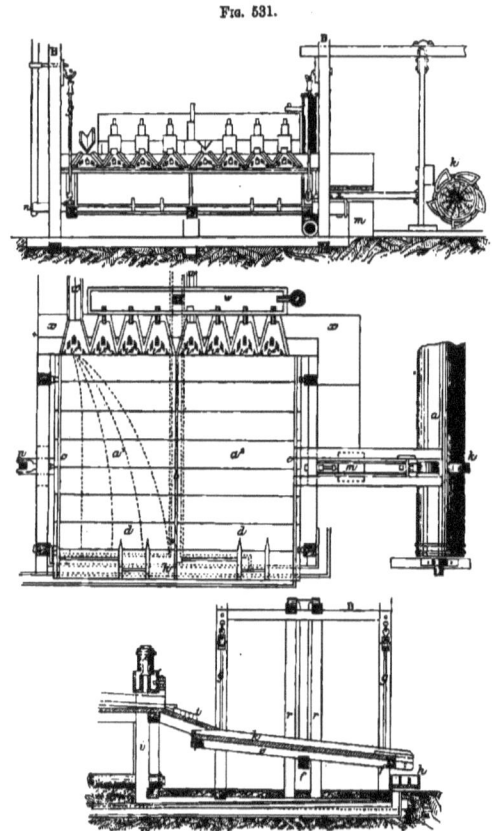

Rittinger's Percussion-table.

sulphuric acid. The boards are carefully laid crossways, and fixed with wooden pegs to the lower floor, made of pine-boards tightly screwed to a stout wooden frame, consisting of 4 or 5 bars *e* lengthways, and 3 *f* across, which are mortised and screwed together and secured by iron angle-braces. The centre crossbar f is nearly double as strong as the others, and projects on both sides a certain distance beyond the platform. It is called the "tongue" or "percussion-bar," as it forms the part to which the side-movement and percussion of the table is imparted. The double table is suspended by 4 iron rods g, having adjusted shackles, and at either end eyes that are connected with hooks; the upper ones screwed into stout uprights, that form part of a strong framing b, braced well together at top and bottom; the lower ones screwed into the sides of the platform frame at about 1 ft. from either end. The arrangement for imparting the side way motion and percussion to the tables consists, in the first instance, of a wooden axle a, furnished with 4 or 5 cast-iron cams k, opposite the centre of the table. This axle is turned by an endless strap from a shaft connected with the axle of a water-wheel. The cams act upon—i. e. push forward—the iron-faced projection of a vertically-suspended wooden lever, which swings at its upper end on small iron pivots between two crossbars connected with the framing, whilst its lower end moves between guiding ledges, nailed to the floor of the building. About level with the frame of the table the ends of 2 wrought-iron bridles are joined to it by means of a screw-bolt. These bridles transfer the forward movement of the lever to the table by being fixed with a screw-collar over a horizontal screw-spindle, that is fastened and adjusted on top of the projecting portion of the percussion-bar or tongue, and by means of which the length of the forward movement—i. e. the side-throw can be regulated. The end of the projection of the percussion-bar, being slightly rounded off in front and strengthened by an iron rim, presses, when at rest, against the bumping-block m—a stout square pillar—which is joined to a foot-piece, 5 or 6 ft. deep in the ground, and well stayed at the back to resist the shocks, and its face, at the place of contact with the percussion-bar is generally covered with leather for the purpose of deadening the blows. The pressure of the percussion-bar against the bumping-block is produced by a stout spring n of iron or some tough wood, attached to its prolonged end at the opposite side of the table, and can be regulated by a screw, that adjusts the tension of the spring. Both prolongations of the percussion-bar move on each side between two uprights r, connected with the outside framing, by which means the transversal movement of the double table is guided.

The following is a somewhat more simple arrangement for moving the table, and is similar to that used for the common percussion-table. The cams of the driver are made to act against the iron-faced top end of a lever, that moves on an axle at foot, and to which is attached, by means of a regulating screw, a horizontal wooden bar. This axle slides between guides fixed on top of the bumping-block, and at the other end is in contact with an oblong block of hard wood, screwed on top of the prolongation of the percussion-bar. In order to prevent the horizontal bar and lever from jumping back too far when the table strikes against the bumping-block, the bar is furnished with a bolt, which ensures the normal position by resting against guiding-ledges on top of the bumping-block. The other arrangements, as regard the spring, regulation of pressure, &c., are the same.

The tables receive their supplies of classified material and of a necessary amount of cleaning water, evenly distributed by means of triangular inclined dividing-planes t^1, &c.—furnished with wooden buttons in the usual manner, from separate troughs w, into which the material and water are conveyed by small launders or pipes u, from the respective classifier and main water-launder. The whole of this arrangement, inclusive of platforms x for the workmen to stand on, rests on a low framing i above the head of the table. There are generally 3 or 4 dividing-planes t^1 to t^4 for each table, one of which supplies the classified material for a breadth of 8 to 10 in. at that side of each table opposite to where the percussion takes place, whilst 2 or 3 others provide the cleaning or washing water, the one nearest the percussion side always in somewhat larger quantity.

The principle upon which the ore-concentration on these tables is based, differs from that of the common percussion-tables mainly in the side action of the percussion on the stuff treated, which produces two movements of the particles, viz. one down the incline, the other forward; and in the mean direction resulting from this—i. e. diagonally downward—they pass off the table. As now the heavier particles—i. e. the ore—are not only thrown farther, but are also, on account of their stronger friction on the boards, more or longer exposed to the forward movement than lighter ones (waste) during the same interval of time, it follows that they travel outside these, gradually separate according to their specific weights into distinct bands, and that this separation becomes more perfect, the nearer they approach the end of the table. The whole of this action is at the same time enhanced by the "cleaning-water." This prevents the stuff from spreading at once all over the table; it cleans the outside streaks of ore, and the stronger portion serves specially for washing the ore finally off the apparatus. For securing the partitions of the different portions of ore, separated on the table, small, pointed, movable pieces of wood d, called "tongues" or "pointers," are screwed upon the table near its foot, by means of which the streaks of ore are divided from that of the waste, and guided through narrow slits in the table, furnished with sheet-iron lips, into separate launders h, underneath, that discharge into settling- or catch-pits; or else the ore passes off the table over small movable strakes into launders placed in front. The waste runs off the table in both cases over a broad strake into a small launder in front, that conveys it to the main waste channel. In this, and in all instances where the discharge takes place over the front of the table, it is, like the slits above mentioned, provided with sheet-iron lips projecting about 2 in., in order to prevent the stuff licking back underneath.

The result produced by the operation of these tables on any of the four previously classified portions of the stuff under notice is, to divide them into five products, viz. :—

1. Lead-ore, containing fine free gold.
2. Lead-ore, poor in gold.
3. Copper- and iron-pyrites, mixed with lead-ore.
4. Poor ore-slime (i. e. imperfectly concentrated); and
5. Waste.

These run in as many well-defined streaks down the tables, and are, as above described, parted by the pointers d and guided, the first three, into launders communicating with separate settling-pits; the fourth, into a strake, that conveys it to a catch-pit, from whence an elevator, or lifting-wheel, raises it into a small pyramidal box for re-classification—the portion issuing from the apex orifice being then conveyed to, and re-treated, on a separate table. The quality of the three first portions of ore collected in the settling-pits is such as to render them fit for direct metallurgical treatment, the auriferous portion being, however, previously submitted to gold extraction, as will be seen hereafter. There are many conditions, as regards adjustment of stroke, supply of material and washing-water, &c., necessary to ensure the satisfactory working of this machine. This is shown by the following table (prepared by Franz Rauen), which gives the results of practical experience in treating the four classified portions of sands and slimes, both of auriferous lead-ores and silver ores.

DRESSING MACHINERY.

ADJUSTMENTS OF RITTINGER'S CONTINUOUSLY-ACTING SIDE-THROW PERCUSSION-TABLE.

Description of Ores.	Inclination of the Table.		Throw or Stroke.			Supply of Material and Cleaning Water per Minute upon a Double Table.			
						Classified Material.		Cleaning Water.	
	In the Direction of the Throw.	Towards the Front.	Constant Pressure of Table against Bumping-block.	Length.	Number of Throws per Minute.	Supply in Measurement.	Weight.	Solid Contents Carried by the Water.	Front Cleaning Water.	Back Cleaning Water.
For auriferous lead-ores.	lines	lines	lb.	lines	No.	cub. ft.	lb.	lb.	cub. ft	cub. ft.
Coarse sand, 1st classifier	..	64	140	24	73	0·392	23·51	2·665	1·023	1·251
Fine ,, 2nd ,, ...	6	58	110	21	85	0·333	20·09	1·763	1·093	1·234
Coarse slime, 3rd ,, ...	7	56	106	18	100—105	0·323	18·94	0·914	0·888	0·977
Fine ,, 4th ,, ...	14	52	100	10	112–130	0·236	14·07	0·750	0·669	0·914
For silver ores.										
Coarse sand, 1st classifier	6	72	212	18	76—78	0·420	26·64	4·956	0·711	1·187
Fine ,, 2nd ,, ...	6	54	183	12	86—88	0·405	24·80	3·410	0·256	0·677
Coarse and Fine slimes } 3rd and 4th classifiers ..	6	30	100	10	100–110	0·261	16·55	3·580	0·313	0·708

Another condition, not less important than those just given, for good concentration of the different classified portions of material, is the proper regulation of the velocity of the throw. For the tables applied to the treatment of the two coarser sizes, it ought, in the average, not to exceed 1 ft. per second, whilst slime-tables require a velocity of stroke of only 0·5 ft. per second; a greater velocity causes the tables to slide, so to speak, from underneath the particles of ore, thus retarding their progress.

The motive power required for working a double table is, by dynamometric experiments, proved to be 0·26 H.P.; and the working effect per hour of a single table is 55 lb. of slimes and 300 lb. of sands. Six continuously-acting double tables are capable of working in 24 hours 10-12½ tons of crushed material, classified by four pyramidal boxes. Four of these treat the four classified portions —i. e. each table is devoted to one particular class—and the two remaining ones are necessary for reworking the intermediate products—i. e. the imperfectly concentrated ore matter. As regards supervision and manual labour required one workman is quite sufficient for attending on two properly-adjusted double tables. His principal work consists in looking after the right position of the "pointers," and the steady and regular supply of material and cleaning-water. On comparing the relative quantities of ore produced and the loss sustained during the same time by this and the old percussion table, the produce of the former is 3 to 4 per cent. smaller than that of the latter, and its loss of ore about 2 per cent. larger i. e. the new table loses 23 to 24 per cent., whilst the old one only loses 21 to 22 per cent. These disadvantages of the new invention are, however, more than compensated by the greater purity of its ore effecting a saving both in transport and smelting expenses, and more especially by the greater amount of free gold contained in the auriferous portion. But it greatly excels the old table on account of its continuous, steady self-action; and consequently its working expenses are, without regard to wear and tear, fully 60 per cent. lower than those of the other.

Ordinary percussion table.—The end shake percussion table shown in Fig. 532 is a simple and inexpensive form of concentrator largely used in the Australian reduction works, and, when carefully handled, gives exceedingly good results with ores which do not offer special difficulties in concentration. It will be seen from the engraving that the wooden box into which the pulps flow is suspended by iron rods from cross-beams carried on the top of the upright standards, and, by the action of a cam driven by the pulley on the cam shaft and working against a cast-iron block which is fixed on the lower end of the box or table, a succession of sharp blows is given, causing the sulphides and heavier portions of the crushed ore to settle to the bottom, while the lighter material is carried away by the flow of the water. Where closer concentration is necessary, special machines are undoubtedly better in every way; but this old-fashioned shaking-table is still preferred in some districts on account of its simplicity and the ease with which it can be repaired.

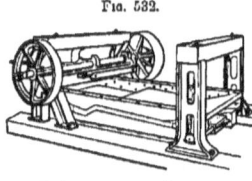

Fig. 532.

End Shake Percussion Table.

Rotating table.—Rittinger's rotating table, Fig. 533, is specially applicable for the concentration of fine slimes, and for this operation is preferred to both common and side thrown percussion-tables. The concentrating portion—i. e. the table proper—may be described as a shallow, inverted conical or flat funnel-shaped ring, consisting of even-grained, well-planed 1-in. pine-boards. The outer diameter is 16 to 18 ft., and the inner 5 to 6 ft., with an inclination of 6 in. to its radial width. It is furnished round the outer periphery with a rim of board 2 to 3 in. high, and is divided radially by narrow battens into 32 equal segments, that are somewhat contracted at the inner periphery by the ends of the battens being split, where, attached to the end of each segment, is a funnel-shaped descending-pipe of wood or sheet-iron v, serving for discharge into receptacles beneath. This ring rests exactly horizontal, or rather the boards which it consists of are fastened horizontally crossways and watertight, upon 16 radial wooden bars or arms r, attached by means of a cast-iron rosette q to a central vertical wooden axle s of 16 in. diameter, to which (and consequently to the table) a slow steady revolving motion is imparted by means of a tangent-screw, from a shaft connected with the water-

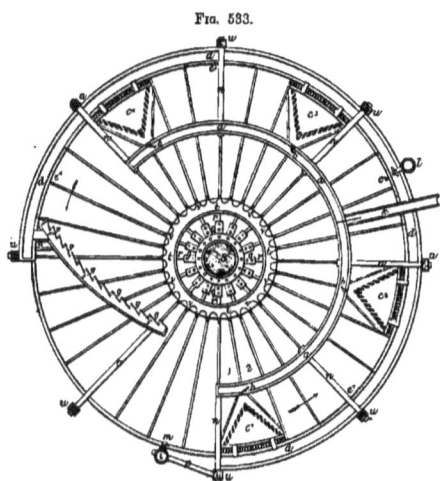

Fig. 533.

Rittinger's Rotating Table.

wheel. The tangent-screw operates upon a finely toothed cast-iron wheel, about 2½ to 3 ft. in diameter, fixed on top of the axle. The lower pivot of the latter turns in strong cast-iron bearings, whilst the upper one revolves within an iron collar; the bearing and collar, being fixed to a strong framing, consisting of 8 radial spars at top and bottom, connected, about 1 ft. outside the circumference of the table, by as many stout uprights u, and strengthened above and below by 4 braces, forming squares, near the centre. To this framing are attached all supplementary portions of the table, that take no part in its motion, such as the troughs for supply of clear water and material, triangular distributing-planes, &c.

The circular trough a that supplies the material consists of boards of sheet-iron, is 3–4 in. wide and extends over 20 segments of the table. It rests horizontally on supports n, fastened to the uprights u of the framing, within about 2½ ft. of the outer periphery of the table, and at such a height above it that the triangular distributing-tables c, that introduce the material close within its rim, have a fall of at least 20°. The bottom of the circular trough, from the point where a launder k introduces the material from the classifier—i. e. just at the centre of the curve—is saddle-backed at a fall of about ¼ in. per ft., which prevents the stuff from settling. Up the centre of the mouth of the launder k a small movable tongue of wood is fixed to distribute the material equally towards both sides of the saddle. There are generally 4—sometimes 5—square openings b through the outer rim at the bottom of the trough, also provided with tongues, to regulate the outflow of the material on to as many triangular distributing-tables $c^{1, 2, 3, 4}$, the relative positions of the openings being such that one is near each end of the trough, and the other 2 or 3, as the case may be, are placed respectively at equal distances from these and between each other. As regards the triangular distributing-planes c, they are fastened with their pointed ends by means of bolts and eyes underneath the trough, whilst their lower edges have hook-like strips of sheet-iron attached, by means of which they hang firmly on the inner rim of the main clear-water trough d, presently to be described. This arrangement, though ensuring a fixed position of the planes, yet permits their easy removal in case of access to the table being required. They are, as usual, constructed of pine-boards, to which oblong wooden distributing buttons are fixed; but they have, for the more equal distribution of the material, attached to their lower edges either pieces of sheet-iron with serrated lips, or are provided harrow-like with a number of thin iron spikes. The width of these lower edges, which are concentric with the table, is $\frac{1}{16}$ of the circumference of the latter—i. e. each supplies 2 of the 32 segments at a time.

The main clear-water trough d, constructed of boards or sheet-iron, is likewise circularly bent, and rests horizontally, 3 or 4 in. above the circumference of the table, on flat iron supports projecting from the uprights u of the framing. Its inner rim lies just within the rim of the table; the outer one touches the uprights, and is attached to them. It commences radially abreast of one of the ends of the feeding-trough a, just level with the first distributing plane c^1, but it extends 4 segments beyond the last distributing-plane c^4 at the opposite end of the trough a, and encompasses in all 24 segments of the table. It is constantly and in regulatable quantity kept supplied with clear water by means of a tap b. The amount required for concentration runs from the trough on to the circumference of the table through open cuts e^1, e^2, e^3, e^4, in its inner rim, 2½ in. deep, and as wide as suffices for the supply of 2 segments of the table. For the equal distribution of the water, sheet-iron plates, deeply serrated on both edges, are fixed in front of these cuts (indicated by dark lines on the sketch). The teeth of the lower edges of the plates stand about 1½ to 2 in. off the surface of the table, while those of the upper edges reach to within 1½ in. of the level of the outer rim of the trough. The number of places of overflow $e^{i, 2, 3, 4}$ corresponds with that of the distributing-tables c furnishing the material, and they are so arranged that if, for instance, there are 4 such tables, 3 overflows e^1, e^2 e^3 occupy the centres of the spaces—i. e. they supply the two middle segments of the 4 between these tables, whilst the fourth e^4 lies 2 segments beyond the fourth table. Adjoining this last place of outflow, and communicating by a short spout h with the main trough d, a peculiarly-constructed special clear-water trough f commences and extends horizontally in a flat curve across 6 segments of the table to near the inner periphery of the latter, where its end rests on a strong support o, projecting from the nearest upright of the framing. The outer rim of this trough is even, but the inner one presents a succession of deep notches or breaks, and the dark lines g signify the places lower than the rest, where the clear water flows over the serrated sheet-iron plates, that extend down to within 2 in. of the surface of the table. And as they have to follow the inclination of the table, whilst the trough lies level, they become gradually longer towards the end of the trough. The special purpose which this trough serves for the material under notice is the separation of the pyrites from the lead-ore. As regards the arrangement for washing this ore off the table, it consists of a vertical pipe i, that by means of a flat mouthpiece m, 10 lines wide and 3 lines high, discharges, under a pressure of 8 to 9 ft. and at a very oblique angle, a stream of clear water on to the table near its rim. This pipe is closed at the bottom, and communicates at the top with a reservoir or launder. It rests on a support p projecting from the nearest frame-upright u, close outside the circumference of the table, and in the centre of the space between the commencement of the main clear-water trough d and the end of the special one f just described, a spot coinciding with the centre of the middle one of the 3 remaining segments of the table.

For the separate reception of the three kinds of material—viz. waste, pyrites, and lead-ore,— that run off the table at its inner periphery (through the funnel-shaped pipes v attached to the contracted ends of the segments), either one circular trough t divided by 2 partitions into 3 compartments, or else 2 separated troughs, one within the other, are used, the outer one of which is devoted to the waste, whilst the inner one receives the 2 kinds of ore into separate compartments by means of 2 short, shallow strakes, resting on top of the outer trough. The positions and lengths of the respective compartments correspond, of course, with the numbers and positions of the segments of the table, from which waste, pyrites, and lead-ore are washed off. The troughs rest on the foundation-bars of the framing, and the compartments for the 3 products communicate by holes in their bottoms with separate launders underneath, that conduct the ore to settling-pits outside the table, and the waste into the main waste-channel.

The mode of action of this table is very simple, and its working requires no manual labour whatever; the machine is, in fact, like the percussion-table—self-acting. The process of concentration on the two segments 1 2 is, for instance, as follows:—Being coated with material by the triangular plane c^1, they pass during the slow rotation of the table underneath the first overflow e from the main clear-water trough d, and the ore deposited on them is cleaned from the waste; but, progressing farther, they receive a fresh supply of material from the second distributing-table c^2, are again cleaned by the second overflow e^2 of clear water, and after having undergone this double operation twice more, they progress underneath the special cleaning-trough f. On account of the inwardly-bent form of this trough, the numerous streams of clear water, issuing from it at g, act gradually on nearly the whole surface of the segments, completely washing off the pyrites and leaving them coated with pure lead-ore only, which, on farther progression, is also removed by the forcible stream of water, discharged from the mouth-piece m of the vertical pipe i. With this operation a full rotation of the table is completed, and the whole process, as regards the 2 segments, commences again. But it will no doubt be understood, that every 2 segments composing the table undergo the same operation in steady regular succession, and that the concentration is thus continuous as long as the table rotates.

The conditions necessary for satisfactory working of the machine are the following :—

(1) The supply of material for 4 to 5 simultaneous coatings ought not to exceed 0·6 to 0·7 cub. ft. per minute.

(2) The solid contents per cub. ft. of supply must not amount to more than 1 to 1·25 lb.

(3) 2·25 to 2·40 cub. ft. of clear water are required per minute, i. e. 0·75 to 0·80 cub. ft. for separating the lead-ore from waste and pyrites and 1·5 to 1·6 cub. ft. per minute for washing it off the table.

(4) The table must not rotate faster or slower than once every 10 minutes, i. e. 6 times per hour.

Comparative experiments touching the relative merits of this machine and the old percussion-table for the treatment of slimes, have proved that, whilst on the latter an amount of material holding 45 to 50 lb. of solid contents can per hour be passed through all stages of concentration to that of pure ore; the rotating-table accomplishes the same result in like time from a supply of similar stuff containing 90 to 120 lb. of solid matter—i. e. it concentrates fully double the quantity in the same time. The concentrated ore from both machines is equal, as regards purity, &c.; but comparing the relative amounts produced, viz. 80 per cent. by the new and 72 per cent. by the old one from an equal quantity of material, it follows that the old table loses 8 per cent. more than the new one.

As regards working expenses, the latter has also an important advantage over the former; for its supervision can be accomplished by a boy, whilst the manipulation on the percussion-table requires a strong and experienced person, whose wages are of course considerably higher, or nearly in the proportion of six to one. The expenses for working an amount of material with 5 tons of solid contents are, for instance, on the old table about 12s., on the new one a trifle over 2s. The motive power required for one of these tables is very small indeed, considering the size of the machine: 1 H.P. is quite sufficient to turn 10 to 15 of them.

There is a further special advantage connected with the construction and mode of action of this rotating table, namely, that with some modification in the arrangement of the supply and clean-water troughs, &c., and some additions, two different classes of material can, if required, be worked separately on one and the same table. For this purpose, 2 of the 4 triangular distributing-planes have to be devoted to each class of stuff. The supply- and reception-troughs have to be properly partitioned off, and there are in addition required one special cleaning-trough and one pipe for washing off the ore (both intermediate between the two pair of triangular planes), also double the number of launders for supplying stuff and conveying the products to their respective settling-pits.

Buddles.—The Convex or Centre-head buddle (Figs. 534, 535) consists of a circular pit, about 22 ft. diameter, and 1-1½ ft. deep at the circumference, with a raised centre 10 ft. diameter, and a floor falling towards the outer circle at a slope of about 1 in 30 for a length of 6 ft. The stuff is brought to the centre of the buddle in launders a, into which a constant stream of water flows; and it is distributed upon the raised centre from a revolving pan b, carrying a number of spouts, so as to spread the liquid stream very uniformly in a thin film, which flows gradually outwards over the whole of the sloping floor to the circumference. In its passage down the slope, the material held in suspension by the water is gradually deposited according to its specific gravity, and the tin ore being the heaviest is the first thrown down, and is consequently in greatest proportion towards the centre of the buddle. The overflow c for the waste and slime from the circumference of the buddle is regulated by a wooden partition perforated with horizontal rows of holes, which are successively

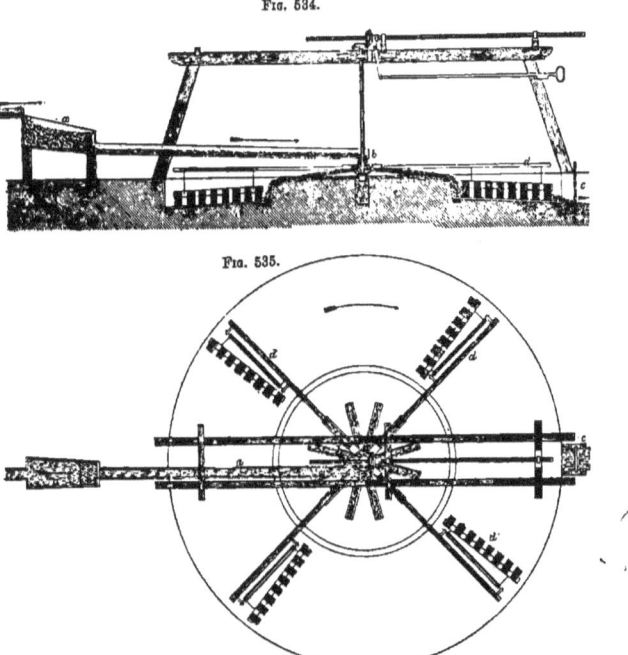

Fig. 534.

Fig. 535.

Convex or Centre-head Buddle.

plugged up from the bottom as the height of the deposit in the buddle rises. To facilitate the uniform spreading of the stuff over the floor of the buddle, and prevent the formation of gutters or channels in the deposit, a set of revolving arms d are employed, from each of which is suspended a sweep carrying a number of brushes or small pieces of cloth, and these being drawn round on the surface of the deposit keep it to an even surface throughout; the distributing spouts and sweeps are driven at about 5–6 rev. per minute. As the deposit accumulates in the buddle, the sweeps are successively raised to a corresponding extent; and the process is thus continued until the whole buddle is filled up to the top of the centre cone, which usually takes about 10 hours. The contents are then divided into three concentric portions, each about a third of the whole breadth, which are called the head, middle, and tail; the head, or portion nearest the centre, contains about 70 per cent. of all the metal in the stuff supplied to the buddle, the middle nearly 20 per cent., and the tail, or portion next the circumference, contains only a trace; the remaining particles of metal are carried off by the water in the state of slime.

In the concave buddle (Figs. 536, 537) the stuff is supplied at the centre, but is conveyed thence direct to the circumference, by revolving spouts that deliver it in a continuous stream upon a

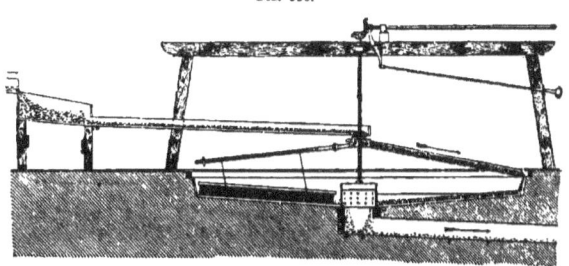

Fig. 536.

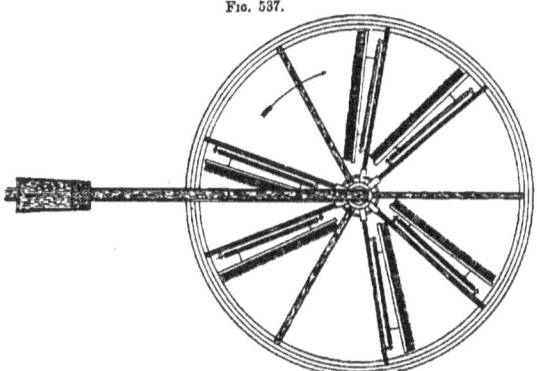

Fig. 537.

Concave Buddle.

circular ledge, from which it flows uniformly over the conical floor, falling at a slope of about 1 in 12 towards the centre; it is kept uniformly distributed by means of revolving sweeps. The greatest portion of the metal is in this case deposited round the circumference of the floor, and the slime and waste flow away through rows of holes in the sides of a centre wall; as the depth of deposit increases, the level of the overflow is gradually raised by plugging up these holes in succession.

MINING AND ORE-DRESSING MACHINERY.

Borlase's concave buddle (Figs. 538, 539) has a mechanical arrangement for adjusting the level of the central outflow, by raising a ring, that slides upon the centre vertical shaft. By this means the height of the outflow is adjusted more gradually and uniformly than by the plugged holes in the ordinary buddles, and there is less liability to waste by guttering. The sliding ring is raised by hand by a rod and lever, provided with double adjusting nuts; and the arms of the sweeps being

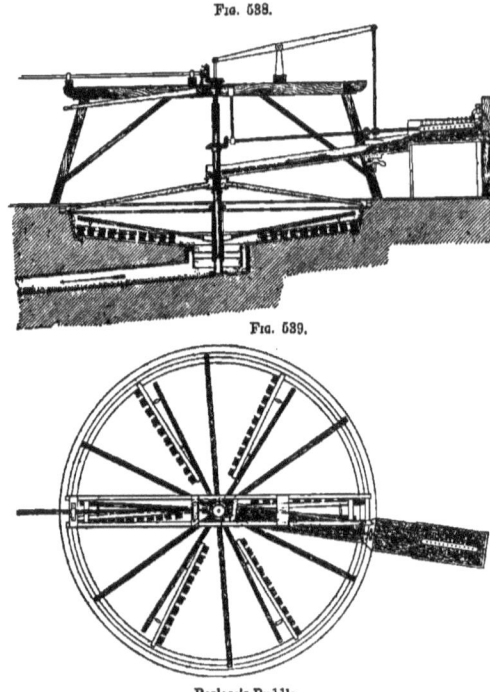

Fig. 538.

Fig. 539.

Borlase's Buddle.

supported upon the rising ring, are kept constantly at the proper height by the same adjustment. A mechanical agitator at the head of the feeding launder stirs up the stuff before entering the buddle.

The Propeller Knife Buddle (Figs. 540 to 542) consists of a cylindrical frame, 9¼ ft. long and 6 ft. diameter over all, rotating on a horizontal axis, and carrying a series of scrapers or knife-blades arranged in spiral lines round its circumference, which revolve close to a cylindrical casing lined with sheet iron, but without touching it; the casing forms the bottom of the buddle, and extends rather less than one quarter round the circumference of the revolving frame. The stuff is supplied at one end of the buddle from the hopper a, and is made to traverse gradually along the whole length to the other end by the propelling action of the revolving knives, which are fixed obliquely and follow one another in spiral lines round the cylindrical frame. A gentle stream of clear water flows down over the whole curved surface of the bottom of the buddle from a trough b along its upper

Fig. 540.

Fig. 542.

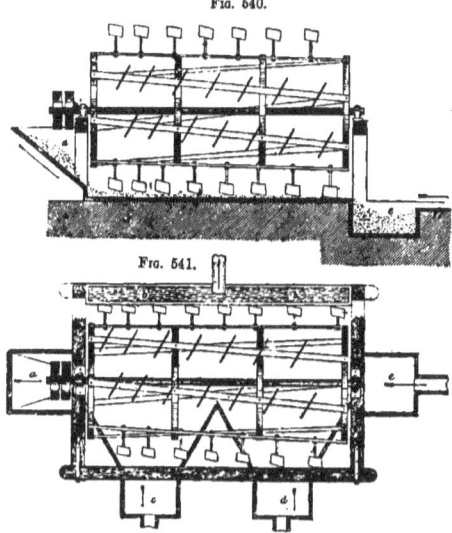

Propeller Knife Buddle.

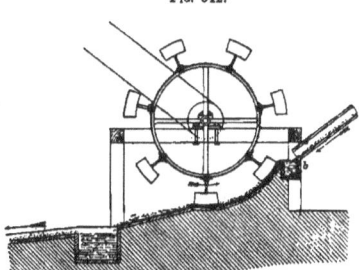

Propeller Knife Buddle.

edge, and washes away continuously into the two side hutches $c\,d$, the lighter materials that are mixed with the ore, whilst the particles of ore remain behind on the bottom of the buddle, and are gradually propelled to the farther end, where they drop over the edge into the receptacle e. The machine is driven at about 20 rev. per minute, giving the knife-blades a speed of about 370 ft. per minute. The action of this machine is found to be very perfect, the whole of the stuff being continually turned over by the knife-blades and pushed upwards against the descending stream of water, which washes out the lighter particles; the result is an unusually complete separation of the ore, in a single operation, with only a small proportion of loss in the waste. The contents of the second waste hutch d are so poor as not to pay for any further dressing; and the waste in the first hutch c containing a small proportion of slime is passed through the buddle a second time.

In Munday's improved round buddle (Fig. 543), the sand enters the receiving trough fixed on the axle, and is thence conveyed through pipes to the rim of the basin, where it is discharged; the heavier portion of the sand thus treated gradually settles down to the bottom of the basin, while the lighter portion washes away. The detention of the heavy sand is facilitated by the scrapers being fixed angularly on the arms and intercepting the sand as it flows down from the edge toward the centre, and causing it to return towards the rim. The detention of the heavy sand is likewise facilitated by the recesses formed by the circular ribs attached to the bottom of the basin. The action of the scrapers is believed to be improved by arranging them in a spiral form, one succeeding the other at distances of about 1½ in. The heaviest portion of the material treated is found to accumulate within 2 ft. of the rim of the basin.

FIG. 543.

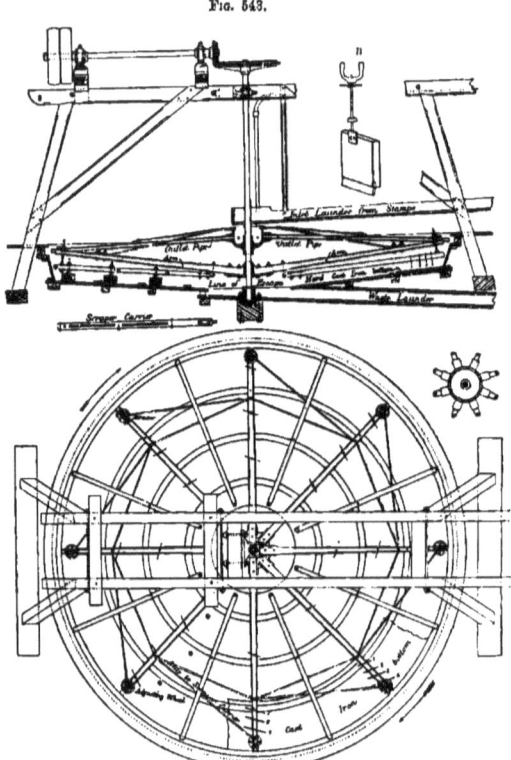

Munday's Round Buddle.

The circular basin may be made of wood, iron, or masonry, and of any convenient size from 12 up to 34 ft. diameter. Motive power may be communicated by belting or other means from existing machinery. The power to drive a 24-ft. buddle may be estimated at from 2 to 4 H.P.; the speed of the buddle is 5 revolutions a minute. Water from the stamps to be regulated according to the nature of the sand. A buddle of 24 ft. diameter will efficiently treat from 20 to 30 tons of sand from gold quartz in a day, taking it from the batteries. A slope of 1 in. to 1 ft. is generally given to the bottom of the basin. The angle at which the scrapers are set with the arms is about 18° from the right angle.

This buddle has been successfully adopted at many gold-mining claims in Victoria. But in some cases slight modifications are adopted. For instance, those used at the Port Phillip works are constructed of brickwork and cement instead of iron, and are then found to be cheaper while equally efficient. The ribs marked * in the plan are not used by Rosales at the Walhalla works; and the figures 1 to 7 show the position and angle at which the scrapers are set, Nos. 4 and 7 being 17 in. long, i. e., 1¼ in. longer in front, and 2 in. longer at the back than the others. Rosales prefers this arrangement to Munday's. The scrapers used by Rosales are of rubber, and their shape is shown at B, Fig. 543.

Alve's Concentrator.—At the Sons of Freedom Gold Mine, Boggy Creek, Victoria, one of Alve's patent concentrators has been at work for some time, and has given satisfactory results. The local mining registrar thus describes the apparatus:—

"There are three lengths of tables, each 6 ft. by 2 ft.; blankets on the tables, and wire screen ⅜ in. mesh firmly fixed on blankets; the pyrites are caught in the wire-netting, the blankets are washed every hour, then taken and put through a reducing table of the same construction with a greater pitch and 6 ft. longer. After leaving the reducing tables, the pyrites are thoroughly clean, and are dried on an iron plate over a slow fire. The blanketing tables have a pitch of about 1 in. to the foot, and the reducing tables 1½in. to the foot. At the end of the blanketing tables there is an amalgamator, which is charged with about 70 lb. of mercury, and is so constructed that everything must come through the mercury; after which there are several ripples to catch any mercury that may escape, which is next to nothing. Previous to Alve's patent being attached to the Sons of Freedom battery, I am informed by Mr. Preston, the mining manager, that the assay of the tailings, both at the School of Mines, Ballarat, and Sandhurst, was 6½ dwt. per ton. Since the fixing of Alve's patent, no particles of gold can be found outside of the machine, and in the opinion of Mr. Preston, it is the most simple and efficient pyrites saving apparatus at the present time."

These machines are also in use at the Rob Roy and Hans companies and at Walhalla. It is said that by their operation tailings containing 3 gr. of gold per ton can be profitably treated, the machines being designed to reduce the cost of working to a minimum. No machinery in motion is required, the whole apparatus being worked by gravitation. It is light, portable, and durable, and works with very little water pressure, catching in the amalgamator any floating gold there may be, and in the concentrator any auriferous grains of sand or pyrites. The cost of a machine capable of treating 100 tons of tailings per week is about 30l.

Dodge's Concentrator.—The concentrator introduced by M. B. Dodge, and made by Malter, Lind, & Co., 189, Broadway, New York, is shown in Fig. 544. The pulp from the batteries flows down the sluice a on to the distributing-board b, which is provided with spreaders. The sluice and distributing-board are inclined in an opposite direction from the concentrating table c. The pulp is divided evenly across the table by the spreader at a point just below the double arrow. Under the spreading-board is a riffle, to arrest the downward movement of the mineral when it leaves the spreader. An end shake is given by the cam d against the lug or tappet e, moving the concentrating-table gently forward, when the springs f draw the table suddenly back, striking on the upper corners of the table against the buffers g, which causes the pulp to move upwards on the machine. The pipe h delivers clear water, through cross pipes provided with small orifices, in jets against the head of the table; the water flows downward in a thin sheet against the pulp, causing the gangue to move down and out of the table at the lower end, as shown by the arrow; while the mineral, being heavier than the gangue, moves upwards, as shown by the double arrow, against the clear water into the depression and through the orifice. This orifice is provided with a wooden plug, with a suitable passage for mineral, and water enough to convey the concentrations down the short sluice into the box provided to receive them; thus forming a continuous discharge of mineral and gangue from the machine. At the lower end of the table is a receptacle to receive any fine mineral and mercury that might be left in the gangue while passing over it, which can be removed

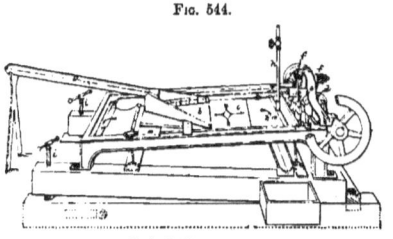
Fig. 544.
Dodge's Concentrator.

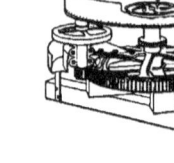
Fig. 545.
Duncan Concentrator.

from time to time, or discharged continuously. This is one of the great advantages claimed for this concentrator over all others. The machine can be regulated for different kinds or grades of ores, by the amount of clear water turned on, and by the length of stroke, which is regulated by jam-nuts on the buffers g; also, by means of the screws i, the machinery may be raised or lowered to regulate the incline of the table. A great deal depends on the pressure put on the springs f.

Duncan Concentrator.—As will be seen from the illustration (Fig. 545), this machine is neither an imitation of, nor an improvement on any of the numerous belt machines or percussion tables; but by an ingenious arrangement of gearing, a combination of movements is obtained, closely resembling the motion given to the prospector's dish in washing and panning, and very similar to that of the vanning shovel. The revolving motion of the pan a causes the sulphides and other heavy minerals to flow towards the circumference and settle to the bottom, the gangue being held in suspension and gradually carried away by the force of the current down the central discharge b, whilst a transverse movement, obtained by an eccentric working on the underside of the pan, materially assists in throwing down the heavier portions of the pulp, and facilitates the speedy and complete separation of the ore from the gangue. Means of adjustment are provided, so that the movement of the pan can be modified to suit various kinds of ore; and argentiferous galena, copper pyrites, &c., are concentrated as successfully as gold or silver ores. It is claimed for this concentrator that it has fewer component parts than other machines; that of those parts only three or four are subject to any special amount of wear; that they can be replaced, when required, at a very moderate cost; that the first cost as well as the weight for transport is considerably less than other machines of equal capacity; and that it is erected more cheaply and quickly; whilst it also occupies less space in the mill. On some classes of mineral it has proved itself superior to all other concentrators tried in competition with it.

Frue Vanner.—The Frue "vanner," made by Fraser and Chalmers, 23, Bucklersbury, and by Calvert, Cornes, & Harris, 76, Cannon Street, London, is shown in Fig. 546. a are the main rollers that carry the belt and form the ends of the table; each roller is 50 in. long and 13 in. in diameter. In order that they may be light, yet strong and durable, these rollers are made of No. 16 sheet-iron, riveted lengthways, and crowned in the centre about ¼ in. The roller is secured at the ends by rivets to light cast-iron frames. The whole is galvanised when finished, so that even the rivets are protected from rust. The roller, when finished, is strong, and only weighs 70 lb. The bolts which fasten the boxes of a to the ends of f, also fasten to f the chilled cast-iron supports of the flat bars of iron n. b and c are of the same diameter, and are made in the same way as a. The belt e passes through water underneath b, depositing its concentrates in the box 4; then, passing out of the water, the belt e passes over the tightening-roller c. b and c are hung to the shaking-frame f by straps p, which swing on the bolts fastening them to f. By means of the hand-wheels, b and c can be swung on either side, thus tightening and also controlling the belt.

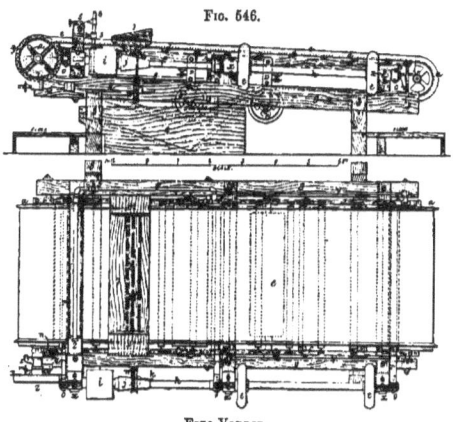

Fig. 546.

Frue Vanner.

The boxes holding a in place have slots, so that by drawing out or shortening, a can be employed as a means of tightening the belt e; and as e sometimes travels too much toward one side, this tendency can be stopped most quickly by lengthening or shortening on one end or the other of a. The swinging of b and c out of line also controls the belt, but neither has influence equal to a. The small galvanised iron rollers d and their support cause the belt e to form the surface of the evenly inclined plane table. This movable and shaking table has a frame f of ash, bolted together, and with a and a as its extremities. This frame is braced by 5 cross-pieces (shown by dotted lines). The bolts holding together the frames pass through the sides close to the cross-pieces; the cross-pieces are parallel with a and d, and their position can be understood by the 3 flat spring connections r o, which are bolted to 3 of them, one to each, underneath the frame. The belt e is 4 ft. wide, 27½ ft. in entire length—being an endless belt of rubber with raised sides.

The stationary frame g is bound together by 3 cross-timbers, which are extended on one side to support the crank-shaft h. g supports the whole machine, and the grade or inclination of the belt is given by elevating or depressing the lower end of g. This is accomplished by means of wedges; for this frame rests on uprights 3, fastened to two sills, which form the foundation of the machines in the mill. f is supported on g by uprights n, 3 on each side. These uprights are of flat wrought iron, drawn to a knife-edge at each end, and case-hardened, with bearings above and below of chilled cast iron; each middle bearing on f has one bolt-hole, and there are two of them, one on each side. The end ones have two bolt-holes, and there are four of them, one on each side. These bolts pass through the frame f, and hold to the frame the bearings of a, which work in a slot. The bearings of the head-roller are higher than those of the foot-roller; i. e. a is a trifle higher than the regular plane of the table, and the first small roller d should be raised a little.

The cross-timbers binding together g and resting on them are extended on one side, and on these extensions rest, with its connections, the main or crank-shaft h, in bearings x; the cranks are $\frac{1}{2}$ in. out of centre, thus giving 1 in. throw. The driving-pulley i forms with its belt the entire connection with the power. j is a cone-pulley on the crank-shaft h. By shifting the small leather belt, the uphill travel of the main belt e is increased or diminished at will. The small belt connects to j the grooved pulley w, which is on the small shaft k, and by means of the hand wheel can be shifted on k and held in place. The two bearings of k are fastened to the swing-box y, a cast-iron shell protecting the worm z and worm-gear l; y turns on a bearing bolted to the outside of g, and thus becomes a swing-box for swinging w and k. The object gained by this is that the weight of w and k (swinging with y) hangs on the small leather belt, and keeps it tight, so that this small belt will last for a year without slipping or breaking. Before this improvement, the small belt was constantly breaking or slipping. In some cases, this movement is accomplished with step pulleys and flat belt. A hand-wheel m is used to relieve the small belt from part of the weight of k and w; by screwing it up, k and w can also be raised, taking all the strain off the small belt, and thus stopping the uphill travel. k terminates in a worm z, which connects with a worm-gear l, travelling in a bearing bolted to the outside of g. z and l are protected from dirt and water by the cast-iron shell y enveloping both.

The short shaft which l revolves terminates in an arm s, which drives a flat steel spring q (which is a section of a circle), connected with the gudgeon of a. r are 3 flat steel spring connections bolted underneath the cross-pieces of f, and attached to the cranks of the shaft h by brass boxes o. These springs give the quick lateral motion—about 200 a minute. t are two fly-wheels. v are two rods passing from the middle cross-timber to the upper bearings of the lower uprights n. The cast-iron washers on the bolts of the cross-timber have lugs cast on them, and so have the bearings of the lower n. v pass through these lugs, and at each end are nuts on each side of the lugs. Thus v prevent the whole movable frame f from sliding either up or down, and by them f is squared. 2 is the clear-water distributor, and is a wooden trough, which is supplied with water by a perforated pipe; the water discharges on the belt in drops by grooves $1\frac{3}{4}$ in. apart. 1 is the ore spreader, which moves with f, and delivers the ore and water evenly on the belt. 3 are upright posts, which are firmly fastened to two sills, forming the foundation for any number of machines. 4 is the concentrates-box, in which the water is kept at the right height to wash the surface of the belt as it passes through. 5 are the cocks to regulate the water from the pipes 6.

The ore is fed with water on the belt e by means of the spreader 1. Thus the feed is uniform across the belt. A small amount of clear water is distributed by 2. A depth of $\frac{1}{2}$ in. of sand and water is constantly kept on the table, and the table should receive about 200 shakes a minute. The uphill travel or progressive motion varies from 3 ft. to 12 ft. a minute, according to the ore; and the grade or inclination of the table is from 4 to 12 in. in 12 ft., varying with the ore. As previously explained, the inclination can be changed at will by wedges at the foot of the machine, these wedges being under the lower end of g, and resting on uprights from the main timber of the mill. The amount of water used, the grade, and the uphill travel, must be regulated for every ore individually; but once established, no further trouble will be experienced in the manipulation. In setting up the machine, everything must be in line, except the tightener-roller c. The tightener-roller not only tightens the belt but regulates it and keeps it in place on the table. This wide belt travels uphill very slowly, so that it takes several minutes to recover its central position on the table, and at times one bearing may necessarily be several inches farther up than the bearing on the opposite side, thus twisting c out of line. In treating ore directly from the stamp, too much water may possibly be used by the stamps for proper treatment of the sand by the machine. In such a case, there should be a box between the stamps and the concentrator, from which the sand with the proper amount of water can be drawn from the bottom, and the superfluous water will pass away from the top of the box; but as mineral will also pass away with this water, there should be settling-tanks for this water, and the settlings can be worked from time to time as they accumulate.

The surface of the belt lasts for 3 months at least. As soon as the belt shows wear, it should be preserved; and the belt should never be allowed to wear to the canvas. The belt is preserved by a coat of rubber paint, prepared expressly for the purpose. Accidental breaks in the belt may be repaired by rubber cement. In renewing the surface of the belt, it must be dry. The paint should be thinned, if necessary, with benzine and naphtha, so that it readily flows from the painter's brush. The surface of the belt should be cleaned with naphtha. Then, a man standing at the lower end a paints liberally across the belt, and for 2 or 3 ft. up, as he can conveniently reach; then revolves the belt in its usual direction, i. e. upwards, for a short distance, and paints another short piece. This operation is continued until the whole surface has been painted. The rubber paint dries almost immediately, and in a very short time the belt is ready for work again. Every two months this should be repeated. The paint should be put on uniformly, but not so hastily but that the portions painted have time to become nearly dry before reaching the tightener c. In using the paint, it must be kept well stirred. The main body of the belt suffers hardly any wear, since it merely drags its own weight slowly around the freely revolving rollers; and the life of the belt is lengthened by this precaution, viz. to keep it clean from sand at every point except the working surface, thus sand cannot come between the belt and the various rollers.

The concentrates-box 4, which is kept full of water, and through which e passes, may be of any size or depth desired; in front of it may be an apron to catch any chance droppings of concentrates from the belt. Though not indispensable, it is best to have a few jets of water playing above and underneath on the belt as it emerges from the water in 4, so as to wash back any fine material adhering to the belt, and as such a method will cause an overflow in 4, the waste water, being full of finely divided mineral, should be settled carefully in a box outside. Every few hours the concentrates may be scraped out with a hoe, into a small box that can be placed under the inclined end of 4, and if this box be on wheels it can be readily run on a track to the place where the concentrates are stored: such a method seems clumsy, but there is comparatively a small quantity to handle. Frequently the sand on the belt forms a corner on each side, and to break up these corners and keep a uniform consistency on the belt, a system of drops or small jets of water can be used on each side to advantage. Such will help to increase the capacity of the machine, and will enable it to do uniformly better work. The latest improvement is to corrugate the large travelling belt; recent trials of this modification have been very satisfactory.

3 B

Halley's Percussion Table.—This machine is shown in Fig. 547. An inclined table about 8 ft. by 4 ft. is slung at the corners, the inclination being adjustable. A 3-toothed cam rotating against a spring gives a reciprocating motion. The ore to be classified is fed on at the upper end with water. By the repeated concussions the light and heavy ingredients are separated, the former flowing away with the water, while the latter are detained in the depressions of the table. Each 5 head of stamps requires one table.

FIG. 547.

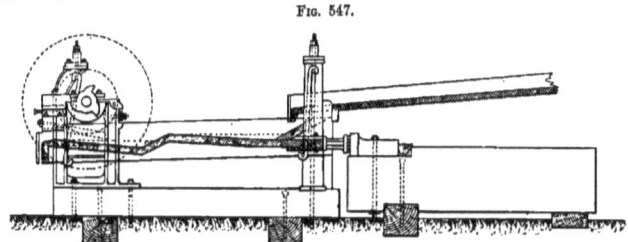

Halley's Percussion Table.

FIG. 548.

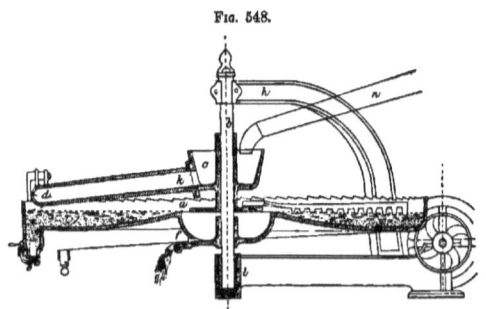

Hendy's Concentrator.

Hendy's Concentrator.—This apparatus, Fig. 548, consists of a shallow iron pan, 5 to 6 ft. in diameter, supported by a vertical shaft in the centre, and made to oscillate to and fro by means of cranks on a shaft at one side, and joined by connecting-rods to the periphery of the pan. The pan turns for a short distance at every revolution of the crank-shaft. A frame supports the central pin and crank-shaft, as well as arched arms h, which rise over the pan and sustain the upper end of the vertical shaft b. The bottom of the pan is raised in the centre around the shaft, nearly to the height of the rim, and thence descends towards the periphery in a parabolic curve, by which the movement of the particles from the centre towards the circumference is facilitated, and their passage in the other direction obstructed.

When placed for operation, the apparatus should be perfectly level. The stuff to be concentrated is delivered by the trough n to the hopper c, whence it is fed through the pipe k and distributor d, into the pan near its outer edge. The feeding extends around the whole circumference, by causing the distributor to rotate around the vertical shaft, accomplished by the movement of the pan. The upper edge of the pan is a continuous ratchet, into which 2 pawls connected with d drop during the motion of the pan from the distributor, and in the return motion give a velocity to the distributor equal to that of the pan. Continued impulses in this way keep the distributor in regular rotation around the shaft. Rake-like arms are bolted to a flange on the bottom of the hopper c, and are carried around with the distributor, serving to separate the compact mass of sand and pyrites as it settles, and breaking the scum that gathers on the surface. The crank-shaft makes 200 to 220 revolutions a minute, thus throwing the pan to and fro an equal number of times, and keeping the materials in a constant state of agitation. The heavier substances, such as the pyrites and any stray particles of mercury or amalgam, settle to the bottom, and accumulate in the lowest parts of the pan, gradually displacing the sand and lighter materials, which, with the excess of water, flow over the raised bottom at the centre, and out of the pan by a central discharge. The accumulated sulphurets are discharged at the gate e, the opening of which is regulated by a small handle at the front of the machine. The pyrites that are discharged may be received into boxes or troughs.

These machines weigh about 1000 lb. each. They are run by a belt, and usually set in pairs. The amount of water required is not large; not more than what flows away from the batteries with the sands to be concentrated. Each machine will receive and concentrate 5 tons of stuff every 24 hours; 8 tons have been put through in that time, but the product was not entirely free from sand, the presence of which is not objectionable in some processes of working, and if clean pyrites are desired, the discharge from 4 machines is delivered into a fifth, and this gives a complete clean concentration.

Fig. 549.

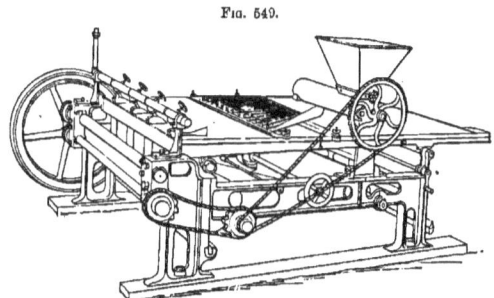

Imlay Concentrator.

Imlay Concentrator.—Favourable accounts are given of the Imlay concentrator and amalgamator, shown in Fig. 549. Described generally, the machine consists of a flat table having a copper surface, with two upturned sides and one similar end, the opposite end being open to permit the discharge of the tailings. This table is set at an incline, varying from $\frac{1}{2}$ to 2 in., the waste discharge end being lowest. At its opposite upper end, the table is provided with outlets for the reception and discharge of the concentrations. The table is supported upon 4 arms, one at or about each corner, which arms project upwardly from two transverse rock shafts at either end of the machine and about 1 ft. below the table; when motion is duly communicated, these arms vibrate to and fro, a longitudinal reciprocating movement, or a lengthwise movement, of the tables being thus effected.

A variable movement, which is the peculiar characteristic of this machine, is obtained by a very ingenious, and yet simple combination of mechanism. The main shaft of the machine is provided with an eccentric gear-wheel, which meshes with a like gear on a counter-shaft, parallel with the main shaft. This counter-shaft is an eccentric or crank-shaft, from which two connecting-rods lead to the reciprocating-table. The eccentric-gears communicate a variable rotary movement to the crank-shaft, and this through the connecting-rods, produces the variable reciprocating lengthwise movement of the table. The effect of this movement is to cause any material above a given specific gravity, laid upon or fed to the table, to travel upwardly upon the latter, while anything below such gravity will be caused to pass down the same. As the pulp almost invariably carries some mercury, the latter soon forms an amalgamated surface on the copper-plates.

Keeve or Tossing-tub.—This is shown in Fig. 550, and measures 30 in. deep, 4 ft. wide at top, tapering to $3\frac{1}{2}$ ft. at bottom; it consists of 2-in wooden staves, bound with three iron hoops, $2\frac{1}{4}$ in.

FIG. 550.

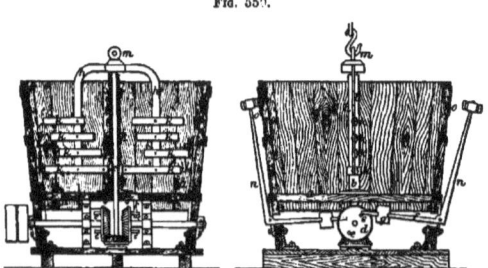

Keeve or Tossing-tub.

wide and $\frac{1}{4}$ in. thick. The staves are carried 2 in. below the floor of the tub, and 4 of them are prolonged a further 10 in., and are bolted to a platform 6 in. high, so that the floor of the tub is 18 in. above the floor of the mill. A hollow cast-iron tube b, 3 in. in diameter, passes centrally through the floor of the tub, and is bolted to it by a flange cast on the tube. This tube is traversed by a 2-in. wrought-iron shaft c, carrying a yoke h, and eye m, to which 8 flat iron stirrers k, 2 in. wide, are bolted. The shaft is rotated by gear-wheels d, to which are attached 2 pins e, which catch an arm f, bearing counterpoises g, which end in hammers n. When the tub is to be charged, a wooden wedge is pushed between the side of the tub and the hammer handle, to prevent the pins e striking the lever f. First the tub is half filled with water, and the stirrers are rotated 48 times a minute; then the tub is almost filled with tailings, which should be introduced near the outer edge.

This accomplished, the stirrers are stopped and lifted out by means of the hook and rope attached to m, and the wedge is withdrawn so that the hammers can play on the sides of the tub at o. When this has proceeded long enough to completely settle the contents of the tub, the water is drawn off at the top, and the settlings are divided into "tops," "middles," and "bottoms." The tops extending about 2 in. deep, are scraped off and discarded. The middles are retossed, and the tops from these again discarded. The remainder of the middles goes back to be buddled, while the bottoms should consist entirely of clean sulphurets.

McNeill's Concentrator.—Mr. J. R. McNeill, engineer of the Long Tunnel Gold Mining Company, Walhalla, Victoria, has made some important improvements on what are known as Brown and Stansfield's concentrating pans. These pans were formerly used at the Long Tunnel battery, but were discarded owing to the amount of wear and tear and breakages. McNeill's improvements consist in minimising the wear and tear by a different method of oscillation and suspension of the pans, but principally in lining the pans with mercury-coated copper plates, which are thickly studded with copper pins similarly coated. The materials treated by the Long Tunnel Company are pyrites and blanketings caught from the tailings after having passed through the different gold-saving appliances inside the battery, and the gold obtained is, therefore, so much absolutely saved from going into the waste heap. Mr. Ramsay Thomson, the manager of the company, saved, in 3 months, by means of the improved pans, 66 oz. 5 dwt. 8 gr. of gold, which would otherwise have been lost.

Mr. William Parker, of the Long Tunnel Extended Company, has also used McNeill's pans, and in reply to inquiries he wrote:—"I have no hesitation in saying that McNeill's is the best concentrator that I have yet seen, as it is not only a good separator of the pyrites from the quartz-sand, but it is also a good amalgamator, and, if worked judiciously, is capable of bringing together a very large proportion of the broken-up floured mercury, and forming it into amalgam. I consider this one of the most important features in connection with the machine, because simple concentration is not much trouble, but the collecting of the floured mercury in the form of amalgam is, and up to the present time has baffled the efforts of our best men."

The following is a description of the machine:—In Fig. 551, a, is the concentrator pan, the working-surface of which is lined with mercury-coated copper plates; b, vertical spindle; and c, supporting frame. d are

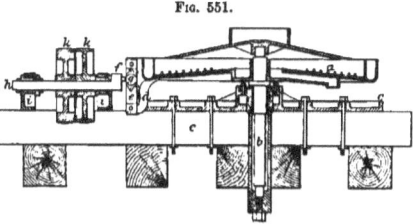

Fig. 551.

McNeill's Concentrator.

brackets formed upon or secured to the outside of the pan, having bolted to them the flat bars e, between which the sliding block f works. Such sliding block is made in halves, bolted together at its end snugs, and bored to fit the ball crank-pin g, which is secured or formed upon the end of the shaft h, which is supported in bearings i in the framing, and upon which shaft is fixed the driving-pulleys k. The concentrator pans have snugs formed on their bottoms, in which are secured ball pins, to which the ends of a connecting rod are attached, while motion is imparted to the connecting rod at its centre, where a slot eccentric is formed for a sliding block, such slot eccentric being formed by the two T ends of the connecting rod being bolted together with a distance ferrule between them, the sliding block in this case working upon a plain crank pin on the end of the crank shaft. Stays are attached to the slides of the pan, and at their upper ends meet at a common centre, which is supported in the overhead framing.

Rew and Jones's Concentrator.—This has a reciprocating motion, and is designed on the principle of the ordinary miner's pan. It is claimed to be successful even with the finest sulphurets.

Treatment of Slimes.—A very simple and effective form of self-acting slime frame or "Rack" is shown in Figs. 552 to 554, by means of which the attendance requisite is so far reduced that one boy is able to attend to 20 frames. The launder a bringing the slimes from the buddles passes

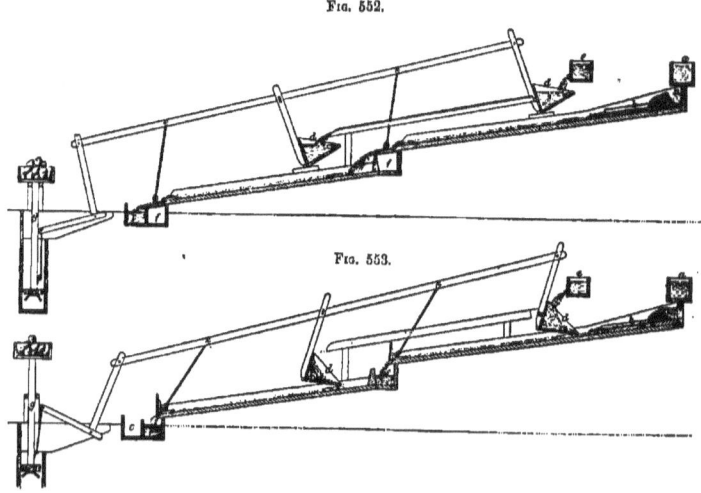

Fig. 552.

Fig. 553.

Self-acting Slime Frame.

between two rows of the slime frames, set back to back, and the delivery to each frame is distributed by a fluted spreader b, and then flows uniformly in a gentle stream over the surface of the frame, which is at a slope of 1 in 7, and is divided at the middle into two halves by a 5-in. step; the waste flows off at the bottom of the frame into the launder c. The stuff deposited on the frame is then flushed off at successive intervals of a few minutes each, by a self-acting contrivance consisting of two rocking troughs d, which are gradually filled with clear water from a launder e; when full they overbalance, and discharge their whole contents suddenly upon the top of each half of the frame. The tipping movement of the troughs opens at the same time the covers of two launders f, one at the foot of each half of the frame, into which the stuff deposited on the frame is washed by the discharge of water, the two halves being kept separate because the greater portion of the tin ore is retained on the upper half of the frame. The readjustment of the whole into the original position is effected by a cataract g of simple construction.

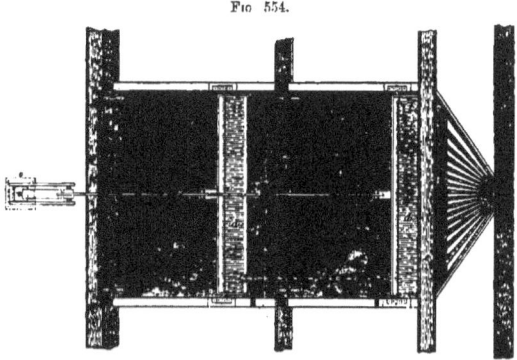

Fig. 554.

Self-acting Slime Frame.

In Fig. 555 is shown a treble stationary slime table with rotary feeder and cleaner, from which greatly improved results are expected.

It is unnecessary to repeat that the greatest difficulties and comparatively heaviest losses in ore dressing are experienced in the extraction of the ore from the slimes. The appliances generally used have been described in the preceding pages. With buddles, the ores are deposited on the table in concentric rings, the heaviest near the centre, the lightest near the periphery, and the barren sand has to be swept from the table by the water. The operation of separating the different kinds of ore and seconds must be executed by hand, by digging out separately the concentric rings. The separation is done in the direction of the radius, and it is evident that, although the buddle can be made of very large diameter (say 30 ft.), there is no distinct classification possible, as the various grades are mixed at their boundary. Moreover, the proper execution of the separation rests entirely upon the ability of the workmen; the surface of the table, when covered with a quantity of ore-slimes, becomes irregular, and the most important condition for good working of this apparatus, viz., a smooth sheet of water running down the table-surface, is not complied with any longer. It must be pointed out that in case the difference of specific gravities of ores and quartz is used for separation, the stuff which is to be treated must be properly prepared. The classification to size of slimes by aid of sieves is impossible; and, in fact, there is no practical means known to execute such classification. Rittinger invented his "Spitzkasten" (pyramidal boxes with a syphon-outlet at the bottom) with a view that, if a current of slimes passes the box, those particles which have the greatest tendency to settle will (in consequence of the reduced velocity of the current in the box) fall to the bottom and pass out sideways, whilst the lighter particles are carried away with the current. And if a series of boxes increasing in size are placed one behind the other, the velocity of the current will decrease, and the bottom-efflux of each following box will consist of particles of less tendency to settle—i. e. of finer grains.

But although the bottom-efflux from the first box will be coarser than that of any one of the following, and that of the last (and largest) box will be the finest, it cannot be denied that the grains of the efflux of any one of them cannot be of uniform size, except all grains have the same specific gravity, which is contrary to experience. If the slime which is to be treated contains galena, pyrites, and quartz, it is evident that the efflux from the bottom of any box must show the above

FIG. 555.

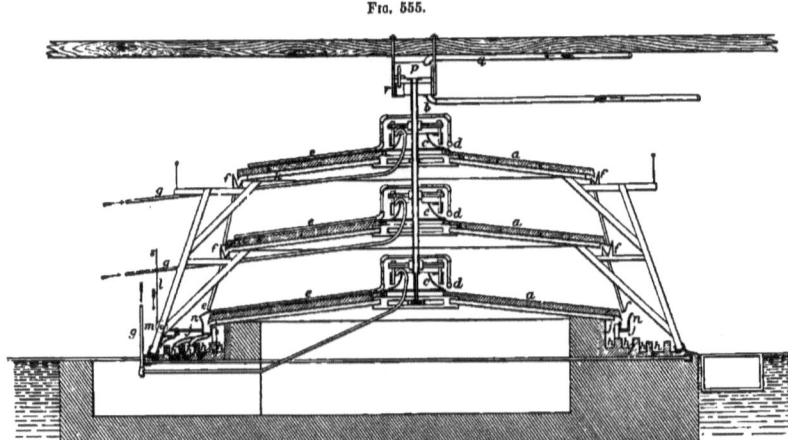

Stationary Slime Table.

minerals in sizes proportionate to the specific gravity, viz., large quartz-grains, medium pyrites-grains, and small galena-grains, and it is impossible to separate them in a jigger. In Germany such grains are called "gleichfällig," which means grains of equal falling speed in water or air; and it is clear that such grains cannot be separated in jiggers which are making use of the difference in the "falling-speeds" of the grains, which must be of uniform size and various specific gravity.

Buddles and those tables which are used for the separation of slimes must be based on a different principle, if a perfect separation is to be obtained. The stuff treated on them, viz., the bottom-efflux of one spitzkasten (at the time) consists of grains of equal falling-speed, but various sizes; the largest grains being those of lowest specific gravity, and, consequently, offering to a horizontal or nearly horizontal current of water not only the largest surface, but also the smallest resistance; and if these large light grains are exposed, together with small heavy grains (of equal absolute weight), to a current of water running over an inclined smooth surface, the large light grains will advance quicker than the others; and if the current is properly regulated, the large grains will run off the incline with the water, whilst the heavy grains are deposited on the table, the heaviest first. This operation requires, of course, a perfectly smooth surface, and all appliances constructed with the intention of retaining the clean ore on the surface can only work effectively during the first rotations of the feeder. As soon as the surface is covered or partially covered with ore, the action becomes irregular.

This was the reason why buddles were replaced by rotary conical tables with fixed feeders and cleaners. In the course of a rotation, a sector of the conical surface first passes a feeder (fixed near the centre) where it receives a certain quantity of slime: it then passes a spray of clean water, ejected from the centre, and strong enough to carry down the quartz and lighter ore-grains whilst the specific heaviest grains are settling. After that, the said sector passes underneath a water-basin fixed at a certain distance from the centre and ejecting a light spray on to the table, sufficient to sweep down the quartz grains only. At a third portion of the table similar arrangements can be made for washing down the ores of lighter specific gravity deposited near the periphery; and before the sector of the table (under consideration) comes again to the fixed feeder, the heavy ore is brushed off. Of course, separate gutters are fixed under the drip, so as to receive the various kinds of ores, seconds, and tailings. The said sector then appears again in front of the feeder perfectly cleaned.

It follows from the above description, that with a buddle, the separation is executed by grouping the various kinds of minerals in concentric rings, and that with the rotary table the different kinds are obtained on the periphery. As a matter of fact, in neither of these cases are the various kinds of minerals sharply divided; if galena is the heaviest, there will be a space on the table where it is mixed with pyrites, and it is easily understood that the dimensions of the table are of great influence. On a small table or buddle, it will scarcely be possible to divide the pyrites from the zinc-blende, and, consequently, large diameters are necessary for both apparatus, and this is somewhat difficult to execute with the rotary tables. If their diameter exceeds 15 ft., they begin to oscillate, and during a rotation the incline of the table is always changing.

Notwithstanding the superiority of rotary tables as separators over buddles, which are in fact concentrators only, the said defect could not be overlooked, and the inventor of the stationary table succeeded in eliminating this defect, and combining the good qualities of both buddle and rotary tables, by fixing the table and making feeder and cleaners rotating. Any diameter can now be used, a perfectly smooth surface (polished cement) is obtained, and the separation of the various kinds of mineral is carried out to perfection.

On the old rotary table the gutter which surrounds it under the dripping-nose is divided into as many compartments as various kinds of stuff are to be obtained, and each compartment is connected by a pipe with a settling pit. The places of discharge of the various kinds of stuff are, in this case, fixed, as they are at certain distances from the fixed feeders and cleaning-pipe. But as, on the new table, the latter are rotating, it is necessary that the said gutter is also rotating, and that as many stationary ring-shaped gutters are surrounding the table as kinds of stuff are to be obtained. The rotating gutter, which, of course, is divided into the same number of compartments, is provided in each of the latter with a spout leading into one of the ring-shaped stationary gutters. Each of the latter is in communication with a settling pit by aid of a clay pipe.

The arrangement is shown in Fig. 555: n is the rotating gutter of sheet iron fitted with wheels, which are running on a circular rail; j are the spouts leading into the fixed gutters h; g shows the feeding-pipe leading from a "Spitzkasten" to the feeding-basin c; only a part of the latter allows the efflux of the slime on the table surface a; the spindle b is hollow, and supplies the pipes d and e with water for cleaning the table; the pipes e are fixed to the rotary gutter n, and the rotating motion is given to the latter, the spindle, the cleaning-pipes, and the feeding-basin c, by an endless chain slung round the gutter n and passing over the two pulleys m to a shaft under the ceiling, which is driven again by belt.

The sprays of water for washing the slime on the table are ejected from perforated pipes, which are movable and can be placed in any position required thereby; and as the table is thoroughly cleaned after each rotation, and its action can be easily controlled, a common labourer is able to regulate the water sprays in such manner as is required for perfect separation of the various classes.

The stationary table is more expensive than a buddle; but finally it is considerably cheaper, because it is a true separator, and the stuff which has been treated on it does not require further treatment; and it works without interruption, whilst buddles work intermittently, because they must be cleaned by hand. During this time, of course, the buddle cannot work, and besides, wages are consumed for removing the deposited ores from the table. The stationary tables are made in three sizes, viz. of 19 ft. 8 in., 23 ft., and 26 ft. 3 in. diameter respectively. A table of 26 ft. 3 in. diameter is able to perfectly separate 7 tons of slime, dry weight, in 10 hours, and requires about 20 gal. of water per minute. The price of the iron parts of a stationary table of 26 ft. 3 in. diameter is, f.o.b. Australian ports, 280l., including patent licence. The cost of the brickwork is (in Germany) about 75l.

George Green, of Aberystwith, is the originator and manufacturer of a complete set of machinery for crushing, dressing, and thoroughly cleaning, in one operation, the ores of copper, lead, silver lead, and blende. A set of his crushing and dressing plant contains one or more of the following machines. Stone breakers, crushing rolls, trommels (or revolving classifiers), jiggers, water current classifiers, and buddles. The lumps of ore and stone, as they come from the mine, are fed into crushing rolls, and crushed at the rate of 16–45 tons a day, according to the size of rolls employed. Beneath these rolls is a revolving cylinder of perforated steel, copper, iron, or wove wire. The axis of this cylinder is inclined, and the crushed ore from the rolls falls into the higher end of this perforated cylinder, and, owing to the inclination and revolving motion of the latter, gradually works its way to the lower end, where such of the ore stuff as has been insufficiently crushed, and has, therefore, not gone through the perforations, is passed into an elevator, which re-delivers it into the crushing mill, whilst all that passes through the perforations is delivered into an iron trough, and so conveyed on to the next operation, which is performed by three or more revolving classifiers, similar to that just described, but with smaller perforations, these being finer in each descending classifier than those in the one above. Perforated pipes are placed inside each classifying trommel, from which a sufficient quantity of water plays on the ore stuff, to wash through the perforated plates of the trommel, all the slimes and particles that are finer than the holes, and all that passes through the perforated plate of one classifying trommel is discharged into the next in succession, whilst a sized product is delivered at the lower end of each trommel into a trough or shoot, which conveys it into a jigging machine to suit. These classifying trommels are arranged end to end in succession, each one being on a lower level than the preceding one, so that the stuff which has passed through the sieve of the higher ones is easily delivered into the inside of the next lower trommels, and each of these trommels having its own jigging machine beside it, but at a lower level than the trommel, the stuff from the latter falls into the jigging machine down a shoot.

In succession, then, each classifying trommel discharges a sized product entirely free from slime, and out of the trough surrounding the last trommel all the slime and particles are discharged into a launder which carries them to be treated apart in a series of patent saddle-back classifiers, which are vessels with four inclined sides, meeting in an inverted pyramidal point at the bottom. What takes place in these saddle-back classifiers is this :—A current of water with the slimes, &c., in suspension, delivered by the last trommel, flows into the first saddle-back classifier at one end, deposits some of its suspended matter, flows off at the other end into a second saddle-back classifier, and then onwards to the others in succession. These classifiers are of graduated sizes, the first in order being the smallest, and the current flows through them at different velocities, so that in the first and smallest, the current being the strongest, the larger particles only are deposited; smaller ones in the next, and so on.

The smaller saddle-back classifiers are provided with water-pipes attached at the bottom to deliver a spray of clean water at a head of 15-20 ft. pressure, and sufficient in volume to carry forward the dead slimes to the last and largest classifier, where the current is very slow and weak, and which has no pipes connected for clean water; the current being almost stagnant in this, all ore worth saving is sure to deposit itself.

The classified stuff from the first three saddle-backs is delivered into three jiggers, and that from the remaining two saddle-backs runs into buddles or other efficient slime washers.

The use of the jigging machines is to thoroughly cleanse the gangue and other foreign substances from the ore which has been sized by the classifiers. A jigger consists of a horizontal hutch, constructed of wood or iron, which is divided transversely into two, three, four, or more compartments. A vertical partition also extends from end to end down the centre of the hutch, along the upper part of the compartments. In each compartment, along one side of the hutch, is a plunger or piston, to produce the jigging motion of the water; in each compartment on the other side there is a sieve. Several standards are fixed on the top of the jigger to carry a longitudinal shaft, on which eccentrics are fixed; the eccentrics being connected by rods to the plungers put the water in motion. The separation is effected by the jigging action of the water with which the hutch is filled, and which is made to work up and down through the sieves by the plungers. A layer of ore is put on the sieves, which has the effect of allowing particles of the same specific gravity as itself to pass through, whilst it keeps back any particles of less specific gravity, which last are gradually washed over the end from each compartment to the next lower one, the light waste from the last compartment finally passing away. A suitable appliance for regulating the stroke of each plunger is attached.

A chief feature in this mode of dressing is the way in which gravity is made to do the principal part of the work, as any stones or slimes composing the ore stuff are floated away by the water, whilst the ore, being of greater specific gravity, is left behind. This is the principle on which the saddle-back classifiers and jiggers depend for their efficiency.

The system that distinguishes a well-arranged modern factory is carried out here, that is, the rough material enters at one end as it comes from the mine, and is delivered without hand labour at the other end, cleaned and classified into sizes ranging from a three-quarter inch cube to dust as fine as the finest gunpowder. The machines being arranged each one lower than the preceding one, gravity, aided by the water required for washing, carries the ore stuff through all the necessary operations. A large number of sets of this dressing machinery are working successfully both in this country and abroad.

Fig. 556 shows the arrangement of a complete set of the machinery. A is a crushing mill, with rollers 26 or 30 in. diameter, into which the ore stuff to be treated is put. This crusher can be driven by a water wheel as shown on the drawing, or by a steam engine, whilst the dressing machinery may be driven by a separate water wheel or steam engine. When started, the whole of the machinery is put in motion, and the work goes on regularly without hand labour. Any machine can be stopped or started by shifting the belt on or off the loose pulley.

B is a revolving classifier, covered with iron perforated plate,—the size of perforation in such iron plate is determined by the nature of the stuff to be treated. The richer it is in mineral the larger the perforation, and *vice versâ*. This classifier receives the crushed ore stuff from the trough

Fig. 556.

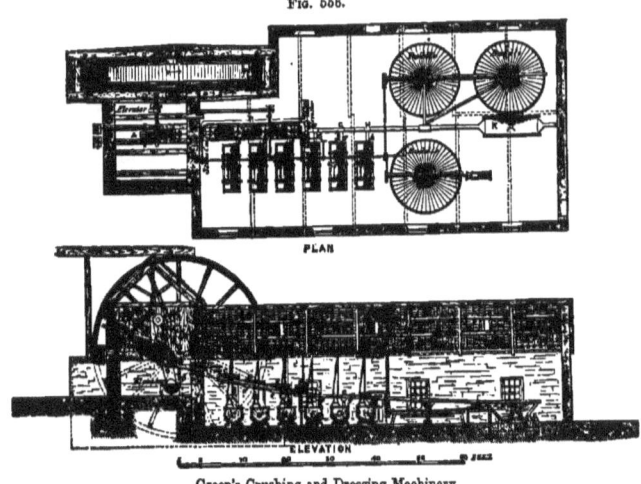

Green's Crushing and Dressing Machinery.

a immediately under the rollers. The ore stuff delivered into this classifier which has been insufficiently crushed, is passed into an elevator, which re-delivers the same into the crushing mill, whilst all that passes through the perforation is delivered into an iron trough *b*, which conveys it on to the next operation, which is performed by

C, D, E,—which are three of Green's Patent Automatic Classifiers and Feeders. Each of these classifiers is covered with perforated iron plate of a suitable sized perforation to suit the first classifier B, each descending one being finer than the one above, so that B, the first, is the coarsest, and E, the fourth, is the finest. Perforated pipes are placed inside of each classifying trommel, from which a sufficient quantity of water plays on the ore stuff to wash through the perforated plates all the slimes and particles which are finer than the holes, thus all that passes through the perforated plate of one classifying trommel is discharged into the next in succession, whilst a sized product is discharged at the end of each into iron troughs or shoots *c, d, e*, which convey it into a jigging machine to suit. In succession, then, each classifying trommel discharges a sized product entirely free from slime, and out of the trough *e* surrounding the last all the slime and finer particles are discharged into a launder, which carries them to be treated apart from the troughs in F, G, H, I, K, which are five

Patent Saddle-back Classifiers and Feeders, which are made with inclined sides meeting in an inverted pyramidal point at the bottom. A current of water, with the slimes, &c., in suspension, delivered by the last riddle, flows into a classifier at one end, deposits some of its suspended matter, and flows off at the other end into a second classifier and then onwards to the others. These classifiers are of graduated sizes, the first in order being the smallest, and the current flows through them at different velocities—so that in the first and smallest, the current being the strongest, the largest particles are deposited, and smaller ones in the next, and so on. The smaller classifiers are provided with water-pipes attached at the bottom to deliver a spray of clean water at a head of 15 or 20 ft. pressure, and sufficient in volume to carry forward the dead slimes to the last and largest classifier, where the current is very slow and weak, and which has no pipes connected for clean water—the current being almost stagnant in this, all ore worth saving is sure to deposit itself. The classified stuff from F, G, H, is delivered by the troughs, f, g, h, into the jiggers, F, G, H, and the stuff from I and K through trough i, k, into either buddles or other efficient slime washers.

C, D, E, F, G, H, are six three-compartment self-acting jiggers, which receive the classified stuff delivered by the classifiers as explained above. The jigger comprises a horizontal hutch, constructed of wood or iron, which is divided into two, three, four, or more compartments, by transverse ends and partitions. A vertical partition extends along the upper part of the compartments; and on one side thereof are a set of plungers or pistons to produce the jigging motion of the water, whilst a series of sieves are placed on the other side. On the top of the partitions there are fixed a number of standards to carry a longitudinal shaft on which the eccentrics are fixed, and which, being connected by rods to the plungers, put the water in motion. The separation is effected by the jigging action of the water with which the hutch is filled, and which is made to work up and down through the sieves by the plungers. A layer of ore is put on the sieves, which has the effect of allowing particles of the same specific gravity as itself to pass through whilst it keeps back any particles of less specific gravity, which last are gradually washed over the end from each compartment to the next lower one—the light waste from the last compartment finally passing away. A suitable appliance for regulating the stroke of each plunger is attached.

Buddles, or other efficient slime machines, are attached to the larger classifiers, and the stuff flowing in a perfectly even current from the bottom of such classifiers on to each separate buddle, makes them quite self-acting, and of course more effective. The finest or dead slimes are worked by an ordinary paddle trunk. The whole is complete and continuous, and worked without labour from the roughest prills to the finest slimes,—each distinct size having a machine suited in speed and action for its treatment. The drawings represent a set of 6 jiggers and 3 buddles, but more or less jiggers and buddles may be used as circumstances direct. Below are rough estimates of the plant required for treating medium quality ores, not including crushing mill, motor, or shafting.

Fig. 557 represents a 4-compartment self-acting jigger, which receives the classified stuff delivered by the classifiers. The jigger comprises a horizontal hutch, constructed of wood or iron, which is divided into two, three, four, or more compartments; and on one side thereof there are a set of plungers or pistons to produce the jigging motion of the water, whilst a series of sieves are placed on the other side. On top of the partitions there are fixed a number of standards to carry a longitudinal shaft, on which the eccentrics are fixed, and which, being connected by rods to the plungers, put the water in motion. The separation is effected by the reciprocating action of the water, with which the hutch is filled, and which is made to work up and down through the sieves by

MINING AND ORE-DRESSING MACHINERY.

Plant consisting of the following Machines.	Approximate Weight.	Approximate Horse-power Required.	Measure for Shipment.	Medium Quality Ore Treated per Day.	Approximate Cost.
	tons cwt. qr.		cub. ft.	tons	£
3 self-acting jiggers (three compartments) 2 Green's patent classifying trommels 1 iron saddle-back classifier 1 wood classifier 2 Green's round buddles (ironwork only) 1 agitator (ironwork only)	4 12 0	3	800	15	220
5 self-acting jiggers (three compartments) 3 Green's patent classifying trommels 2 iron saddle-back classifiers 2 wood classifiers 3 Green's round buddles (ironwork only) 1 agitator (ironwork only)	7 14 0	4	1300	25	363
7 self-acting jiggers (three compartments) 5 Green's patent classifying trommels 2 iron saddle-back classifiers 2 wood classifiers 4 Green's round buddles (ironwork only) 2 agitators (ironwork only)	10 16 0	5¼	1800	35	506
9 self-acting jiggers (three compartments) 6 Green's patent classifying trommels 3 iron saddle-back classifiers 3 wood classifiers 5 Green's round buddles (ironwork only) 2 agitators (ironwork only)	13 14 0	7	2550	45	652

the plungers. A bed or layer of ore is put on the sieves, which has the effect of allowing only particles of the same specific gravity as itself to pass through, the speed and length of stroke being adjusted according to size of stuff delivered to jigger whilst it keeps back any particles of less specific gravity, which last are gradually washed over the end from each compartment to the next lower one; the light waste, which contains no ore, from the last compartment finally passing away. A suitable arrangement for regulating the stroke of each plunger is attached.

	Cost.	Working Capacity.
	£ s. d.	
4-compartment self-acting jigging machine	52 0 0	..
3 ,, ,, ,, ,,	42 10 0	..
3 ,, ,, ,, ,, (large)	52 0 0	..
2 ,, ,, ,, ,, ,,	42 0 0	..

DRESSING MACHINERY.

Fig. 558 represents a side view of one of a series of trommels, the number of which, as well as the size and mesh of perforators, is always adapted to suit the class of ore to be operated upon. They are arranged so as to effect a perfectly automatic classification of the granular portions, which are

Fig. 557.

Green's 4-Compartment Self-Acting Jigger.

Fig. 558.

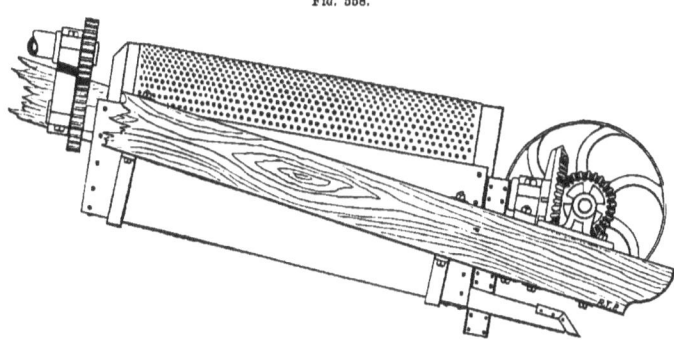

Green's Trommel.

received bodily from the crusher or pulveriser, as also to give a continuous and uniform feed to the jiggers, which are placed to receive the ore so classified direct. They are constructed with a view to the utmost durability and ease of access, and can be covered with steel, copper, or iron perforated sheet or wove wire.

	£	s.	d.
Prices of trommel with trough and shoot, but without gearing	12	10	0
" " shoot and gearing	14	0	0

GALENA AND BLENDE.

The dressing of galena and blende was made the subject of a paper contributed by E. du Bois Lukis, A.M.I.C.E., to the 'Proceedings' of the Institution of Civil Engineers (James Forrest, Esq., Secretary), and as this paper records detailed observations made whilst preparing galena and blende for market, it is replete with most valuable information. Following is a summary of the main facts observed.

The ores dealt with were from the mines of Sentein, in the Pyrenees. They were intimately mixed in the proportion of 8-10 per cent. galena, 15-20 per cent. blende, and gangue consisting of hard quartz, quartzose rock, schist, &c. The market lead-ore obtained included 16-20 oz. silver per ton. The blende did not contain sufficient silver for valuation. By experiments in the laboratory, it was found that the galena lost very little silver by fine crushing and washing.

The machinery, supplied by George Green, of Aberystwith, was erected in existing scattered buildings. A method of arranging the whole of the required plant under one roof, with slight

FIG. 559.

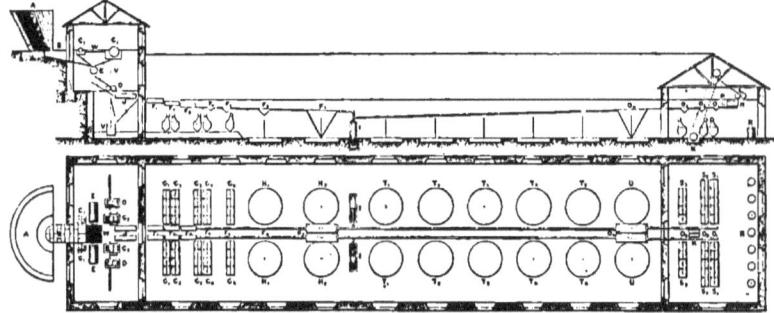

Galena and Blende Dressing Machinery at Sentein.

modifications, that would render the dressing-floors more efficient, is given in Fig. 559, the substantial structure indicated being necessary on account of the climate of the Pyrenees, where protection from frost and snow is indispensable. Water-power was used, being abundant.

The first point of importance is the size of the ore-stuff. The operations should be so conducted as to separate the marketable minerals in as large grains as possible. The reduction to absolute fineness should be gradual, and intermediate dressing operations resorted to, for the finer the particles through subdivision the more difficult and costly become the dressing operations, and the greater the loss of minerals. In the present case the ore-stuff is crushed to pass through a riddle having square holes of 0·18 in.

Jiggers are used as far as possible to separate the minerals from their gangue, as they entail only one-tenth the cost of buddle-work as regards labour, and less loss of mineral. Ore-stuff that would pass through a riddle with holes 0·02 in. diameter, could be jigged perfectly well, if freed from slimes. Ore-stuff that cannot be jigged is divided into two classes, "fines" and "slimes," and is dressed by buddling. The rich heads of "buddles," when concentrated to about 60 per cent. of metallic lead for galena, or blende of about 42 per cent. of metallic zinc, are worked by tossing and packing in a kieve or dolly, so as to obtain marketable ores of about 69 per cent. lead and 48 per cent. zinc respectively.

This operation of ore-dressing is divided into eight sections : (1) picking for prills ; (2) breaking and crushing ; (3) sizing and classifying ; (4) jigging ; (5) buddling ; (6) re-crushing, pulverising, and dressing chatts ; (7) dollying, or tossing and working the flat-buddle ; and (8) treating and collecting slimes.

(1) *Picking for Prills.*—At the Sentein mines the ores were too intimately mixed to render picking of practical value ; but at mines whose ores are rich in silver this process is most serviceable. The ore-stuff is tipped into a large masonry hopper A (Fig. 559), at the bottom of which cast-iron plates B are placed, and a revolving picking-table might be also used. The ore-stuff is washed by a jet of water from a hose, enabling the workmen to quickly distinguish and pick out the prills or pieces of virgin galena or blende, as they rake the ore over the plates towards the grating W. The plates B are so arranged as to allow the water to run off into a launder below, carrying with it fine particles of ore and slimes. This water passes through the double trommel E, and thence to the dressing-floors, where its contents are treated. The prills being put to one side, are again picked over before sampling for market:

(2) *Breaking and Crushing.*—The ore-stuff is next raked over the grating W, made of flat-iron bars, 3 in. deep and ½ in. thick, set on edge 1·57 in. apart. The ore that passes between these bars is conducted by a launder, in which water flows, to the double trommel E, which has an inner sieve, with holes 0·79 in. square, and an outer sieve, with holes 0·18 in. square. The inner sieve is merely to protect the outer sieve from unnecessary wear. The ore that will not pass through the outer sieve is conducted in a launder to the crushing-rolls D. The fine stuff that passes through the outer sieve goes at once to the dressing-floors.

The rock and stones that remain on the grating W are put into the stone-breakers C_1 C_2, where they are broken into fragments that will pass through a ring 1·57 in. diameter. These fragments also go into the double trommel with the small stuff that has passed through the grating, the fine ore-stuff going to the dressing-floors, the coarse to the crushing-rolls. The stone-breakers at Sentein are of two sizes. The smaller one, C_1, for medium sized stones, has a mouth 9·84 in. long by 5·91 in. wide. The larger one, C_2, for large stones, has a mouth 19·68 in. by 9·84 in. The faces of the jaws are of cast iron, chilled to a depth of 1·18 in. ; the wearing edges of the toggles, and the bearings in which they work, are also chilled. Only two such stone-breakers were used at Sentein, but they were insufficient for the work. The large stone-breaker was driven by belting from a water-wheel, 14 ft. diameter and 3 ft. breast, and it often had to be worked day and night to keep one-half of the floors supplied with ore-stuff during the day. With an additional pair of stone-breakers such dressing-floors would be amply supplied with ore-stuff.

The quantity of stuff that may be crushed by rolls in any given time depends upon the size to which the ore-stuff has been first reduced by the stone-breakers, and it was found that the fragments should be able to pass through a ring 1·57–1·97 in. diameter.

The crushing-rolls at Sentein were three in number, one 24·02 in. diameter by 15·95 in. wide, and two others, each 14·96 in. diameter by 12·99 in. wide. It was found that, with the assistance of the small stone-breaker, the large rolls could do nearly as much work as the two small, one of which was assisted by the large stone-breaker.

In the proposed dressing-floors, rolls 26·77 in. diameter by 18·11 in. wide are designated as being more efficient. These rolls consist of three parts: the shafting, 5·91 in. square, of wrought iron; the core, which should be well keyed on to the shafting and the same width as the roll, is of cast iron, not chilled; and the ring, of cast iron, with the face chilled to a depth of 1·18 in., about ·15 in. thick, with grooves 0·79 in. deep and 0·79 in. wide diagonally across the face, half-way across; six grooves in one half alternating with six grooves in the other half of the face. These grooves do not continue to the edge of the face, but to within 1·18 in. of it. The core is made about 1·18 in. less in diameter than the inside diameter of the ring, so that the space between may be wedged up with dry deal wedges, driven in from both sides, which are then keyed up with small soft iron wedges. To keep the rolls tightly pressed together when working, levers and a balance-box were found to answer better than springs or rubber. Rubber cushions soon deteriorate, and workmen do not pay enough attention to them. The rolls worked better when only one roll was connected with the driving-shaft, the second roll working by friction on the first. By adopting this method more work was done than when either equal or differential gearing connected the two rolls, and less driving power was required. The driving-roll made 10 revolutions per minute to 8 revolutions of the second roll, and consequently wore away faster; but by occasionally changing the relative positions of the rolls the ill effects of unequal wear were obviated. The speed of the driving-rolls at the periphery is about 60 ft. per minute. The ring lasted 5–6 months, working 6 days a week.

The crushed ore-stuff is conducted by a launder below the rolls to the riddle J, covered with sieving having holes 0·18 in. square, equal to circular holes about 0·20 in. diameter. The stuff passing through goes to the dressing-floors; the coarser grains are returned to the crushing-rolls by the elevator V. This elevator consists of a rubber belt 5·91 in. wide and 0·39 in. thick, passing over and under two pulleys, fixed at different levels. The top pulley is worked by a small-toothed wheel and pinion. Small buckets are bolted on to the belting, at intervals of about 4·92 ft., which take up the ore and discharge it at the upper level as they turn over the top pulley. The elevator is inclined at about 80° with the horizontal plane.

(3) *Sizing and Classifying.*—The ore-stuff having been crushed small enough to pass through the sieve J, consists of particles of all sizes, from fine dust to the largest grains that could pass through the sieve. To permit the separation of the particles of different densities by dressing operations, those of equal volume must be collected together, and others eliminated as much as possible, by mechanical means. To do this, riddles and classifiers are used. The riddles F_1 to F_3 are cylindrical, and covered with copper plates, pierced with circular holes of varying diameters, and they make 10 revolutions per minute. The first riddle, F_1, has holes 0·16 in. diameter; the second, F_2, holes 0·12 in.; and the third, F_3, holes 0·08 in. diameter. The first is 7·22 ft. long, the second and third are 6·23 ft. All three are 23·62 in. diameter.

The first classifier or spitzkasten F_4 (Fig. 560), has a depth of 5·91 in. below the level of the bottom of the launder; the second, F_5, a depth of 7·87 in.; the third, F_6, a depth of 11·81 in., the inclined planes making an angle of 45° with the horizontal line. A pipe, of 0·98 in. bore, enters

Fig. 560.

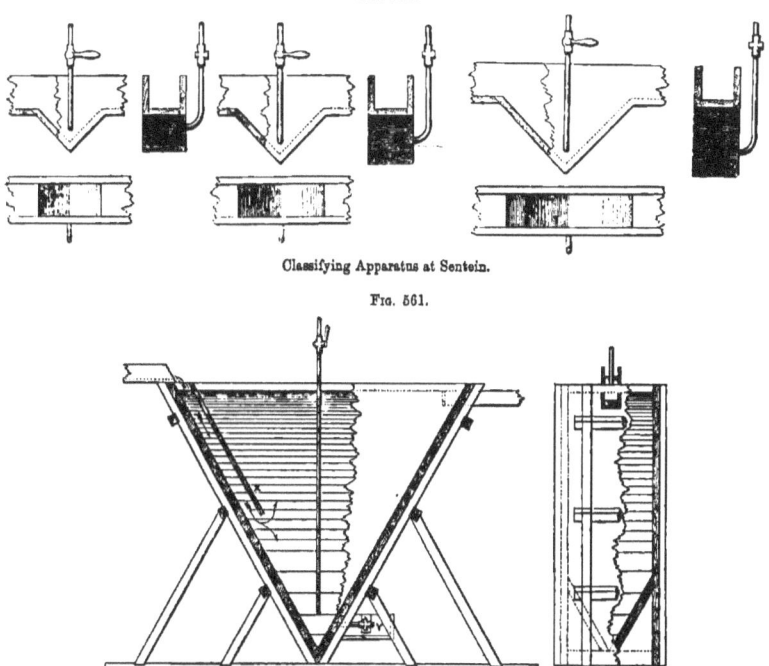

Classifying Apparatus at Sentein.

Fig. 561.

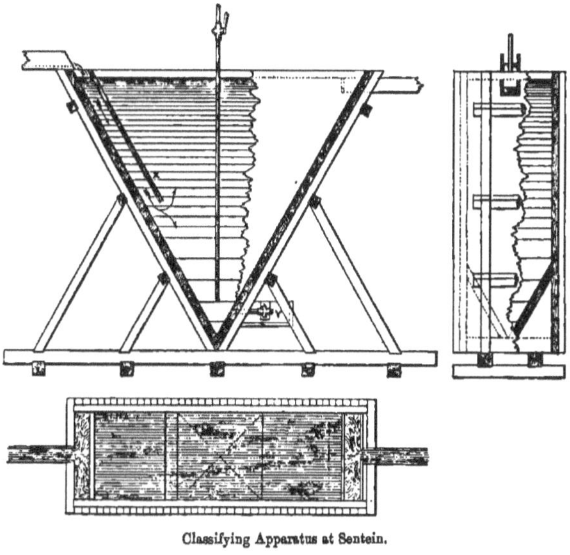

Classifying Apparatus at Sentein.

the side of each classifier about 1·97 in. from the point, and is connected with a water-main of 2·95 in. bore, under pressure of a head of water of about 6·56 ft. The pressure to each classifier is regulated by a tap fitted to each small pipe. The large classifier, F_7 (Fig. 561), for slimes, is 9·84 ft. deep, and 3·28 ft. wide, the inclined planes making an angle of 60° with the horizontal line. A straight pipe, with a bore of 0·98 in., fitted also with a tap, and connected with the water-main, passes down the centre, reaching nearly to the bottom. Holes are made in the small classifiers, in that side opposite to which the hydraulic pipe enters, which can be partly closed with wooden plugs, so as to regulate the feed of ore to the dressing-machines. A tap is fixed to the bottom of the large classifier, through which the thick slimes are drawn off and supplied by a launder to the buddles.

The principle upon which the action of these classifiers depends is as follows:—When particles of matter of varying densities are carried along by a stream of water in a launder, the heaviest flow in the stratum of water nearest the bottom; those of the next lower density in the stratum of water immediately above, and so on. Further, when particles of varying densities are simultaneously immersed in a column of water, and allowed to subside freely, the heaviest reach the bottom first. Thus when the crushed ore-stuff is carried along the launder to the first classifier, both these principles come into action, and the heaviest particles can be drawn off from holes in the bottom of the classifier, while the lighter ones, further assisted by the upward flow of water from the hydraulic pipe, flow on to the second classifier, and so on.

The ore that does not pass through the first riddle consists of particles 0·16–0·20 in. diameter and is supplied by a launder to the first jigger G_1; that not passing through the third riddle, 0·12–0·16 in. diameter, goes to the second jigger, G_2; that not passing through the third riddle, 0·08–0·012 in. diameter, goes to the third jigger, G_3. An iron trough, under each riddle, receives the stuff and conveys it to the classifiers which are fixed to that launder. On reaching the first, F_4, the heaviest and largest particles, 0·04–0·08 in. diameter, fall to the bottom, and are drawn off through the holes in the side to supply the fourth jigger, G_4; and the pressure of water in the small pipe connected with the bottom of the classifier is so regulated that the whole of the slimes, with much fine stuff, rise and flow on to the second classifier, F_5, where the same action is repeated, no slimes being allowed to reach the fifth jigger G_5, which takes stuff from about 0·02 to 0·04 in. diameter. The third classifier F_6, in the same way supplies "fines" to the first buddle H_1, with but little slimes. The water in the launder, now charged with only very fine ore-stuff and slimes, passes over a straight-edge for the whole width of the large classifier F_7, and under the board X, Fig. 561. The liquid from the bottom flows through the tap Y to the buddle H_2, and, as the water becomes free from muddy matter suspended in it, rises to the surface, and flowing over another straight-edge in a thin film almost clear and limpid, is used to work a wheel I, 9·84 ft. diameter and 19·68 in. breast, and supplies the motive force for the buddles. The holes in the riddles are kept clean by a spray of water, under pressure, from a perforated pipe which plays upon them from the outside along their whole length.

(4) *Jigging.*—The jiggers are five in number, each consisting of four compartments; the compartment or hutch is equally divided into a jigger-case and a piston-case, Fig. 562.

They are made of pine deals 2·95 in. thick, laid on a keel, Fig. 563; all longitudinal joints are tongued with dry oak 0·98 in. by 0·04 in. The structure is fastened together by five 0·59 in. bolts vertically through the cross-heads of cast-iron, Fig. 564, and across by bolts A, Fig. 562, passing through the divisions.

The cases thus made are supported on stands, Fig. 565 to which the shafting-stands are fastened by two bolts. Light rods are bolted between the stands near the head, through holes X X. A

DRESSING MACHINERY. 389

turned shaft 1·73 in. diameter runs through the stand-heads working in brasses well lubricated. On to this shafting the eccentrics are keyed, so that the piston-rod attached may be plumb over the centre of the piston-cases. The piston is shown in Fig. 566. Fast-and-loose pulleys are also put on the shafting in a convenient position for the belting from the driving-shafting.

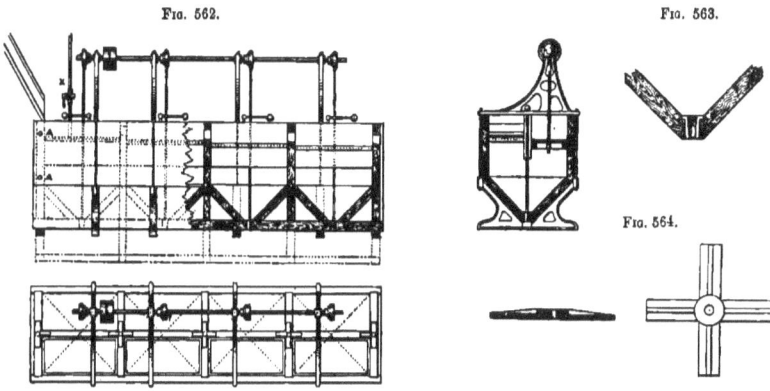

Fig. 562.   Fig. 563.

Fig. 564.

Classifying Apparatus at Sentein.

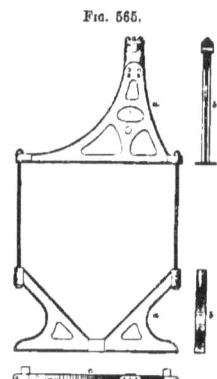

Fig. 565.

Classifying Apparatus at Sentein.

The sieves or bottoms of the jiggers are put on a grating of cast-iron, Fig. 567, which rests on planking screwed on all round the jigger-case. A similar grating over the sieve is kept in place by planking screwed on in the same manner as that below, but with copper screws to facilitate changing the jigger-bottoms.

Each compartment of the first jigger G_1 has a depth from the lip at the overflow to the top of the sieve of 2·76 in.; the compartments of the second jigger have a depth of 2·56 in.; those of the third jigger of 2·36 in.; those of the fourth jigger of 2·17 in.; those of the fifth of 1·97 in.

The lip of each jigger-case has a fall of 0·79 in. from one compartment to another, and is covered with a cast-iron plate, Fig. 568, to prevent wear.

The eccentric, which drives the plungers of the jigger, Fig. 569, is made in three parts. One of these, shown in back elevation at b, is keyed to the shaft, and has two bolts projecting from it; the others are the eccentric and eccentric-strap. A is the eccentric which can be moved laterally for the length of the slots Y (a), through which pass the bolts of the fixed part b. These slots allow a displacement of the eccentric of 0·79 in. from the dead centre, which is equal to a stroke of 1·57 in. of the piston. By loosening the nuts on the bolts, and giving a slight blow to the side of the eccentric, the distance from the dead centre to the centre of the eccentric will be slightly altered, which distance is equal to half the difference made in the length of the stroke of the piston. Thus, suppose the eccentric to be at the dead centre, by moving it 0·04 in. out of the centre a stroke of 0·08 in. is obtained.

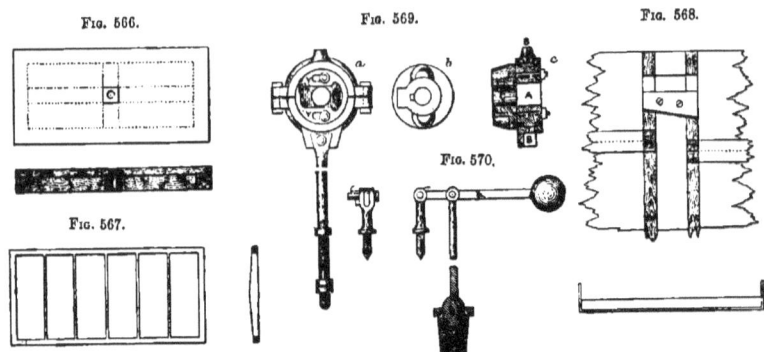

Classifying Apparatus at Sentein.

Various experiments have been made to find a metal that will wear the least, of which the eccentric may be constructed, and close grained strong cast-iron has been found to answer as well as any thing, besides being cheapest. All working parts should be accurately fitted and well lubricated. The sieves are of copper plates punched with conical holes; the rough side is uppermost. The four compartments of the first jigger, G_1, have plates with holes 0·22 in. diameter; those of the second jigger, G_2, holes of 0·18 in.; those of the third, G_3, of 0·14 in.; of the fourth, G_4, 0·12 in.; and of the fifth, G_5, 0·10 in.

A valve, valve-rod, and lever, Fig. 570, complete the jigger, which is placed over a long trunk of deal 2·95 in. thick. having four compartments, corresponding with the four compartments of the jiggers. When the valve is raised the ore is received in these compartments, the overflowing water being conducted to the slime pits, &c.

To begin operations, a bedding of ore 0·79-1·18 in. deep, galena being used in the first compartment, is placed on the plates of the jigger, mixed galena and blende on the second, and blende on the plates of the remaining compartments. After some weeks' work, chips from miners' drills accumulate on the bottoms of the jiggers, and form a better bedding than galena or blende, for these latter are too brittle. It would therefore be better to use small chippings from a fitting-shop, or disks from punched iron plate, to commence with. These should be a little larger than the holes in the bottoms of the jiggers, so as not to pass through them.

The jiggers are filled with water from the tap X, Fig. 572, and the ore-stuff is supplied through launders from the several riddles and classifiers. At the bottom of each launder, there should be a distributing-plate, Fig. 571, made of cast-iron, so that the ore may enter at the head of the first compartment without disturbing the bedding. This was not done at Sentein, where the want of it caused some inconvenience. As the ore-stuff is supplied, it travels onwards towards the outflow at each stroke of the piston, assisted by a continual flow of water supplied from the tap X; the heavy particles percolate through the sieves into the hutches below according to their densities. Thus

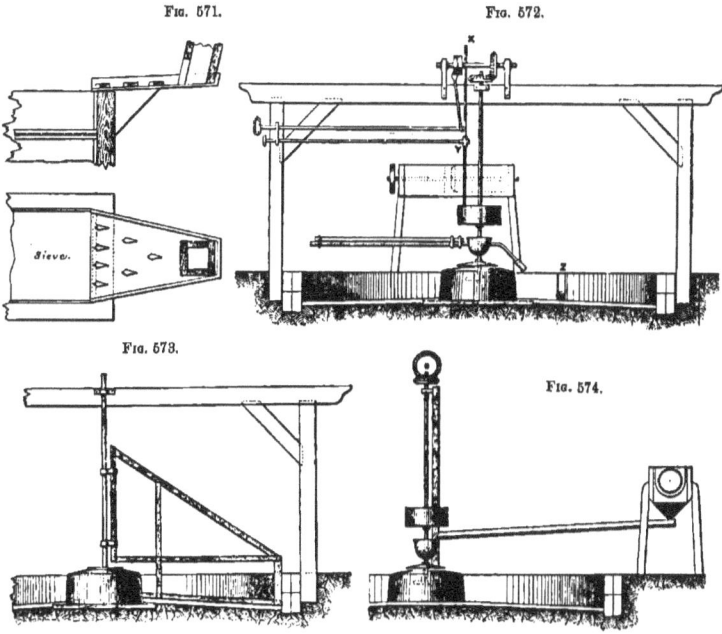

Fig. 571.   Fig. 572.
Fig. 573.   Fig. 574.

Classifying Apparatus at Sentein.

galena passes into the first compartment; mixed galena and blende into the second, and blende into the third and fourth compartments, of each jigger. The waste passing over the lips of the fourth compartment is almost free from mineral. It may sometimes be necessary, in order to prevent loss of mineral in the waste, to allow a little gangue to remain in the fourth compartment with the blende.

The mixed ores of the second and fourth compartments of each jigger are again treated by separate machinery, being further crushed, sized, jigged, and buddled, &c., until the waste is free from mineral, and the galena and blende are ready for market. About 80 per cent. of all the ore-stuff is treated by jigging, the remainder goes to the buddles. The fourth jigger, however, does more than one-third of the work, and requires special attention. The results of the assays, as shown in the following Table, demonstrate where modifications should be made.

ASSAYS OF RESULTING ORES IN THE HUTCHES.

Ore Stuff. Average Sample.	1st Jigger.	2nd Jigger.	3rd Jigger.	4th Jigger.	5th Jigger.
Per Cent. { 9½ Pb 21 Zn	6 Pb Zn ?	7½ Pb Zn ?	9½ Pb 20½ Zn	24½ Pb Zn ?	6 Pb 22¾ Zn
1st compartment or hutch ..	per cent. 67½ Pb	per cent. 72 Pb	per cent. 76 Pb	per cent. 77¾ Pb	per cent. 77¾ Pb
2nd " " ..	36½ Pb 22½ Zn	49½ Pb 17½ Zn	30½ Pb 32 Zn	30½ Pb 33 Zn	28 Pb 36 Zn
3rd " " ..	7 Pb 39½ Zn	9 Pb 45 Zn	6½ Pb 45 Zn	8 Pb 44 Zn	7½ Pb 45½ Zn
4th " " ..	6 Pb 46 Zn	5 Pb 50 Zn	6 Pb 46 Zn	6½ Pb 45 Zn	5 Pb 48 Zn
Waste	0 Pb 12 Zn	0·3 Pb 7½ Zn	0·5 Pb 4½ Zn	0·7 Pb 4 Zn	0 Pb 4 Zn

The ore-stuff supplied from the crushers contained about 9½ per cent. Pb and 21 per cent. Zn; this being classified showed that the galena and blende, not being so hard as the gangue, were crushed finer than could have been wished, but it was not to be prevented. The first hutch of the first jigger only gave 67½ per cent. of lead, which was too low, and both the bedding and the stroke had to be altered to improve the result, bedding being added and the length of the stroke being slightly diminished, as some grains of gangue percolated into the hutch. The blende was too rich in lead, so some bedding was taken out of the second compartments and more mixed ore was produced; this also assisted in diminishing the loss of blende, which, especially for the first jigger, was enormous. The results sought were to obtain galena containing 75 to 78 per cent. of metallic lead; with blende at 47 to 49 per cent. of metallic zinc, not more than 3 per cent. of metallic lead, and waste to contain not more than 0·5 per cent. of Pb, and 1 to 1½ per cent. Zn, and this was done. By frequently testing the resulting ores on a vanning shovel, and rubbing the samples very fine with a hammer, the relative percentages of lead and of blende can be easily ascertained. Such tests should be verified by assays, and a little practice will enable an ore-dresser to arrive at estimates within ½ per cent. of the truth in the case of lead. This more especially refers to the blende ores. If the proper result is not obtained, it must be sought by altering the length of stroke of the piston, and adding or removing some of the bedding on the jigger-bottoms.

The length of stroke of the piston should be just sufficient to lift the mineral on the surface of the bedding to a height equal to the diameter of the particles under treatment. Thus, for the first jigger, the grains of 0·16–0·18 in. (square sieve), need a stroke of about 0·35 in. to raise them to a height of 0·16–0·18 in.; and in the fifth jigger it requires a stroke of about 0·12 in. to raise the grains a height of 0·02–0·04 in. No rule can be given, but practice will soon show what length of stroke is necessary.

The number of strokes per minute of each piston is the next thing to attend to. The grains should be allowed sufficient time after each stroke to fall through a distance equal to their diameters; the strokes being given in quick succession allow the heaviest grains just to settle in the bedding when the next stroke further tends to free the descending particles from those of less density which surround them, and thus by degrees permit them to reach the plate and pass through into the hutch below. The number of strokes per minute for the first jigger should be about 200; for the second about 220; for the third, 240-250; for the fourth, 260-270; and for the fifth jigger, 280-300.

The ore should be frequently drawn off from the hutches to allow sufficient space inside the cases for the proper working of the piston and the water. The galena is taken from the trunks below the jiggers to the flat-buddle, where it is freed from any slimes or fine blende, and then put to pile ready for market. The blende is ready for sampling without treatment on the flat buddle.

The loss of lead in the waste is accounted for by mere specks of galena on grains of gangue, and in the blende to the lamellar fractures which cannot be saved without more cost than profit.

(5) *Buddling.*—The " fines " and " slimes " supplied by the classifiers, F_6, F_7, are treated by round buddles. The first for very fine-grained stuff; the second for the slimes H_1 H_2.

These buddles are circular (Fig. 572), 14 ft. in diameter, and 13·98 in. deep.

They are built of stone or brick, preferably the latter, and are well cemented. In the centre, the cast-iron cone A is placed on firm ground, two pieces of wood being bolted to the base. Small broken stones or bricks beaten down form a bottom on which a layer 1·97 in. thick, of a mixture of cement, hydraulic lime, and sand, is evenly laid with an inclination of 2·17 in. from the outer circumference of the cone to the inner circumference of the brick or stonework. The vertical shafting being fixed in position, a gauge is adjusted to it, Fig. 573, which, being turned round the shafting, regulates both the circle of masonry and the level of the cemented bottom. The vertical shafting is turned 1·50 in. diameter, and rests on a footstep, F (Fig. 574), which can be lubricated by a small hole in the cap X. This hole is closed by a wooden plug when the buddle is at work, to prevent sand from running in. No brushes are used with these buddles, but a " hose " and a " rose " are substituted. The " hose " D is a zinc pipe of 1·97 in. bore, soldered on to an iron pipe that passes through the side of the centre-piece C, and is screwed to the perforated pipe H, shown in detail in Fig. 575. The zinc pipe is pierced with holes about 0·04 in. diameter, and 1·18 in. apart in three rows. A fourth row may be added if the ore under treatment needs much water. The " rose " is a copper cylinder E, 1·97 in. bore. It is screwed on a bent iron pipe of 0·98 in. bore, and also passes through the side of the centre-piece opposite the " hose," and is screwed to the perforated pipe H. The hose and rose are supplied with water through this perforated pipe from the trough above B, to which it is keyed. A continuous flow of water from a supply-pipe X, 0·98 in. bore, fitted with a tap at Y, regulates the supply and keeps the requisite quantity in the trough. In using a cast-iron centre-cone with a smooth surface, runnels do not form in the ore-stuff in the buddle as is the case when wooden centre-cones are employed. The hose supplies water to the buddle, the supply increasing as the radius of the buddle from the cone to the circumference, and does excellent work even with very fine slimes, when the quantity of water used is properly regulated.

The ore-stuff is supplied by a launder to the centre-piece C, Fig. 576, and passes through the holes at the bottom to the cap, which distributes it evenly all round the cone as the shafting revolves. Layer by layer the ore-stuff covers the bottom of the buddle, the heavier particles remaining at the head, and the lighter ones being washed down the inclined plane to the tail. A small ring of water is kept at the tail of the buddle to prevent the ore from escaping with the water, and, as the stuff rises in the buddle, pieces of wood are placed in the slot Z, to keep the water at the requisite level. When the deposit reaches the top of the cone, the work is stopped, a groove is cut from head to tail with a shovel, and samples are taken, which must be crushed and washed on a vanning-shovel to judge where the divisions should be made; for at the head the ore is rich in

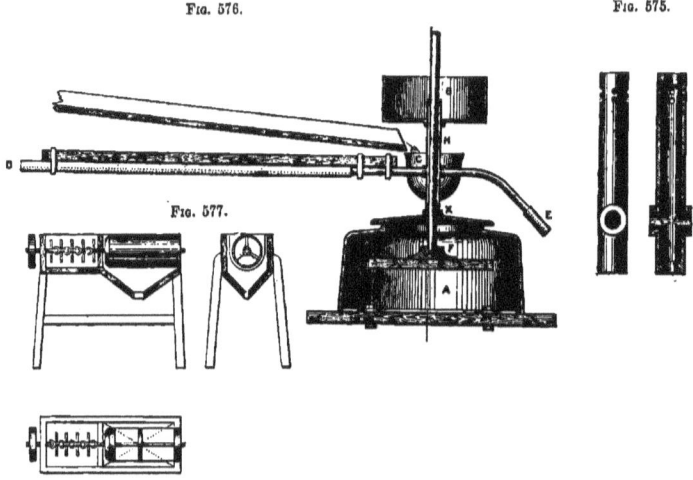

FIG. 576.   FIG. 575.

FIG. 577.

Classifying Apparatus at Sontein.

galena, then follow two qualities of mixed ore of galena, blende, and gangue, and lastly poor tailings. Rings are marked round, and the different qualities are taken away for further treatment in other buddles, T_1 to T_5. The heads, after being once reworked, will be ready to go to the dolly; but the mixed and the poor middles must be treated several times if the waste is to be made as free from mineral as possible.

The buddles T_1 to T_5, are fed by hand. The ore is put into the trough of the mixing-machine (Fig. 577); each buddle being furnished with one, water is supplied by a pipe regulated by a tap, and as the mixing machine revolves, the ore passes through a sieve with holes 0·18 in. square to a launder and thence to the buddle. These buddles are of the same construction as those that have already been described; but they may be made a little larger, namely 17·06 ft. diameter, and 16·54 in. in depth.

The ore must be regularly supplied to ensure the proper working of the buddles. When full, these are emptied like the others, the different classes of ore being treated over again with other ores of approximate richness and size, until all the gangue is extracted, and the galena separated from the blende. The galena is enriched to 50–60 per cent. Pb; the blende to about 42 per cent.

Zn, and 3 per cent. Pb, and then tossed and packed in the dolly. The waste from the buddles contains 0·25-0·5 per cent. Pb, and from 1 to 1·5 per cent. Zn.

The motive power for the buddles is furnished by a water-wheel I, driven by the overflow from the large classifiers F_7, O_4.

(6) *Re-crushing, pulverising, and dressing chatts and ragging.*—The mixed product of the jiggers in the second and fourth compartments, called chatts or ragging, must be separately treated. The chatts from the first three jiggers are raised by the elevator K, and conducted at a higher level by a launder to a pair of crushing-rolls L, to be further crushed to pass through a riddle covered with a copper plate having holes 0·08 in. diameter. These rolls are 14·96 in. diameter by 12·99 in. wide, and are driven at a speed of about 50 ft. at the periphery per minute. The rolls suggested in the plan of proposed dressing-floors have a diameter of 19·68 in., and a width of 13·39 in., as likely to be more efficient, for those in use at Sentein in this department did not do enough work. The surfaces of the rolls are chilled to a depth of 1·18 in., but they are not grooved. The construction is the same as that of the rolls already described, excepting the mode of keeping them pressed together. Instead of levers and a balance-box, a spring, formed of layers of thick rubber, was used, which could be tightened when required. It was considered that the rubber spring yielded too much, and that more rigidity and better work would be done by levers, and a balance-box weighted to suit requirements. Precautions should be taken that crushing rolls should be always supplied with ore-stuff, otherwise the external ring and even the levers are liable to be broken.

The chatts from the fourth and fifth jiggers are pulverised in one of Hall's grinding-mills P. It consists of two renewable cast-iron grinding plates with chilled faces B, Fig. 578, bolted to two permanent driving-plates C D, within a casing of cast iron. These grinding-plates are slightly concave, and have races cast in them; representing those cut in millstones. The upper one is 20·67 in., and the nether one 21·65 in. diameter. Their axes are set 1·18 in. out of centre, so as to produce an eccentric motion between them when set in rotation. The nether plate is directly connected with the driving-motor by gearing, and makes about 200 rev. per minute. The axis of the upper driving-plate C, Fig. 578, is truncated, and projects for some distance beyond the casing of the mill. The projecting part carries a worm-wheel gearing into a screw. To produce the necessary grinding action, the upper plate is driven at a much lower speed than the nether one. This may be effected either by a pinion on the driving-shaft, communicating motion to a larger pinion on the axis of the endless-screw by a driving-chain, as in Fig. 578; or else by a frictional brake. This brake consists of a conical gland, in three or four parts, inserted between the bearing of the axle of the endless-screw and the axle itself. A screwed collar forces the gland between the bearing and the axle which is adjustable, and the required speed of the upper plate is obtained by the friction produced. The plate can be prevented from turning, but this should be avoided, as unequal wear of the grinding face would ensue. The upper plate is kept pressed upon the nether plate by levers with movable weights which can be raised or lowered by the screw F, Fig. 578. The ore is regularly and gradually supplied to the plates through the central projection, but is not reduced to extreme fineness in one operation. Repeated grinding is resorted to, so that between each operation particles of galena and blende may be separated by dressing, of as large size as possible.

The crushed and pulverised ores are conducted by a launder to four classifiers O_1 to O_4, of the same construction as those previously described, but of different depths. The first classifier, O_1, has a depth of 4·72 in.; the second, O_2, of 6·30 in.; the third, O_3, of 7·87 in.; but the fourth, O_4, is similar to that shown by Fig. 561. These divide the grains of ore according to their respective sizes. The first two classifiers supply ore-stuff to two five-compartment jiggers S_1 S_2; the third classifier feeds a four-compartment jigger S_3. The fourth classifier supplies "fines" and "slimes" to a round buddle U. The jiggers and buddle are of the same pattern as those previously described.

Fig. 578.

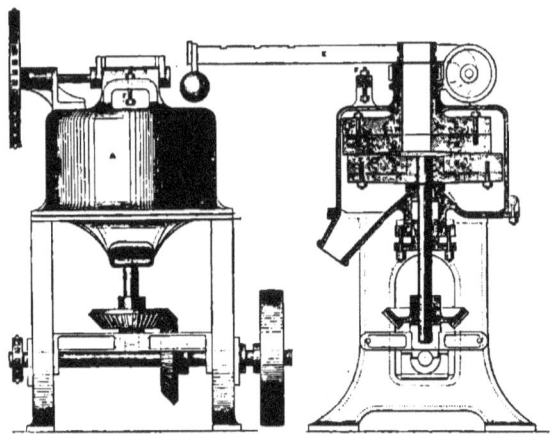

Classifying Apparatus at Sentein.

The depth of each compartment of the first jigger, from the lip at the overflow to the top of the sieve, is 2·36 in.; of the compartments of the second jigger 2·16 in.; of those of the third jigger 1·97 in. The plates of the first jigger have holes 0·12 in. diameter; those of the second and third jiggers 0·10 in. The piston of the first jigger gives about 250 pulsations per minute; that of the second about 270; and that of the third about 300 pulsations.

The results directly obtained from these jiggers depend on the quality of the chatts under treatment, whether they are rich or poor; but the principles upon which they are worked are the same as those of the jiggers G_1 to G_5. It was very difficult to completely free the galena in this department from the blende, and second-class lead ores were only obtained averaging 69 per cent. Pb. The blende contained as much as 4 per cent. Pb. to 42 per cent. Zn. To obtain these results the chatts were re-crushed several times, and treated again and again. It was estimated that 8–10 tons of chatts were passed through this pair of rolls and the grinding-mill per day.

(7) *Dolly-work, or tossing and packing.*—The different classes of fine ore having been enriched by buddling, to 50–60 per cent. of lead for galena, and 39–40 per cent. of zinc for blende, are further enriched by dollying or tossing.

The dolly is a tub made of oak 1·77 in. thick, strongly bound round with iron hoops. In the tub is a fan A (Fig. 579). The dolly rests firmly on the flooring, but should never be packed round the bottom. Manual labour is used at Sentein to work the dollies, but mechanical means should have been adopted. In Cornwall a lighter fan is driven by overhead motion, which can be easily thrown out of gear, and the fan removed. Lukis suggests a plan, as shown in Fig. 579, to work the jig dolly by mechanical means so far as the striking is concerned. D represents the main shafting upon which a bevel-wheel can be put into gear by a clutch and lever (not shown). The toothed wheel E under the dolly drives three pinions, one of which is shown at F keyed on the vertical shafting of the striker, supported by the stand O. At the top of the shafting is a cam K working against a stop H, fixed on the square bolt M, at the end of which is a striker L of about 8 lb. weight. A strong spring is placed between the striker and the head of the stand, capable of giving a blow of about 30 lb. when the striker is pulled back 1·57 in. The stop H can be adjusted by a screw, so that it may give lighter blows if necessary.

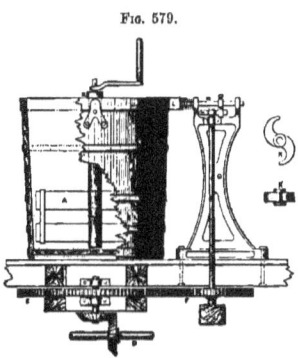

Fig. 579.

Dolly Tub at Sentein.

The ore-stuff is treated in the following manner: Water is put into the dolly to the level of the top of the fan, and the fan is made to revolve whilst a man throws in the ore-stuff, until the ore and water nearly fill the dolly. The fan is made to revolve for a few minutes longer, and then removed from the tub as quickly as possible without stopping the rotary motion of its contents, and the strikers are at once set to work. The heaviest particles subside to the bottom of the tub, the lighter ones rising to the surface.

The number of blows, and their power, depend upon the coarseness or fineness of the ore-stuff. The finer the ore the lighter the blows and the quicker in succession; 80-150 blows per minute are required, and the knocking is continued for 40-50 minutes until the ore has "packed" or settled in the tub. The water is then drawn off from a plug-hole in the side of the dolly, and the ore examined with a vanning shovel. At the top will be found a stratum of sand and a little galena and blende, then a stratum of mixed galena and blende, and lastly galena ready for market. When blende is treated, the top stratum contains sand and blende, the middle stratum is put to pile ready for market, and a little at the bottom of the tub is treated again for the lead in it. Ore-stuff containing 60 per cent. of metallic lead, when finished in the kieve, was divisible into three layers, of which No. 1 assayed 5 per cent., No. 2, 41 per cent., and No. 3, 74 per cent. of lead. Ores of different sizes should on no account be mixed before treatment in the dolly. R shows the position of the dollies in the proposed floors.

The flat buddles (Fig. 580) are erected outside the floors and covered with a light shed. Two would do all the work of the floors. About 1½ cwt. of galena from the jiggers is put on one side of the water-supply X; the water is turned on, and with a hoe-shaped tool the mineral is passed little by little across the stream, which washes out slimes and small particles of blende from the galena. Blende is not submitted to this operation. The slimes are deposited in the trunk at the end of the buddle. Some lead ores containing about 60 per cent. Pb, can be enriched to about 78 per cent. Pb by this means. The flat buddle is a simple wooden structure with an iron plate fixed at X, upon which the ores are worked.

(8) *Treating slimes, &c.*—Unfortunately the automatic means of treating the ores at Sentein did not extend to the slimes. These were collected in pits, which were occasionally emptied, and the accumulated stuff was put aside for future operations. For the economical treatment of slimes they should, however, not be allowed to dry and cake. Exposure to the atmosphere for any length of time decomposes the ores, and particles that were once free adhere to others, and it is then very difficult and costly to so mix them in water as to separate the valuable mineral from the gangue. The whole of the thick water from the dressing-floors should pass over a large classifier like that shown in Fig. 561. The concentrated ore-stuff drawn from this classifier could be treated directly by various means, such as shaking-tables, or self-acting Cornish frames, or even buddles.

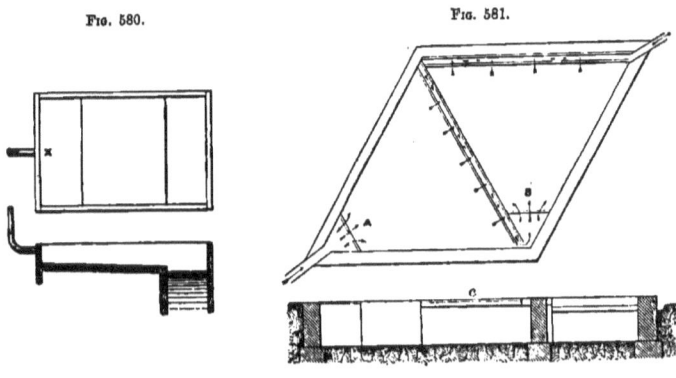

FIG. 580.   FIG. 581.

Classifying Apparatus at Sentein.   Classifying Apparatus at Sentein.

Finally, the water is conducted to triangular slime-basins (Fig. 581); the stream flowing over a straight-edge A at one of the angles of the first triangle, spreads out as it advances towards the opposite side, losing its velocity, and depositing the particles held in suspension, passes in a thin film over a straight-edge extending along the base of the first triangle into a parallel launder below, which carries it to the head B of the second triangle, and so on, to others, until the water is clear enough to be returned to the river.

The cost of dressing 8235 tons of ore-stuff was at the rate of about 2s. 6¼d. per ton, from which 879 tons of market lead ore, and 2720 tons of market blende, were obtained. About 30 persons were employed, men being paid 2·25-2·75 francs per day, and lads and women 1·25-1·50 franc per day.

It was found that one small crusher, one large pair of rolls, a set of 5 four-compartment jiggers, with the necessary trommels and elevator, could be worked by an overshot water-wheel 22 ft. diameter by 4-ft. breast, supplied with 42·3 cub. ft. of water per minute. This is equal to 17·6 H.P., but taking the effective at 70 per cent., the power utilised would be about 12·3 H.P. The addition of a large stone-breaker would need about 4·6 HP. extra, say 17 H.P. for one-half the department, treating crude ore. The one-half of the department treating chatts and ragging, that is, one pair of rolls, one grinding-mill, two five-compartment jiggers, one four-compartment jigger, elevators, &c., required about 8 H.P. Therefore about 50 H.P., as obtained from water-power, would be needed to work the whole dressing-floors to treat 65 tons of crude ore per 12 hours. As already mentioned, the overflow from the classifiers supplies the motive power for the round buddles.

The clear water was supplied to the various machines for dressing purposes through a main 2·95 in. internal diameter, extending the entire length of the building under a head of about 6 ft. pressure. Rubber belting connected the various jiggers and buddles with the main driving shaftings.

TIN.

The Cornish system of tin dressing has received much attention from the local Inspector of Mines, R. J. Frecheville, who remarks that the loss by the processes to which the tin-stuff is subjected at the mines is very considerable. This is proved by the fact that during 1884, 1326 tons of black tin, sold to the smelters for 41,055l., were obtained from the tin streams in the parishes of Camborne, Illogan, and Redruth alone. Even then the sands and slimes escaping from the mines were not perfectly untinned, as the dressing operations carried on at Gwithian and Portreath beaches plainly indicate.

From a number of samples taken and tests made, Frecheville computes that the mines save 84·37 per cent. of the quantity and 89 per. cent. of the value of the tin in the stone treated. This cannot be regarded otherwise than as a good result, but Frecheville believes it is possible in some degree to further increase the efficiency of the process.

Having gone very carefully into the matter of cost at two leading Cornish mines, he finds that, including every charge from the time that the tin-stuff is delivered to the stamps until the ore is ready to be sent to the smelters, the cost amounts to 5s. per ton of tin-stuff dressed. This includes a charge for repairing the floors, but not for depreciation in value of machinery. Adding 2d. per ton for this item, would make the total cost 5s. 2d. per ton. This is the weak part of the process. The cost, owing to the large amount of manual labour employed, is too high, and, in these days of improved machinery of all descriptions, should most certainly be reduced; though, as labour is cheap and plentiful in Cornwall, there is not the same necessity as exists in some other countries, the United States for instance, of introducing automatically working machinery.

The Cornish system of tin dressing is of native growth, and naturally, in many ways, eminently suitable to local conditions, but certain details and appliances can be grafted on to it, that will tend both to increase its efficiency and diminish its cost. In this connection the following suggestions are made:—

(1) Constant assays should be made, by an independent man, of the sands and slimes leaving the floors, as a check on the dressers.

(2) Abundance of clean water is essential for good dressing; where the supply is deficient it should be supplemented by constructing reservoirs in suitable localities.

(3) Stone-breakers should more generally replace the muscular arms of Cornish maidens.

(4) Great economy in stamping would result from the employment of Husband's Oscillating Cylinder Stamps.

(5) The treatment of the stamped stuff should invariably be preceded by classification, that is, not only should the slime be separated from the sand, but the latter should be sorted acccording to the different sizes and weight of the grains.

(6) When the stuff is discharged from the stamps direct into a buddle, much of the slime tin passes to the tail, and is not saved by the subsequent operations to which this is subjected.

(7) As the "strips" that formerly were universally used in front of the stamps to a certain extent classified the stuff, their abandonment is decidedly a retrograde step in dressing. These "strips" are, however, by no means to be compared in efficiency with "pointed boxes," and besides, the stuff deposited in them has to be shovelled out by hand, while from the pointed boxes it is delivered without expense, and at any point required.

(8) For the treatment of the coarser sand delivered by the pointed boxes, the employment of Rittinger's double side-blow percussion table is recommended. It would give three products, namely, ore fit to be sent to the calciner, ore associated with vein-stuff for the pulverisers, and valueless waste. It is the best continuous working machine yet invented for dealing with the coarser portions of stamp work. Borlase's buddle is well adapted for the middle fine sands; and for dressing slimes, the Cornish frame is an excellent appliance, especially when carefully constructed and arranged like those to be seen in the stream-works of John Williams at Tuckingmill.

(9) A great deal has been said of late years about the wonderful results that would be obtained if jiggers were used in dressing tin-stuff. With the object of ascertaining whether these assertions are well founded, Frecheville took at several mines samples of the stuff just as it passed through the stamp grates, which in each case were of the size known as No. 36, the perforations of which are ·028 in. diameter. Results proved that these samples of stamped tin-stuff contained 6 to 45 lb. of black-tin per ton, which, owing to its physical condition, would be very unsuitable for concentration by jiggers.

Jiggers could no doubt be applied with advantage for the treatment of stuff where the grain of the black-tin is coarse, such for instance as that produced by Mulberry and Drakewalls Mines, but for the bulk of the ore yielded by the principal tin lodes they are not likely to prove satisfactory machines.

(10) In all departments of dressing, the object aimed at should be to attain the highest possible degree of efficiency consistent with the employment of the least possible amount of manual labour; not only on account of the greater cheapness of the work performed by self-acting machines, but also because when the back of the master is turned the quality of the work remains the same.

SILVER.

Fig. 582 shows an elevation of a complete dressing plant designed by Commans & Co 52, Gracechurch Street, London, E.C., for the Ravenswood Extended Silver Mining Co. in Queen.t. nd, and is said to be the most perfect arrangement of continuous ore dressing machinery in the colony. The ore from the mine is delivered direct to the top of the building by means of a hoist. Later on it is intended to employ an endless aerial ropeway some 720 yd. in length, on the Otto system. The ore to be treated is a rich argentiferous lead, somewhat finely disseminated, and requiring careful sizing. As the ore arrives on the top floor, it is tipped over a coarse screen; the lumps are passed on to the stone-breaker, and after being crushed rejoin the ore that falls through the screen, the whole passing on to a large revolving trommel where a preliminary sizing takes place. The finer particles go direct to a series of sizing trommels or classifiers, over the jiggers, the coarser being delivered, if rich, on to a picking table, or otherwise direct to the crushing rolls (Fig. 583). The ore, after passing through the crushing rolls, falls into the elevator pit, and is raised to the

Fig. 582.

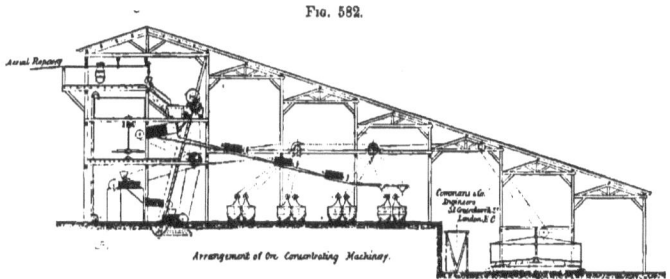

Commans' Dressing Plant.

Fig. 583.

Commans' Crushing Rolls.

sizing trommels. Two pairs of crushing rolls are used, so as not to unnecessarily reduce the ore, and to retain the same in a granular condition, and so avoid a loss by production of slimes. The rolls are fitted with forged steel shells, and are of the most modern design, the upper pair being fitted with an automatic feed to ensure a regular supply of ore. The sieves of the sizing trommels, as with the jiggers, have a gradually reducing size of hole, the smallest being $1\frac{1}{2}$ mm. diameter. The ore below these trommels is sized in pointed boxes, the sand passing to the fine jiggers; and the very fine slimes, which cannot be effectually treated on the jigging machine, after flowing over large

3 F

V-shaped boxes, are concentrated on patent Linkenbach buddles (Fig. 584), they being the best form of concentrator for the purpose.

This buddle is so arranged that the slimes and wash water are distributed over the bed or table of the buddle, which enables a very large concentrating surface to be secured, the latter being no less

FIG. 584.

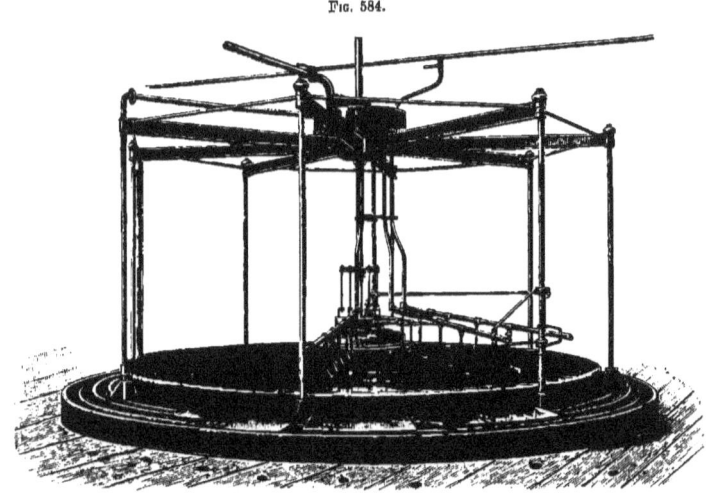

Linkenbach Buddle.

than 26 ft. in diameter; a diameter which would be impossible to obtain by employing an ordinary continuous revolving buddle. The wash-water pipes are supported from the vertical shaft by means of a light framework, and are arranged in such a manner that the concentrates can be washed off at any desired point into the channels round the bed. The slimes are delivered from the pointed boxes by means of a pipe, and fed into the spreader or distributor at the centre of the buddle. The washwater is conveyed to a small tank attached to the upper part of the framework (or it may be delivered through the vertical shaft, which can be made hollow for this purpose), and from thence it passes to the horizontal pipes; the flow over the bed being regulated as the circumstances require. The finished products are washed off the bed over aprons into annular troughs, from which the minerals are delivered into their respective settling pits. These aprons are secured by means of angle-iron rings to the framework supporting the water pipes, and revolve along with the latter. The tailings flow direct into the channels set apart to receive them. Water is also passed through these channels, the products being thereby carried off and deposited in settling pits, out of which they are dug as the pits fill up. By using these aprons in connection with a series of troughs circulating round the bed of the buddle, the ore under treatment can be separated into any number of component parts varying in richness as desired. This, combined with the very large concentrating surface, gives to the Linkenbach huddle advantages for the concentration of rich slimes possessed by no other concentrator at present in use.

COPPER.

Copper ore is raised in the same manner as tin ore, but it presents a marked contrast to tin ore in being very much less finely disseminated throughout the lodestuff with which it is associated; the coarser spots or patches in which it is met with necessitate consequently a very different treatment from that adopted in dressing tin ore. The most abundant ore of copper is yellow copper ore, also called "copper pyrites," which has a bright yellow colour, much like good brass; it is a sulphide of copper and iron, containing when pure only 34·6 per cent of copper, with 30·5 per cent. of iron, and 34·9 per cent. of sulphur. The other principal ores of copper are the red, black, grey, purple, and green ores. The red and black ores are oxides, containing when pure 89 and 80 per cent. of copper respectively; red, which is the more common of the two, is quite brittle, and is easily broken up into a red powder. Grey copper ore is a sulphide, containing when pure 80 per cent. of copper; it has much the appearance of metallic lead, but may be broken up by a hammer. Purple copper ore, also called "horseflesh ore," is a sulphide, but not so rich as the grey, part of the copper being replaced by iron; when pure it contains nearly 70 per cent. of copper. Green copper ore, or "malachite," is a carbonate, and is much less common than any of the others; it contains when pure 57 per cent. of copper. None of these ores of copper are very hard, all being readily scratched with a knife.

The ore as raised from the mine is tipped into spaces called "slides," in quantities averaging from five to twenty tons in each slide. The larger stones having been separated, and "ragged" or broken up into smaller pieces by hand hammers, the whole is passed through two revolving riddles of different mesh, and then handpicked by children and sorted into three qualities. These are called "prills" or best, consisting of pieces of very nearly pure ore; "dradge" or second quality, in which the ore is more or less interspersed with matrix; and "halvans" or leavings. As much of the best as will pass through a riddle of $\frac{3}{4}$-inch mesh is taken at once to the pile ready for market, and the rest goes to the crushing rolls to be crushed down smaller. The second quality has to undergo both crushing and jigging.

The ore is tipped from a tram waggon into a hopper above the rolls, and after passing through them it falls into a shoot below, by which it is conveyed to an inclined revolving screen or riddle, having holes $\frac{3}{8}$-inch square, and making thirty-two revolutions per minute. The pieces that are too large to pass through the screen are delivered from its lower end into the rim of the revolving raff wheel, the cups of which raise the stuff to the upper floor, where it falls over an inclined plane into the hopper, and is again crushed by the rolls until all are reduced to a size small enough to pass through the screen. The best and second quality ores are crushed separately; the former does not require any further treatment, and is ready for the market. The second quality is taken to the jigging machines for further separation.

COAL WASHING.

The several distinct operations connected with the washing of coal by machinery, and the machinery used therewith, especially the manner in which the operation is carried out, and the machinery used, at the washing establishment at Dowlais, have been fully described in a paper read by T. F. Harvey, before the Institute of Civil Engineers (James Forrest, Esq., Secretary).

Tipping.—Two kinds of tip are available for discharging heavy waggons of coal with facility; namely, the power-tip, actuated by mechanical means, such as steam or water; and the self-acting tip, which is worked by the loaded waggon itself. The power-tip requires somewhat less height; but where, as is generally the case, the necessary height is obtainable, a well-designed self-acting tip is equal to the power-tip in facility and rapidity of discharge, and has decided advantages in simplicity of construction and in economy of first cost, and therefore is almost universally used in connection with coal-washing.

At the Denain collieries, a self-acting tip is used for discharging 10-ton waggons sideways. The cradle is supported at each end upon a gudgeon fixed at about the rail-level and between the rails, but on the opposite side of the centre line of the rails to that on which the coal is discharged, so as to give the waggon a turning moment in that direction. As the axis of rotation is much below the centre of gravity of the waggon, the moment of the latter will increase as the angle of inclination increases; the rising side of the cradle is therefore connected to a series of graduated counterweights, which are lifted successively as the tipping moment increases, and which not only retard the undue velocity of descent of the waggon, but also bring the tip back to its normal position when the waggon is empty. During the process of tipping, the action of the apparatus is controlled by a brake, and the waggon is prevented from leaving the rails by being held firmly between strong adjustable brackets fixed to the cradle.

The tipping-cradle at Dowlais (Figs. 585 to 587) consists of a rectangular timber framing, 13 ft. long, 7 ft. 8 in. wide, and 12 in. deep, covered with timber planking 3 in. thick, upon which a pair of rails with curved-up ends is securely fixed by bolts and brackets. When in normal position, it is placed horizontally over an oblong pit with strong side walls, upon the top of which are two cast-iron plates, one on each wall, well bolted to the masonry. To each side of the framing of the cradle is secured a strong cast-iron bracket, with a curved flange convex downwards projecting from its side. These curved flanges act as gudgeons to the cradle, and bear upon the plates fixed to the top of the side walls. When the cradle partially revolves, as in the process of tipping a waggon, the flanges roll upon the plates which are straight, being prevented from slipping by projections or teeth cast on their underside, fitting into corresponding grooves cast in the plates. In order that a loaded waggon, when on a tip of this description, may turn it from the horizontal position to the required angle for discharge, and that the tip may bring the waggon back from that angle to the horizontal when the waggon is empty, it is necessary that the centre of gravity of the whole mass, tip and loaded waggon combined, should in the former case be some distance in advance of (or nearer to the descending end of the tip than) the centre of gravity of the whole mass in the latter case. This distance must be such that the moments of the weights or forces acting at the centres of gravity in respect to the middle point of the straight line which joins those centres of gravity, shall be somewhat in excess of the frictional resistances of the cradle with its load. The distance between these centres of gravity may be readily adjusted by varying the length of the cradle, or the position of the turned up or stop ends of the rails thereon. It may also be altered by the introduction of weight at the rear or rising end of the cradle when the length is constant and the position of the rails unaltered, or by the three methods combined.

In determining the distance between the centres of gravity mentioned above, a liberal allowance

DRESSING MACHINERY. 405

has to be made for the difference between the weights of the waggons, and also of their loads, as well as for the position of the load in the waggon, which is frequently unsymmetrical, and by making that distance sufficiently great, and attending to other points in the design, it is possible with facility to discharge by the same tip waggons varying between 7 tons and 10 tons load.

Fig. 585.

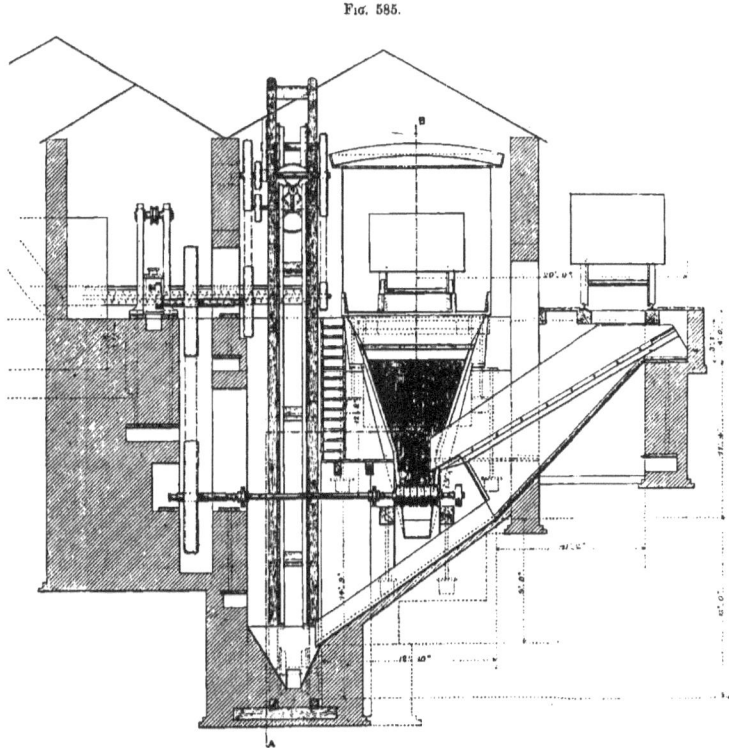

Tipping-cradle at Dowlais.

The segmental gudgeon on which the tip rocks, and upon the position of which its efficiency so much depends, is struck from the middle point in the straight line between the before-mentioned centres of gravity, with a radius the length of which depends upon structural considerations. The balance of the cradle would of course be unaltered if a gudgeon of small radius were struck from the same centre, but it would have to be connected by a tall bracket or other similar means, and would have to rest upon a similar bracket secured to the side walls of the pit. Such brackets would, however, be not only costly but extremely inconvenient. It is therefore preferable to use a gudgeon of large radius rolling upon a suitable bearing near the level of the road, which bearing is made

Fig. 586.

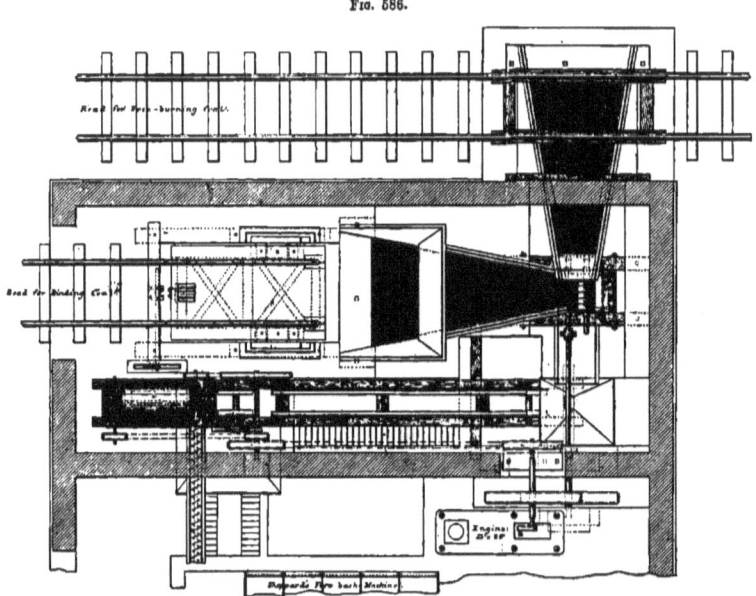

Tipping-cradle at Dowlais.

straight in order to get rid of the excessive friction that would occur with a concave bearing of so large a radius. To control the tip in the horizontal position, or at any angle, a powerful brake is provided and fixed at its rear end, by which the cradle with a waggon thereon can be held horizontally, and the speed of turning regulated, and when the necessary angle of discharge has been attained, the cradle can be held in that position for the coal to run out. The waggon being discharged, the descent is controlled by the same means.

The process of tipping or of discharging a waggon is performed in the following way: a loaded waggon with the door forward, having been run into position on the cradle while the latter is lying horizontally and securely held by the brake, the iron keys holding the door closed are removed, and a wooden one is inserted. The brake is then released, and the cradle with its load automatically assumes the required angle of discharge, which in this case is about 50°, the whole being brought quietly to rest and held in position by the re-application of the brake, the buffers of the waggon having descended upon blocks fixed in the top of the screen. During the latter portion of the descent the coal presses against the door sufficiently to break the temporary wooden key, when the door flies open, allowing the coal to run out on the screen. When the waggon is empty, the

FIG. 587.

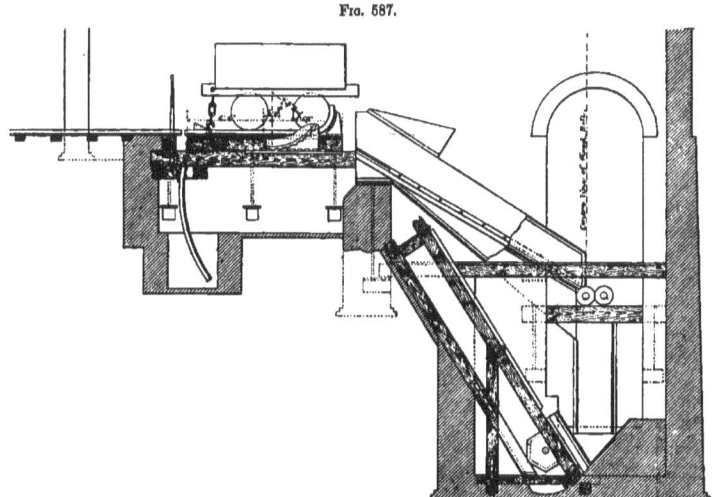

Tipping-cradle at Dowlais.

brake is again released, and the cradle and waggon return to the horizontal position. A cradle of this description may be designed so as to tip a waggon through an angle of 90° if required, but an angle of 50° is sufficient to discharge completely the most unfavourable coals. When this angle or any greater one is adopted, it is necessary to attach the waggon to the cradle, to prevent the waggon from being thrown over by the velocity it acquires in tipping. The attachment may be simply effected by placing a strong hook with a T head in the end link of the coupling chain, and sliding the T head into a suitable bracket fixed to the tip. This arrangement allows the backward motion communicated to the empty waggon during its descent after tipping to propel it from the cradle on its return journey from the tip towards the empty-waggon siding.

Screening.—A great portion of the coal which requires washing is too large to be effectually separated from its impurities, or to be raised by the elevators or conveyed by the creepers. This is especially so, when the whole output of a colliery has to be washed, or when the colliery screen-bars are set with a wide space between them, with the view of taking out only the bigger lumps for household purposes. Coal which has passed between the flat bars of a screen 1 in. apart is washed by machinery at Ebbw Vale, and at other places. This may be regarded as the superior limit as to size of coal to be washed, when even moderate efficiency is to be expected, and when it is neither interstratified with shale, nor impregnated with sulphur in the form of iron pyrites. All larger pieces should be crushed; and to avoid reducing that which is already sufficiently fine, it is separated from the coarser material by screening.

Of the various classes of screens used for this purpose, the inclined flat-bar screen, which for simplicity and durability is unsurpassed, is almost universally adopted at the collieries and coal-shipping ports of this country. But it has the serious defect of allowing long wide pieces of shale to pass between the bars; these pieces when very thin are difficult to separate from the coal in the process of washing, and when thick give trouble with the machinery, especially the creepers. If, instead of the bars, were substituted plates perforated with holes of a suitable size, the difficulty with the long wide pieces of shale would be avoided, and a screen almost as effectual in other respects would be obtained.

Guinotte uses, at the Mariemont Colliery in Belgium, several screens consisting of flat bars; but instead of the bars being fixed and placed at an angle of 25°-30°, as with the class mentioned above, they receive a swinging motion from eccentrics at one end, while the other is guided in a straight line. The bars are driven in two sets, every alternate bar receiving motion from the same eccentric, or from eccentrics fixed in the same line; while every adjacent bar is driven by eccentrics placed diametrically opposite. The motion imparted to the bars in this way is not only well suited for screening, but is suitable, and is frequently used, for travelling coarse coal in a nearly horizontal direction. But the passage of long wide pieces between the bars is inseparable from this system.

A third type of screen is the reciprocating, which may be placed at an angle or horizontally. These screens are suspended by links, or supported on friction wheels, and are driven by a releasing cam. When placed horizontally, the return portion of the stroke is performed by springs; but when inclined at an angle no springs are required, as the screen is brought back by its own weight.

The last class to be noticed is the revolving screen, which is extensively used. Practice differs as to form and construction of these useful screens. They are made either conical or cylindrical. The conical have always their axis horizontal, the difference between the diameters of the two ends giving the necessary inclination to the bottom side of the screen; but the cylindrical are made with the axis either horizontal or inclined at an angle varying between 1 in 16 and 1 in 10. When the axis of a cylindrical screen is horizontal, it is necessary to fix inside the periphery a spiral or screw to travel the coal onwards, as the screen revolves. Revolving screens are frequently made with two shells, both perforated, having an annular space of 6-8 in. between, the object of this arrangement being to save length. For the coarser coals, shells of perforated plate, generally of wrought iron, are used; but when the finer descriptions, below $\frac{1}{4}$ in. cube, have to be extracted, the shells are of woven wire or gauze. These screens are made of lengths varying between 6 ft. and 20 ft., and of diameters from 3 ft. to 12 ft., those of the largest diameter being used for extracting the finer coal. The speed of the circumference varies between about 120 and 200 ft. per minute.

One of the greatest difficulties encountered in screening coal arises from the choking or filling up of the spaces of the screen, owing to the clogging nature of the coal when damp. This is especially so when fine coal is extracted. Attempts have been made, in the case of revolving screens, to overcome this difficulty by causing blows to be delivered automatically by hammers along the top of the screen, so as to dislodge the coal which fills up the spaces. Another arrangement for attaining the same object is to press a wire brush against the periphery of the screen as the latter revolves.

Water is also sometimes delivered into the screen with the coal, which renders the material in the more liquid state less liable to choke the spaces.

A method also employed is to place a small pipe immediately over, parallel with, and extending the whole length of the screen, having its underside perforated with small holes. Through this pipe steam at a high pressure is occasionally blown. This method effectually clears the screen, but the condensation of steam upon it aggravates rather than reduces the evil it is intended to remove. Compressed air similarly used would probably reduce the evil to a minimum. Very little in the way of clearing the screen can be effected with a reciprocating screen, because the whole screening surface is covered when working. Such screens may therefore be regarded as unsuitable for separating the finer classes of coal.

At the Dowlais new establishment two screens (Figs. 585 to 587) are used, one for the free burning, and one for the binding coal, which meet at right angles over a pair of crushing rolls, to which the screens deliver the coarser coal. These screens are of the flat-bar type, similar to, but much larger and stronger than, the ordinary colliery screens, as they have to receive the load of a railway waggon, whereas the colliery screens are supplied by trams. The screens taper in plan, to deliver the coal more conveniently to the crushing rolls. Their principal dimensions are: length, 24 ft.; breadth at the upper end, 10 ft., at the lower end, $2\frac{1}{4}$ ft.; depth of sides at the upper end, 3 ft., at the lower end, 2 ft.; space between the bars, 1 in.; cross section of the bars, 4 in. deep, $\frac{3}{8}$ in. wide above, $\frac{5}{16}$ in. below; angle of screen 30°.

To add to the efficiency of the screen for the binding coal, the top is formed into a kind of hopper, by fixing at about 8 ft. from the upper end of a strong plate 5 ft. deep across the screen, allowing a space of 18 in. between the bottom of the plate and the bars; this both prevents the coal from being thrown too far down on the screen by the impulse it receives in the tipping, and also causes it to slide along the whole length of the screen bars. A sliding door across the lower end of each screen regulates the supply of coal to the rolls. Under the sloping bottom of each screen is placed a pocket or shoot, into which falls the fine coal that passes through the screen bars. These shoots meet at right angles, and deliver the fine coal underneath the rolls, in the same way as the screens meet over the rolls. All the coal is thus brought to one place, both the free-burning and the binding coal, and is consequently well mixed—a consideration of some importance when it is used for coking purposes. From this point it is conveyed by a short shoot, fixed at an angle of 35°, down to the foot of the elevator.

Crushing.—Before dealing with the question of crushing, it may be well to suggest that the lumps of shale mixed with the coarse coal which has passed over the screen may be advantageously picked out by hand previous to its delivery to be crushed.

On the Continent, where less pure coal seams are worked than is generally the case in this country, and where the small coal only is washed, the whole of the coarse is usually prepared for the market by subjecting it to a careful process of handpicking, and the arrangements for the purpose are frequently very complete and convenient. The chief object to be kept in view in arrangements for this purpose is to pass the coal to be picked slowly and in a thin layer before the operatives; which may be effected in several ways. The two following methods are perhaps the best, and are most generally practised. First, by means of a circular revolving table of wrought iron, upon which the coal is delivered from the screen and carried slowly round, while persons stationed round the table examine the stuff and pick out the shale. The other method is to stretch over the circumference of two parallel cylindrical drums, which are in the same horizontal plane at a suitable distance apart, a long, wide, flat, endless belt or band of hemp. This belt receives a slow motion from one of the drums, and the whole is so arranged that the coal is conveniently delivered from the screen on to the belt near one of the drums, and is slowly carried before the operatives, who pick the shale out, and is then delivered over the other drum on to a floor, to be filled into waggons.

It has already been noticed that when coarse coal requires washing it is necessary to prepare it for that purpose by crushing. Not only is it difficult to deal with large lumps by the machinery usually employed at a coal-washing establishment, but it is impossible to give the necessary agitation to large lumps in the washing bashes. There is, however, another reason why crushing should be resorted to, previous to washing. Reference has been made to the occurrence of coal in the mine in an interstratified condition with shaly and pyritic impurities. The adhesion between these impurities and the coal, when they occur in this manner, cannot be overcome by the washing process. It accordingly necessitates crushing.

The degree of fineness to which the coal should be reduced depends, therefore, partly upon the capabilities of the machinery, partly upon the amount of agitation that can be given to the water, and partly upon the manner in which the impurities are associated with the coal. Lumps of unmixed coal and of unmixed shale, which will pass through a hole 2 in. diameter, can be separated with facility by washing : but when the shale and coal are interstratified, or when pyrites exist in large quantities in the coal, crushing must be carried to a greater degree of fineness. No particular limit as to fineness can be stated generally, because that would vary with different coals; but crushing may be safely carried on until no one piece shall consist of both coal and shale.

Coal intended for conversion into coke to be used in blast furnaces is frequently pulverised to a state of coarse powder of about $\frac{1}{16}$ in. cube, in order to increase the hardness and density of the coke. This may be, and generally is, done after washing.

For reducing coarse lumps of coal to smaller pieces, the most suitable class of machine seems to be a roller crusher, having one pair, two pairs, or three pairs of rolls, according to the size of the lumps to be reduced; but when small lumps are to be reduced to a coarse powder, experience favours those machines which act by percussion rather than those which act by direct crushing. Of the percussive class of machines, which is extensively adopted for the purpose, Carr's disintegrator and Hall's may be mentioned.

A roller-crusher having one pair of rolls 18-20 in. diameter and 3 ft. long, well fluted with semi-circular flutes, about $1\frac{1}{4}$ in. deep, and rounded ridges $\frac{1}{2}$ in. wide, and running at 100 revolutions per minute, will reduce coal, which has passed through a $2\frac{1}{2}$ in. flat-bar screen, and over a similar screen with 1 in. space between the bars, to a convenient size for washing, at the rate of 30 tons per hour. When coarser coal has to be reduced, containing lumps of 6—8 in. cube, such as the output of an ordinary bituminous seam in the South Wales district, two pairs of rollers should be used, one pair over the other; the upper pair, between which the coal passes first, being set about $3\frac{1}{2}$ in. apart and made with coarser flutes so as partially to reduce the coal for the lower and closer pair, which are similar to those of a one-pair crusher. For coal of still larger dimensions, three pairs of rolls are employed.

Instead of fluted rolls, tooth rolls are sometimes adopted when the coal is very large or of the hardest descriptions. In such rolls the teeth, which are pointed, may be cast with the rolls, or they may be made of steel and welded in by fixing them in the mould and running the metal around

DRESSING MACHINERY. 411

them, or the steel teeth may be turned and fitted into sockets formed in the rolls. This kind of roll is used in pairs so geared together as either to have the teeth of one roll opposite those of the other, or to have the spaces of the one opposite the teeth of the other; or one toothed roll may be used with a plain cylindrical roll, the action in this case being somewhat similar to that of a pick on a floor. A pair of plain cylindrical rolls, varying between 2 and 3½ ft. in diameter, and placed almost in contact, is sometimes employed for the purpose of reducing coal to a coarse powder preparatory to coking. Ordinary edge-runners as used in mortar mills are also applied to this purpose; but these are by no means the most suitable machines for this class of work.

To prevent the rolls, or the gearing driving them, from being broken by pieces of iron, parts of old tram-wheels, or sleepers, which sometimes find their way with the coal into the rolls, one of the pair revolves upon bearings placed in slides, and the bearings are held in position by an elastic plunger or buffer, or by a weighted lever (which latter can hardly be so good an arrangement), with sufficient force to crush the hardest coal or shale, but sufficiently yielding to allow a piece of harder material to pass through. The teeth of the gearing should be made of extra length, to permit the necessary lateral movement without becoming disengaged, as that would be likely to result in the fracture of some of the teeth.

Another point requiring attention in designing roller crushers is the protection of the journals from coal-dust and consequent abrasion. By carefully enclosing the rolls in a casing of wrought or cast iron, having holes fitting accurately around the axis between the journals and the body of the rolls, this object may be tolerably well attained. A reduction of the torsional strains to which the roll-shafts are subject may be obtained by fixing toothed gearing at each end, instead of at one end only, as is the usual practice.

A principle too frequently neglected, and which should be regarded in arranging crushing machinery of any description, is that no coal that does not require reducing should be passed through the crusher. In order to carry out this principle, it is necessary to extract by efficient screening the whole of the coal of a finer size than that requiring to be crushed. When two or three pairs of rolls are adopted, screening and crushing should be performed alternately.

The crushers used at Dowlais are of the roller class; each consists of one pair of fluted rolls 18 in. diameter by 3 ft. long, running about 50 revolutions per minute at the new establishment, which is found amply fast enough, as only that portion of the 300 tons washed which requires reducing is passed through the rolls. At the old establishment the rolls run 100 revolutions per minute. The brass steps in which the journals revolve are divided diagonally, and are fitted into cast-iron standards, with steel riders turned down at each end, so as to embrace the tops of the standard, over which they are well fitted, thus giving assistance in cases of unusual strain upon the standard.

A 4-inch shaft is coupled to and drives one of the rolls, motion being communicated from this to the other by a pair of pinions fastened outside the journals. The steps of the roll which receives its motion direct from the shafting are fixed in the standards; but those of the other roll are permitted a lateral motion, being held in position by a strong elastic plunger of rubber springs, such as are used on the draw-bars of railway waggons. The journals are made large, the rolls are well encased, and no trouble is experienced from heating of the journals.

Transmission.—Of the various methods by which the coal is conveyed or transmitted from one part of the washery to another, that by inclined troughs or shoots is the simplest. For allowing small coal to gravitate freely down these shoots, which are preferably made of wrought-iron plates, and thus to avoid the necessity of handling, they should be placed at an angle of not less than 35°. Experience shows that, unless the coal is very dry and hard, the adoption of a smaller angle, in order to save height, is false economy.

This mode is almost universally used where the distances are short, even though the coal has to be raised for the purpose, it being often found better to prolong an elevator a few feet higher, rather than to complicate arrangements by adding a separate conveyance; but its application is, in all except very rare cases, limited to short distances, owing to the great height required.

Sometimes coal is conveyed into, and away from, the washing machine by water running in troughs, in which case, instead of being at an angle of 35° they have an angle of 3°–5°. In this way coal is conveyed from the elevator to the machines, a distance of about 18 ft. at the old establishment at Dowlais and at Ebbw Vale, the troughs in both instances being of cast iron, about 6 in. square inside.

Coppée, in the many extensive establishments designed by him, and erected on the Continent, uses this means of transport freely, especially with the finer coal below $\frac{1}{4}$ in. cube, which he conveys from the screens to the washing bashes, and again from the latter to the depositing tank, through distances which, in the aggregate, often amount to 200–300 ft. The water for the conveyance of the coal by this method is also used for the washing process, and, as no machinery is required nor extra water consumed, this system seems peculiarly adapted to the applications above referred to.

Vehicular transmission, although not much used within the precincts of a washing establishment, is, nevertheless, occasionally adopted. One case of its adoption has already been described in connection with the washing arrangements at Tredegar, where the coal is hauled in trams to a bunker at the top of the washing troughs. An interesting and extensive application of vehicular transmission has been made by Marsaut, at a coal-washery in connection with the Bessèges and Molières collieries in France, in a communication by him, which was substantially reproduced in *Engineering*, vol. 29. Conveyance by machinery seems, however, to be the most suitable to the requirements of the washery, as it is not limited in its application to short distances like shoots, neither is it intermittent in action like vehicles.

There are several useful machines for transmitting coal horizontally. Of these may be mentioned the Creeper, or Archimedean screw, revolving in a trough; a belt or band similar to that described for handpicking coal; and a carrier, consisting of a plated-chain, of which there are several varieties. For raising coal from a lower to a higher level only one machine has received extended application, namely, the Elevator or Jacob's Ladder.

The creeper is probably the most suitable machine, and certainly the one most extensively applied, for conveying small coal in a horizontal direction, and is used not only for dry or damp coal but frequently for conveying coal or shale in water. Although not so well adapted for raising the material, still there are places where it is used for this purpose. At the Marie Anne and Steinbank Collieries in Germany, a creeper about 50 ft. long is placed at an angle of about 10°, raising washed coal into a bunker. The object of this arrangement is not so much to elevate the coal as to permit any water mixed with it to drain back out of it.

A creeper consists of a square shaft of wrought iron, having journals turned in it at intervals. Upon the shaft is a screw of coarse pitch and very thin thread, in this case called a blade, working in a wrought or cast iron trough of slightly greater width than the diameter of the screw, having a

DRESSING MACHINERY.

semi-circular bottom. The end journals revolve in steps fixed in ordinary plummer-blocks, provision being made for resisting the longitudinal thrust of the screw; and the intermediate journals revolve in steps fixed in hangers suspended from the top of the trough. The screw is driven by a pulley or toothed wheel, fixed on the end of the shaft outside the trough, and conveys the coal introduced at one end of the trough to its other end. Although this is a very good contrivance, its use is attended with much friction when dealing with fine coal; and when conveying shale in water the screw-blade wears out rapidly. With a given kind of coal, the quantity conveyed by a creeper depends upon the diameter, pitch, and velocity at which it is driven; but a given creeper will transmit a much greater quantity of coarse coal above $\frac{1}{2}$ in. cube than of fine coal below $\frac{1}{4}$ in. cube. In actual practice a creeper, 14 in. diameter, 12 in. pitch, running 48 revolutions per minute, has conveyed 40 tons per hour of coarse coal; while a creeper 10 in. diameter, 7 in. pitch, making 90 revolutions per minute, has conveyed only 10 tons per hour of fine coal or duff.

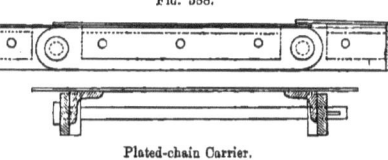

Fig. 588.

Plated-chain Carrier.

The belt or band has been already described in connection with the process of hand-picking, and therefore needs no further mention.

The last kind of machine for horizontal transmission, of sufficiently extended application to require notice, is the plated-chain carrier. Fig. 588 represents a pair of links in one pattern of this chain, with the plate between them. The endless chain is stretched over a pair of polygonal drums at each end of a frame, and is supported between the drums by rollers or slides, by which it is kept in a nearly straight line. This apparatus cannot be worked in water like the creeper, but it can be used for raising coal at a considerable angle; indeed, the elevator is but a modification of this machine.

The elevator, for elevating or raising the coal, seems to be an indispensable accessory to a washing establishment, and, when well proportioned to the amount of work required of it, leaves but little to be desired, as it does its work in an efficient manner without any attendance, is durable, and neither difficult nor expensive to construct. The coal is raised in buckets of cast or wrought iron, capable of holding 20–100 lb. each, put between and secured at equal intervals to a pair of endless chains formed of flat links, which, for convenience of fixing to the buckets, and for attachment to each other, as well as for giving greater stiffness to each individual link, are placed on edge. Although these chains are sometimes constructed of single links, they are preferably made of single and double links alternately, united by bolts, cottar pins, or rivets, which are turned to fit accurately into holes bored near the extremities of the links. The chains, with the buckets attached, are strained over a pair of polygonal drums fixed on shafts, which revolve in suitable pedestals secured to a strong framing. One of the drums is set at a higher level than the other, the difference between their levels depending upon the height to which it is necessary to elevate the coal; the inclination of the elevator is generally 60°–70° to the horizontal. The uppermost drum is driven slowly by toothed gearing at a speed which should not exceed about 12 revolutions per minute. In some cases the lower drum is used as the driver, but this arrangement necessitates great tension being placed upon the chain in order to prevent the drum from slipping, because the tension produced by the weight of the chain with its load is not here available for keeping the chain in contact with the lower drum; whereas, when the upper drum is made the driver, the weight of the chain with its load is sufficient without additional tension. For facility of fixing the chain, and for taking up any slack caused by wear at the joints, the pedestals of one of the drums have provision for adjusting the distance between the centres of the drums. Where possible, the upper pedestals should be the movable ones, keeping the position of the lower ones fixed, so that the buckets may always pass close to the bottom of the elevator pit, and clear out the coal as it descends from the shoots. From the neglect of this simple precaution great trouble is often experienced in feeding the elevator with coal. In case the gearing makes adjustment at the top objectionable, the plate forming the bottom of the elevator-pit should be movable, so that it may follow the bottom drum when that is adjusted. Throughout the distance between the drums, the chains are supported upon angle-iron slides fixed to the frames, or at intervals by rollers; and to prevent the buckets from coming into contact with the slides or the drums, and thus being speedily worn, the links project about $\frac{1}{4}$ in. below the buckets. This machine is adapted for raising coal or other granular material to heights varying between a few feet and 100 ft., and in quantities suiting the requirements of the most extensive washing establishment; for, although the speed is limited, the buckets may be of any size likely to be required.

Resuming the description of the New Coal Washery at Dowlais, the elevator for raising the coal after it has been crushed is 45 ft. long from centre to centre of the drums, and is placed at an angle of 60°. The buckets are of wrought iron (a material which is generally to be preferred) $\frac{3}{16}$ in. thick, and hold, when full, 80 lb. of coal. The drums are of hexagonal form, and make about 6 revolutions per minute, the upper one being driven by a toothed wheel, 4 ft. diameter, fastened on its axis, into which is geared a pinion 9 in. diameter. On the pinion shaft is fixed a pulley 5 ft. diameter, driven from a pulley of the same diameter, fastened on the engine shaft through a flat belt 8 in. wide. The chains are composed of flat links 18 in. long from centre to centre, single and double alternately; the single links are 3 in. by $\frac{3}{4}$ in., with thickened ends, and the double links are 3 in. by $\frac{1}{2}$ in. united together by bolts $\frac{7}{8}$ in. diameter; and the chains are supported between the drums by angle-iron slides $3\frac{1}{2}$ by $3\frac{1}{2}$ by $\frac{1}{2}$ in., during both their ascent and their descent. This elevator is capable of raising 400 tons of coal per day of 10 hours at the slow speed stated above.

The elevator delivers the coal into the upper end of a cylindrical revolving screen, 3 ft. 6 in. diameter, 7 ft. long, having an inclination of 1 in 10 from the horizontal, and driven at about 26 revolutions per minute. This screen separates the coal into two sizes, the holes in the shell being $\frac{1}{4}$ in. diameter. All coal which passes through the holes in the shell, amounting to about 25 per cent. of the whole, is delivered into a creeper 10 in. diameter, 7 in. pitch, and 60 ft. long, driven at about 90 revolutions per minute, and is thereby conveyed to a bunker outside the building, to be taken direct to the coke ovens without having been washed. The coarse coal, delivered from the lower end of the screen, runs down a shoot into a bunker, whence it is raised by a second elevator and delivered into a horizontal creeper placed over the washing machine. This creeper conveys the coal and distributes it into the different bashes through openings fitted with slides in the bottom of the trough. The buckets of the second elevator are of cast iron, $\frac{3}{8}$ in. thick, and carry, when full, 25 lb. of coal. They are attached to chains of single and double links alternately, the single links being $\frac{3}{4}$ in. thick, and the double ones $\frac{1}{4}$ in. by $2\frac{1}{2}$ in. deep. In this case the links are joined by turned rivets $\frac{5}{8}$ in. diameter, enlarged in the middle, where they pass through the single link, to

$\frac{7}{8}$ in. diameter. The drums are hexagonal, and the chains with their buckets are guided upon angle-iron slides. The length from centre to centre of the drums is 25 ft., the angle 60°, and the speed about 20 revolutions per minute. The distributing creeper is about 18 ft. long, 20 in. diameter, 14 in. pitch, and is driven at the rate of 25 revolutions per minute.

Washing.—The process of washing, or rather of separating the coal by machinery from mechanical impurities, may be effected either by causing an upward current of water to pass through the coal, or by allowing the coal to fall through a great depth of still water. The feasibility of both these methods depends upon the difference between the densities of the materials requiring separation.

Attention to the following elementary considerations will make clear the principles upon which the separation is effected by these methods. If a quantity of coal and shale, consisting of particles of equal dimensions, be thrown into a deep vessel full of still water, the particles of shale will, owing to their greater weight, descend at a higher velocity, and will soon be separated from the particles of coal. Again, if in a vessel having an upward current of water, material of the same description be placed, as all the particles are of equal volume, each will be buoyed upwards by an equal force, which force will communicate a greater velocity to the lighter coal than to the heavier shale, and thus cause them to part company. On the other hand, neglecting the influence of gravity, if the materials to be separated be placed in a downward current of water of sufficient strength, the reverse of the above will take place; that is, instead of the coal being on the top of the shale it will be found underneath, a result due to the greater velocity imparted to the lighter material by the force of the current. When particles of different sizes and of the same density are subjected to these conditions, they will, if similar in shape, be arranged according to their respective volumes, the larger pieces behaving in the same way as the denser pieces mentioned above. It is therefore evidently impossible to effect the perfect separation of coal from the shale by either of these methods, when the sizes of the constituent particles differ greatly, because the same velocity may be imparted in still water by means of gravity, or in an upward current by the force of that current, to a smaller piece of shale as to a larger piece of coal, and they will consequently still continue associated.

Hence it becomes necessary to separate the particles of coal according to their sizes, by some system of screening, previous to the process of washing, which latter, it is to be observed, effects a separation according to their densities, as before intimated. It should also be borne in mind that a downward current, having an opposite effect on the mixed bodies to that which an upward current has, will, if used in conjunction with an upward current, more or less neutralise the effect of the latter, and should consequently be avoided as much as possible.

The manner in which these principles are applied in practice will be understood by reference to Fig. 589, which shows in longitudinal section an ordinary washing machine in which the coal is treated by an intermittent upward current of water. Theoretically, the action of an upward current presents no advantages over that of a fall through still water, in separating the coal from the shale ; and although Marsaut and others have proved that the practical solution of the problem can be effected by the latter system with as much simplicity and efficiency as by the former, still the larger part of the coal-washing machinery, both in this country and on the Continent, is constructed on the principle of the intermittent upward current. Marsaut's machine, for a fall through still water, consists of a cage suspended from the piston of a hydraulic cylinder, the cage having a perforated bottom, upon which the coal to be cleansed is placed to a depth of about 4 ft. The cage is caused to descend by a succession of short drops through still water in a rectangular tank about 6 ft. long, 10 ft. wide and 24 ft. deep, within which the cage fits with tolerable accuracy. When the cage performs a short descent or drop, it leaves the coal behind, momentarily suspended in the water; the coal then descends by gravity on to the cage, the velocity of each individual particle depending on its density, inasmuch as sizing has of course been effected previously to the washing.

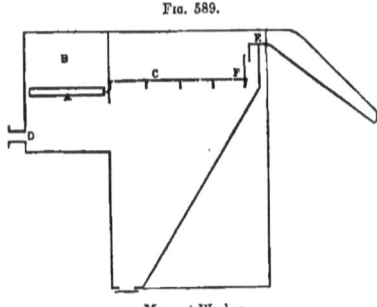

Fig. 589.

Marsaut Washer.

Fig. 589 shows the essential parts of an ordinary machine with intermittent upward current.

The piston A works upwards and downwards in the compartment B, and during the downward stroke forces the water up through the perforated plate C with sufficient velocity to lift the coal and shale to a considerable height. The velocity imparted to the coal is greater than that imparted to the shale, for the reasons already stated; and the tendency therefore is for them to separate. After the piston has reached the bottom of its stroke, the upward current produced by its descent soon ceases; and the particles of coal and shale then fall through the water in the order of their densities, the heavier shale falling more rapidly than the lighter coal. Their complete separation is produced by repeating the pulsation; and when the stroke of the piston is short the pulsations are often repeated to the extent of 250 times. When heavy shale alone is mixed with the coal, it may be easily separated; but when the specific gravity of the impurities approaches more nearly that of the coal, separation can only be accomplished by a long-continued repetition of the operation. To prevent a downward return current from being produced while the piston ascends—which would carry the coal down again among the shale on the sieve, and some of the smaller particles of coal even through the shale—water is frequently admitted at D under a head of several feet to supply that displaced by the piston. With a view to secure the same object, valves opening downwards are sometimes inserted in the piston, so as to prevent the formation of a vacuum underneath as the piston ascends. An alternative method is to make the piston perform the upward stroke more slowly, by driving it either with a suitably formed cam, or by means of a crank and a block sliding in a slotted lever, similar to that used for giving a quick return motion to slotting and shaping machines, and known as the slot-bar motion. The pulsations imparted to the water not only separate the coal from its impurities, but carry both of them gradually forwards along the whole length of the compartment, and ultimately effect their discharge—that of the coal over the dam E, and the shale through the valve F. Moreover, with a view to assist the motion of the shale towards the valve F, the perforated bottom C is often inclined downwards in that direction. It is the general practice to construct the compartment C—or, as it is called, the bash, in which the washing process is performed—about 5 ft. long, from the rear to the dam, 3 ft. 3 in. wide, and 1 ft. 3 in. deep below the top of the dam, the piston having an area of 5—6 sq. ft., with a stroke varying between 18 and 3 in., and being driven at the rate of 26—120 double strokes per minute; such an apparatus is capable of cleansing about 5—6 tons of unwashed coal per hour. When a long stroke of the piston is adopted, the number of strokes is proportionately less; and of late years the long-stroke piston seems to be gaining favour for washing coarse coal.

It is important that the current of the water be of equal intensity throughout the whole length of the bash, a condition which limits the length to about the above dimension of 5 ft., when the agitating piston is placed at the rear of the bash, as in Fig. 589; but if the piston were placed along the side of the bash, the length would not be so restricted. By adopting the latter arrangement of piston, and adding to the length of the bash, a proportionately greater number of pulsations would be given to the coal during its travel along the bash. This plan might be adopted with advantage in the case of coal largely mixed with light shale difficult to separate; and would be simpler and more economical than resorting to double washing, as is frequently done in such instances.

Endeavours have been made to determine, from theoretical considerations, the limit of the differences between the sizes which may be treated together, so as to be compatible with good results; but, owing to the great diversity of forms of the particles, some being nearly cubical while others are thin and flat, no rule of universal application can be established.

The following table shows the extent to which classification by screening is carried out in practice at different places, when due attention is paid to efficient washing. These limits of sizes of the material treated in the same bash have been arrived at in each case after an extended experience, and give satisfactory results with the particular class of coal treated.

Meier Iron Works, United States.	Bochum, Germany.		Coppée, Belgium.	
inch	millimetres	inches	millimetres	inches
0·06 to 0·10	0 to 10	0·00 to 0·40	0 to 8	0·00 to 0·32
0·10 ,, 0·20	10 ,, 15	0·40 ,, 0·60	8 ,, 14	0·32 ,, 0·56
0·20 ,, 0·30	15 ,, 30	0·60 ,, 1·20	14 ,, 24	0·56 ,, 0·96
0·30 ,, 0·42	30 ,, 45	1·20 ,, 1·80	24 ,, 40	0·96 ,, 1·60
.. ..	45 ,, 70	1·80 ,, 2·80	40 ,, 70	1·60 ,, 2·80

Below $\frac{1}{8}$ in. cube (0·12 in.) the coal cannot well be washed in a machine of the kind above described, neither can it be easily classified by any of the processes of screening already mentioned; and consequently in many instances that portion of the coal less than $\frac{1}{16}$ in. cube (0·06 in.) has been thrown away as valueless, being unsuitable for fuel, and considered incapable of economical improvement by washing.

By a system known in Belgium and France as Coppée's, and in Germany as Lübrig's, the finest description of coal may be effectually cleansed. The coal to be washed is mixed with water sufficient to carry it freely down a slightly inclined trough, and is caused to pass through a series of inverted pyramidal vessels on its way to the washing machines. In these vessels the water deposits the coal with a considerable amount of regularity in respect of size, the largest particles falling into the first vessel, and the smallest into the last, whilst the intermediate vessels receive the intermediate sizes. The different sizes are then washed in separate bashes, as with the larger coal. The washing machines used are of the intermittent upward-current description; the bottom of the bash, instead of consisting of a perforated plate, is in this case formed by a layer of felspar about 4 in. thick on a coarse net of wire cloth, the sizes of the pieces of felspar varying between 1 in. and 2 in. cube, according to variations in the sizes of the coal. Through this porous bottom, which is kept open by the pulsations imparted to the water by the pistons, the shaly impurities descend into a hutch underneath, and are thence discharged through a valve at the bottom of the machine. The pulsating pistons are of short stroke, from about $1\frac{1}{4}$ in. down to only $\frac{1}{4}$ in., and run at a speed of 120–200 double strokes per minute.

Although extensively used on the Continent, this excellent system is comparatively unknown in Great Britain, and, it is believed, has not been adopted in a single instance. It is, as already stated, the only existing system suitable for washing fine coal; and, it may be added, the only one by which coal interstratified with shale, or containing iron pyrites, can be washed effectually, inasmuch as these kinds of coal must be finely pulverised before it is possible to remove their impurities by washing.

Sheppard's machine erected at Dowlais, and set to work in the beginning of 1881, has five bashes, and deals with about 300 tons of unwashed coal per day of ten hours. By means of a creeper 20 in. diameter, placed over the machine, the coal is fed into the rear end of the bashes, which are 5 ft. long by 3 ft. 3 in. wide; and by the pulsations imparted to the water it is conveyed forwards along the whole length of the bashes, and is discharged over the dam, whence it falls to the bottom of the lower hutch, along which the coal is conveyed by a second creeper to one side of the machine, and is raised thence by an elevator, which delivers it into a bunker outside the building. The shale, with the other impurities separated from the coal, passes through valves immediately under the dam; these valves, extending the whole width of the bashes, are lifted at intervals, and discharge the shale into an inner compartment extending the whole width of the machine; the shale is thence conveyed by a third and smaller creeper along the bottom of this compartment to the side opposite that to which the washed coal is carried; it is thence raised by an elevator and delivered into a shoot, from which it is taken by trams to the rubbish tip. The water in each bash receives pulsations from a rectangular piston, 2 ft. by 3 ft. 3 in., having a stroke of 1 ft.; the piston is driven by a crank at the rate of 32 double strokes per minute, the velocity of the up and down strokes being equal. The bottom of the bash is formed of a copper plate, 19 B.W.G. = 0·04 in. thickness, perforated with holes of $\frac{3}{64}$ in. diameter, and about 60 to the sq. in.; and supported by a cast-iron grating, the bars of which are $\frac{3}{4}$ in. wide, 2 in. deep, and form spaces 15 in. wide by 6 in. long. Angle-irons are riveted to the side plates of the bashes for the whole length of the bash, and upon these the grating with the perforated plate is bolted, having a downward slope from the rear to the dam.

This machine, illustrated in Figs. 590, 591, is of the intermittent upward-current type; but instead of a portion of the water being discharged from the machine at each pulsation, it is delivered into the lower compartment with the washed coal; the piston, in ascending, draws the water up from this compartment through a foot-valve beneath, and at the top of the stroke the valve closes; in descending, the piston forces the water upwards again through the unwashed coal in the bash. In this way an intermittent circulation of the water is kept up in one direction only, with but little downward return current through the sieve. Formerly a valve was used in the opening from the piston hutch into the washing bash, with the view of preventing return of the water as the piston ascends; but this valve has been found unnecessary. The machine is driven by a pair of vertical engines with cylinders 10 by 18 in. stroke, making 72 revolutions per minute. A continuation of the crank-shaft is carried upon a suitable frame across the whole width of the machine, and has fixed upon it the necessary pulleys and toothed wheels for driving the elevators, creepers, and piston cranks. The elevator raising the washed coal out of the machine is similar to that previously described for delivering unwashed coal into the machine, except that the cast-iron buckets are here made long and shallow, and are perforated with numerous holes to allow the water to drain from

FIG. 590.

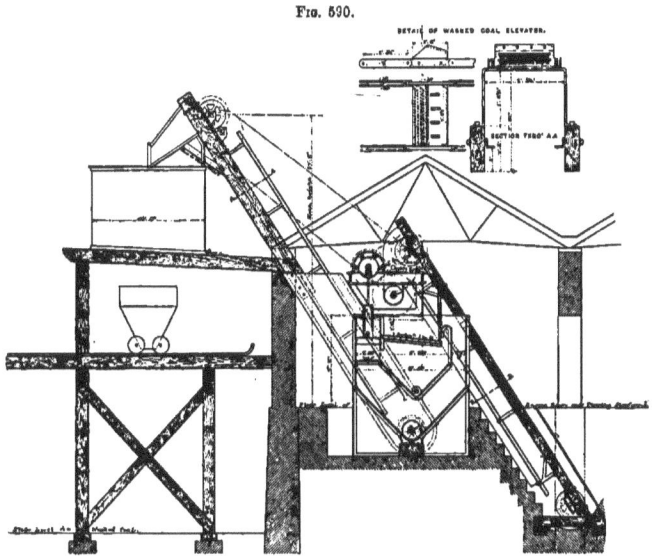

Sheppard's Washer.

the coal as it ascends. It is driven from a 4-ft. pulley fixed on the crank-shaft, connected by a 6-in. leather belt to a 3-ft. pulley at the head of the elevator frame; on the same shaft with the latter is a pinion 7 in. diameter gearing with a 3-ft. wheel on the drum-shaft of the elevator, which therefore makes about 19 revolutions per minute. The bottom hexagonal drum of this elevator is fastened on the end of the axis of the large creeper which works in the bottom hutch of the machine, and to which motion is thereby imparted. The shale elevator is similar to, but smaller than, the washed-coal elevator, and drives in the same way through its lower drum the shale creeper working in the bottom of the upper hutch. The shale apparatus is large enough to carry away shale amounting to 25 per cent. of the unwashed coal delivered into the machine.

This machine seems to be well adapted for washing coarse coal containing little or no "duff," and occupies less space for a given capacity than any other, and in the matter of cost compares advantageously with other systems. With suitable coal the water is not changed, except when the machine is cleaned out, which would be at intervals of from once a week to once a month; and consequently no settling ponds are required.

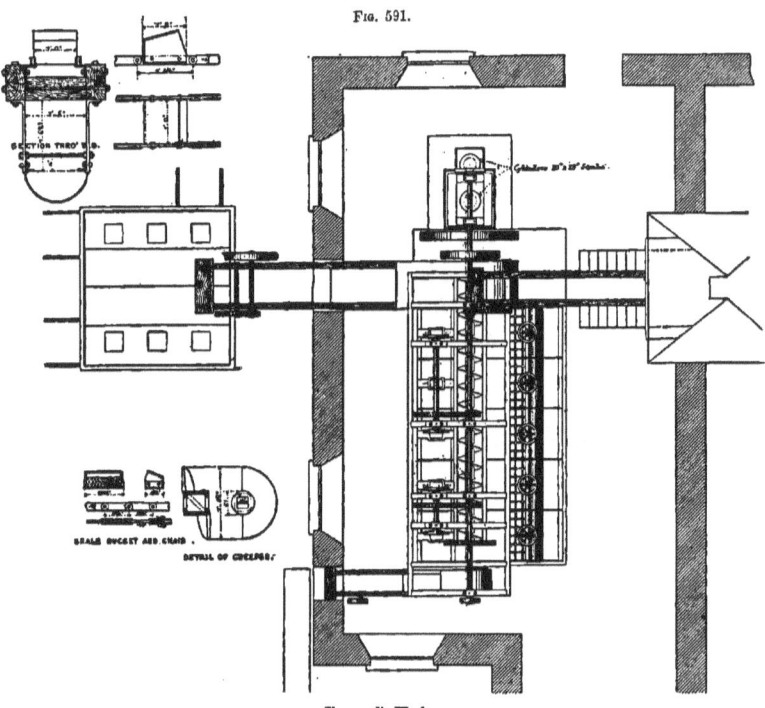

Fig. 591.

Sheppard's Washer.

At Dowlais, however, settling ponds were made, owing to the large amount of fine stuff in the coal. The fine is now extracted by screening previous to washing; but the ponds are still used to some extent, and a considerable quantity of pyrites is deposited in them. With respect to this machine it must be stated that a considerable reduction of the speed at which some of its parts are driven, especially the elevators, would be an improvement, and would diminish the wear and tear.

There is also at Dowlais another washing establishment (Fig. 592) which was erected some fifteen years ago, and at which there are two "Bérard" machines (Figs. 593–597) of four bashes each (making eight bashes altogether), placed back to back, with the elevator between them, the two

DRESSING MACHINERY.

machines together being capable of treating 480 tons of unwashed coal per day of 10 hours. The same descriptions of coal are dealt with as at the new establishment, and in a similar manner, except that the whole of the fine is washed here. A vertical engine, 16 by 16 in. length of stroke, making 80 revolutions per minute, drives the whole of the machinery, which consists of the two washing machines, a pair of crushing rolls, and an elevator for raising the coal. On the crank-shaft of the engine are keyed three pulleys, of which one pulley 8 ft. diameter drives the rolls at 100 revolutions per minute, and by means of a 3 ft. 6 in. pulley attached to one of the rolls, and reducing toothed

Fig. 592.

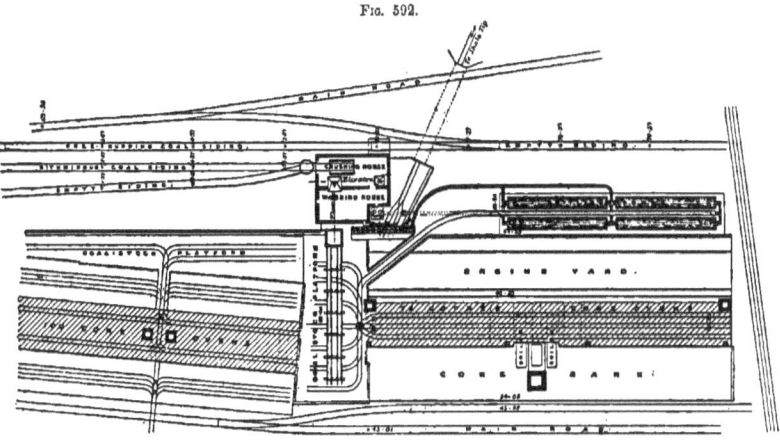

Dowlais Washery.

gearing, gives motion to the elevator, the hexagonal drums of which make 5 revolutions per minute. The other two pulleys are each 5 ft. 6 in. diameter, and are connected by belts to pulleys 4 ft. in diameter on the agitator shafts of the washing machines, which they drive at 95 revolutions per minute. Motion is imparted to the pistons from the shafts by straps and eccentrics of cast iron, and wrought-iron forked rods, the length of stroke of the pistons being 3 in.

The coal to be washed is discharged from the waggons into shoots having short screens, which take out some of the small. From these shoots it passes through the rolls, where it is crushed; and it is then elevated by an ordinary bucket-elevator between the machines. The elevator raises and delivers the coal into the upper limb of a ⋏-shaped shoot, which divides it into two portions, and conveys it into troughs fixed at a slope over the machines, down which the coal is conveyed by water and delivered into the rear end of the bashes. The coal, having been washed, is discharged over the dam of the machine with the water into a short inclined shoot, the bottom of which is made of wire cloth, to allow the water to drain from the coal, while the latter runs downs into a tram under the end of the shoot. A large quantity of fine coal and slimes pass with the water through the wire cloth; the fine coal is intercepted as completely as possible by sieves fixed in wooden

MINING AND ORE-DRESSING MACHINERY.

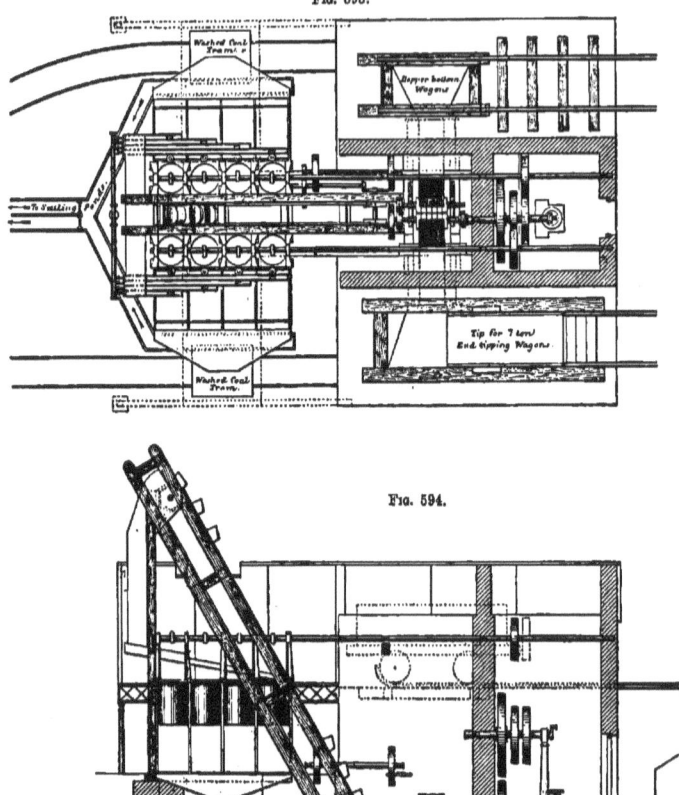

Fig. 593.

Fig. 594.

Bérard Washer.

troughs, along which the water passes on its way to the settling ponds, where the slimes are deposited. There are three settling ponds, 40 ft. long by 25 ft. wide and 3 ft. deep; they are used in succession.

In this machine the shale leaves the bash through a valve under the dam, and falls into a hutch beneath; from thence it is periodically run out into trams below, through an opening in the bottom fitted with a slide, to be taken to the rubbish tip.

About eight years ago the Ebbw Vale Company erected at their works extensive machinery (Fig. 598) and plant for washing coal to be converted into coke for use in blast furnaces. The coal

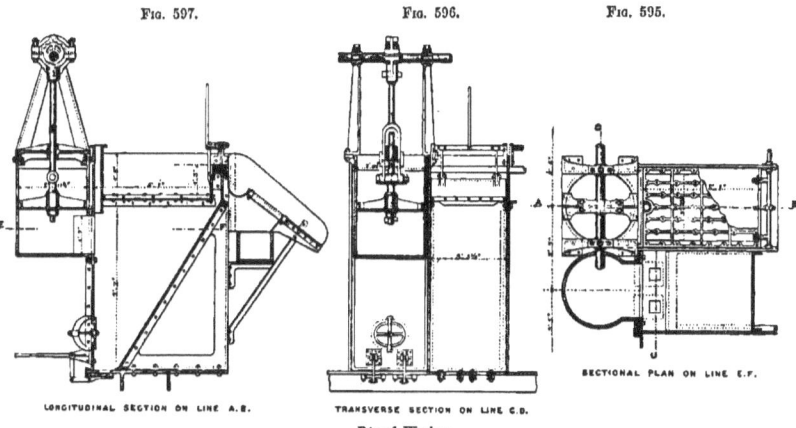

Bérard Washer.

is brought to this establishment, after having passed through flat-bar screens with 1-in. spaces, in end-tipping railway waggons of various capacities from 4 to 10 tons load. These are discharged into a bunker automatically by a tipping cradle similar to that described in connection with the new washery at Dowlais, except that the cradle has in this case to be brought back to the horizontal position by a winch when the waggon has been emptied.

The coal is raised from the bunker, into which it has been tipped from the waggons, by a bucket-elevator of ordinary construction, the drums making 12 revolutions per minute; the elevator delivers the coal into inclined troughs, along which it is conveyed by water and fed into the machines in the same way as at the old establishment at Dowlais. The washed coal is discharged from the machines on to a fixed inclined screen composed of wire $\frac{1}{8}$ in. diameter, and $\frac{1}{8}$ in. apart. The screen is 7 ft. long, and extends the whole width of the machines, which allows the water to drain fairly well from the coal. Under the lower end of this screen and parallel with the ends of the machines is a creeper 14 in. diameter, of 12 in. pitch, and 24 ft. long, making 48 revolutions per minute, into which the coal falls from the screen, and by which it is conveyed to a second elevator similar to the former. This elevator raises the washed coal, and delivers it into a large cast-iron bunker outside the building; through openings in the bottom of the bunker, fitted with slides, it is run out into trams and conveyed to the coke ovens as required. From the bashes the shale is delivered continuously through an opening under the dam in each bash, fitted with two sliding valves, into a creeper in front of the machines, which conveys it to the bottom of an elevator at one side of the machines, to be raised and delivered into trams.

At this establishment there are four single-bash machines, the bashes being each 6 ft. 10 in. long, by 4 ft. 10 in. wide, and the four together capable of treating 400 tons of unwashed coal per day. The pistons are rectangular, 4 ft. by 2 ft., having a stroke of 18 in., and making 26 double strokes per minute. Water is admitted into the machines under the pistons, through two 4-in. pipes in each machine, under a head of about 27 ft., from a large tank covering the whole of the top of the building; there is accordingly a continuous upward current of water through the coal in the bashes, and the current is intensified by the descending stroke of the pistons, which have rectangular holes without valves to prevent the formation of a vacuum underneath as they ascend. The bottom of the bash in which the coal is washed consists of flat bars on edge $\frac{3}{16}$ in. apart, their ends resting upon ledges cast on the back and front plates of the bash. This is a more durable and less expensive bottom than the perforated copper plate on a cast-iron grating of the other machines described previously; and where, as in this case, there is a continuous upward current of water to prevent fine coal from being drawn down through the bars, there seems to be but little objection to its adoption. It should be observed, however, that the proportion of clear spaces to the solid bars is less than with the perforated copper plate, and that the friction of the water passing up between the comparatively deep bars must be greater.

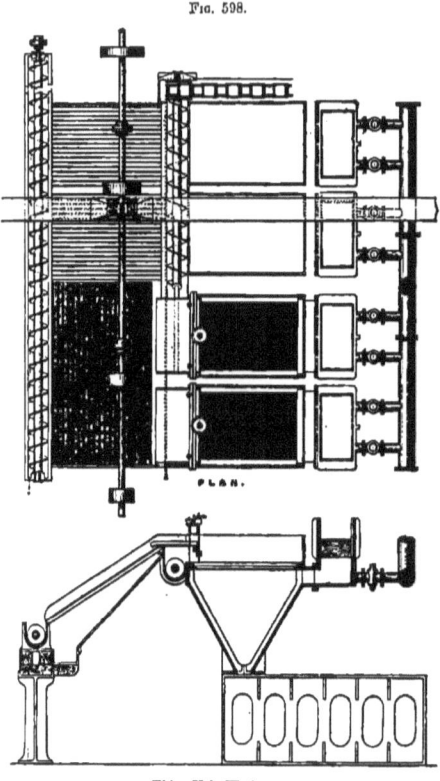

Fig. 598.

Ebbw Vale Washer.

Each of the systems above particularly described as in use in South Wales is equal to the extraction of about 6 per cent. of ash by one process of washing. This is, however, by no means the limit to which the separation may be effected; for, in Continental practice, coals containing 17 or 18 per cent. of ash have, in many instances, as much as 14 per cent. removed by washing.

Rathbone holds that the chief cause of the perfection of Continental systems is the irregular character of the coal seams, which renders the coal very brittle, and also the large intermixture of "dirt," or shaly matter, which is most difficult to get rid of. In the Lührig and the Coppée systems, the great point is the successful treatment of the fine "smudge" coal, which at English collieries is often wasted. In one Continental establishment the treatment of the entire coal output is of such an efficient character that the waste of coal is reckoned at 2 per cent. only. At one of the establishments where Lührig's system is at work, Rathbone was assured that the coal, which before washing was associated with 25 per cent. of impurities, afterwards contained only 5-6 per cent., the cost of the operation being very moderate. He considers that some modification of the systems of Lührig and Coppée might be introduced with advantage into districts in England where the coal is impure, and would certainly be far superior to anything known in England at the present time.

Bewick's experience of Sheppard's machine leads him to believe it is an excellent one. In one colliery in the county of Durham, where a considerable quantity of small coal was washed to make coke (without washing it was impossible to make saleable coke), the cost, including labour, wear and tear, interest on capital, &c., was only $1\frac{1}{2}d$. per ton. Another advantage in Sheppard's machine is that it is compact, and uses a very small quantity of water, which is an important consideration, because at many collieries water is scarce, and where plentiful, in flowing off it interferes with the stream into which it falls, and then proceedings are likely to be taken by the river conservators.

Cochrane thinks that the question of washing is one only to be considered where there is a large percentage of ash and shale. When coal contains 3-4 per cent., as in the North of England, he questions whether the washing process is required, and whether the process of abstracting the shale by handpicking is not the best. On working out the economy of washing such coal, he does not think the result would be found to repay even the small cost of $1\frac{1}{2}-2d$. a ton. The elaborate process of washing such coals would not be required if more care were taken in the separation of the shale upon the screens. Cochrane adopts the principle of endless bands, formed of steel plates carried by an endless rope about 30 ft. long, passing nearly horizontally before 6 or 8 operators, the coals being made to travel in a very thin stratum, so that the dirt can be well picked out before it goes to the crushers. He thinks that method is attended with the largest economy in the case of coals used for the production of coke.

Marten once had occasion to inspect some coal-washing machines in operation at Wigan, where a large amount of "smudge" had accumulated at one of the principal collieries. This "smudge," in consequence of the impurities contained in it—consisting of dirt, pyrites, and shale—had, prior to the introduction of the washing-machines, not only no commercial value, but the accumulations of it had become a positive encumbrance. An enterprising man, who was familiar with the process and advantages of coal-washing as conducted on the Continent, purchased these accumulations of smudge, set up washing-machines similar to some of those described, removed by these means the great bulk of the impurities, and was rewarded by producing a clean "residuum," from which a coke of fine class was manufactured. Marten's object in inspecting the washing-machines at Wigan, was the application of the same process to the small coal, or "slack," of the South Staffordshire district. The practical result of his experiments with that material was not satisfactory. Large quantities of shale and pyrites were undoubtedly removed from the fine "slack" or "smudge," but the attempt to convert the clean "residuum" into coke failed, as it was not sufficiently bituminous to coke, and burnt away into ash. He considers there are numerous descriptions of bituminous coals to which the system of machine washing may be advantageously applied, but there is no corresponding advantage from its application to non-bituminous, free-burning small coal. It should also be remarked that, in bituminous small coal, the washing produces a much more satisfactory result where the sulphurous element is concentrated in the shape of pyrites, than where it is chemically diffused through the entire substance of the coal, as in the latter case only a fractional portion of the sulphur is removed. This is the case with one of the inferior measures of coal known in Staffordshire as the "stinking" coal, which is so loaded with diffused sulphur that its fractured side, when exposed to the air, frequently becomes covered with all the colours of the rainbow. Within the limits named, the washing of bituminous "smudge" is undoubtedly of great commercial advantage, as by that means a waste and cumbersome by-product is converted into one capable of inaugurating and profitably sustaining important industrial enterprises.

Marsaut thinks the different sorts of coal-washing machines embody a limited number of principles common to all, and which have to be considered when judging of the intrinsic merits of individual machines. Mechanical separation may be effected either by a stream of water ascending through the charge, or by the free descent of the material through still water. These two modes achieve the relative motion of water and of coal, which form the basis of all the systems. Substances always fall through water, whether the water ascends or descends, because their density is greater. The only effect of a descending current is to quicken the process of descent, without in any way affecting the work of separation. It does not neutralise the effect of the ascending current. Further, the descending current, instead of being injurious, is very beneficial. In Marsaut's opinion, it constitutes the most valuable means of preparing certain coals containing much admixture of impurities. It is this downward current that, with certain coals, reduces the yield of ash to 2-4 per cent. For example a piston-jigger, having a valve to prevent the descent of water with the charge, will under certain conditions leave 3-4 per cent. more ash in coal subjected to the same number of piston-strokes than will be the case with a similar jigger without the valve. That was what he found at Bessèges, and at that moment he was constructing machines founded specially on the principle of the recoil of the water, which was jealously conserved, in order to utilise this effect. The merits of the machines of Lührig and of Coppée rest entirely on this alternate upward and downward movement, but especially on the motion of descent. There are two principal types of washing-machines; the piston-jigger, with alternate motion of the water, and the basket-washer, or English jigger; both being dependent on the relative movement of the water and of the material to be washed. Apparatus based on the alternate movement of coal and of water, realised what he calls the suction-phenomenon, which is extremely useful in respect of the degree of cleansing. It certainly augments the loss, but this is not sufficient to counterbalance the superior amount of cleansing. These machines are the nearest to perfection. In continuous-current apparatus, which altogether prevents the return of the water with the charge, and which are furnished with overflows, the only effect obtained is that of sorting by equivalence. The loss is diminished, but there remains in the washed product all the impurities of duff and slack. In apparatus which only partially prevent the return of the water with the charge, and which may be called "mixed," intermediate results are obtained. Such machines are those of Bérard and of Sheppard, which latter is only an intelligent modification of Bérard's. By abolishing in Sheppard's washer the valve in the passage, a very useful effect has been produced from the point of view of cleansing. This machine thus comes under the denomination "mixed" before-mentioned. In all machines working by a stream of water, either continuous or alternate, the relative movement of coal and of water remains uniform for every particle of material. This constitutes classification by equivalence, which is at the same time greatly improved by the effect of suction in the return-current apparatus. On the other hand, the washers based on a free descent of the material in still water, realises the "differential relative movement." At the beginning of the descent, advantage is taken of the action of gravity, which is independent of the size of the pieces. This is known as the initial stage of the descent. It favours classification by density, and if it could be managed so as to constitute the entire process, it would effect perfect separation of all the particles of all sizes, that is to say, without any preliminary sorting.

The washers of Lührig and of Coppée, which are nothing more than the old Hartz riddle, take advantage of suction, and even utilise it for the total separation of all the impurities. This kind of washer is at the present time in great favour in France and Germany. It remains to be proved that it washes better than the ordinary piston-jigger with return-current. It has the serious disadvantage of requiring a considerable stream of water, which acts injuriously in the process of cleansing. Particles of $\frac{1}{2}$ millimetre, and even larger, were carried off by this quick current, and without being sufficiently cleansed. The favour actually enjoyed by this washer appears to Marsaut to belong rightfully to the general disposition of the apparatus, and to the water-carriage of the product, which lessens the amount of manual labour, replacing it by a mechanical power, of which the cost might be neglected.

To sum up :—

(1) The return-water piston-jigger effects the cleaning in a superior manner. It even dispenses with preliminary sizing, but it entails more waste.

(2) The same type of jigger, with a valve to prevent the return of the water, lessens the waste, but does not cleanse so effectually.

(3) Free-fall washers give results equal to No. 2 type, if this coal be first sized, a result which is, moreover, increased by the effect of the initial period of the fall.

(4) Free-fall washers leave nothing to be desired for coals well-sorted as to size, as in the case of all the other types of washers not returning the water with the charge.

The initial period of fall and suction constitute the two elements of difference between all systems of washing by water ; accordingly, as either of these elements preponderates, so will the washing be more efficient in the different types. Unfortunately, suction entailed a slight waste of useful material, but that is far from counterbalancing all its advantages. By well considering these theoretical principles it is quite possible to arrive at a correct valuation of the different forms of washer. The other part, viz., the machinery, has to be considered as between manual and mechanical methods. Both have their peculiar advantages, and, consequently, their advocates. It must also be remarked that the special conditions of each case and the results aimed at go for much, and will probably influence the sort of machine chosen.

Harvey has pointed out that if a piece of light coal and a piece of heavy shale of the same form be exposed to the action of the same current of water, the velocity communicated to the coal by the force of that current will be greater than the velocity communicated to the shale. Marsaut admits the truth of this statement with regard to the upward current, but seems to deny the same effect to the downward current, as he states that "the only effect of a descending current is to quicken the process of descent, without in any way affecting the work of separation." Harvey maintains that the tendency of the force is to separate them, as well as to accelerate their descent; but instead of conducing to bring the heavier shale below the coal as the upward current does, and as also does the force of gravity in still water, the tendency of the force of this downward current is to bring the lighter coal below the shale, and thus to some extent it neutralises the effect both of the upward current and of gravity, and as a consequence it involves waste.

Harvey cannot agree with Marsaut's statement that the merit of Coppée's machines rests "especially on the motion of descent." Coppée, in his machines for washing coarse coal, uses the slot-bar motion referred to. This has the effect of causing the upward stroke of the piston to be performed more slowly than the downward stroke, and thus reduces the velocity of the return current. Moreover, he admits the water into the machine under the piston with a head of several feet, and he lays so much stress on the uniformity of head, that he prefers supplying the water by a centrifugal pump, driven at a given velocity, to drawing the supply from a reservoir with the risk of a variable head. With Marsaut, Harvey considers Coppée's coal-washing establishments well arranged, but he does not think the favour they enjoy depends to any appreciable extent on the adoption of water as a means of transport; for this means of transport he uses but little in the case of coarse coal.

Coal-washing without previous sizing appears to Harvey to be established on no very rational basis, and can probably be effected as efficiently, and perhaps more so, in a machine with a strong return-current, as in a machine in which that current is prevented. But his observations lead him to favour preliminary sizing, and then separation according to density in an intermittent upward current of water. He has known establishments where sizing is carefully effected, and where separation is afterwards performed in an upward current at which 12 per cent. of ash is removed, and the washed product is only 17 per cent. less in weight than the unwashed coal. On the other hand, he has seen at a Bérard machine, where no means are used to prevent downward current, and where no sizing is done, a loss of 20 per cent. by weight sustained by the removal of only 6 per cent. of ash.

In July, 1886, various coal washing machines were inspected by David Cowan, manager for the Carron Company, and reported on by him substantially as follows. Samples of coal and dross were taken, both before and after washing, also of the rubbish washed out of dross, with the object of having these analysed, in order to ascertain the actual work performed by each machine. The 12 machines inspected were in operation at the following collieries:—

No.	Name of Colliery.	Maker.
1	Tannochside, Uddingston	Boll, Wishaw.
2	Woodend, Bathgate	Coltness Iron Co.
3	Devon, Sauchie	M'Culloch, Kilmarnock.
4	Kinnoil, Bo'ness	Sheppard.
5	Turedalo, Durham	Boll & Ramsay.
6	Howle, East Durham	Ramsey, East Howle.
7	Binchester, Bishop Auckland	Robinsons, Bishop Auckland.
8	Auckland Park, Bishop Auckland	Robinsons, Bishop Auckland.
9	Black Boy, Bishop Auckland	Robinsons, Bishop Auckland.
10	Great Western Co., South Wales	Sheppard.
11	Dowlais, South Wales	Coppée, Brussels.
12	Carronhall, Falkirk	J. Clelland, Carronhall.

DRESSING MACHINERY. 429

The above washing machines may be divided into four distinct classes, viz. :—
(1) The Bash Washer, of which Nos. 1, 2, 3, 4, 10, and 11 are examples.
(2) The Open Trough Washer, of which Nos. 5, 6, and 12 are examples.
(3) The Rotary Washer, of which Nos. 7, 8, and 9 are examples.
(4) The Feldspar Washer, No. 11 being the only example. This is a modified or improved form of Bash.

The Bash Washers are supplied by different makers, but all are much alike in construction, the main difference being in the bashes or cylinders for agitating purposes, some of which are vertical and others horizontal. The first costs of these machines are much alike, and to erect any one of them complete, capable of washing 200 tons of dross per day, should cost about 500l.

From the results of the laboratory experiments, the Sheppard machine (which comes under this class) in the process of washing removed only 39·9 per cent. of the total rubbish in the dross. The Woodend machine removed 45 per cent. of the total rubbish. The rubbish taken out of the dross by these machines contained dross to the extent of, at Kinneil, 4·25 per cent.; Devon, 30·13 per cent.; Woodend, 23 per cent. In comparing these figures it has to be borne in mind that the dross unwashed contained 15·37 per cent., 39·26 per cent., and 20 per cent. of rubbish respectively.

As regards the percentage of dross mixed with the rubbish taken out by the washers, the Sheppard machine gives fairly satisfactory results, but none of them can be considered satisfactory so far as removing the total quantity of rubbish in the dross is concerned. The cost of washing by these machines varies from 2d. to 3d. per ton on the washed dross, not including cost of water, which averages 200 gal. per ton of dross washed.

The Trough Washers are perhaps the oldest form of washer in existence, and are, up to the present day, largely in use. There are many forms and types, but the experimental washer at Carronhall—a trough with a stream of water and dams across—is perhaps the oldest form. Many improvements have been made from time to time, one of which is the method of agitating the dross as it passes through the trough, notable instances of which are those in use at East Howle and Tursdale Collieries. The cost of these machines to wash an output of 200 tons of dross per day varies from 250l. to 400l.

Some of these washers give very fair results, the Tursdale washer removing 75·44 per cent.; Carron, 68·92 per cent.; and East Howle 46·80 per cent. of the total rubbish contained in the dross. It has to be noted in making comparisons of these percentages that the dross at Carronhall contains about 10 per cent. less rubbish than the others.

Cowan is of opinion that much better results would have been obtained from the machine at East Howle if it had been kept working on a smaller quantity of dross, as there is little doubt the machine was washing up to something like 20 per cent. over its capacity. This is further shown by the fact that in the rubbish delivered from the machine no trace of dross was found. The percentage of dross left in rubbish taken out by the Tursdale machine was 13·38, and by Carron 9·14.

The cost of washing by these machines was given by the parties using them as follows:—East Howle, 1·8d. per ton of coke, equal to 1·1d. per ton of washed dross. Tursdale, 1·5d. per ton of coke, equal to 0·9d. per ton of washed dross. The cost at Carron is 7d. per ton, no doubt due to the crude state of the plant, and that manual labour is employed in agitation. These costs include enginemen and all other labour and stores. The quantity of water used is about 350-400 gallons per ton of dross; depreciation and establishment charges are not included.

The "Rotary" Washer, patented by Robinson of Howlish Hall, is in operation at Binchester, Auckland Park, and Black Boy Collieries, in the County of Durham, and at the Bothwell Pits of Messrs. Baird of Gartsherrie. This washer is in the form of an inverted frustum of a cone, 8-10 ft. diameter at top, tapering to 2 ft. diameter at bottom, and fitted with necessary outlet valves at bottom of cone. In the centre of the cone revolves an agitator, the washed coal flowing out at the top of the cone and the rubbish being removed from the bottom.

The whole of this plant is very compact and takes up little space, and can be fitted up complete for an output of 200 tons per day at a cost of about 250*l.*, including pulsometer pump. This form of pump is preferred as being less liable to get out of order. This washer, according to the samples taken, has not given very good results, removing only 33 per cent. of the total rubbish in the coal. There is, however, very little coal left in the rubbish, the percentage being 3·80. It should be noted that this is a very dirty dross, the rubbish being 27·23 per cent. The cost of washing with this machine is said to be fully ½*d.* per ton of dross washed. This is a low estimate, and it might be taken at 1*d.* per ton. The quantity of water used is a constant one of about 10 gal. per minute, or about 30 gal. of water per ton of dross washed. From the compactness of this washer, its small first cost, and the low working expenses, it is worth consideration. Cowan has been informed that better results have been got from this machine than those stated above.

The Feldspar washer, patented by Coppée, and until recent years not in operation in this country, is made in two forms—one to wash coal of sizes from 5 in. to ¾ in., and the other, called the Feldspar washing machine, to wash coal of sizes from ¾ in. down to impalpable powder.

This machine was inspected at Dowlais Iron Works, where it is fitted up to wash 1000 tons per day, and comprises two sets of plant, viz.—one for washing bituminous coal, and one for steam coal. The coal in both plants is divided into five different sizes; this being accomplished by the use of revolving screens, reciprocating tables, and pointed boxes. Each size is washed in a separate compartment. These are constructed of timber, and divided into two divisions, the one being the washing compartment and the other the piston compartment.

The piston has a speed of 60-75 strokes per minute, and the length of stroke is varied from 1¾ to 4 in., according to size of material to be washed, by an arrangement of a cranked lever, similar to that used in an iron shaping machine. The speed of the up stroke is much less than that of the down stroke. The rubbish is drawn off continually and automatically from the machine. The Feldspar machine is used for washing the sizes of coal from ¾ in. downwards. These are much of a similar construction to the foregoing washer, but are fitted with screens, which are covered with a bed of feldspar 3-4 in. thick, through which the rubbish is drawn.

This machine is the only one in which the division of coal into different sizes has been carefully gone into, and that there is a decided advantage in this there is no doubt. The cost of such a plant for washing 200 tons per day has been estimated by Coppée at 1900*l.* There will, however, be the whole of foundations, buildings, &c., which are not included in this tender; the total cost for the whole erected complete at Carron may be about 2500*l.*

The best results were got from this washer. In a dross containing 22·64 per cent. of rubbish, 85·69 per cent. of the total was washed out, which contained 5·70 per cent. of coal. Coppée

DRESSING MACHINERY.   431

guarantees his washer to remove rubbish in the coal to within 4 per cent. of the total. The experiments were even a little better than the guarantee, viz. 3·24 per cent. They, however, show that instead of 2 per cent. of coal in the rubbish there was 5·70 per cent. The cost of washing by this machine is said to be about 3d. per ton of washed dross. The quantity of water used is about 300 gal. per ton of dross washed.

The total cost of washing by the several processes is calculated to be as follows:—

Makers.	Where in Operation.	Pence per Ton.
Bell, Wishaw	Tannochside	4·6
M'Culloch	Devon	4·48
Carlton Iron Co.	East Howle	1·24
Bell & Ramsay	Tursdale	1·27
Bolchow, Vaughan, & Co.	Binchester	1·48
Coppée	Dowlais Co.	5·3
Sheppard	Kinneil Iron Co.	2·5
Carron Co.	Carron	7·6

TABLE SHOWING THE QUANTITY OF RUBBISH IN DROSS PREVIOUS TO WASHING AT THE FOLLOWING COLLIERIES, AND THE RUBBISH LEFT IN DROSS AFTER WASHING; ALSO COAL IN RUBBISH REMOVED BY WASHING, AND PERCENTAGE OF TOTAL RUBBISH WASHED OUT.

Name of Colliery.	Name of Washing Machine.	Rubbish in Dross Coal previous to Washing.	Rubbish left in Washed Dross Coal.	Quantity Rubbish removed by Washing Operation.	Coal in Rubbish removed by Washing.	Total Rubbish and Coal taken out in Operation of Washing.	Percentage of Total Rubbish in Dross taken out in Washing.	Weight of Dross lost as being mixed with Rubbish taken out in Operation of Washing on every 100 Tons Washed.	Weight of Rubbish left in every 100 Tons of Washed Dross.
		per cent.	per cent.	per cent.	per cent.	per cent.	per cent.	tons cwt. qr.	tons cwt. qr.
Dowlais	Coppée	22·64	3·24	19·40	5·70	25·10	85·69	1 8 2	0 16 0
East Howle	Ramsay trough	25·35	13·51	11·84	trace	11·84	46·80	trace	1 11 3
Binchester	Robinson	27·23	18·24	8·99	3·80	12·79	33·00	0 9 2	2 6 2
Tursdale	Bell & Ramsay trough	13·84	3·40	10·44	13·38	23·82	75·44	3 3 2	0 16 0
Kinneil	Sheppard	15·37	9·06	6·31	4·25	10·56	39·90	0 8 3	0 19 0
Devon	M'Culloch	39·26			38·13				
Carron	Trough	14·32	4·45	9·87	9·14	19·01	68·92	1 14 2	0 16 3
Woodend		20·00	11·00	9·00	23·00	32·00	45·00	7 6 0	

MINING AND ORE-DRESSING MACHINERY.

TABLE SHOWING THE RELATIVE FIRST COST OF THE DIFFERENT WASHING MACHINES, THE COST OF WASHING PER TON OF DROSS, THE QUANTITY OF WATER USED, THE COST PER TON OF WASHED DROSS TO REDEEM THE CAPITAL OUTLAY AT 10 PER CENT. PER ANNUM, CALCULATED ON AN OUTPUT OF 30,000 TONS; ALSO LOSS PER TON IN DROSS WASHED OUT WITH RUBBISH, AND ANNUAL VALUE OF DROSS LOST IN WASHING.

Name of Washer.	Total Cost.	Quantity of Water used per Ton.	Loss per Ton in Dross left in Rubbish, calculated at 2s. per Ton.	Cost of Washing per Ton.	Cost per Ton of Dross to Redeem Capital at 10 per Cent., 30,000 Tons per Annum.	Total Cost of Washing per Ton.
	£	gal.	d.	d.	d.	d.
Bash Machine—						
Bell & Sons, at Tannochside	500	200	1·7	2½	·40	2·90
M'Culloch's, at Devon	600	200	say 1·0	3	·48	3·48
Trough—						
East Howle	300	350 to	nil	1	·24	1·24
Tursdalo	250	400	·7	1	·20	1·20
Rotary—						
Robinson	250	30	·2	1	·20	1·20
Feldspar—						
Coppée	2500	300	·3	3	2·00	5·30
Sheppard	500		·1	2	·40	2·40
Carron—						
Experimental	..	..	·4	7	..	7·40

* Labour, cost, and depreciation only included; power and other establishment charges not included.

Lührig's method of coal washing consists in the use of continual action jiggers, Figs. 599, 600, 601. A medium size of these machines, which, according to the degree of impurity and coarseness of the coal, washes 150–200 cwt. of coal per hour, is found to give favourable results. With coal giving on assay 24 per cent. of impurity, they wash about 150 cwt. per hour, but on the percentage of impurity being reduced to 10 per cent. they wash 200 cwt. The following is a description of this medium-sized jigger:—The jigger box is held together by uprights 2½ in. square and is about 9 ft. 10 in. long, 3 ft. 1½ in. broad, and 4 ft. 7 in. deep. The top part of the box is divided by a partition board a, which reaches to a depth of 2 ft. 2¾ in. This partition, therefore, divides the whole box into two parts, namely, one for the piston and the other for the sieve. The inside of the box is lined with wood 1¼ in. thick, so that there still remains an area of about 4 ft. 6 in. × 2 ft. 10 in. = 12·75 sq. ft. for piston and sieve. The bottom b, the lowest point of which should be exactly under the middle of the partition, is curved so as to cause, through the up and down motion of the piston c, a flow of water favourable to the sorting and delivery of the coal. The mesh of the sieve d depends on the size of the grain washed; that is, for example, ½ in. for large size coal (cobbles) and ¼ in. for nuts or smalls. The sieve is fixed so as to be 19 in. below the top edge of the front side of the box, and 21 in. below the top edge of the partition, so that there is an inclination of 1 in 24 from the front to the back. This is done in order the better to retain the heavier pieces of dirt. Under the sieve are fixed five wooden bars e, to resist the strain on it. The feeding takes place by means of a trough, which is fixed on to the back side of the jigger box, so that it just

DRESSING MACHINERY. 433

comes over the deepest part of the sieve. The coal, being lifted by the upward pressure of the water, is carried out through the slit f, which extends across the whole front side of the jigger box, while the dirt sinking on the sieve gradually falls out through the opening g (4 in. × 10 in.), which is fixed 2 in. above the level of the sieve, into the dirt trough h. The slide or valve i, which regulates

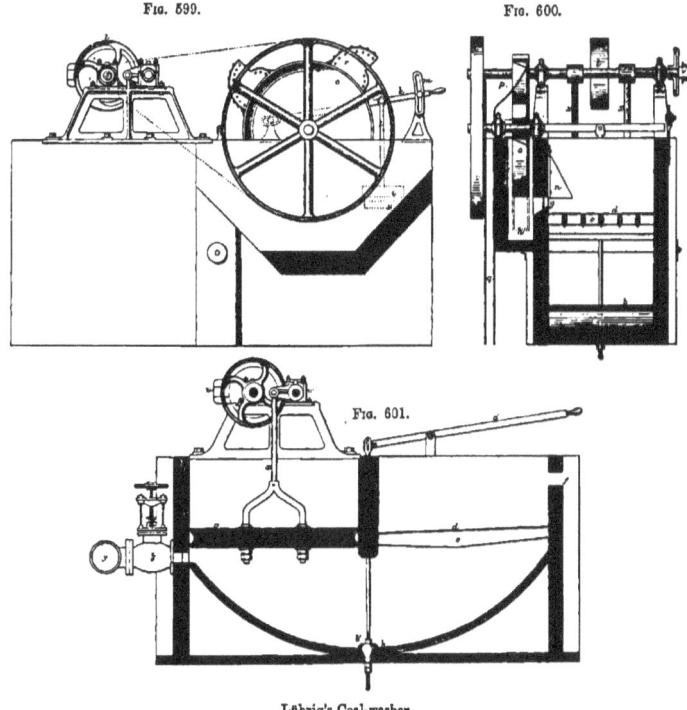

Fig. 599.   Fig. 600.   Fig. 601.

Lührig's Coal-washer.

the size of the opening, is acted on by a lever k and a connecting rod l, and is kept in its place by the slot m. But in order to prevent the coal when entering from passing through this slit, a shield or screen n is fixed over the same, which can be regulated by screws, but as a rule when once it has been adjusted it does not require to be again altered. The dirt passing through the opening g is caught up by the scoop-wheel o, and delivered by it on to the funnel p, over which it falls into the spout q, from the bottom of which it is removed in a manner to be described hereafter. The separate delivery of the dirt from each jigger by the scoop-wheel possesses one great advantage, namely,

3 K

there is always a safe control over the amount (*i.e.*, the quantity and quality) of work which each jigger is doing. In order to watch this control a sort of spoon is used by which samples can be brought up from time to time, thus it can immediately be detected if any one machine is delivering coal with dirt, or otherwise neglecting its duty. Each separate scoop *r* is so pierced with holes that the water can easily drain off. The motion of these scoop-wheels is transmitted through bands from the principal working axis *s*, the axis receiving its motion from the pulley-wheel *t*. The motion thus transferred to the principal axis is again transferred through the slot *u* with a slide block *v* to the axis *w*, and from there by two forked piston rods *x* to the piston, which is therefore fixed at four different points, whereby a regular and steady motion is ensured to the course of the piston. In order to replace the loss of water caused by the continual outflow, a water pipe *y* is connected to each machine, the supply for which is regulated by the valve *z*. The coarse sizing jiggers make 60 strokes per minute, those washing the next largest or middle size 65, and all the rest 70. The scoop-wheel makes 10 revolutions per minute. Lastly, the very finely divided refuse which sinks through the mesh of the sieve is collected at the bottom of the jigger box, and is from time to time cleared out by means of raising the valve *b'* by the lever *a'*, and is thus allowed to flow out into a trough in connection.

Lührig's fine coal jigger is on the whole similar to the one known amongst ore dressers as the continual fine grained jigger, and is shown in Figs. 602, 603. The inside of the box is lined as

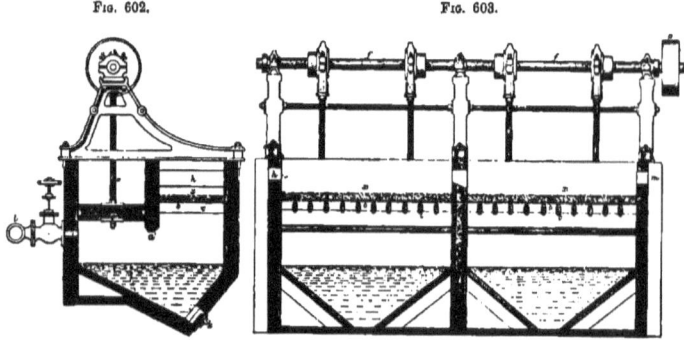

FIG. 602.   FIG. 603.

Lührig's Fine Coal Jigger.

before with wood, and has a length of 8 ft. 3 in. at the top, and a breadth of 3 ft. 3 in. At the bottom the back has a length of about 2 ft. 6 in., but increases in length on the front side to 4 ft. 4 in., the bottom of each jigger being thus pointed. The partition *a* divides the whole into two separate jigger boxes, and the partition *a'* divides this again into two equal parts, so that there are two pistons and two sieves in each jigger box. This partition extends to a depth of 19 in., the area taken up by the piston and sieve being about 5¼ sq. ft. The sieve *b* is fixed 11¾ in. below the top edge of the jigger box, and is strengthened as before by wooden bars *c*. The piston *d* is constructed out of two massive pieces of wood, has a thickness of 3½ in., and is fixed as shown to the

DRESSING MACHINERY.

piston rods e, which derive their motion from two eccentrics fixed on to the axis f. The length of the stroke varies according to the size of the grain treated, from $\frac{1}{4}$ to $1\frac{1}{2}$ in. The axis is rotated by means of a pulley-wheel g, the number of strokes varying between 130 and 175 per minute. A layer of 3 in. of felspar x is placed on the top of the sieve. The stream of coal which has previously been subjected to a separation in spitzkästen, or grading boxes, flows on the first sieve through the entrance h, and is there immediately acted on as follows :—The water in the jigger box below the sieve is subjected to short and quickly-repeated blows of the piston; the lighter particles of coal are thus kept continually in suspension, and are therefore washed by the force of the entering stream into the next sieve compartment, while the heavier particles of dirt sink and pass through this bed of felspar and through the sieve into the pointed-formed part of the jigger box below. From here it is led off by a trough in connection with it through an opening i with a lid k. The second jigger compartment does the same work over again, in order to cleanse the whole satisfactorily, namely, to separate more perfectly any fine dirt particles which may have washed over with the coal. A stream of clear water runs in continually by a pipe l, which can be regulated by a valve as shown. The washed coal flows out of the jigger box by the outlet m.

At the Barrow Colliery, dry cleaning is performed as follows :—The coal is filled underground with the ordinary close shovel and sent to bank as triping, and there prepared for market. The hutches are made of wood and have close ends, the boxes being about 4 ft. long, 3 ft. wide, and 2 ft. deep; capacity, 9–10 cwt.; wheels, 8 in. diameter, with flanges to run on edge rails laid to a gauge of 24 in.

The tumbler is of the backward tumbling kind, which turns the hutch upside down, as shown in Figs. 604, 605, and delivers the coal on the top plate of the screen. At the bottom end of this

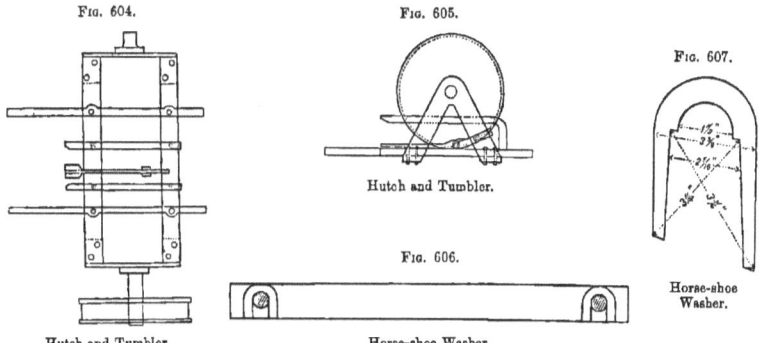

Fig. 604. Fig. 605. Fig. 607.

Hutch and Tumbler.

Fig. 606.

Hutch and Tumbler. Horse-shoe Washer. Horse-shoe Washer.

plate, and at the top of the screen bars is a shutter or door, to which a boy constantly attends, his chief duty being to regulate the flow of coal down the screens.

The several screens are arranged all at the same inclination of about 15 in. per yard, one above the other, and only a few inches apart. The top screen is fixed and formed of bars about 12 ft. long, supported on round rods, the distances apart being regulated by washers of a horse-shoe pattern,

(Figs. 606, 607). This form enables the washers to be placed on and removed from the cross-bearers supporting the screen bars with facility, and the spaces between the bars varied at will with as little labour as possible.

The under screens are riddles, varying in the mesh from $\frac{3}{8}$ in. to $1\frac{3}{4}$ in., and vibrating 80–100 strokes per minute, each of which delivers the coal passing over it into shoots, discharging on travelling picking tables. The number of under screens or riddles corresponds with that of the grades of coal prepared for the market. The bottom riddle passes only gum, and delivers into a hopper which empties itself into a cast-iron trough, about 12 in. wide, and in which moves a creeper that rakes the gum towards the stokehole of the colliery boilers.

FIG. 608.

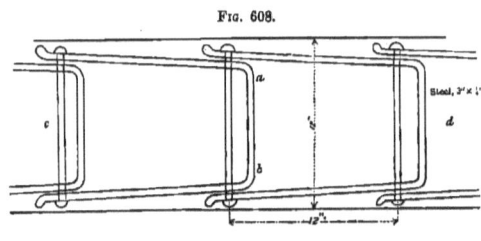

Creeper at Barrow.

The construction of this creeper is shown in Fig. 608. Two pitch chains carrying a cross-piece ab work in a trough cd. The motion of the whole is from a common shaft with belts.

The picking tables are endless bands of wire netting made by suitable mechanism to travel at a speed of 30–35 ft. per minute. In some the meshes are closed with strips of hoop iron inserted so as to make practically a close sheet. The picking tables are arranged one on each side of a line of rails, and at right angles thereto. One travels towards the right and the other to the left, both discharging on a trimming shoot suspended above the railway waggon. When tender coal has to be dealt with, these trimming shoots are sometimes automatic balance plates, arranged so as to gently lower the coal into the waggon. The outer ends of these travelling tables can be elevated so as to allow the coal to be delivered into tubs and then run out to the bing when stocking is necessary. All screens and picking belts are started and stopped by levers and clutches as required, and are driven by belts from the main shaft. The refuse consists of stone and shale got along with the coal, and is loaded into hutches and taken to the spoil bank. About 1·25 per cent. of dirt is taken out in the process of dry cleaning.

Guinotte & Briart's device for allowing a practically instantaneous enlargement of the apertures to any extent is effected by a very simple differential grid, shown in Figs. 609 and 610, where A represent the bars which it is desired to open out. These bars are carried by spindles B, threaded with right- and left-hand screws, which take the sleeves C, having threads in a contrary direction. These sleeves are traversed by a shaft D, common to the whole system, and holding the sleeves by the key E; consequently the rotation of the shaft D widens or narrows the spacing of the bars as desired. This is effected by a handle fitted to the square end of the shaft. The degree of widening is indicated by figures on the shaft D, which serves as a scale. It may also be shown on a dial-plate

DRESSING MACHINERY. 437

by a needle attached to the shaft. In order to effect the symmetrical divergence of the bars, the middle sleeve of one of the systems is keyed to the shaft, while the bars of the other system are attached to cross-bars immovably fixed to the framework of the apparatus. This arrangement, which

Fig. 609.

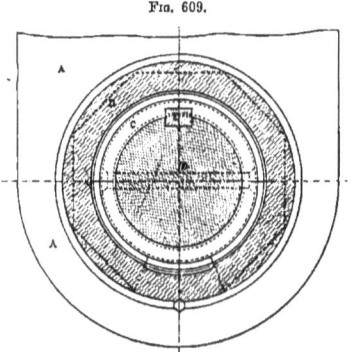

Guinotte & Briart's Differential Grid.

has been adopted in all their new machines for the last three years, gives entire satisfaction. Not only is the loss of time referred to avoided, but any degree of sizing can be obtained by the same machine within the limits of traverse of the bars, and without any expense.

Fig. 610.

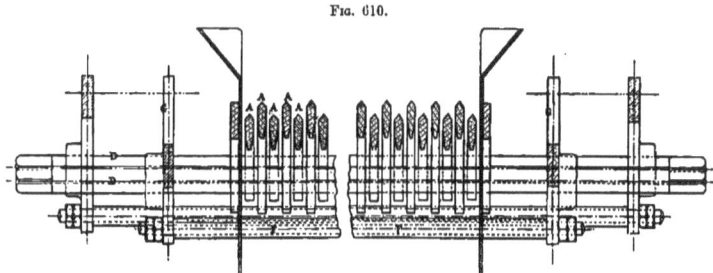

Guinotte & Briart's Differential Grid.

For dry cleaning at the Aldwarke Main Colliery, the hutches are made of wood, weigh 3¾ cwt., and carry 8–9 cwt. of coal. The dimensions are 3 ft. 6 in. long, 2 ft. 10 in. broad, and 2 ft. deep, with wheels about 8 in. diameter, adapted to run on edge rails. All the hutches have close ends. The tumblers in use are those known as "Rigg's," shown in Figs. 611, 612. They tip forward and turn the hutch over.

438   MINING AND ORE-DRESSING MACHINERY.

The general arrangement is shown in Fig. 613. The several screens are indicated by the letters $B^1 B^2 B^3$, and the picking bands by the letters $C^1 C^2 C^3$, the tippers being fixed at the point A. These tippers are fitted with a heavy hinged flap for regulating the flow of coal. No attendants are necessary for this purpose. But in addition to these flaps, sometimes a heavy piece of iron or wood

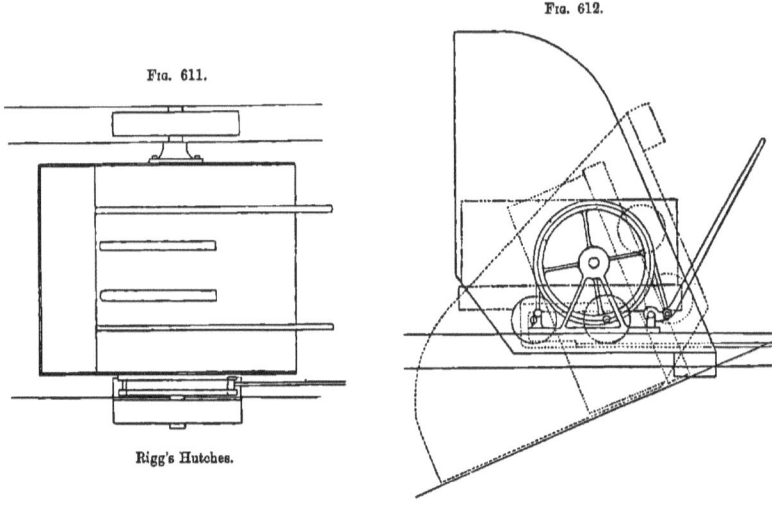

Fig. 612.

Fig. 611.

Rigg's Hutches.

Rigg's Hutches.

is suspended by chains from the top of a screen or shoot B^1, and lies across the floor of it at about one-third of its length from the top. Further across the inner end of the picking table there is fixed a board, adjustable vertically like a sluice. This sluice board effectually serves its purpose, the layer of coal delivered on the travelling belts being exceedingly regular. To each screen there is a corresponding picking table. The bands or tables are formed of plates about 14 in. broad, fixed at each end to endless chains working over polygonal drums. The bands are about 60 ft. long, and travel at the rate of 40–50 ft. per minute. At the delivery end of each band is an arrangement of short screens for removing any gum made on the tables.

The *modus operandi* is as follows :—Screen B, which is inclined at an angle of 18°, has a riddle B^1 with mesh $1\frac{1}{2}$ in. by $\frac{7}{8}$–$1\frac{1}{4}$ in. fixed over it, and vibrated at 106 strokes per minute by an eccentric of 7 in. stroke. The riddle B^1 delivers the lump coal on the picking table C^1, and the dead shoot D delivers what passes through riddle B^1 of screen B on to B^2.

Screen B^2 is $\frac{3}{4}$ in. mesh. What passes over it on to the picking table C^2 are nuts, what passes through it falls on to the screen B^3, the mesh of which is $\frac{3}{16}$ in. What pass over B^3 on to the picking table C^3 are peas; what passes through B^3 is smudge, and it is delivered either into a waggon L^3,

DRESSING MACHINERY. 439

direct, or into a creeper which conveys it to an elevator in connection with the washing apparatus. Fine dross taken out by the short screens H and G^1 is conveyed by the lower or return side E^2 of creeper xy to the waggon L^4.

FIG. 618.

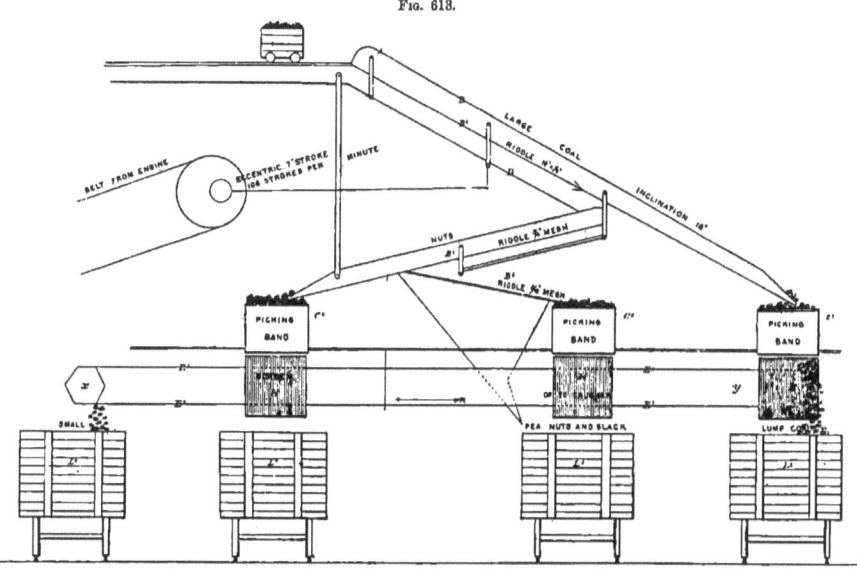

Dry Cleaning at Aldwarke Main.

These picking tables are fixed on a platform above the several lines of rails, and instead of being at right angles to the railway, as at Barrow, are parallel thereto. There is a line of rails corresponding to each grade of coal produced.

To facilitate the mixing of the different grades (as is done to produce No. 2 and 3 gas mixtures), by-pass doors F, forming the dead plate of screen H, are raised so as to direct the coal over screen G^1 into the creeper E^1. The mode of operating is as follows:—A waggon L^1 is placed on the outer line of rails, on the right and under the lump coal band C^1, into which is delivered the lump coal. The by-pass doors F, at the end of the nut or pea bands, or both, as the case may be, are raised, and nuts or peas, or both, are diverted into the creeper E^1 (instead of into waggons), which is travelling towards the right hand, and arranged to deliver these along with the lump coal into the waggon L^1; the gum being always taken out and carried back by the creeper E^2, to the waggon L^4. No balance plates are used at this colliery.

A convenient arrangement for delivering coal, from say the middle of a picking table into a shoot, may be here mentioned. It consists of a fluted roller working obliquely across the table

(Fig. 614), the roller A is made to revolve, and is adjusted so as to work either the whole or a part of the coal from the band B, into the shoot C. These rollers can be fixed anywhere and in any number along the line of tables, and in certain cases are a very great convenience in sorting coal.

The screens and the three picking tables or bands above described, deal with 600 tons of coal in 8 hours. On the band C^2, which deals with nuts, or about 15 per cent. of the total output, 6 boys are employed. On band C^3, over which passes about 10 per cent. of the output in the shape of peas, 2 boys are employed. On band C^1, which deals with large coal, or about 60 per cent. of the output, 6 young men are employed. The dirt picked out is about 5 per cent. of the gross output, and the dust or gum amounts to 12–14 per cent. The refuse consists of stones and shale, and is conveyed to the spoil bank.

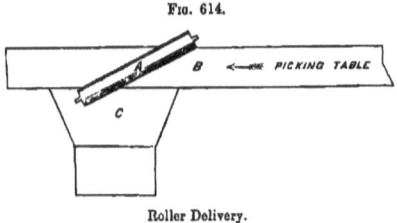

Fig. 614.

Roller Delivery.

Fig. 615, 616 shows the arrangements for screening and picking the coking coal, and washing the dross at Flimby Colliery. These consist of a jigger screen, a revolving table, and a trough washing machine. The jigging screen is 14 ft. long by 8 ft. wide, and has an inclination of 1 in 5.

Fig. 615.

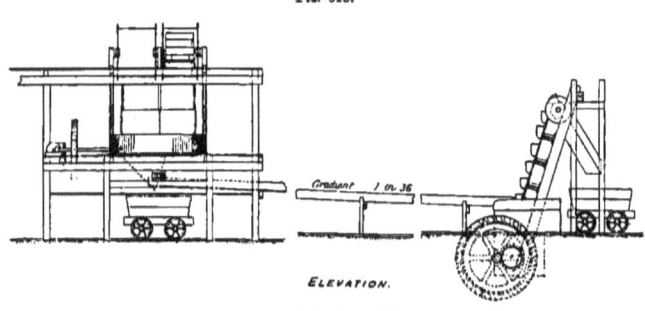

Screening and Washing at Flimby.

It has a stroke of 4½ in., and makes 90 strokes per minute, deriving its motion from cranks on a transverse shaft driven by spur gearing from the crank shaft of a small horizontal engine. The jigger has steel bars 12 ft. long and ¼ in. apart, and of the same section as those in the common screens. It is hung by four chains, which are said to have an advantage over rods in giving a slight hitch every stroke, but this hitch is so slight as to be imperceptible. The jigger receives the coal from two of Cook's tumblers, which are placed side by side, and each of the screens is fitted with a "kepper" at the top of the bars to regulate the descent of the coal on to the screen. This kepper is in use at nearly all the collieries, and the best form is shown in Fig. 618. When the kepper is shut, the crank and connecting rod are in a straight line, so that no pressure of coal will

DRESSING MACHINERY. 441

open it, while the slightest pressure on the lever is sufficient to do so. These keppers are of great use in regulating the supply of coal to the jigger, or to the bars of the screen, so as to prevent overcrowding, and also for keeping hutches separate for fining purposes, there being room for a

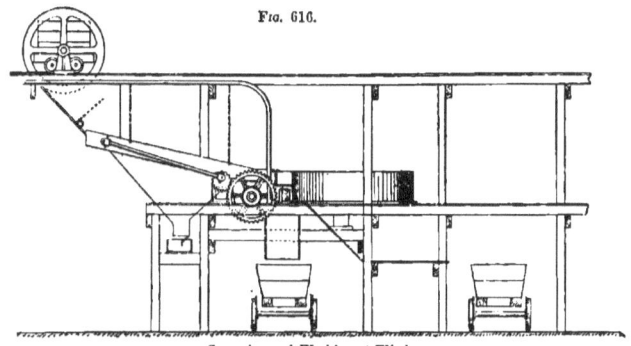

Fig. 616.

Screening and Washing at Flimby.

hutch or two above the kepper. Of course, the screen must be so steep that the coal will start to run down after being stopped.

The jigger lands the coal on the revolving table, from which, after being picked, it is discharged into the waggon. This table is 10 ft. diameter, with a space in the centre 6 ft. 8 in. diameter, leaving a ledge 20 in. broad all round on which the dirt is picked. The table revolves 3 times per minute, and the whole output of 350 tons triping, yielding 250 tons of round coal, is picked by 6 women, who throw all dirt, and pieces of coal and dirt sticking together, down through the hole in the middle of the table, whence it passes by a shoot to a small platform, where it is further picked by 3 women, who break up the coaly pieces and fill the whole of the dirt into hutches, in which it is conveyed to the sea. About 9 tons of dirt are picked at this table per day, so that the cost of picking 250 tons will be:

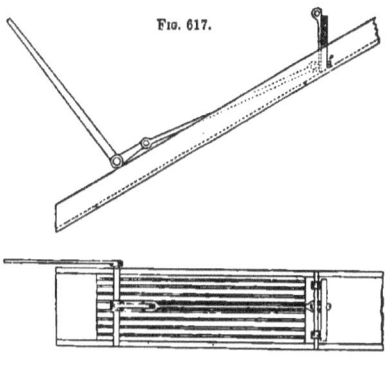

Fig. 617.

Screen and Kepper.

	s.	d.
9 women at 1s. 6d. per shift	13	6
1 boy working engine and regulator at 1s. 4d.	1	4
Removing 9 tons dirt, at 3d. per ton	2	3
Total wages cost for picking 250 tons	17	1

which is at the rate of 0·82d. per ton of round coal picked.

3 L

The house coal dross is sold as it comes from the screen without further treatment; and the dross from the coking coal is washed, crushed, and coked. The system of washing adopted here is the trough system, as shown on Fig. 617. The dross falls directly from the screen into a trough, along which a stream of water is passing, by which it is carried down to the washing trough. At the point where the dross first meets the water, the trough or rhone is divided into two by a longitudinal partition extending down to within 1–2 in. of the bottom. This always leaves a clean water-way on one side, and so prevents flooding in the event of an extra rush of dross, while, at the same time, it causes the dross to become thoroughly saturated as it is washed away from the tail. The inclination of this part of the rhone is also steeper than the washer proper, so as to carry both dirt and dross down to the upper end of the washer. The washing machine consists of two parallel rhones, each 150 ft. long by 17 in. wide and 13 in. deep, lying at an inclination of 1 in 36. These rhones are used alternately, one washing while the other is being cleaned. The velocity of the water is about 300 ft. per minute, and the quantity used is said to be about 400 gal. per minute for 100 tons per day of 10 hours. As the current passes along, the larger and heavier pieces of dirt are deposited near the head of the rhone, while the coal and lighter pieces of dirt are carried farther down. To assist in settling the small pieces of dirt, there are two stoppers about 2 in. deep, one at the foot of the rhone and one about 20 ft. farther up. While the washing is going on, the attendants occasionally stir up the dirt which has settled, so as to relieve any pieces of coal which may have become covered. When the dirt begins to show through the coal in the trough, the water is turned into the other one, and the dirt is shovelled out into hutches which stand alongside. The coal is washed directly into a large wooden box, from which it is elevated to the top of the waggon by a bucket elevator driven by a 6 ft. water-wheel worked by the overflow water from the wooden box. There are one man and two boys in attendance on the washing machine. In wet weather the water from the pit is sufficient, but in dry weather the water is pumped up by a 3 ft. centrifugal pump running at 250 revolutions, and used over again. The cost for washing 100 tons is as follows:—

	s.	d.
1 man, at 2s. 8d. per shift	2	8
2 boys, at 1s. 4d. „	2	8
Removing about 18 tons of dirt, at 3d. per ton	4	6
Total wages cost for 100 tons ..	9	10

which is at the rate of 1·18d. per ton of dross washed.

The washed dross is taken in waggons to the ovens, where it is crushed in a 4 ft. 6 in. Carr's disintegrator before being coked.

At the Robin Hood Pit, a washing machine by Bell & Ramsay is in course of erection. The whole arrangements as proposed are shown in Fig. 618. The coal will be tipped on a jigger screen, from which the dirt will be picked and the coal landed directly into the waggon, while the dross will fall into a pit, from which it will be elevated to the washer. The washer is essentially a trough washer, with the addition of mechanical stirrers to help to float the coal, and so save water, and an almost automatic arrangement for cleaning out the dirt. Each rhone is hinged at the upper end, and has a shaft running along the whole length. This shaft is hung independently of the rhone, and is provided with stirrers, as shown in Fig. 619, which have a transverse oscillating motion. There are also stoppers distributed along the length of the rhone, some being fixed to the

DRESSING MACHINERY.

framework near the low end and others near the top end, being movable up and down by means of levers. For catching the fine gum floating with the water there are one or two scum plates fixed to the rhone. These stoppers and scum plates are shown in Fig. 620. When the rhone is to be cleaned, the lower end is lowered clear of all the stirrers and fixed stoppers, and the movable stoppers are raised so as to allow the rush of water to carry off the whole of the dirt to a hutch placed to receive it. Fig. 621 shows the travel of the prongs.

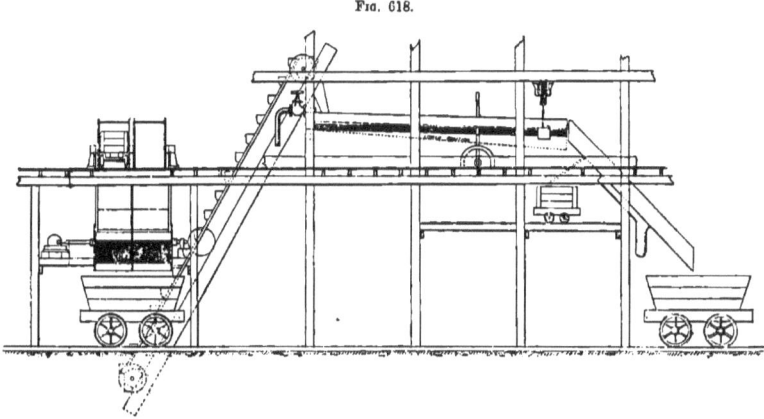

Fig. 618.

Bell and Ramsay Washer at Robin Hood.

The dross-washing machine at Glenclelland Colliery was erected in the summer of 1883, from drawings supplied by Kerr & Mitchell to John Bell, engineer, Wishaw. A considerable number of machines of the same type have since been made by Bell, and the machine is now known as "Bell's." Several alterations have been made on the machine at Glenclelland since it was erected, and Fig. 622 shows it practically as at present working. In its general construction it is very similar to other forms of "Bash" washers. The following is a description of the working:—

The motive power is supplied by a 12 in. cylinder horizontal engine, working with a 2 ft. stroke, and running 40 revolutions per minute. The dross is elevated from the dross hole D by the elevator E^1, which is worked by means of a belt from the crank shaft of the engine, and which has 32 buckets, each bucket holding 25 lb. of dross. The dross is conducted from this by a shoot into the tanks of the washing machine T, and falls upon the perforated plates B, where it is agitated by the water alternately rising and falling through the holes in the plates. As the dross is agitated, the impurities having the heaviest specific gravity sink to the bottom and rest on the perforated plates, the good dross rising to the top until there is a thick layer of dirt above the valve V^2. When this accumulation has taken place, the valve is opened, and the dirt falls to the bottom of the tank where again the valve V^1 is opened, and the dirt falls into the chamber containing the dirt elevator E^2.

444   MINING AND ORE-DRESSING MACHINERY.

Fig. 619.   Fig. 620.

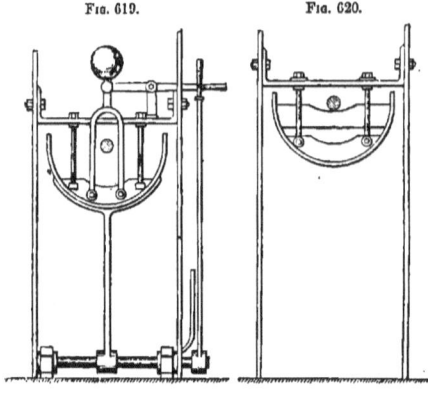

Fig. 621.

Fig. 622.

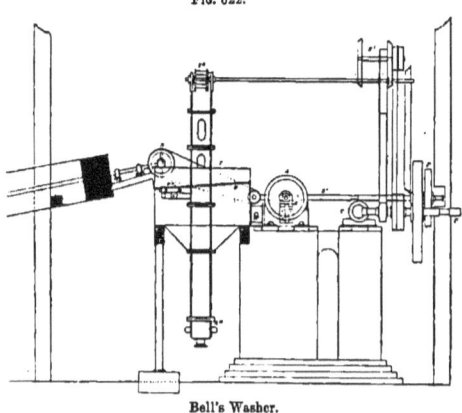

Bell's Washer.

This elevator has 30 buckets, each holding about 8 lb.; and by it the refuse is lifted and discharged into a hutch to be conveyed to the dirt bing.

The valve V^2 stretches the whole way across the plate B, and is in the section of the shape shown. The valves V^1 are simply half hollow balls fitting into circular holes, and are lifted and lowered by means of levers at the sides of the tanks within reach of the man attending the machine. The good dross is pushed over from the front of the tanks by the skimmers S, and passes down the shoot into the revolving riddle R. This riddle is 11 ft. 6 in. long, and is divided into meshes to suit the separation of the dross into peas and nuts. The first division—2 ft. in length—where the water strikes the riddle, is of fine mesh $\frac{3}{16}$ in. The water is removed by this, and carrying with it the smallest of the dross, it passes along a rhone to the front of the boilers, where there is a small revolving riddle of very fine mesh, which removes the dross and passes it to the front of the boilers ready for firing.

The special feature of this machine, and that by which it differs from other washers, is that the agitation is produced by means of a piston working in a horizontal cylinder A. The working part of the cylinder is bored out, and is 33 in. diameter. The motion to the piston is communicated by a parallel eccentric motion, worked from the shaft S^1. It is geared to work at 80 strokes per minute. The length of the stroke is $4\frac{1}{2}$ in. The water is introduced at the ports W, one for each tank, between the agitating cylinder and the tanks. The water comes from a water tank, the bottom of which is about 12 feet above the level of the top of the dross tanks, and is pumped up to it by means of a 9 in. ram, with a 12 in. stroke, worked off the crank of the engine. As the piston moves backwards and forwards in the agitating cylinder, the water in each tank alternately rises and falls, the water in the one tank rising while it falls in the other. Strong agitation is thus produced, the amount depending on the size of the cylinder and the length of the stroke.

The horizontal cylinder was adopted for the sake of both economy and efficiency. The construction is less costly, as there is only one cylinder to bore, and gearing required to produce motion for one piston instead of for two. The power required to drive for the same agitation should also be less, as there is less friction and complete reciprocity of motion. The results of the working of the machine have been very satisfactory, and although it was erected more than 6 years ago, when there was very little attention paid to washing, it probably still works as well as one of the ordinary Bash machines, the great point in its favour being strong agitation, which, it is hardly necessary to say, is required for clean washing.

There was some difficulty found at first in keeping the piston tight, owing to its weight. To help this, a bracket K was fixed to bear up the outer end of the piston rod, and a system of gutta-percha rings round the piston has been adopted. These rings are made out of old pump-bucket gutta-percha, and measure 2 in. × 1 in. section. They are put in place of the ordinary metallic piston rings, and, according to the condition of the cylinder, they last 1–6 months, and require to be renewed to keep the piston tight.

Experience has shown that the chief defect of this washer, as of all the ordinary Bash machines, is that in spite of constant renewal of the perforated plates B, the holes through which the water passes wear so quickly that it is impossible to prevent some of the very fine dross from passing down through the plates and being carried away with the dirt. The only way to get completely over this difficulty is to size the dross before washing, and to wash the small sizes in a more suitable machine, such as Robinson's, or, perhaps, better, a Bash washer with the use of felspar on the top of the plates. The defect is increased by the sucking down of the water through the holes in the course of agitation. This might be partially overcome by giving a slow downward motion and a quick upward motion to the water, which might be arranged by means of some form of link motion, although not in conjunction with a horizontal cylinder. A further defect is the valve arrangement for dropping the dirt from the perforated plate. Its efficiency depends too much on the care of the attendant, and it would be a great improvement to have some form of automatic dirt separator. A very simple form of such a separator is that of a wheel with little buckets working at the side of the tanks in communication with an outlet for the dirt. This form is used in Continental machines, and works well, there being an arrangement by which the communication can be adjusted to prevent any dross passing through. The machine will wash rather more than 100 tons daily. In Fig. 622 C is a 12 in. engine supplying the motive power; E^1, dross elevator; F, fly-wheel on agitating shaft; S^1, agitating shaft; A, agitating cylinder; W, ports for water to enter tanks; T, tanks for washing; B, perforated bottoms in tanks; V^2, valves to let dirt fall from perforated plate; V^1, half-ball valves to let dirt down to dirt elevator; S, skimmers to push washed dross towards riddle; R, revolving riddle for screening dross; E^2, dirt elevator; P, position of 9 in. ram, working from crank of engine; K, bracket for bearing up end of piston rod.

The coal washing machinery at the Cwm-Avon works is on the felspar system, and is used to wash dross for coke-making. It was constructed and erected by Evence Coppée & Co., of Cardiff, and is shown in Fig. 623. The small coal arriving from the mines is delivered into the pit A. An elevator B takes it up and delivers it to the screen C. This screen is provided with round perforations of ¾ in. and ⅜ in. diameter; therefore, the sizing is as follows :—From ¾ in. upwards; from ¾ in. to ⅜ in.; from ⅜ in. downwards. The first two of the above sizes are washed in the "Bash"

Fig. 623.

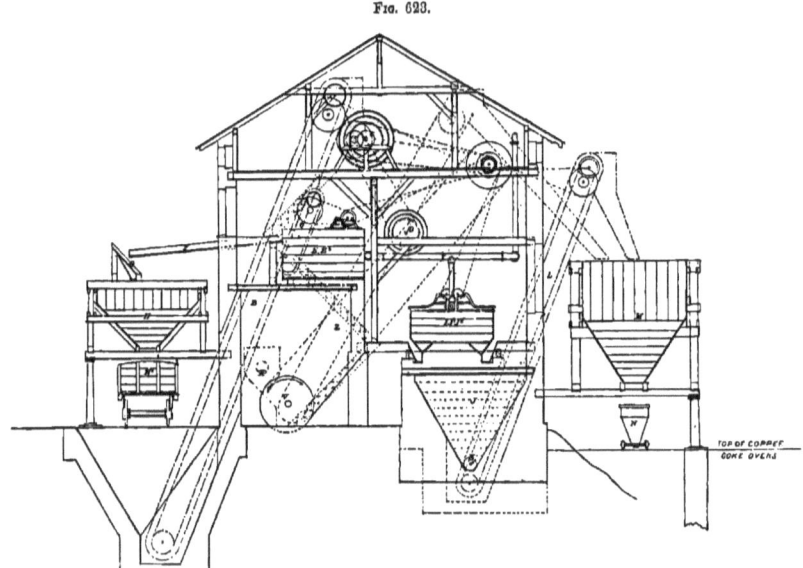

Coppée Washer at Cwm-Avon.

nut coal washers E E¹, and after being washed are brought by means of water through the shoot F, upon the draining screens G, after which they accumulate in the bunker H, from which they are loaded into the truck H¹. The water passes through the screen G, and goes through shoots to the basin J. The refuse or dirt of the nut coal washers is taken a few feet up by the small elevator O, which delivers it by a shoot into the shale basin. The coal of ⅜ in. downwards is sent by water into a screen D, which is provided with perforations of ¼ in. and ⅛ in. diameter, so that the coal is sized as follows :—From ⅜ in. to ¼ in.; from ¼ in. to ⅛ in.; from ⅛ in. downwards. These are washed separately in four felspar washers I, I¹ I² I³, intermediate being most abundant. The coal, after being washed over the felspar, accumulates in the coal basin J, in the bottom of which there is a screw K, which delivers the coal into the bunkers M, whence it is taken to the coke ovens. The rubbish from the felspar washers is carried by shoots to the shale basin, from which an elevator

DRESSING MACHINERY. 447

raises the whole of the rubbish and delivers it into a bunker, where it is taken by trucks to the tip.

At X is shown the position of the coal crushing machine, and at V that of the elevator for taking up the crushed coal. This part is not yet erected, but provision is here made for it if required.

A centrifugal pump sends the water to the upper screen C, and to all the washers. And it must be here noticed that the dross is damp before being delivered to the washer, and in this state would not size in the screen; and it is for this reason that a jet of water is sent into the screen, so that the dross may be thoroughly separated. The overflow water from the shale basin runs into a basin, then into J, in which the dirt is deposited. It then flows into a well, from which it is raised by the centrifugal pump, and led by means of cast-iron pipes back into the washers.

The steam engine V is horizontal, and drives the whole system : cylinder, 21 in. diameter; stroke, 42 in. There is sufficient power in the engine to drive the crusher when put up. There are three boilers, each 28 ft. long and 5 ft. 6 in. diameter, and supplying steam at 40 lb. The boilers are fired by the waste gases from the coke ovens, and the feed water is heated to about 212° F.

FIG. 624.

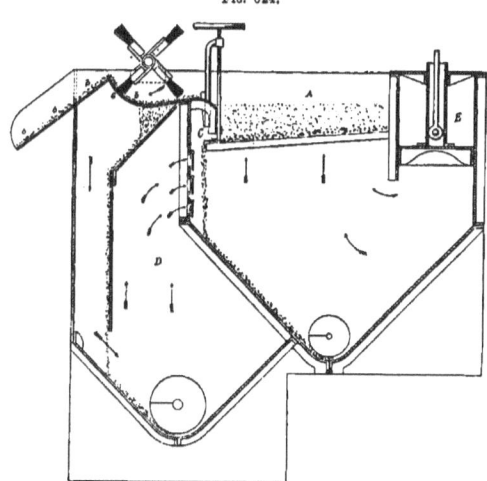

Sheppard Washer.

The centrifugal pump is 18 in. diameter, and works at a speed of 845 revolutions per minute. This pump circulates the water from the well, and replenishes the waste that escapes with the washed dross and shale. The delivery pipe is 7 in. diameter.

The screen C at the top is made of plates with perforated holes, and a 2 in. malleable iron pipe for supplying water to wet the dross is led into the shoot that delivers the coal from the elevator to the top end of the screen. The diameter of the screen at the top is 5 ft. 4 in., and at the opposite end 6 ft. 8 in., the shaft being level. The total length of the screen is about 12 ft., and the speed 12 revolutions per minute. The screen D is also about 12 ft. long, and is formed of malleable iron plates with perforated round holes. The diameter next the shoot is 3 ft. 3 in., and at the delivery end 4 ft. 6 in.; speed, 8 revolutions per minute.

Three tons of washed dross produce two tons of coke, costing about $7d.$ per ton for labour at the works. The coal-washing plant is situated at a considerable distance from the mines and close to the coke ovens and Cwm-Avon furnaces. The coal is tipped over simple screens, having bars 20 ft. long, and $1\frac{1}{4}$ in. apart; and 60–75 per cent. passes through, the remainder being used without further handling as house coal. The average quantity of dirt in the mixture of the dross sent to the washing machine is 15–18 per cent. The large nuts are used in gas producers, and the small nuts and gum for coke making. The two sizes of nuts, the gum, and the dirt are all delivered into their respective hoppers before being dropped into waggons or bogies. The dirt is taken away in ordinary waggons, and emptied by hand. This latter arrangement will probably be improved on and the

FIG. 625.

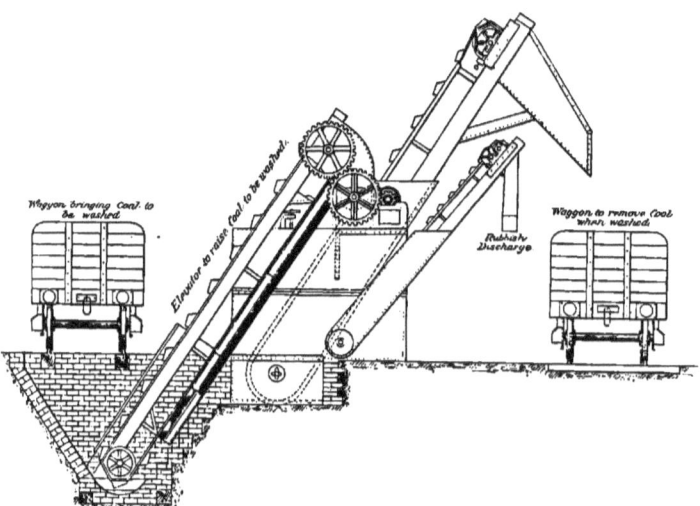

Sheppard Washer.

labour saved, when the washing plant is longer in use. The following analyses show very good results:—Ash in nuts over $\frac{3}{4}$ in. screen, $1 \cdot 85$ per cent., not coked; ash in nuts over $\frac{1}{4}$ in. screen, $2 \cdot 35$ per cent. coked; ash in gum, $5 \cdot 23$ per cent. coked; ash in coke requiring 30 cwt. raw coal per ton of coke made = 6·38 per cent. = 4·26 per cent. ash in raw coal. The dirt contained an appreciable quantity of small coal. The number of men required to work the machinery, exclusive of emptying or filling waggons or disposing of the dirt, is as follows :—1 man attending engine, centrifugal pump, &c. ; 1 man attending on the upper floor, that is on the first revolving screen and bash washer for the larger sizes of coal ; and 1 man attending the felspar machines in which the smaller sizes are treated, or, in all, 3 men. The cost per ton for labour is under ¾d.

The usual form of Sheppard washing machine is shown in Fig. 624, and a general view of the machine and elevators in Fig. 625.

The water is agitated by means of the piston and cylinder E, driven from the crank shaft of an engine. The dross falls into the tank at A, and, as the agitation proceeds, the dirt gravitates to the inclined bottom of that division, and, finding its way out at C, falls into the principal division of the tank, at the bottom of which is a screw to convey the dirt to the elevator, where it is raised to the discharge spout. The washed dross passing into the space B is swept by the revolving brush H, and the larger pieces fall over the shoot into waggons, while the smaller particles pass through the perforated plate under the brush, and fall into the tank D. A screw at the bottom of this tank passes it on to an elevator, by which it is raised to a convenient height to be loaded. The arrows indicate the direction of flow of the water in the machine.

Robinson's washing machine (Fig. 626) consists of a truncated cone, inverted, measuring 8 ft. diam. at top, 1 ft. 10 in. diam. at bottom, and 6 ft. 6 in. deep. The plates are steel, ¾ in. thick. A strong shaft is fixed vertically, to run in the centre of the cone ; on this shaft is a cast iron cross head, to which are bolted 4 cross arms of oak, 8 in. by 6 in. To each of these cross arms are bolted 3 heavy malleable iron bars, projecting down so as almost to touch the sides of the cone or washer, and on the bottom of the driving shaft are bolted shorter arms.

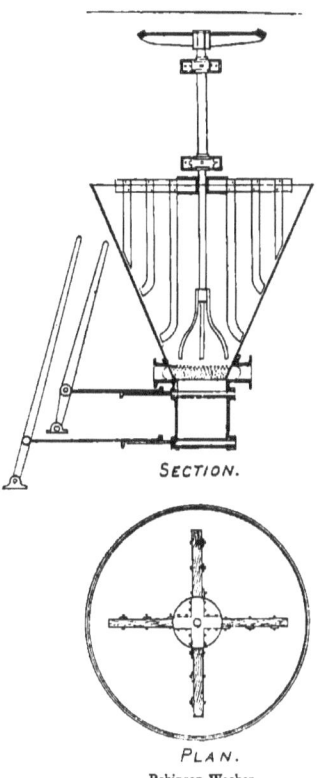

Fig. 626.

SECTION.

PLAN.

Robinson Washer.

At the same time as the dross is falling into the cone, a stream of water is being forced in at the bottom, through the cast iron bottom piece, which is covered on the inside by the side plates, having 8 rows of ½ in. holes. The shaft, with crosshead and depending arms, revolves inside at the rate of 14 revolutions per minute. The dross coal is carried by the water over the side of the washer, and the heavier particles—brasses and stones—

fall to the bottom. Fixed on the side of the washer, with a suitable inclination, is a sheet of finely perforated zinc, 6 ft. by 3 ft. by 0·55 in., with 72 holes per sq. inch. In passing over this zinc sheet, the water is separated from the dross, which is again elevated and carried along a shoot into a hopper placed conveniently for loading the washed dross into waggons. Before this dross passes into the hopper, an arrangement is made in the shoot, by perforated plates, for drawing off part of the finest dross for the boiler fires.

The brasses and stones which gravitate to the bottom of the washer are removed by means of the valve arrangement, the upper valve being always closed before the lower one is opened. The débris is also elevated into a hopper, where it is drawn into a hutch and run out to the dirt heap. The water separated from the dross by the perforated zinc sheet carries with it several tons per day of very fine particles of coal, but this is caught in the following manner:—The water is run back in wooden rhones to the tank in connection with a pulsometer pump. The rhones are 6 ft. higher than the tank. The last rhone is deeper than the others, and is 18 in. wide, with a fall of 5¼ in. per yard. When the water enters this rhone, it runs on a sheet of zinc, having 232 holes per sq. inch. The water passes through the holes, leaving the very fine dross to fall over the end of the rhones. It is then filled into waggons, carried to the boilers, and used there for firing.

A No. 9 pulsometer is used for pumping the water, which is forced direct into the washer, in preference to pumping the water into a tank 30 ft. above the washer, and regulating the flow into the washer by a valve. A certain quantity of fresh water is added daily to that in use, but no record of the quantity is kept; but the oftener the water is changed the better it is for the fuel washed. There are two settling ponds in connection with the washer, but they are now very seldom cleaned out. On the side of the last pond there is placed a No. 4 pulsometer pump, which pumps all the dirty water that is allowed to run away from the washer on to the top of the dirt heap. A considerable quantity of fine sludge is thrown on to the top of the heap with the water, but the heap is a perfect filter, the water leaving the bottom of it being as clear as it was before it first entered the washer. The washer was erected to wash 20 tons per hour. The engine for driving the washer, three elevators, nut jigger, nut picking table, dross conveyor (60 ft. by 3 ft.), and the coal crusher, has a 14 in. cylinder and 2 ft. stroke, running 56 strokes per min.; pressure of steam in boilers, 50 lb. All the different parts of the machinery described are driven by shafting connected by bevelled gearing, except the crusher, where belting is used as a much simpler mode of getting up the speed necessary.

DRESSING MACHINERY.

STATEMENTS CONTRASTING THE PRACTICE AT THE SEVERAL COLLIERIES IN THE SHEFFIELD, NOTTINGHAM, AND MANCHESTER DISTRICTS.

I. Dry Cleaning.

Name.	Speed per Minute of Picking Bands.	Number of Belts.	Total Length of Bands.	Total Length of Band available for Operating on.	Capacity.	Vibration of Screens per Minute.	Inclination.
	feet		feet	feet			
Barrow	30 to 35	..	..	..	1200 to 1500 tons daily ..	80 to 100	1 in 2·4
Aldwarke Main	40 ,, 50	3	180	180	600 tons in 8 hours ..	106	1 in 3
Nunnery Main	60	3	120	120	..	..	1 in 4
,, Nut	40	1	62	60	..	..	1 in 4
Annesley ..	30 to 35	1	250	250	700 tons in 8½ hours ..	115	1 in 4
Clifton	15	2	60	10	350 to 400 tons per day of 10 hours	All fixed	1 in 1·8

II. Wet Cleaning.

Name.	Type of Machine.	Nominal Capacity per Day of 10 Hours.	Actual Capacity per Day of 10 Hours.	Dirt in Coal.	Taken out.	Left in.	Smudge Tanks. Capacity.	Settling Ponds. Capacity.	Tank Capacity per Ton washed per Hour.	
									Smudge.*	Settling.*
		tons	tons	per cent.	per cent.	per cent.	cub. ft.	cub. ft.	cub. ft.	cub. ft.
Barrow ..	Robinson ..	300	240 to 280	..	8 to 10	4	6,500	21,540	2·17	7·76
Aldwarke ..	Trough ..	..	180	12	4	20,115	Nil	11·27	Nil
Nunnery ..	,, ..	..	220	..	10	..	5,000	16,000	..	..
Clifton ..	Coppée ..	400	300	..	15 to 18	..	12,150	..	3·04	..

* These are calculated on maximum capacity of the apparatus for want of more reliable data.

III. Trough Washing.

Name.	Dimension of Double Trough Washer.	Washing Capacity.	Inclination of Troughs.	Dirt Removed.	Dirt Left in.	Remarks.
				per cent.	per cent.	
Aldwarke Main	50 yards, 16 by 9	150 tons in 12 hours	1½ in. per yard; washer same inclination	12	4	Crushed.
Nunnery ..	36 yards, 16 by 9	220 tons in 10 hours	From coal hopper to troughs 2 in. per yard; washer level	10		Smudge and peas not crushed.

3 M 2

IV. *Hutches.*

Name.	Box.			Weight.	Capacity.	Diameter of Wheels.	Gauge.	Wheels.	Doors.	Ends.
	Length.	Breadth.	Depth.							
	ft. in.	ft. in.	ft. in.		cwts.	in.	in.			
Barrow	4 0	3 0	2 0	..	9 to 10	8	24	Flange	None	Fast
Aldwarke	3 6	2 10	2 0	3¾	8 to 9	8	..	,,	,,	,,
Nunnery	4 0	3 0	2 0	..	9 to 10	10½	28	,,	,,	,,
Annesley	4 0	3 0	2 0	..	11	..	26	,,	,,	Open
Clifton	3 6	3 0	1 8	..	7	8	22	,,	,,	Fast

NOTE.—All are made of wood.

V. *Costs (Labour and Repairs only) per Ton of Coal.*

Name.	Dry Cleaning.	Wet Cleaning.	Type of Washer.	Condition of Coal previous to Washing.
	d.	d.		
Barrow	1½	2½	Robinson	Crushed.
Aldwarke Main	..	2¾	Trough	,,
Nunnery	3	1·40	,,	Not crushed.
Annesley	2½ to 3	No cleaning here.	..	..
Clifton	3 to 4	6 to 7	Coppée	Not crushed.*

* Machine not in complete effective working.

Much of the foregoing information has been derived from the Report of the Coal Cleaning Committee of the Miners' Institute of Scotland (1890, price 8s. 6d.), which closes with the following general conclusions:—

" The methods and appliances in use in any one district can seldom be adopted as a whole in a similar form in another. This applies in many instances to collieries in the same district, and even to different seams worked by the same shaft. The nature of the coal, the associated and interbedded strata, the skill, customs, and prejudices of workmen, the markets to be supplied, the varying requirements of competition, and the caprice of the public, have all to be taken into account when designing plant for classifying and cleaning coal.

" While coal with marked characteristics can with care be selected underground so as to be filled separately, no process can be profitably applied underground for effectually removing refuse, especially the smaller particles. To clean coal properly, it must be treated on the surface.

" As a considerable percentage of dross is made in transit from the cage to the railway waggon, it is evident that the best results are got where attention is paid to the form of hutch and tumbler, the inclination of screens, and the drop into waggons; and this is specially important in the case of soft coals. A number of contrivances to lessen breakage are mentioned in the report. The careful hand-packing of large coal into the waggons, as practised in the Nottingham district, has advantages.

" For effective screening, especially when a large output has to be dealt with, there appears to

DRESSING MACHINERY.

be no better contrivance than the single or double jigger, or shaking screen, going at 90–100 strokes per minute, and having an inclination suited to the class of coal to be dealt with. There is a preference for wire-meshing for such screens at some collieries, and at others bars or perforated plates are preferred.

"For picking, the shaking screen just referred to, or the travelling band, or both combined, is the most effective and economical—the band being about 4 ft. wide, 40–60 ft. long, and moving at a speed of 30–60 ft. per minute, according to the quantity of coal to be passed. Ample length of band allows large coal to be sized and loaded into separate waggons by hand with despatch and economy.

"In every case it is necessary that the coal be delivered regularly from the tip hopper to the jigger or travelling band. This can be accomplished by regulating sluices worked by an attendant, or automatically by the intervention of a slow-motion band.

"Good light is essential to efficient picking.

"A rough rule for deciding the number and length of picking tables may be stated as follows:—
One picking table for every 30 tons per hour of triping output, travelling at the rate of 40 ft. per minute, and having an effective length of 10 ft. for every 3 per cent. of material to be picked off, plus 15 ft.

"The cost for labour of this system may be taken at about $1\frac{1}{2}d.-2d.$ per ton of round coal for every 5 per cent. of material picked out of that coal.

"For round coal, say above $1\frac{1}{2}$ in. cube, the dry process is universally employed, and this process can be successfully applied to nuts from say $\frac{3}{4}$ in. upwards where the refuse does not exceed 2–3, or even 4 per cent.; and the table capacity required, judging from the examples in the report, is about one table for every 20 tons per hour, travelling at the rate of 30 ft. per minute, and having an effective length of 15 ft. for every $1\frac{1}{2}$ per cent. of material picked off. The cost for labour will probably be $\frac{3}{4}d.-1\frac{1}{4}d.$ for every 1 per cent. picked off. Balanced screens, on which the coal is picked, are available only when the amount of material to be picked off is very small, say $1-1\frac{1}{2}$ per cent. For all small under $\frac{3}{4}$ in., and for dross from $1\frac{1}{2}$ in. downwards, with more refuse than 2–4 per cent., the wet process is most applicable.

"In the wet process it is desirable to have the arrangement so that the small coal can be delivered direct from the screens into the washing tanks without the intervention of waggons. In all the systems of washing, the best results are obtained by sizing the small coal before it reaches the machine. This can most conveniently be done by passing it through revolving screens with meshes of varying size. The supply and degree of pulsation or agitation of the water require careful adjustment to suit the various sizes of coal to be treated, and the relative specific gravity of coal and impurities.

"To remove the refuse from the smaller sizes, say under $\frac{3}{8}$ in., the felspar washer is the most effective. The felspar system is the most valuable where the coal is crushed before washing and is to be used for coke making.

"Where the coal and the refuse approach one another in specific gravity, it appears that in some cases the trough washer gives the best results. It is applicable for small quantities only, and requires a large flow of water and extra labour, but it has the recommendation of simplicity and small capital cost. It may also be sometimes utilised as a means of transport where the distance from the pit to the waggons or coke ovens is considerable.

"The Robinson washer is cheap as regards first cost and upkeep, and requires little water. It largely depends for its efficiency on the attention and skill of the man in charge, who may often be tempted to pass more through it than it can effectually clean.

"Speaking generally, more elaborate machinery is effective in avoiding waste in proportion to its cost; but the capital charges and upkeep are also high in proportion.

"Other things being equal, coal will be washed best with an abundant supply of clean water; but the more water used, the greater the risk of fine coal being lost, and the greater the difficulty of filtration. Water to wash coal for coking should not be often used over again, as dirty water dulls the coke.

"The particulars furnished as to settling ponds do not give sufficient data to justify any definite conclusion as to their capacity in relation to the quantity of coal washed. In most cases no record was kept of the quantity of water used; but settling ponds are a necessity, and their capacity will depend on the special circumstances of each case.

"There seems no better way of filtering the foul water, after it has passed through the settling ponds, than pumping it on to the rubbish heap, and allowing it to percolate through, as at Earnock.

"The washed gum of coal not suited for coking is meantime used almost entirely for firing colliery boilers. Briquettes are made of it to a small extent, but new outlets are required for this product.

"The large quantity to be treated daily, and the varying nature and proportions of the coal and dirt to be separated, render washing, at most collieries, a troublesome process; and unqualified satisfaction is seldom expressed as regards any machine in use. In some cases the machine may not be quite adapted to the peculiarities of the coal treated, or it may be over-driven, or not have a sufficiency of water, or be allowed to get out of repair, all or any of these causes leading to disappointment as to results. A separate siding for each class of coal is a desirable arrangement."

CHAPTER XV.

MISCELLANEOUS.

SIGNALLING.—A simple and effective signalling apparatus, or indicator, for use at the winding shaft, is shown in Fig. 627. It is placed in the engine-house, within view of the engine-driver. It consists of a board on which is marked to scale the positions of the several levels and the brace of a mine. Above the brace and below the bottom level—on the board—are placed wheels over which runs an endless chain; on the chain are fixed wooden blocks representing the cages and their relative positions in their shaft. The motive power is obtained by a band from an axle of the winding gear connected with the wheels of the indicator. It rings a bell to warn the engine-driver when the cage is within 10 ft. of the surface, and when the cage has reached 10 ft. above the brace of the shaft.

Another arrangement is shown in Fig. 628, which works as follows:—

When the knocker line is pulled, the catch presses in the lever b, and lifts the spring c, which

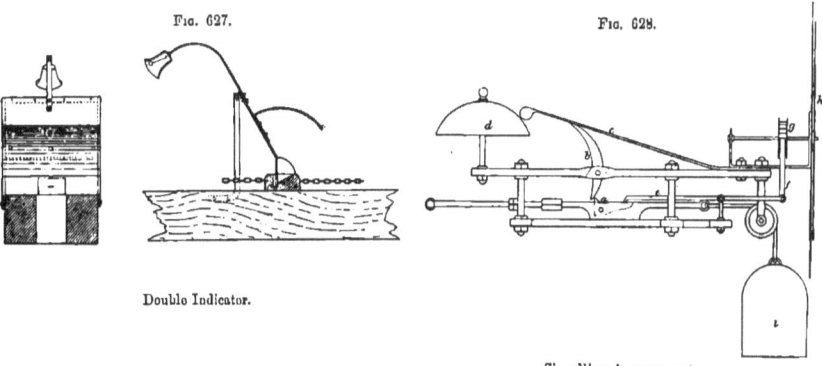

Fig. 627.

Fig. 628.

Double Indicator.

Signalling Arrangement.

strikes the bell d. The catch a then, returning to its original position, lifts the lever e, which is connected with the spring f. This latter turns the ratchet-wheel g, the handle h on the dial being moved forward one number for every knock given. The ratchet is kept from slipping back by a

Fig. 629.

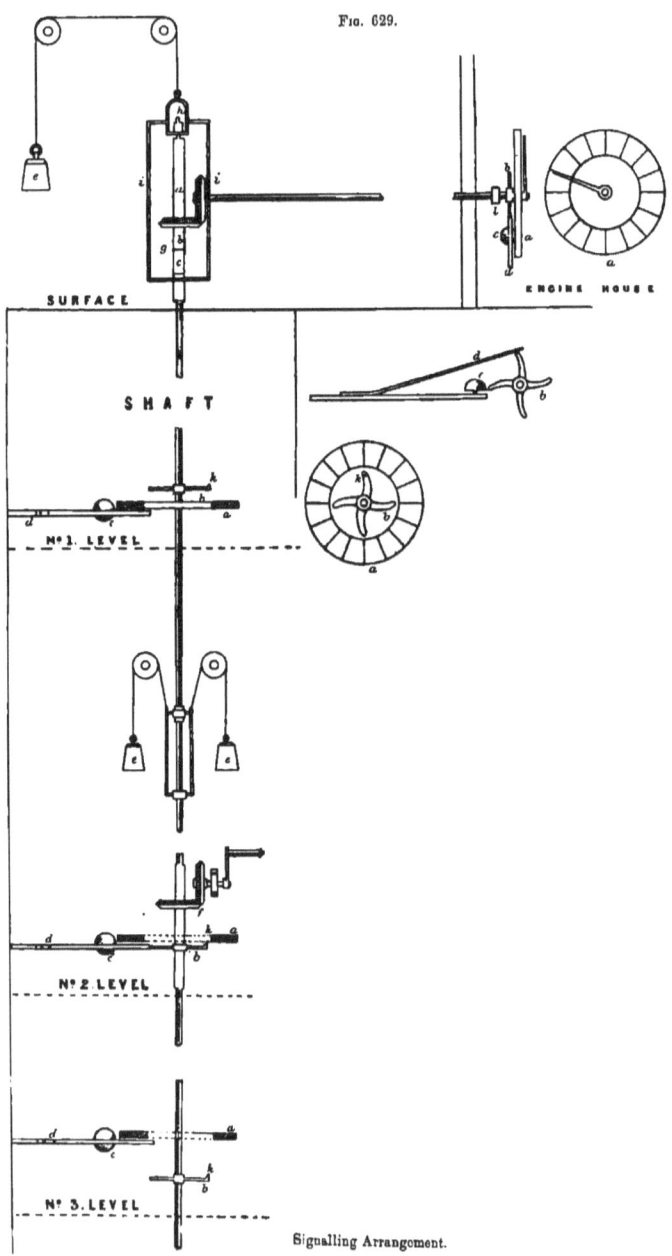

Signalling Arrangement.

small spring. By a weight i the catch a is brought back into its first position ready for use. A spring could easily be substituted for the weight if it were considered better. The engine-driver, when he has occasion to leave the room, observes what number on the dial the pointer is standing at, and, should any knocks have been given during his absence, the pointer will indicate how many. For example, if the pointer had stood at 2, and 5 knocks were given, it would have shifted to the number 7 on the dial. It is useful for preventing mistakes when knocks are given. Of a set of these signals, one is supposed to be in the engine-house, the second on the brace, the third on the surface, and the fourth in the plat below. A copper wire is continued right through, so that, upon one indicator being moved, the others change position in a similar manner. The short hand on the dial will indicate all the ordinary signals now in use in mines, while the long one is intended for other purposes. The pointers are of different colours, and the letters are coloured to correspond.

Fig. 629 illustrates another system: a, dials; b, ratchet wheels; c, bells; d, springs which strike the bells; e, balance for raising or lowering the rod; f, bevel wheels and crank for turning the rod; g, square bar which slides up through bevel wheel showing No. of level; h, swivel; i, guides to prevent swivel from turning; k, index point on ratchet wheel; l, small ratchet wheel to prevent rod from twisting.

The increased depth of shafts, and the introduction of rapid winding gear, have rendered the method of mechanical signalling known as the hammer and plate system quite obsolete. After repeated trials of many kinds of wire for transmitting an electric current down shaft, &c., No. 4 (0·238 in.) galvanised telegraph wire was found by Bagot to be the most suitable for shafts, and No. 8 (0·165 in.) for inclined planes. The shaft conductors were hung vertically from shackles on the pit-frame to the bottom of the shaft without any intermediate support, the depth in some cases being 600-700 yd. To the lower end of each wire, which hung free in the sump, was attached a 20-lb. weight, to act as a compensator. The wires in the planes (or drives) were supported by the ordinary stoneware insulators, spiked into the props, or into the overhead cross timbers. The insulated copper wires connected with the battery were made of No. 16 B.W.G. (0·065 in.) with a covering of gutta percha, making them equal to No. 7 B.W.G. (0·180); were then bound with tape, and covered with Stockholm tar. The 12-cell large-sized Leclanché batteries were found to be most suitable, and the outsides and insides of the glass cells down to the level of the exciting fluid were well brushed with paraffin oil, to prevent efflorescence, evaporation, &c. For the transmission of ordinary signals, a single-stroke bell circuit for operating on 9-in. electric gongs of special make was used, and the number of strokes would correspond with the number of knocks on an ordinary knocker plate. For special orders a 12-order dial circuit was employed, the current for working this circuit being supplied from the batteries required for the bell service. The order transmitted to the person to receive it is also shown on the dial of the transmitter's instrument, and he only can alter it, &c. It is said that in practice twelve separate orders can be transmitted by this system in ten seconds.

Many electric signal bells have been fixed in mines by John Davis & Son, of Derby, London, and Cardiff, one of whose bells is shown in Fig. 630.

SURVEYING, &c.—A number of instruments are used in mine surveying, the chief of which may be briefly referred to. The principal makers are W. F. Stanley, of Great Turnstile, Holborn, London; John Davis & Son, of Derby, London, and Cardiff; and W. H. Harling, 47, Finsbury Pavement, London.

MISCELLANEOUS.

Davis's clinometer, shown in Fig. 631, is capable of doing the work of the dumpy level and the Hedley dial approximately, although it is not intended to take the place of either. Where great accuracy is not required it will save time and a more expensive instrument, and may be used where a level or dial cannot, on account of its extreme portability, its outside dimensions being 6¼ in. long, ½ in. wide, 3 in. deep. Price complete, in case, with portable tripod, 3*l*. 10*s*. The 2 in. compass on pivots is shown at A, the portable tripod screwing on to clinometer at B, and folding sights at C.

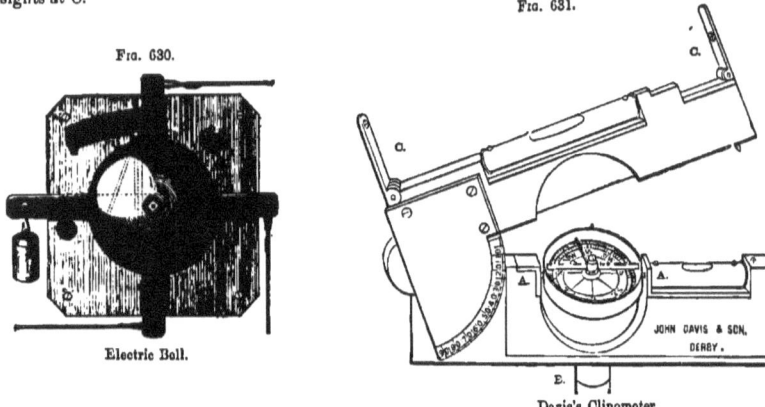

Fig. 630.

Electric Bell.

Fig. 631.

Davis's Clinometer.

In Louis's improved Davis clinometer (Fig. 632), the compass pivots are carried on a brass arc capable of revolving in the lower portion of the clinometer frame, so that the compass can be placed horizontally, and therefore read whatever be the position of the lower limb. The arrangement therefore allows both the amount of dip and the exact strike of strata, the amount and direction of inclination of an inclined shaft, &c., to be read simultaneously on the instrument. The best way of determining the strike of strata being by ascertaining the direction of their maximum dip, this can readily be done by turning the compass until it is horizontal, whilst the lower limb is resting on the strata in the desired position. The improvement also allows the compass to be instantly reversed, so that the same end of the needle may be used for all dial readings in running survey lines up and down hill; this cannot be done with any of the other forms of clinometer, and the instrument in its new shape may be used for all purposes, and will be found sufficiently accurate for most of the requirements of the miner or prospector in metal mining. A further improvement consists in mounting the bubble of the lower limb on a swivel, so that the clinometer may be levelled both ways without being reversed. The size of the clinometer is 6¾ in. long × ½ in. wide × 3 in. deep, weight 1 lb. 2 oz.; tripod with ball-and-socket joint, length 3 ft. 10 in., weight 1 lb. 8 oz.

Fig. 633 shows one of Stanley's most popular instruments. It is adapted to the telescope as shown, but there is also supplied with it a pair of open-sights to attach in its place. In the diaphragm of the telescope two very fine platino-iridium points give index of the reading. These remain in permanent adjustment, and are not liable to any derangement common to spiders' webs, which are generally used for the purpose. The angular displacement of the telescope reads in rather bold lines on the outside of the compass-box, which is divided to half degrees with two verniers. The vertical arc may be placed upwards or downwards. It reads to half degrees only by an index arm carried from the axis of the compass-box. The mounting of the telescope is carried upon a rocking ring upon the Hedley plan. The verticality of the axis is effected by three adjustments, which may be used concurrently or separately, according to the irregularity or inclination of the floor

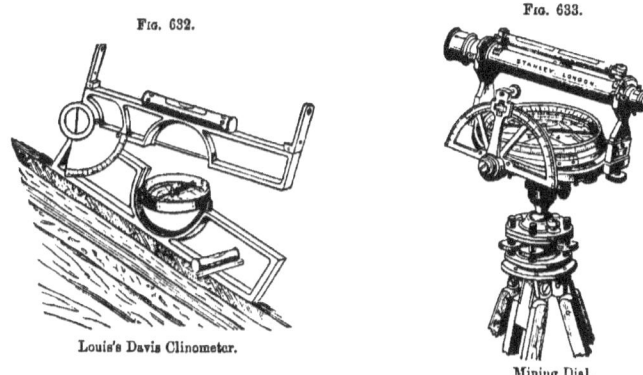

Fig. 632.
Louis's Davis Clinometer.

Fig. 633.
Mining Dial.

surface. The first adjustment employed by any one using the instrument is made by the tripod. This is not jointed in separate screw-on pieces in the ordinary manner, but the lower half of each leg is made to slide up between the limbs of the upper half, so that it may be shortened to half its length or set out at any intermediate position. Fig. 634 shows a perspective view of the upper part of tripod, and jointed part of one leg. After the legs are set up, the instrument may be set approximately level by the ball and socket, which is clamped in different instruments in two ways, depending partly upon the plan of final adjustment. It is clamped by a nipping-plate with a thumb-screw if applied to parallel adjustment, as shown in Fig. 633; or by rotation of the upper plate, which carries a screw covering the ball, if the adjustment is with tribrach screws, as shown in Fig. 634. This plan of separately clamping the ball is a great advantage over the method sometimes used of clamping it at the same time as making the adjustment by the parallel-plate screws, for the reason that when these two operations are performed simultaneously the ball becomes tight or loose and unsteady, according to the inevitable irregularity of pressure in parallel plate adjustment. Otherwise it is much more pleasant to attend to these setting-up adjustments separately as here shown.

Stanley has recently patented (No. 12,590, Aug. 1889) a new form of mining dial, for working in very close seams or veins (Fig. 635). It is said it will work well in as small a height as 15 in. The principle of the instrument is founded upon the prismatic compass, but the dial is of larger size (5–6 in.), and is made transparent. The divisions to half degrees are read through a prism, as with the prismatic compass. A second prismatic arrangement is placed under the compass-box, by which light is thrown from a lamp to illuminate the transparent dial. The back-sight may be extended, as shown in the engraving, upon an arm, to give greater precision of sighting. The compass-box adjusts by the stand, which is arranged upon an entirely novel plan. The legs may be placed at an angle as great as 50° to the floor of the mine if desired, without risk of slipping.

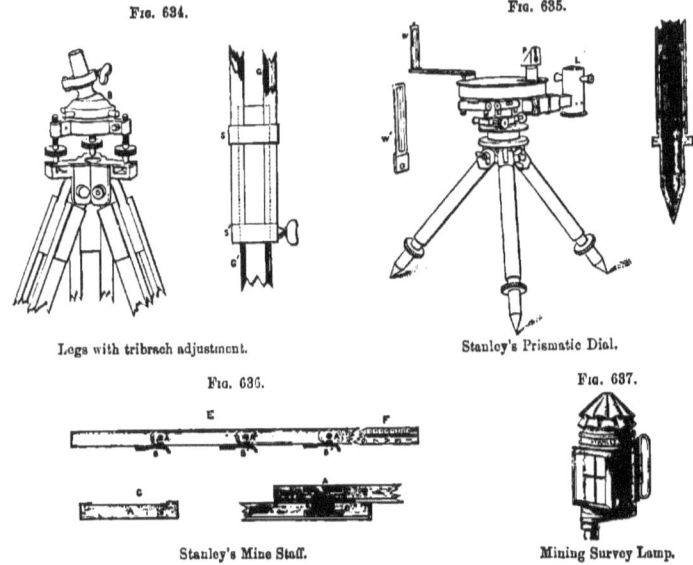

FIG. 634.

Legs with tribrach adjustment.

FIG. 635.

Stanley's Prismatic Dial.

FIG. 636.

Stanley's Mine Staff.

FIG. 637.

Mining Survey Lamp.

Each leg is formed of a pair of brass tubes, in the centre of which is a coarse-thread screw, of about half the length of the leg. This screw is moved by a large milled head near its point, so that either leg may be adjusted by shortening or lengthening as required for irregularity of floor.

Stanley places two stadia points on the back-sight, so that distances may be read by the angle subtended, as with the tacheometer. In this case the instrument is read for distance by his mine-staff, which is also patented.

Stanley's mine-staff may be used either as a stadia for measuring distances by the angle subtended as a telescope or sights, or be used for levelling It is illustrated in Fig. 636. It folds up as a French rule, but each of the joints has a spring clip, so that one, two, or more lengths may be opened out at once, according to the height required. The staff is lighted by a bull's-eye lantern. Details are shown in figure : E, staff in four lengths, fully extended ; A A' A'', joints ; section at A lower figure ; B B' B'', holding clips ; C, B, joint ; E, a piece of the front of the staff; G, section with hollow front.

Stanley's mining survey lamp (Fig. 637) was made originally for Kilgour of Westminster. It is constructed with fittings exactly corresponding with the dial to be used with it, so that by taking the dial off its tripod the lamp may be placed exactly in the same position. The lamp has double glasses; the inner glass has a distinct cross enamelled upon it. The centre of the cross corresponds both with the axis of the telescope of the dial and with the vertical axis of the instrument. These lamps have each a tripod identical with that of the dial, so as to reciprocally be changed, the one for the other, without any change of adjustment of the dial.

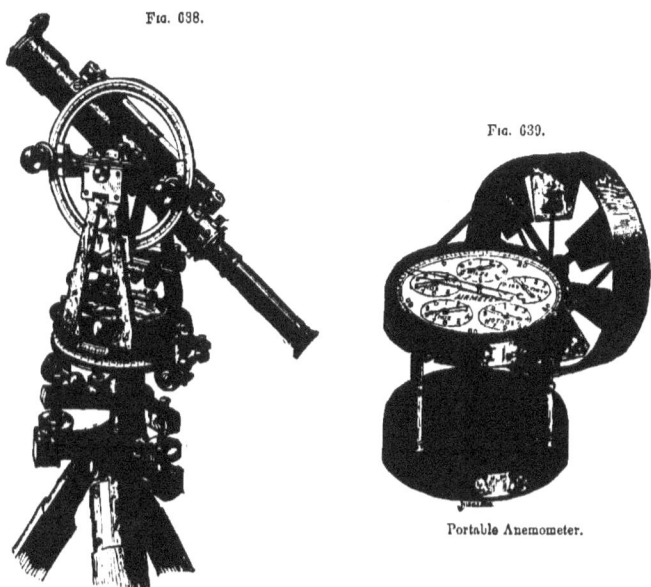

Fig. 638.

Fig. 639.

Harling's Theodolite.

Portable Anemometer.

In Fig. 638 is shown Harling's 6-inch transit theodolite; it is of best workmanship and highest finish, bell-metal centres, &c., with achromatic telescope, erect and inverted eye-pieces, vertical and horizontal circles, divided on silver to 20 in., microscopes, and two verniers to each circle, clamp and tangent screws, and parallel plates, complete in case with tripod stand, price, 26l. 10s.

Fig. 639 shows a portable anemometer by Harling, for the measurement of currents of air through mines, tunnels, &c., packed in box 4 inches square, with a universal jointed socket holder, TWO DIALS reading to 1000 FEET, with disconnector, price, 2l. 5s.

INDEX.

A.

ABEL fuses, 100
Accidents in blasting, 102
Adams on pointed boxes, 348
Air, compressed, for transmitting power, 26, 30
——, compressed, in shaft sinking, 133
——, compressed, meter, 31
——, compressed, motors, 32
—— compressing machines, 32
—— compressors for coal-cutting machine, 141, 144, 151
—— condults, 85
—— cooler, 152
—— pipes, 85
—— power, 82
—— receivers, 85
——, storing, 85
—— supply to mines, cost of, 175
Aldwarke dry cleaning, 437
Alve's concentrator, 365
American magneto firing machine, 94
Anemometers, 182, 460
Angle of windmill sails, 1
Area of windmill sails, 2
Austrian fuse, 101
Automatic sampler, 75
Axles of tubs, 194

B.

BAILEY on endless rope haulage, 245
Baird's coal cutter, 137
Band wheels, 65
Barrels for raising water, 155
Barrow coal cleaning, 435
—— creeper, 436
Bash wheels, 429, 432, 443
Batteries for blasting, 91, 98
Battery for electric lighting, 193
Beches, 80
Bell & Ramsay washer, 429, 431, 432, 442
Bell coal washer, 431, 432, 443
Bells, 456
Bell socket, 61
——, ventilating, 175
Bérard's washer, 420
Beringer on transmission of power, 26
Bewick on coal washing, 425
Bidder's coal falling machine, 147
Hiram anemometer, 183
Bits, 77
Blasting, 88

Blasting, accidents in, 102
—— gear, set, 81
——, precautions, 103
—— sticks, 92
Blende ores, dressing, 378, 384
Bodies of tubs, 197
Bonneted Marsaut lamp, 190
Bore holes, tubes for lining, 61
Borers, 59
Boring by diamond drills, 73
—— by hand, 52
—— by steel drills, 52
——, cost, 72
—— frame, 52, 53
——, machine, 62. 71
——, percussive, 52
—— tackle, with steam winch, 72
—— tools, cost of sets, 73
—— ——, extracting when jammed, 68
Burhse's concave buddle, 362
Bornhardt's firing machine, 95, 99
Burt, 75
Bottom pick, 111
Box, ventilating, 175
Brass lamp, 186
Breaking ores, 395
Breast wheels, 9
Breguet's exploder, 93
Buckets of water wheels, 7
Bucking iron, 115
Buddles, 359, 381, 393, 397, 402
Buddling, 393
Bulling shovel, 109
Bulls, 80
Bull wheels, 64

C.

CABLE boxes, 94
—— for electric hauling, 259
—— for electric transmission of power, 38
Cables of firing machines, 94, 100
Cages, 213
Calow's safety cage, 221
Candlestick, 188
Capels, 219
Capstan for raising and lowering pump rods, 162
Carbons, 75
Carrett, Marshall, & Co.'s coal-cutting machine, 145
Carron coal washer, 431, 432
Carr's disintegrator, 410
Cars, 194
Cartridge, gelatinous, 103

Centre bits, 67
—— head buddle, 359
Centrifugal pumps, 171
Chain carrier, 413
Chair and sleeper, improved, 206
Chalk, sinking through, 116
Chatts, 395
Chavatte on shaft sinking, 134
Chisels for rock boring, 59
Clamps, 213
Clanny lamp, 180
Clarke on coal-cutting machines, 148
Classifiers, 346
Classifying, 386
Clausthal jigger, 341
Claying irons, 80
Clay spade, 108
Clinometers, 457
Clutches, 240
Coal blasting gear, set, 81
—— Cleaning Committee's Report, 452
—— creepers, 436
—— crushing, 409
—— cutting by machinery, the modus operandi, 148
—— ——, hand and machine, cost compared, 152
—— —— machinery, 135–154
—— —— machines, air compressor for, 141, 144, 151
—— —— ——, advantages over hand, 144, 151, 154
—— ——, cost, 152
—— ——, electric, 152
—— ——, upheaving bottom coal, 142
—— —— , working capacity, 141, 142, 147, 149, 151, 153, 154
—— dross removed by washing, 431
——, dry cleaning, 435, 437, 450
—— falling machines, 147
—— getting, cost by hand and machine, 154
——, hewing, 135
—— hutches, 435, 437, 451
—— picking tables, 436, 439
—— screens, 436, 440
—— transport, 411
—— tumblers, 435, 437
—— washers, cost and efficiency, tables, 431, 432
—— washing, 403, 415
—— ——, cost, 431, 441, 442, 443
—— ——, principles, 426
Cobbing hammers, 115
Cochrane on coal washing, 425
Colladon's wheel, 10
Collom's jigger, 343

INDEX.   463

Commans & Co.'s dressing plant, 400
Compensating joint, 86
Compressed air for transmitting power, 26, 30
—— —— in shaft sinking, 133
—— —— meter, 31
—— —— motors, 32
Compressing air, 82
Comstock hydraulic draining, 165
Concave buddle, 361
Concentration, 351
Concreting shaft, 131, 134
Conical drum for regulating load in hoisting, 270
Connections, 212
Convex buddle, 359
Cooke's drum, 174, 180
Coppée's washer, 417, 425, 431, 432, 445, 451
Copper ores, dressing, 378, 403
—— wire for electric transmission of power, 38
Cornish duck engine, 176
—— jigger, 339
—— pump, 160
—— pump in slant workings, 163
—— shackles, 219
—— skips, 205
—— tin dressing, 399
—— water whim, 239
—— winding engines, 271
Corves, 215
Cost of coal washing, 431, 441, 442, 448
—— of sets of boring tools, 73
—— of trial borings, 72
Cotton powder, 103, 106
Coulson on shaft sinking, 134
Counterweights for hoisting loads, 269
Cowan on coal washing machines, 428
Cradles, 203
Creepers, 412, 436
Crow's foot, 61
Crushing coal, 409
—— ores, 385
—— rolls, 401
Cutting picks, 111
—— stone by wire, 46
—— tools in boring, 59
Cwm Avon washery, 445

D.

Daglish on shaft sinking, 123
Davey's adjustment for pumping engines, 169
Davis-Ashworth Muesler lamp, 190
Davis's anemometer, 183
—— dynamo tension exploder, 96
—— magneto exploder, 96
Davy lamp, 188
Delivery of coal, 436, 439
Denain tipping cradle, 404
Dials, 458
Diamond drill boring, 73
Direct winding, 272
Dodge's concentrator, 365
Dolly stamp, 75
—— work, 396
Donaldson on hydraulic transmission of power, 28
Dowlais screens, 409
—— tipping cradle, 404
—— washery, 420
Drag twist, 80

Draining machinery, 155–173
Drawing cages, 213
Dredging, 50
Dressing, cost, 308
—— hammers, 115
—— machinery, 338
——, objects of, 338
——, power needed, 398
Drifting pick, 109
Drilling machines, 81
—— rope, 66
—— tools, 77
Drills, percussive and rotary compared, 87
Driving pick, 112
—— pipes, 65
Dross removed from coal by washing, 431
Dry cleaning coal, 435, 437, 450
Duck engine, 176
Dumping cradles, 203
Duncan concentrator, 366
Dynamo for electric haulage, 258
—— for electric lighting, 193
—— for electric transmission of power, 38
—— for working coal cutter, 152
—— tension exploder, 96

E.

Ebbw Vale washery, 423
Electric bells, 456
—— coal-cutting machine, 152
—— fans, 183
—— fuses, 90, 100
—— haulage, 256
——, cost, 262
—— lighting, 191
——, battery, 193
——, cost, 191
——, engine and dynamo, 193
—— motors, 43, 184
—— portable safety lamp, 193
—— pumps, 172
—— transmission of power, 26, 37
——, efficiency, 38, 41, 42, 44
Electrical firing machine, 95
—— machines, 91, 93
Elevators, 413
End-shake percussion-table, 356
Endless chain haulage, 245, 250
—— rope haulage, 211, 245
—— ——, tightening, 211
—— wire for quarrying stone, 46
Engine for electric haulage, 257
——, oil, 21
Engines, pumping, adjusting, 169
Excavating machinery, 77–115
Excavators, 49
Expansion gear for winding engines, 271
Exploders, 93
Explosives, 102, 103
——, firing, 88
Extracting bore tubes, 70
—— tools, 61

F.

Fabry's wheel, 178
Fans, electric, 183
——, hand, 177
Feeders, walling out, 118
——, weighing off, 117
Filling shovel, 108

Firing explosives, 88
—— machines, 91, 93
Firth's coal cutting machine, 143
Fisher & Walker's friction clutch, 240
Fisher's pulley, 249
Fishing-up jammed borers, 68
Flat chisel, 50
Flimby washery, 440
Flow of water in pipes, 29
Forster on cost of supplying air to mines, 175
Foster on stone quarrying, 45
Fowler's clip pulleys, 209
—— hydraulic loading, 272
Frecheville on tin dressing, 399
Freestone quarrying, 45
Friction clutch, 240
—— firing machines, 93, 99
—— rollers, 200
Frongoch classifier, 349
—— jigger, 344
—— separator, 349
—— skip, 204
Frue vanner, 367
Frying-pan shovel, 108
Furnaces, ventilating, 175
Fuses, 89, 100
——, construction, 91, 100
——, lighting, 105

G.

Gads, 115
Galena, dressing, 378, 384
Galloway on pneumatic water-barrel, 157
Gas lighting, cost, 192
Gelatinous cartridge, 103
German horse whim, 237
—— pyramidal boxes, 346
Gilkes's turbines, 15–18
Gillot & Copley's coal cutter, 139
Girard turbines, 16, 20
Goffint blower, 174
Goolden's coal-cutting machine, 152
—— electric pump, 172
Governor for windmills, 3
Grafting spade, 109
Green's dressing system, 378
—— water wheels, 10
Grids, 436, 440
Guibal fan, 174, 180
Guinotte & Briart's screen, 436
—— screen, 408
Gunpowder, 102
Günther's turbines, 20
Gwynne's electric engine and dynamo, 193
—— pumps, 171
—— turbines, 18

H.

Hacks, 109
Halley's percussion table, 370
Hall's crusher, 410
Hammers, 78, 115
Hand-boring, 62
—— drilling tools, 77
—— fans, 177
—— lever jigger, 339
Hausa shaft, 117
Harvey on coal washing, 404

INDEX.

Harvey and Co.'s air compressor, 84
Harvey's Cornish skips, 205
—— steam capstan for raising and lowering pump rods, 162
Harz water whim, 240
Haulage at South Duffryn, 245
——, cost, 245, 205, 266
——, underground, 245
Hauling by electric-transmitted power, 40, 42
—— engines, 244
—— machinery, 194-283
Hayward Tyler's Cornish pump, 163
—— —— pump for varying levels, 163
—— —— steam pump, 164
Head gear for hand boring, 52
—— —— for hoisting, 225
Heath & Frost's cartridge, 103
—— —— safety lamp, 105
Helicoidal wire, quarrying by, 46
Hemp ropes, 227, 231
Hendy's concentrator, 370
Hewing coal, 135
Hickie's air-cooler, 182
Hoisting machinery, 194-283
——, regulating load, 268
Holing pick, 111
Horn socket, 61
Horse shoe washer, 435
—— whims, 237
Huet & Geyler's jigger, 340
Hurd & Simpson's coal-cutter, 140
Husband's water safety balance valve, 170
Hutches, 435, 437, 451
Hutching, 339
Hydraulic dmining, 165
—— loading, 272
—— power, 5
—— transmission of coal, 412
—— —— —— power, 26, 28

I.

IMLAY concentrator, 371
Impulse breast wheels, 9
Indicators, 454
Insulating wires of firing machines, 91, 100
Iron lamp, 186
—— ropes, 230
—— stone trial borings, 72
—— wire for ropes, 233
—— —— rope, 24

J.

JARS, 67
Jiggers, 378, 383, 388, 441
Jigging, 339, 388
Joining air pipes, 85
Joint, compensating, 86
Jonval turbines, 20
Junctions, 207

K.

KEEPS for cages, 224
Kieve, 372
Kennedy on pneumatic transmission of power, 30
Keppers, 440
Kerr, Stuart, & Co.'s waggon, 203
Kibbles, 237

Kind-Chaudron shaft sinking, 122
Kinder on electric blasting, 96
Kind's plug, 70
Kitto & Paul's classifier, 340
——, Paul, & Nancarrow skip, 204
Körting blower, 175
Kutter's formulæ for flow of water in pipes, 29

L.

LABYRINTHS, 346
Ladd's frictional exploder, 99
Lamps, oil, 186
——, safety, 188, 108
——, surveying, 460
Lead ores, dressing, 378, 384, 400
Leading wires of firing machines, 94, 100
Lebreton on electric haulage, 264
Lemielle drum, 174, 179
Lever boring machine, 71
Lifting dogs, 57
Lighting, 186-193
——, cost, 191
——, electric, 191
—— fuses, 105
Lining bore holes, 61
Linkenbach buddle, 402
Lippmann's cutter, 133
Load, regulating in hoisting, 268
Lowering pump rods, 162
Lührig's coal washer, 432
—— fine coal jigger, 434
—— washer, 417, 425
Lukis on galena and blende dressing, 384
Lupton on overwinding, 221
—— on shaft sinking, 133

M.

McCULLOCH coal washer, 431, 432
Machine boring, 62
—— drills, 81
—— fuses, 91, 92, 100
Machines for blasting, 91, 93
McNeill's concentrator, 373
Magnetic firing machine, 93
Magneto exploder, 96
—— firing machine, 94
Mandrils, 109
Marble polishing, 49
—— quarrying, 45
Marsaut coal washer, 416
—— lamp, 190
Marsden shafts, 123
Marten on coal washing, 425
Mining tools, 107
Mixed breast wheels, 9
Motive power, 1-22
Motors, compressed air, 32
——, electric, 43, 184
—— for electric transmission of power, 38
——, oil, 21
——, pneumatic, 82
——, water, 5
——, wind, 1
Mowatt on self-acting endless chains, 250
Mneseier lamp, 190
Multiple wedge, 107
Mulvany on shaft sinking, 116
Munday's round buddle, 363
Murgue on ventilating machines, 183

N.

NIPPING forks, 57
Nixon blower, 174
North Skelton, drill trials at, 87

O.

OIL engine, 21
—— lamps, 186
Overshot water wheels, 6, 10
Overwinding, 220

P.

PACKING, 396
Paraffin lighting, cost, 192
Percussion tables, 351
Percussive and rotary drills compared, 87
—— boring, 52
Permanent way, 206
Philips's safety cage, 223
Picking for prills, 385
—— tables, 436, 439, 440
Picks, 109
Pipes, air, 85
——, dimensions for transmitting power by fluids, 29
——, flow of water in, 29
Pit-head frames, 225
—— ropes, 227
Plated chain carrier, 413
Platinum fuses, 101
Pneumatic motors, 32
—— power, 1, 82
—— pressure in shaft sinking, 133
—— transmission of power, 26, 30
—— water barrel, 157
—— wheels, 178
Pointed boxes, 348
Polishing stone, 49
Pull pick, 113
Power, motive, 1-22
—— of water wheel, estimating, 6
——, pneumatic, 82
——, steam, cost of transmission, 27
——, transmission, 23-44, 244
——, ——, by wire rope, 23
——, ——, comparison of systems, 26
——, ——, cost, 26
——, ——, efficiency, 25, 31
——, ——, electric, 26, 37
——, ——, hydraulic, 26, 28
——, ——, pneumatic, 26, 30
——, water, 5
——, cost of transmission, 28
——, wind, 1
Preservation of ropes, 234, 236
Prickers, 88
Priestman's excavators, 49
Prills, 385
Propeller knife buddle, 362
Prospecting, 52-76
—— borings, cost, 72
—— stamps, 75
Pulleys, 208, 249
Pumping by electric-transmitted power, 41, 43
—— engines, adjusting, 169
——, safety valve, 170
—— in shaft sinking, 117, 118
—— machinery, 155-173
Pumps, 160-173
——, centrifugal, 171

Pumps, Cornish, 160
——, dealing with, during shaft sinking, 160
——, direct acting, 164, 172
——, electric, 172
—— for varying levels, 163
—— rods, raising and lowering, 162
——, steam, 164, 172
——, —— capstan for raising and lowering rods, 162
——, water-wheel, 164
Pyramidal boxes, 346

Q.

QUANTITY fuses, 91, 92, 100
Quarrying, 45–51
Quick winding, 272
Quicksand, sinking through, 116, 133

R.

RACK, 374
Ragging, 385
Raising pump-rods, 162
—— water, 155–173
Rammers, 80
Ramsey washer, 429, 431, 432
Rathbone on coal washing, 425
Reels, 208
Reid & Jones's concentrator, 374
Regulating load in hoisting, 268
—— tools, 68
Retarding apparatus for pumps, 169
Ribbons of blasting machines, 92
Rigg's tumblers, 437
Rittinger's jigger, 340
—— rotating table, 356
—— sidethrow percussion table, 351
River-beds, dredging, 50
Robin Hood washery, 442
Robinson washer, 430, 431, 432, 449, 451, 453
Rocking lever for hand boring, 54
Rods for hand boring, 58
Roebling on splicing wire rope, 23
Roller delivery, 439
—— mills, 385
Root's blower, 174, 175, 182
Rope for transmitting power, 23
——, iron wire, 24
——, load, 234
——, maintenance, 234, 236
——, pit, 227
——, qualities, 230, 231
——, steel wire, 24
——, strength, 230
——, wire, splicing, 23
Rotary and percussive drills compared, 87
Rotating table, 356
Royalties at stone quarries, 45

S.

SAFETY cages, 216, 220
—— catches, 216, 220
—— fuses, 89, 100
—— hooks, 220
—— lamp, 105, 189, 193
—— valve for pumps, 170
Sampler, automatic, 75
Sampling, 75
Sand pumps, 67
Sawyer on gelatinous cartridge, 103

Schiele's fan, 174, 181
Schütz on cost of haulage, 266
Scoop shovels, 75
Scrapers, 80
Screening coal, 407, 417
Screens, 436, 440
Screw plugs, 70
Self-acting inclines, 250
—— planes, 208
Self-tipping waggons, 203
Sentein dressing works, 384
Separation of minerals, 338
Settlers, 345
Settling pits, 346
Shackles, 210
Shaft, concreting, 131, 134
Shaft-sinking, 116–134
——, accumulation of air in, 134
—— at Hansa, 117
—— at Marsden, 123
—— at Shamrock colliery, 116
—— at Zollern, 118
—— by compressed air, 133
—— by telescopic cylinders, 134
——, conditions that determine method, 122
——, cost, 132, 134
——, dealing with pumps during, 169
—— in quicksand, 133
——, Kind-Chaudron, 122
——, Lippmann's cutter, 132
——, pumps in, 117, 118
——, safety pipe for accumulated air, 134
——, tools used in Kind-Chaudron method, 124
——, treatment of sinkers, 121, 134
——, walling out feeders, 118
——, wedging-off feeders, 117
Shafts, shutting out water, 116
——, tubbing, 121, 129
Shaking tables, 351
Shamrock shaft, 116
Sheaves, 208
Sheet-iron lamp, 186
Sheppard's washer, 418, 425, 429, 431, 432, 449
Shippey's electric fan, 183
Shothole wires, 92, 100
Shovels, 107
Side-tipping cradle, 206
Siemens' dynamo-electric firing machine, 93, 96
Signalling, 454
Silver lead dressing, 400
Silverton battery, 58
Sink walls, 116
Sinkers, treatment of, 121, 134
Sinking, German methods, 116
—— on Westphalian coalbeds, 116
—— shafts, 116–134
—— shovel, 108
—— through chalk, 116
—— —— quicksand, 116
Sizers, 340
Sizing ore, 386
Skips, 204
—— water, 155
Sledges, 78
Sleeper and chair, improved, 206
Slime labyrinth, 346
—— frame, 374
—— table, 375
Slimes, treating, 396
Slitter pick, 109
Slow breast-wheel, 9
Sludgers, 53, 59
Smith on underground rope haulage, 246

Snell on electric transmission of power, 37
Sockets, 219
Spalling hammers, 115
Specific gravities of minerals, 338
Spider candlestick, 188
Spitzkästen, 346, 387
Spitzlutten, 350
Splicing wire rope, 23
Staffs, 459
Stahl on wire rope transmission of power, 23
Stamping by electric transmitted power, 42
Stamps, prospecting, 75
Stanley's coal-heading machine, 150
Steam capstan for raising and lowering pump rods, 162
—— power, cost of transmission, 27
—— pump, 164
Steel boring, 52
—— for ropes, 234
—— ropes, 230
—— wire, quarrying by, 46
—— rope, 24
Stemmers, 80
Stephenson's lamp, 190
Stone picks, 111
—— polishing, 49
—— quarrying, 45
Storing air, 85
—— water, 86
Straight bit, 59
Stream, adjusting wheel to, 6
Stripping, 49
Struvé piston, 174
Surveying instruments, 456
Surveyor's lamps, 187

T.

TAILROPE haulage, 245
Tallie on electric haulage, 256
Tamping, 102
—— irons, 80
Teeming cradles, 203
—— waggons, 202
Tension fuses, 91, 92, 100
—— in electric firing machines, 90
Theodolites, 460
Tigers, 57
Tin dressing, 399
Tipping cradles, 203, 404
—— tubs, 202
Tipplers, 202
Tonite, 103, 106
Tools for extracting tubes, 70
—— for hand boring, 53
—— —— drilling, 77
—— —— machine boring, 62
—— —— shaft sinking, 124
Top stripping, 49
Tossing, 396
—— tub, 372
Traigneaux quarry, 47
Train blasting, 88
Tramway, 206
—— junctions, 207
Transmission of power, 23–44, 244
—— —— by wire rope, 23
—— ——, comparison of systems, 26
—— ——, cost, 26
—— ——, efficiency, 25, 31
—— ——, electric, 26, 37
—— ——, hydraulic, 26, 28
—— ——, pneumatic, 26, 30
Transport, 411

3 O

INDEX.

Transport of coal, 436, 439
Trial borings, cost, 72
Triangular double troughs, 350
Trommels, 378, 383
Trough washers, 429
—— washing coal, 451
Tubbing shafts, 121, 120
Tubes, air, 85
—— for lining bore-holes, 61
Tubs, 194
—— for raising water, 155
Tumblers, 435, 437
Turbines, 11
——, adjusting, 16, 19
——, advantages, 11
——, classification, 14
——, finding dimensions of, 13
——, —— power of, 13
——, Gilkes's 15–18
——, Girard, 16, 20
——, Günther's, 20
——, Gwynne's, 18
——, Jonval, 20
——, proportions, 12
——, speed in relation to fall, 16
——, vortex, 15
—— working at half-power, 19
Turntables, 207
Two sieved continuous jigger, 341

U.

UNDERGROUND haulage, 245
Undershot wheels, 10

V.

VANNERS, 367
Velocity of windmill sails, 4
Ventilating, 174–185
——, anemometers, 182
—— bell, 175
—— box, 175
——, Cornish duck engine, 176
——, cost of air supply, 175
——, electric fans, 183

Ventilating furnaces, 175
——, hand fans, 177
——, Hickie's air cooler, 182
——, water blast, 176
Ventilators, mechanical, 174
Vertical windmill, 1
Vogel on cost of haulage, 266
Vortex turbine, 15

W.

WADDLE fan, 174
Wadhook, 61
Waggons, 194
Walker on cost of coal-getting by hand and machine, 154
Walling out feeders, 118
Washing coal, 403, 415
Water as a motor, 5
—— blast, 176
—— flow in pipes, 29
—— power, 5
——, ——, cost of transmission, 28
——, raising, 155–173
——, reservoirs, 86
——, shutting out of shafts, 116
—— skips, 155
——, transmission of power by, 26, 28
—— wheel, breast, 9
—— —— buckets, 7
—— ——, Colladon's, 10
—— ——, comparison of, 11
—— ——, cost, 11
—— ——, effective power, 6
—— ——, estimating power, 6
—— —— for varying volume, 8
—— ——, Green's, 10
—— ——, impulse breast, 9
—— ——, mixed breast, 9
—— ——, overshot, 6, 10
—— —— pump, 164
—— ——, slow breast, 9
—— ——, turbines, 11
—— —— under pressure, 11
—— ——, undershot, 10
—— ——, Wesserling, 8
—— —— with vertical sluice, 7
—— whim, 239

Wedges, 107, 114
Wedging off feeders, 117
Wesserling wheel, 8
Westphalian coal beds, 116
Wet cleaning coal, 451
Wheels for transmitting power by wire rope, 23
—— of tubs, 194
Whims, 237
Winch for raising and lowering pump rods, 162
Wind as a motor, 5
—— force, adjusting, 3
—— power, 1
Winding by steam from shallow shafts, 242
——, direct, 272
—— drums, 208, 286
—— engines, 271
—— machinery, 194–283
—— precautions, 236
——, rapid, 272
—— ropes, 227
Windmills, 1
——, governor, 3
——, regulating velocity, 3
——, sails, angle, 1
—— —— area, 2
—— ——, velocity, 4
——, turning tower, 3
Winks, Cowling, & Hosken's safety cage, 221
Winstanley & Barker's coal-cutter, 136
Wire for electric transmission of power, 38
—— ropes, 235
——, insulating, 91, 100
——, quarrying by, 46
—— rope for transmitting power, 23
—— ——, iron, 24
—— ——, splicing, 23
—— ——, steel, 24
—— ——, transmission of power, efficiency, 25
Woodend washer, 429

Z.

ZOLLERN shafts, 118

LONDON: PRINTED BY WILLIAM CLOWES AND SONS, LIMITED, STAMFORD STREET AND CHARING CROSS.

一

.

14 DAY USE
RETURN TO DESK FROM WHICH BORROWED

LOAN DEPT.

This book is due on the last date stamped below, or on the date to which renewed.
Renewed books are subject to immediate recall.

ICLF (N)

OCT 8 1966

RECEIVED
OCT 29 '66 -12 AM
LOAN DEPT.

APR 16 1993
REC. CIR. NOV 16

LD 21A—60m-8,'65
(F2336s10)476B

General Library
University of California
Berkeley

YF 00221

www.ingramcontent.com/pod-product-compliance
Lightning Source LLC
Chambersburg PA
CBHW051854300426
44117CB00006B/388